Coral Reefs of the World

Volume 6

Series editors
Bernhard Riegl, Dania Beach, Florida, USA
Richard E. Dodge, Dania, Florida, USA

More information about this series at http://www.springer.com/series/7539

Dennis K. Hubbard • Caroline S. Rogers
Jere H. Lipps • George D. Stanley, Jr.
Editors

Coral Reefs at the Crossroads

Springer

Editors
Dennis K. Hubbard
Department of Geology
Oberlin College
Oberlin, OH, USA

Jere H. Lipps
John D. Cooper Archaeological and Paleontological Center
Santa Ana, CA, USA

Museum of Paleontology
University of California
Berkeley, CA, USA

Caroline S. Rogers
Wetland and Aquatic Research Center
Caribbean Field Station
US Geological Survey
St. John, VI, USA

George D. Stanley, Jr.
Department of Geosciences
The University of Montana
Missoula, MT, USA

ISSN 2213-719X ISSN 2213-7203 (electronic)
Coral Reefs of the World
ISBN 978-94-017-7565-6 ISBN 978-94-017-7567-0 (eBook)
DOI 10.1007/978-94-017-7567-0

Library of Congress Control Number: 2016942791

Printed on acid-free paper

This Springer imprint is published by Springer Nature
The registered company is Springer Science+Business Media B.V. Dordrecht

Paul L. Jokiel (1941–2016)

Paul Jokiel's career as a reef scientist spanned half a century. His interests were broad, and he always seemed to be ahead of the wave – whether it was his considerations of thermal stress in the 1970s, U/V light and photo-inhibition in the 1980s, the myriad roles of water motion in coral ecology, or the growing impact of humanity and climate change. His interests spanned the spatial spectrum from global (coral dispersal and connectivity) to microscopic (the topic of his chapter in this volume). More recently, his efforts broadened to include monitoring throughout the Hawaiian Islands, and he was an active participant in discussions of climate change, reef decline, and environmental management at scales ranging from local to global. We are grateful for his excitement about our project and regret that he did not get to see his significant contribution in print. However, we hope that his spirit of mentoring and his interest in science across broad disciplinary boundaries will be well served by this volume.

Preface

To every thing there is a season and a time to every purpose under heaven (Ecclesiastes 3:1).

I should have written this book in the early 1980s when I was at the West Indies Laboratory (WIL) on the Caribbean island of St. Croix. But, it was not the time; I was busy learning about coral reefs. Too late for the glory days of Discovery Bay in Jamaica, I had come to St. Croix for a year in 1977 right after graduate school and stayed for 22.

I should have written this book in the early 1990s. I had learned so much through the generosity of others and the benefits of living on the reef year-round for well over a decade. But, the West Indies Lab had just been destroyed by a hurricane.

For the next decade, I applied what I had learned at WIL to help frame a coastal-zone management process that was in its infancy in the US Virgin Islands. I learned tremendously as my interests broadened to include the management of already declining reef systems and how science fit into a larger social framework. I could have written this book then, but I was too busy making a living as a marine consultant outside the cloister of academia.

In the 12 years I spent at WIL, I worked with so many talented scientists and colleagues, but more important, I talked with them... after dives... over coffee... on vessels of all shapes and sizes. For me,WIL was a mecca of reef science for that brief time—and all the people who I would have traveled great distances to seek out were coming to me. Their generosity not only taught me about science, but also showed me how much more we can learn when we do it together. My students helped me to set out on what would be a life of not just teaching them but learning with and from them.

Marine labs are special places. Their ethos is difficult to understand until you have spent some time there. So much insight is derived not from carefully designed experiments but from just seeing the same places over and over again... over seasons and decades... and with different people who bring different perspectives and ask different questions. In the 1960s through the 1980s, marine labs were places where many of us came together not just to quantify a particular process or to answer a specific question, but to just try and understand how all the pieces fit together within an incredibly dynamic system—the coral reef. We were all "specialists" with our own sets of priorities. But we soon realized that as we talked to one another, we started asking new questions that were far grander than any of those we had framed on our own. We weren't just fisheries biologists, marine geologists, geochemists, seagrass or coral ecologists, and physical oceanographers. We were a group of researchers living in a special time and at a special place.

Much of our understanding of how these systems worked was serendipity. But, the significance of even fortuitous observations would have been lost without the perspective of time. And, they would have soon faded without someone to tell about it... not in a journal article, but on the way back from a dive, or over breakfast, or at a local watering hole—many people simply observing the same places—again and again.

Great laboratories and large research institutions have existed over the centuries. And marine labs are certainly not limited to the Caribbean nor have they disappeared. But, there was something indefinable, at least to me, about the small marine labs of the 1960s, 1970s, and 1980s. They were perfectly positioned between the broad expeditions of the previous century and the exploding technology that followed. So much of what we take for granted today came out of these small oases. They were also positioned at a perfect time to document the decline that was already occurring.

In the Caribbean, the Bellairs Marine Lab flourished during the 1950s and beyond. Discovery Bay Marine Lab, on the north coast of Jamaica (Fig. 1), the creation of Tom Goreau (Fig. 2), opened in 1965. For decades, it provided a place for reef scientists to gather and ask questions about how the reef functioned as a physical, biological, and chemical system. Ecologists like Judy Lang and paleontologists like Jeremy Jackson started their careers there, and so many of the principles that have stood the test of time were developed at this facility. Lynton Land, a geologist, "excavated" into the reef (Fig. 3) to understand the nature of its interior and to quantify how calcification by organisms living on the surface built the edifice that supported them. He used the "carbonate budget" concept developed by Colin

Fig. 1 The initial Discovery Bay Marine Lab on Jamaica (Photo by Eileen Graham, archived at the Natural History Museum in London. Courtesy of Ken Johnson and NHM)

Fig. 2 Tom Goreau observing the reef he knew best near the shelf edge off Discovery Bay, Jamaica (Photo by Eileen Graham. Courtesy of Ken Johnson and NHM)

Fig. 3 Lynton Land (*left*) and Reg Purdon excavating into the reef at "Lynton's Mine" (Photo by Eileen Graham. Courtesy of Ken Johnson and NHM)

Fig. 4 View looking north over the West Indies Laboratory (*lower right*); Tague Bay and Buck Island National Underwater Monument are in the background (Photo by D Hubbard©)

Stern and Terry Scoffin on Barbados to think about the myriad processes that tie the community on its surface to the physical structure within.

For me, and so many others, it was the West Indies Laboratory on St. Croix (Fig. 4) that provided this opportunity. This brainchild of H. Gray Multer (Fig. 5a) was the unique blend of a strong undergraduate teaching program (Fig. 5b) and a well-respected research facility. While most others focused on the reef surface, some of us thought "deeper." Using smaller drills, and eventually the SCARID drilling system, we cored through quaternary reefs in both deeper (Fig. 5c) and shallow water (Fig. 5d). In the latter 1970s and 1980s, WIL operated NOAA's *Hydrolab* underwater research laboratory (Fig. 5e) that allowed research teams of four scientists to work at depths of over 45 m for a week at a time. Using it as a base, we cored

Fig. 5 (**a**) Founder and first director of WIL, H Gray Multer (*left*), coring the reef in Nonsuch Bay, Antigua, with Heinrich Zankl (University of Marburg), Ian Macintyre (Smithsonian Institution), and Kenny Burger (undergraduate, Fairleigh Dickinson University) (Photo by D. Hubbard). (**b**) Undergraduate students diving in WIL marine science classes. (**c**) SCARID drill in 15 m of water on Lang Bank, east of St. Croix. (**d**) SCARID drill near the reef crest of Buck Island National Monument, St. Croix. (**e**) Hydrolab underwater laboratory in Salt River Submarine Canyon on St. Croix. (**f**) Drilling horizontally into the west wall of the canyon (Photos by D Hubbard©)

into the walls of Salt River Submarine Canyon (Fig. 5f), spending over 12 h a day at depths up to 30 m. We also made many of our early observations of storm-related sediment transport from this facility, first in the Bahamas and later on St. Croix.

Marine labs have created myriad opportunities for reef scientists that would have otherwise been impossible. But more important, long-term measurements and observations provided a framework that benefitted every researcher who briefly passed through, each one adding texture to the picture that had developed over long years of repeated observation and sampling. It is always interesting to watch the discussions that follow some new paradigm as it catches on. The first contrarians are often those who have lived on a reef for an extended time. In most instances this is not due to superior intellect, but just having the benefit of going back to the same place with so many different people.

I never imagined it would take so long for the concept of this volume to emerge. However, it is fitting that it is not a collection of my own biases but rather a variety of views from valued friends and colleagues who have simply spent a lot of time looking at and thinking carefully about reefs. It is dedicated to those who created the marine labs and those who worked at them—as scientists, staff, and colleagues. I will never forget that, after Hurricane Hugo, the staff and scientists of West Indies Lab were back at work within 24 h, even though so many had lost heir homes in the storm. We were in the water within 2 days making observations and taking post-storm measurements as dedicated staff, graduate students, and visiting researchers started the long process of bringing WIL back to life while also tending to over 30 students trapped after the storm. Collectively, they and others like them have fostered the interdisciplinary spirit so deeply engrained in the early International Coral Reef Symposia. Hopefully, this volume will encourage the next generation of reef researchers to not view the details of reef function solely through their own disciplinary lens, but to think more about how we might ask questions together that we cannot even imagine alone.

Oberlin, OH, USA Dennis K. Hubbard

Acknowledgements

It is impossible to adequately acknowledge all the people that have brought this volume to this stage. At the great risk of forgetting so many, this includes colleagues at West Indies Lab—Bob Dill, Bill and Betsy Gladfelter, John and Nancy Ogden, Ivan Gill, Randy Burke, Percy and Wilmot, Hank Tonnemacher, Rich Berey. . . . and all those who brought me balance – my mom, Karla, Martha and Amy. And finally, I would be remiss if I did not thank the editors at Springer for all the help they have given us along the way. Martine van Bezooijen, in particular, encouraged us to undertake the effort in the first place. Unfortunately, her tragic passing will not let her see the end product. However, we have been fortunate to have had the continued support of Judith Terpos and Alexandrine Cheronet, without whom this would have been impossible.

Contents

Contributors

Richard B. Aronson Department of Biological Sciences, Florida Institute of Technology, Melbourne, FL, USA

Wolf-Christian Dullo Ocean Circulation und Climate Dynamics, Helmholtz-Zentrum für Ozeanforschung, Kiel, Germany

Alastair R. Harborne Department of Biological Sciences, Florida International University, North Miami, FL, USA

Marine Spatial Ecology Laboratory, School of Biological Sciences and ARC Centre of Excellence for Coral Reef Studies, University of Queensland, Brisbane, QLD, Australia

Dennis K. Hubbard Department of Geology, Oberlin College, Oberlin, OH, USA

Paul L. Jokiel Hawaii Institute of Marine Biology, University of Hawaii, Kaneohe, HI, USA

Christopher P. Jury Hawaii Institute of Marine Biology, University of Hawaii, Kaneohe, HI, USA

Ilsa B. Kuffner U.S. Geological Survey, St. Petersburg Coastal and Marine Science Center, St. Petersburg, FL, USA

Halard Lescinsky Department of Biology and Earth Science, Otterbein University, Westerville, OH, USA

Jere H. Lipps John D. Cooper Archaeological and Paleontological Center, Santa Ana, CA, USA

Museum of Paleontology, University of California, Berkeley, CA, USA

Jeff Miller South Florida/Caribbean Inventory and Monitoring Network, US National Park Service, St. John, VI, USA

Chris T. Perry Department of Geography, College of Life and Environmental Sciences, University of Exeter, Exeter, UK

William F. Precht Division of Marine and Coastal Programs, Dial Cordy and Associates, Inc., Miami, FL, USA

Caroline S. Rogers Wetland and Aquatic Research Center, Caribbean Field Station, US Geological Survey, St. John, VI, USA

George D. Stanley, Jr. Department of Geosciences and Paleontology Center, The University of Montana, Missoula, MT, USA

Janie Wulff Department of Biological Science, Florida State University, Tallahassee, FL, USA

Smithsonian Tropical Research Institute, Panama

Abbreviations and Acronyms

[X]	Concentration of X, e.g., $[H^+]$
^{14}C	Amount of radioactive carbon relative to stable carbon; used for age dating
AGRRA	Atlantic and Gulf Rapid Reef Assessment
AMS	Accelerator mass spectrometry; for age dating
A_T	Total alkalinity
ATP	Adenosine triphosphate
az-coral	Azooxanthellate corals
Ba	Billion years ago (age)
BBD	Black band disease
BBL	Benthic boundary layer
BL	Boundary layer
By	Billion years (time interval)
CA	Carbonic anhydrase
CalBP	Calendar years before present (1950)
CLIMAP	Climate: Long-range Investigation, Mapping, and Prediction
COTS	Crown-of-thorns starfish
DBL	Diffusion boundary layer
DIC	Dissolved inorganic carbon
DOM	Dissolved organic matter
ΔSL	Difference in sea level from today to some time in the past
ENSO	El Niño Southern Oscillation
FST	Falling systems tract
GBR	Great Barrier Reef
GCRMN	Global Coral Reef Monitoring Network
Gnet	Net calcification
GPS	Global positioning system
HST	Highstand systems tract
HTM	Holocene thermal maximum
ICRS	International Coral Reef Symposium
IPCC	Intergovernmental Panel on Climate Change
ISRS	International Society for Reef Studies
K-Pg	Cretaceous-Paleogene
K_{sp}	Solubility product
LEC	Light-enhanced calcification
LGM	Last Glacial Maximum
LIDAR	Light detection and ranging (a high-precision altimetry system)
LIT	Line intercept transect
LST	Lowstand systems tract
Ma	Million years ago (age)
MBL	Momentum boundary layer
MPA	Marine protected area
MWP	Meltwater pulse

My	Million years (time interval)
MYA	Millions of years ago
NASA	National Aeronautics and Space Administration
OA	Ocean acidification
OAE	Ocean anoxia event
P	Pressure
P-TR	Permo-Triassic
pCO_2	Partial pressure of CO_2 (in air or water)
PETM	Paleocene-Eocene Thermal Maximum
Pg	Gross (total) carbonate produced on a reef
pHT	pH at a stated temperature (in Chap. 2 = 25 °C)
PIT	Point Intercept Transect
P_n	Net amount of carbonate remaining on a reef after sediment is exported
P_{net}	Net photosynthesis
S	Salinity
S_R	Amount of sediment removed from a reef
SST	Sea surface temperature
SW	Seawater
T	Temperature
TAZ	Taphonomically active zone
TST	Transgressive systems tract
U/Th	Ratio of uranium to thorium; used for age dating
WBD	White band disease
ybp	Years before present
z-coral	Zooxanthellate coral
ZC	Zone of rapid calcification
ZP	Zone of rapid photosynthesis
μatm	Microatmospheres: gas pressure, equal to its mixing ratio in parts per million (ppm)
ΩX	Saturation state of X

Coral Reefs at the Crossroads – An Introduction

1

Dennis K. Hubbard

Abstract

Coral reefs are complex systems that are difficult to fully understand when viewed from a single perspective. As we have separated ourselves into increasingly smaller and more specific disciplines, we often lose sight of important connections between physical and biological factors and how they can change over different spatial and temporal scales. As stresses on these robust yet fragile systems broaden and deepen, it is becoming increasingly important that we break down artificial disciplinary barriers and ask questions that are difficult to frame from a single scientific perspective. This chapter provides a jumping-off point to examine coral reefs – sitting at both a disciplinary and a temporal crossroads.

Keywords

Interdisciplinary • Multidisciplinary • Scale • Perspective

Perhaps more than any other earth system, coral reefs sit at the crossroads of science. While organisms largely provide the raw materials, reef building responds to a wide range of non-biological processes. Water temperature confines today's scleractinean coral reefs to a narrow equatorial belt. Waves & currents control regional patterns of coral dispersal, dictate where larvae might or might not survive and create the critical linkages between reefs across individual ocean provinces. The underlying edifice provides a physical structure upon which wave energy, light intensity, sedimentation and chemical cues conspire to create zonation, the fundamental underpinning of reef characterization both today and in the geologic past.

The feedbacks among all these processes are myriad; spatial complexity is built by calcifying organisms that are in turn dependent on both surface topography and the vast network of environments on and in the reef for their success. More than 85 % of the reef's surface area exists within cryptic spaces that house roughly half of the fish species (Chap. 10) and similar proportions of other organisms living within the ecosystem boundaries. Understanding the nature and significance of these complex interactions is critical whether we are considering modern reefs (Chaps. 2, 4 and 5), their forebears throughout geologic time (Chaps. 3, 6, 7, 8, 9 and 10), or their descendants in an increasingly stressful world (Chaps. 9, 11 and 12).

Most recently, coral reefs have reached another crossroads. They evolved over the past 500 million years (Chap. 8) in response to large-scale changes in ocean processes (Chap. 9). While truly "pristine" reefs may not have existed for a very long time (Pauly 1995; Jackson 1997; Pandolfi et al. 2005), the past three decades have witnessed fundamental changes that have been far greater than anything documented over just the previous century (Wilkinson 2008; Jackson et al. 2014). Providing effective strategies for documenting and mitigating recent decline (Chap. 12) will require an approach that embraces a variety of disciplines that can transform scientific understanding into social will and political implementation. This volume will hopefully provide a starting point for reconnecting the different perspectives from which we view reef development.

D.K. Hubbard (✉)
Department of Geology, Oberlin College, 52 W. Lorain St., Oberlin, OH 4074, USA
e-mail: Dennis.Hubbard@oberlin.edu

D.K. Hubbard et al. (eds.), *Coral Reefs at the Crossroads*,
Coral Reefs of the World 6, DOI 10.1007/978-94-017-7567-0_1

The persistence of this critical natural system may depend on our ability to tie together seemingly disparate views. In any event, this will certainly make our individual understanding richer.... and a lot more fun.

1.1 Coming Together

The proceedings of the first International Coral Reef Symposium in 1969 contained only 22 scientific papers. However, they covered 11 broad topics ranging from reef distribution and ecology to the geology of uplifted islands, research methods, and even the history of reef science – while spanning nine separate ocean regions. The number of papers from the second meeting increased fivefold and the topics were broadly distributed among biology, geology, chemistry, physical oceanography and management across an even wider geographic range. Presentations included some of our earliest discussions of coral biogeochemistry (Smith 1974), reef controls by sea-level rise (Hopeley 1974) and reef accretion in both the Atlantic (Land 1974) and the Pacific (Tracey and Ladd 1974).

Just as important was the consistent intersection of disciplines focusing on a single theme – coral reefs. The idea in the earlier meetings was to encourage the movement of participants freely and often from one topic or session to another. Their most unique element was that participants were drawn by a single interest in reefs and not the discipline that they occupied for the other 360 or so days of the year. More so than many that followed, the earlier ICRS meetings reminded us that we could learn more in a diverse group asking what we don't know than with close colleagues discussing what we think we do.

Our perceptions of coral reefs vary both spatially and temporally. Biologists can observe and manipulate complex processes in ways that are impossible for geologists to reproduce. However, these are generally limited to small areas and short intervals of time. Geologists have traditionally relied on biological models to understand the past, but are increasingly using their longer perspective to provide a view of reefs absent the stresses of climate change and other manifestations of human proliferation. Chemists can look at both minute and broad scales, providing what would be otherwise overlooked processes. Modelers can take seemingly disparate observations and combine them into simulations that can test existing ideas and generate new ones begging for field data. More recently, monitoring and management have increasingly relied on information generated from scientific studies to make wise decisions while, at the same time, asking questions that science has not yet realized are "important".

As we ponder the recent and dramatic changes on coral reefs, it is difficult to quantify factors beyond the scale of a single experiment or an individual researcher's career. Conversely, geologists consider broad expanses of time and space, but too often forget that this larger view is the cumulative result of small and short-lived events. The collective record was probably controlled more by these day-to-day factors than we acknowledge, and millennial scale processes alone cannot explain the time-averaged fossil record. Chemistry, physics, oceanography and a host of other related disciplines likewise contribute to the overall picture but, like biology and geology, each has its own unique perspectives, priorities and limitations. And, as short as the biological time-scale might seem to geologists, the election cycle of politicians and policy-makers can render coral reefs as little more than distractions. However, when all the perspectives of too-often disparate groups are combined effectively, they can provide insight that is impossible within any single discipline. This realization was at the core of the early reef symposia where much time was spent just talking about "how reefs work". This volume hopes to rekindle interests in viewing common problems from different perspectives.... together.

1.2 Our Changing View

On the morning of May 5th, 1961, Alan Shepard left Cape Canaveral on America's first manned flight into space, a journey that would last only 15 min and 22 s. Four minutes after launch, he deployed his periscope (windows were not added until two flights later) and reported, " What a beautiful view.... I can see Okeechobee.... identify Andros Island.... identify the reefs" (Hammack et al. 1961). Within a minute, he fired the first of three retro-rockets that would bring him back to the surface just 500 km from where he began.

Shepard's view from *Freedom 7* (Fig. 1.1) represents our earliest remote observation of Earth from space[1].... and what he saw were the reefs along Great Bahama Bank. Since then, we have seen photos of Earth from the moon ... and even farther as Voyager 1 departed the solar system in August of 2012. As a result, we have become accustomed to the spectacular images generated by manned spacecraft and orbiting satellites (e.g., Fig. 1.2). Students who have easy access to images on "Google Earth" and "Google Maps" on their cell phones take these for granted, failing to appreciate the limited perspective from low-flying aircraft in the latter twentieth century, just as we failed to appreciate challenges on the deck of the Beagle and other nineteenth

[1] Three weeks earlier (April 12, 1961), Yuri Gagarin had completed a single orbit around Earth. However, the small viewing port beneath his feet was configured to view Earth only for a final alignment during re-entry.

Fig. 1.1 Photograph taken by the automated camera system of *Freedom 7* on Alan Shepard's first US space flight in 1961 (Courtesy of NASA)

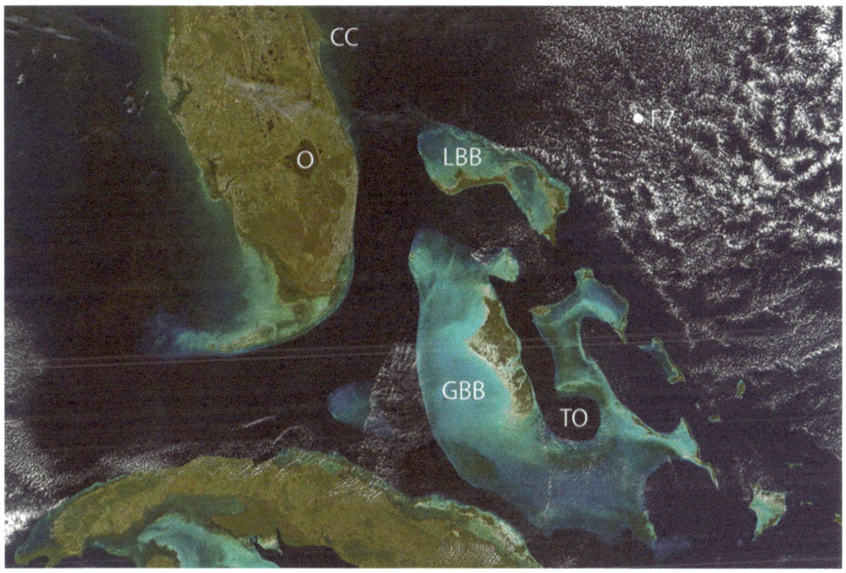

Fig. 1.2 NASA image of Florida and the northern Bahamas. The flight of *Freedom 7* lasted less than 16 min and covered only 500 km from Cape Canaveral (*CC*) to the "splashdown" site (F7) northeast of Little Bahama Bank (LBB). *O* Lake Okeechobee, *GBB* Great Bahama Bank, *TO* Tongue of the Ocean (Courtesy of NASA)

century sailing vessels. We have access to terabytes, petabytes, or even zettabytes of data and increasingly rely on satellite images, huge banks of remotely collected data and complex computer models of natural systems to conceptualize processes operating at scales ranging from microscopic to global.

However, this ever-broadening view has been accompanied by a narrowing of our individual focus. At the time of Darwin, natural philosophy blurred the boundaries between biology, geology, chemistry, physics and even the humanities. His seminal ideas on the role of subsidence in the evolution of Pacific reefs from narrow fringes along volcanic slopes to atolls were born not from the lofty perspective of orbiting satellites. Rather, they were the logical explanation for patterns revealed in early maps and the logs of observant seamen. According to Darwin, the solution was so obvious that "the whole theory was thought out on the west coast of S. America before I had seen a true coral reef".[2]

Today, the evolving scientific landscape has increasingly "organized" us into rigid disciplines or even sub-disciplines. Meaningful discussions still occur, but the goal is more often to seek validation or clarification of specific concepts than it is to question what we might be missing by staying in a familiar intellectual space. Even scientists working on large vessels that are funded by multi-disciplinary programs too often have separate research agendas and work on deck at

[2] Barlow N (1958) The autobiography of Charles Darwin 1809-1882, with the original omissions restored: http://darwinonline.org.uk/content/frameset?itemID=F1497&viewtype=text&pageseq=1, page 98.

different hours to maximize costly ship time. It is more common for different research groups to focus on their own piece of the larger puzzle than it is to look for questions that can only grow out of interactions within a broader group. There are exceptions, but they are too few.

Past attempts to bring different perspectives together in a single volume have still tended to focus largely on one discipline, perhaps adding a thoughtful contribution from another for context. For example, treatments of reefs through time have tended to use a description of modern reefs as a backdrop against which a primarily historical discussion of deep time can be set. Conversely, a volume might start with a broad-brush treatment of reef controls or evolutionary changes in reefs over time to introduce the largely biological themes that follow. Even the seminal *Biology and Geology of Coral Reefs* was organized in four volumes that tended to treat the two disciplines separately.

The goal in the following pages is to focus on a few broader themes, using contributions by different authors to highlight alternative ways of thinking about each. Obviously, this cannot be exhaustive either within or between topics. The main strategy is to group contributions that address a few important areas from different perspectives. Our hope is that readers drawn to chapters written by experts in their discipline will also examine related chapters that consider the same topic from a different viewpoint.... and that this will inspire them to look elsewhere in the volume and in the annals of reef studies with a new eye.

1.3 A Brief Look Back

In the nineteenth century, naturalists struggled to understand both the structure of coral reefs and the distribution of organisms that inhabited them. Much of the early understanding of coral zonation came from sounding leads and dredging. However, even with the crude methods available at the time, the depth limits for most modern corals were surprisingly well constrained to between 20 and 30 m, seemingly at odds with suggestions that reefs appeared to have built from significant depths.

The solution came from geologists in the form of subsidence. Charles Lyell (1832) suggested that atolls might have formed atop the rims of volcanic craters. As they sank, reef building offset subsidence, resulting in accumulations much thicker than the depth range across which corals were known to occur. Charles Darwin (1842) considered the specific tie to crater rims to be "a monstrous hypothesis"[3] and suggested instead that reefs evolved from fringes along the flanks of subsiding volcanic cones to barrier reefs and atolls as the central landmass sank beneath the waves. Darwin's (1842) answer for "the coral reef problem" spurred a heated debate that would last for over half a century.

Mojsisovic (1879) similarly argued that the well-developed coral reefs preserved in the Dolomites were the result of major tectonic uplift rather than a biblical deluge or any other upward excursion of sea level. His argument benefitted from the general acceptance of uplift as an important geologic process. However, the evidence for Darwin's subsidence hypothesis lay beneath the water, making his ideas just as problematic for empiricists like Alexander Aggasiz as they had been for biblical literalists like his father, Louis. The debate continued for over half a century until cores on Funifuti (Royal Society of London 1904) and Bikini Atolls (Emery et al. 1954; Tracey and Ladd 1974) revealed their volcanic ancestry.

The link between tectonic forces and carbonate island building remained the "important" question of the day... so much so that William Morris Davis (1928) characterized an overnight stay on the reef off Cairns as, "entirely fruitless as far as the origin of the reef is concerned".[4] But, the "age of reef ecology" would soon come. A host of marine biological labs can trace their roots back to at least the nineteenth century, but anything akin to modern coral-reef ecology had to wait for the development of the demand regulator (aka scuba) in 1943. This opened an era of intense exploration that allowed us to closely observe, measure and photograph marine systems. In the early 1950s, Tom Goreau visited Discovery Bay where he would eventually create a small marine lab in 1965. For decades, it grew and attracted scientists from different disciplines who repeatedly demonstrated the value of interdisciplinary study focusing on a specific natural system – the coral reef. Another notable Caribbean example was the West Indies Laboratory in the U.S. Virgin Islands, the brainchild of H. G. Multer and Fairleigh Dickinson Jr. Many marine scientists, some of them contributors to this volume, benefitted from the thoughtful discussions among mentors and peers brought together at these two facilities. Marine labs and field stations have come and gone, but the latter part of the twentieth century marked what was arguably an unparalleled growth of interdisciplinary, field-based, coral-reef studies, much of this owing to such places.

[3] Letter to Caroline S. Darwin dated 29 April, 1936: http://www.darwinproject.ac.uk/entry-301#mark-301.f2

[4] See the discussion of W.M. Davis' support of Darwin's subsidence theory by Hopley (1982).

1.4 Where Are We Now?

Until the 1980s, researchers spent most of their time documenting and explaining complex interactions among reef organisms and the edifice where they lived. The explosion of predators like *Acanthaster* on the Great Barrier Reef and the rapid decline of coral cover on Atlantic and Caribbean reefs suddenly expanded the "mundane" and underappreciated activity of coral-reef monitoring. The focus was more on decline than it was on function, and arguments over the relative importance of top-down (mostly overfishing) versus bottom-up stresses (e.g., nutrients) reflected a perception that impacts were to be found on this side of the horizon.

Arguments over methodology were often as heated as those over the dominant causes of decline and the solutions that might reverse it. What level of decline do we want to detect? Do we measure using fixed quadrats, rigid linear transects or irregular ones that take into account the spatial complexity of the reef surface? Direct measurement of coral abundance was the standard but required long and expensive hours spent underwater. Photographs and video were more efficient in the field but too often could not resolve small organisms, especially in hidden cryptic spaces. As field costs soared and image resolution improved, the scales gradually tipped toward photographic or video surveys. While we might think of this as a unique evolution of methods in response to new and specific needs, we should consider that the earliest practitioners had already had this discussion. While neither of the cameras in Fig. 1.3 enjoyed widespread use, the principles inherent to both underwater documentation and monitoring owe their origins to these and other early attempts.

Most recently, rising temperatures and changing ocean chemistry have broadened the discussion. The inadequacy of any single discipline to fully understand these and other problems we face should come as no surprise. Chemists and biologists have combined forces to address possible impacts of ocean acidification. Detailed genetic studies have revised our taxonomy and have provided unique ways to track evolutionary patterns of reef organisms and pathogens that threaten them. Remote sensing has evolved a complex alert system for bleaching and disease. Huge databases provide valuable repositories for information that can be combined to address critical problems, some never envisioned by their creators. And modeling can combine this information with new field and laboratory data to better constrain the controls of observed patterns and address future scenarios that we are yet to experience. Nevertheless, problems are growing faster than resources to study them and we need to figure out how to better triage the growing list of impacted species and systems. The answer is arguably the greater rigor and power of questions and protocols generated by groups of diverse investigators with related interests.

1.5 Where Are We Headed?

Wendell Berey (1987) reminded us that, "in order to understand what we are doing, we need to understand what nature would be doing if we were doing nothing." Observations and

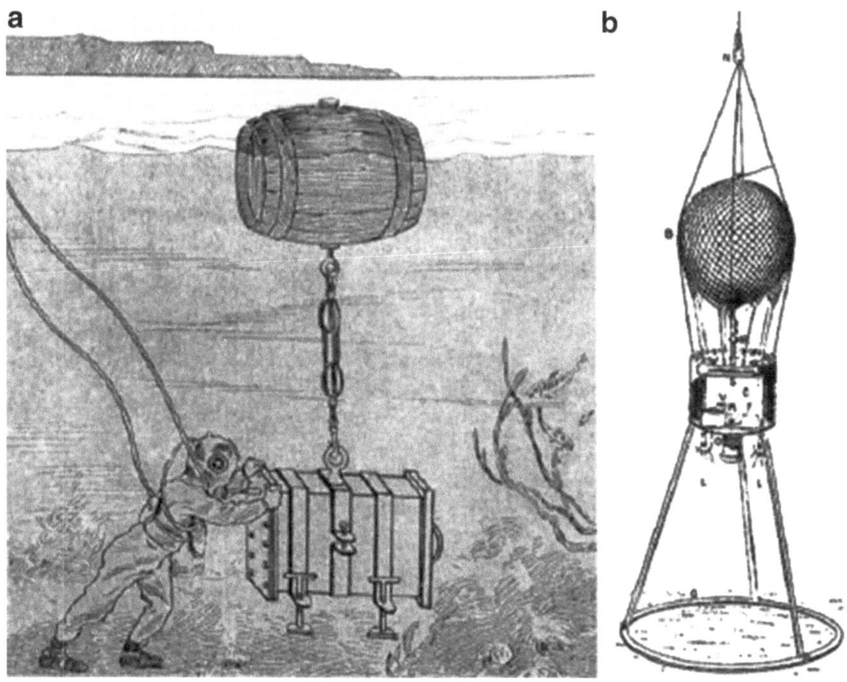

Fig. 1.3 Early underwater cameras. (**a**) Underwater camera system designed by Louis Boutan in the late nineteenth century for recording general underwater scenes. Low light levels and the insensitivity of early photographic plates required exposure times of up to 30 min (From Boutan 1900, p. 198). (**b**) Proposed camera for photographing the seabed. This apparatus, conceived by Regnard (1891, p. 72), was never put into practical use

measurements on modern reefs provide important information that allows us to better understand the nature and the magnitude of recent change. However, they have often been short-lived and all of them record changes that occurred long after the first human stresses were applied. Historical records predate our most ambitious monitoring efforts, the lifetime of a scientist, and especially the attention span of the political body. However, they still fail to accurately record the accumulated stresses that were already contributing to environmental instability – even if their impacts remained hidden until recently. The geological past provides an opportunity to more-realistically consider reefs when "we were doing nothing", but the record is both intermittent and incomplete due to selective preservation and time averaging. However insightful any approach might be by itself, when combined with others, it can help us with the triage we are currently undertaking – whether we are looking to the past, trying to predict the future, or just want to understand how a reef works.

References

Berey W (1987) Home economics. North Point Press

Boutan, L (1900) La photographie sous-marine et les progrès de la photographie. Paris: Schleicher Fréres. available on line at gallica.bnf.fr/ark:/12148/bpt6k10250505/f21.image.r=

Darwin C (1842) The structure and distribution of coral reefs. London, Smith Elder and Co. 215 p

Davis WM (1928) The coral reef problem. Amer Geogr Soc Spec Pub No 9, 596 p

Emery KO, Tracey JI, Ladd HS (1954) Geology of Bikini and nearby atolls, Geol Surv Prof Paper 260-A, U. S. Government Printing Office, Washington D.C. 345p

Hammack JB, Smith N, Hodge JD, Slayton DK, Augerson W, Donnelly PC (1961) "Postlaunch Report for Mercury-Redstone No. 3 (MR-3)", NASA Project Mercury Working paper No. 192

Hopeley D (1974) Investigation of sea level changes along the Great Barrier Reef coastline. Proc 2nd Intl Coral Reef Symp 2: 551–562

Hopley D (1982) The geomorphology of the Great Barrier Reef: Quaternary development of coral reefs. John Wiley and Sons, New York, 453p

Jackson JBC (1997) Reefs before Columbus. Coral Reefs 16: S23-S32

Jackson J. Donovan M, Cramer K, Lam V (2014) Status and trends of Caribbean coral reefs: 1970-2012. Global Coral Reef Monitoring Network, IUCN, Gland Switzerland, 304 p

Land LS (1974) Growth rate of a West Indian (Jamaican) reef. Proc 2nd Intl Coral Reef Symp. 2: 409–412

Lyell C (1832) Principles of geology, being an attempt to explain the former changes of the Earth's surface, by reference to causes now in operation. London: John Murray 2: 286 p

Mojsisovic E (1879) Die Dolomit-riffe von Südtirol und Venetien: Beitrage zur Bildungsgeschichte der Alpen. Alfred Holder. Vienna. 551 p

Pandolfi JM, Jackson JBC, Baron N, Bradbury RH, Guzman HM, Hughes TP, Kappel CV, Micheli F, Ogden JC, Possingham HC, Sala E (2005) Are U.S. Coral Reefs on the Slippery Slope to Slime? Science 307:1725–1726

Pauly, Daniel (1995) Anecdotes and the shifting baseline syndrome of fisheries. Trends in Ecology and Evolution, 10:430.

Regnard P (1891) Recherches expérimentales sur les conditions physiques de la vié dans les eaux. Paris: G. Masson. available on line at http://www.scielo.br/scielo.php?pid=S0104-59702014000301029&script=sci_arttext&tlng=en

Royal Society of London (1904) The atoll of Funafuti: borings into a coral reef and the results. Harrison and Sons, London, 428p

Smith SV (1974) Coral reef carbon dioxide flux. Proc 2nd Intl Coral Reef Symp 1:77–86

Tracey JI, Ladd HS (1974) Quaternary history of Eniwetok and Biukini Atolls, Marshall Islands. Proc 2nd Intl Coral Reef Symp 2:537–550

Wilkinson C (2008) Status of the Coral Reefs of the World: 2008. Global Coral Reef Monitoring Network and Reef and Rainforest Research Centre, Townsville, Australia 296p

Coral Calcification and Ocean Acidification

2

Paul L. Jokiel, Christopher P. Jury, and Ilsa B. Kuffner

Abstract

Over 60 years ago, the discovery that light increased calcification in the coral plant-animal symbiosis triggered interest in explaining the phenomenon and understanding the mechanisms involved. Major findings along the way include the observation that carbon fixed by photosynthesis in the zooxanthellae is translocated to animal cells throughout the colony and that corals can therefore live as autotrophs in many situations. Recent research has focused on explaining the observed reduction in calcification rate with increasing ocean acidification (OA). Experiments have shown a direct correlation between declining ocean pH, declining aragonite saturation state (Ω_{arag}), declining $[CO_3^{2-}]$ and coral calcification. Nearly all previous reports on OA identify Ω_{arag} or its surrogate $[CO_3^{2-}]$ as the factor driving coral calcification. However, the alternate "Proton Flux Hypothesis" stated that coral calcification is controlled by diffusion limitation of net H^+ transport through the boundary layer in relation to availability of dissolved inorganic carbon (DIC). The "Two Compartment Proton Flux Model" expanded this explanation and synthesized diverse observations into a universal model that explains many paradoxes of coral metabolism, morphology and plasticity of growth form in addition to observed coral skeletal growth response to OA. It is now clear that irradiance is the main driver of net photosynthesis (P_{net}), which in turn drives net calcification (G_{net}), and alters pH in the bulk water surrounding the coral. P_{net} controls $[CO_3^{2-}]$ and thus Ω_{arag} of the bulk water over the diel cycle. Changes in Ω_{arag} and pH lag behind G_{net} throughout the daily cycle by two or more hours. The flux rate P_{net}, rather than concentration-based parameters (e.g., Ω_{arag}, $[CO_3^{2-}]$, pH and $[DIC]:[H^+]$ ratio) is the primary driver of G_{net}. Daytime coral metabolism rapidly removes DIC from the bulk seawater. Photosynthesis increases the bulk seawater pH while providing the energy that drives calcification and increases in G_{net}. These relationships result in a correlation between G_{net} and Ω_{arag}, with both parameters being variables dependent on P_{net}. Consequently the correlation between G_{net} and Ω_{arag} varies widely between different locations and times depending on the relative metabolic contributions of various calcifying and photosynthesizing organisms and local rates of carbonate dissolution. High rates of H^+ efflux continue for several hours following the mid-day G_{net} peak suggesting that corals have difficulty in shedding waste protons as described by the Proton Flux Model. DIC flux (uptake) tracks P_{net} and G_{net} and drops off rapidly after the photosynthesis-calcification maxima, indicating that corals can cope more effectively with the problem of limited DIC supply compared to the problem of eliminating

P.L. Jokiel (✉) • C.P. Jury
Hawaii Institute of Marine Biology, University of Hawaii, 1346Kaneohe, HI 96744, USA
e-mail: jokiel@hawaii.edu

I.B. Kuffner
U.S. Geological Survey, St. Petersburg Coastal and Marine Science Center, 600 4th Street South, St. Petersburg, FL 33701, USA

D.K. Hubbard et al. (eds.), *Coral Reefs at the Crossroads*, Coral Reefs of the World 6, DOI 10.1007/978-94-017-7567-0_2

H^+. Predictive models of future global changes in coral and coral reef growth based on oceanic Ω_{arag} must include the influence of future changes in localized P_{net} on G_{net} as well as changes in rates of reef carbonate dissolution. The correlation between Ω_{arag} and G_{net} over the diel cycle is simply the result of increasing pH due to photosynthesis that shifts the CO_2-carbonate system equilibria to increase $[CO_3^{2-}]$ relative to the other DIC components of $[HCO_3^-]$ and $[CO_2]$. Therefore Ω_{arag} closely tracks pH as an effect of P_{net}, which also drives changes in G_{net}. Measurements of DIC flux and H^+ flux are far more useful than concentrations in describing coral metabolism dynamics. Coral reefs are systems that exist in constant disequilibrium with the water column.

Keywords
Calcification • Corals • Ocean acidification • Seawater CO_2-carbonate system • Aragonite saturation state • Boundary layers • Phase lag

2.1 Introduction

Reviews have recently been published on coral calcification (Allemand et al. 2011), on the effects of ocean acidification on coral calcification (Erez et al. 2011) and on the geological record of ocean acidification (Hönisch et al. 2012). These documents provide a wealth of background information. This chapter provides an updated synthesis including new insights on coral physiology and calcification relevant to the geology and paleo-ecology of coral reefs.

2.1.1 Basic Coral Anatomy and Physiology

Reef corals are coelenterates formed by an outer body wall and a basal body wall that enclose a space called the coelenteron. Terminology used here follows that of Galloway et al. (2007). The outer body wall in contact with sea water consists of two tissue layers – an outer epidermis and an inner gastrodermis separated by a jelly-like substance called mesoglea (Fig. 2.1a). Likewise, the basal body wall is a mirror image that consists of the calicodermis and a gastrodermis separated by mesoglea. The space between the two body walls is a cavity called the coelenteron, which interconnects the polyps of the colony and opens to the external seawater through the polyp mouths. The intracellular symbiotic zooxanthellae reside mainly within the cells of the gastrodermis of the surface body wall. The zooxanthellae are photosynthetic and are capable of providing all of the energy needed for basic metabolism of the coral (Muscatine et al. 1984). However, heterotrophic food inputs are still important. Well-fed corals exhibit higher growth rates and greater stress tolerance compared to less-fed colonies (Ferrier-Pagès et al. 2003; Grottoli et al. 2006; Edmunds 2011; Connolly et al. 2012). Calcification occurs in the calcifying fluid located between the calicodermis and the skeleton. A presumed proton transfer process increases the pH and saturation state of the fluid to a point where $CaCO_3$ crystallizes onto the skeleton as aragonite (Furla et al. 2000a, 2000b; Cohen and McConnaughey 2003; Allemand et al. 2004; Cohen and Holcomb 2009; Venn et al. 2011). Energy is needed to drive this process with up to 30 % of the coral's energy budget devoted to calcification (Allemand et al. 2011).

The contemporary four cell-layer structure with metabolic pathways as proposed by Furla et al. (2000a, 2000b) and Allemand et al. (2004) is shown in Fig. 2.1a. This model requires neutralization of the H^+ produced by calcification using OH^- produced by photosynthesis. However, there is a contradiction. The distal areas of the corallum that are growing most rapidly lack gastrodermal cells and their contained zooxanthellae (Gladfelter 1982; Brown et al. 1983; Gladfelter 1983; Tambutté et al. 2007). Jokiel (2011a) hypothesized that H^+ is released directly into the water column in rapidly calcifying areas of the coral (Fig. 2.1b). An alternative explanation is that OH^- is transported from areas of the coral undergoing rapid photosynthesis to areas of the coral undergoing rapid calcification. McConnaughey and Whelan (1997) proposed that calcification at branch tips could discharge protons into seawater within the coelenteron. This water could be transported by ciliary currents to the abundant photosynthetic zooxanthellae in the lateral polyps.

Most studies involve incubation of corals in static containers under controlled conditions with extrapolation of the changes measured in the carbonate-CO_2 chemistry of bulk seawater to precipitation of $CaCO_3$ in the calcifying fluid adjacent to the coral skeleton. These results must be viewed with caution because there is an organism located between the calcifying space and the bulk water being measured as well as a boundary layer (BL) between the organism and the water column. Calcification is under biological control and mediated by organic tissue that separates the calcifying surface from overlying seawater. Therefore calcification occurs in a medium (i.e. the

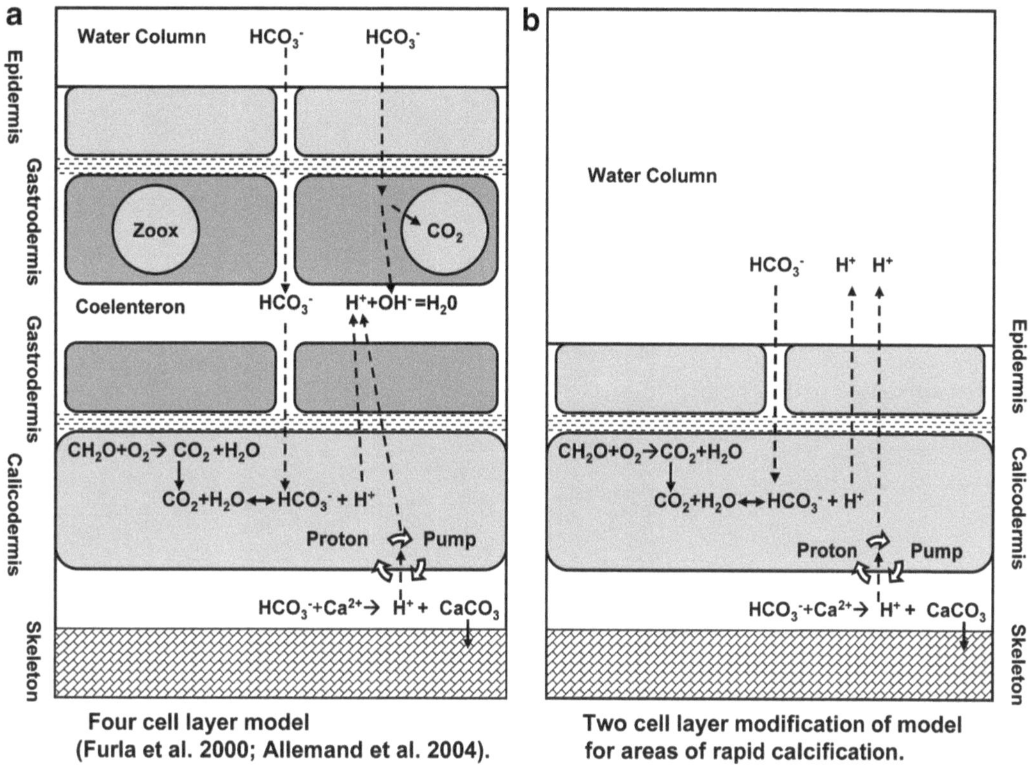

Fig. 2.1 Classic four cell-layer model of calcification (**a**) compared to two cell layer structure of rapidly calcifying areas of the corallum (**b**) as described by Tambutté et al. (2007). Note that protons generated by calcification in (**b**) are shown being released directly into the water column rather than being neutralized by photosynthesis as proposed by Furla et al. 2000a, 2000b; and Allemand et al. 2004 (Figure from Jokiel (2011b) used with permission from the Journal of Experimental Marine Biology and Ecology)

calcifying fluid) that has different carbonate-CO_2 chemistry than the bulk seawater as materials are exchanged through the BL. Additional information on processes occurring within the coral tissues and the BL has been provided through use of microprobes (Kühl et al. 1995; Al-Horani et al. 2003a, 2005a), isotope chemistry (Goreau 1977; Allison et al. 1996; Al-Horani et al. 2005a) and direct measurement of pH within coral tissues (Venn et al. 2009, 2011, 2013). Most of the models have focused on rates of biological processes that occur at the interface between the calicodermis and the coral skeleton (Fig. 2.1a). More recently, Jokiel (2011a, 2011b) has developed a model based on physical control of material flux through the BL and into the water column (Fig. 2.1b).

2.1.2 Coral Morphology

The growth forms of reef corals (Fig. 2.2) are extremely varied (Veron 2000), which has confounded understanding of basic metabolic processes and patterns of calcification. How can a simple organism consisting of only two tissue layers with a total of four cell layers produce so many intricate growth forms? The key to understanding lies in the observation (Fig. 2.3) that all coral growth forms can be reduced to the topological equivalent of a hemisphere containing the photosynthetic polyps and/or tissues containing dense concentrations of zooxanthellae (zone of rapid photosynthesis or ZP) surrounded by a hemisphere dominated by calcification polyps and/or tissues devoid of zooxanthellae (zone of rapid calcification or ZC). Cells and polyps located in the distal portions of a colony (ZC) have few or no zooxanthellae, giving these areas a white appearance (Figs. 2.2 and 2.3).

2.1.3 Models of Light Enhanced Calcification (LEC)

The discovery that calcification in reef corals is accelerated in the light (Kawaguti and Sakumoto 1948) led to the conclusion that photosynthesis by zooxanthellae must somehow be involved in the biochemical pathways of calcification. Experimental evidence was eventually developed by Vandermeulen et al. (1972) who showed that blocking photosynthesis results in a marked reduction in calcification. A number of LEC models have been presented (reviewed by Gattuso et al. 1999; Cohen and Holcomb 2009; Allemand et al. 2011). Goreau (1959) proposed that calcification is

Fig. 2.2 Variation in coral morphology (**a**) branch tip of *Acropora palmata*, (**b**) colony of branching *Acropora humilis* (**c**) plate-like colony of *Montipora capitata* (**d**) encrusting *Porites rus* (**e**) foliose *Turbinaria* sp. (**f**) solitary coral *Fungia scutaria* (**g**) cross-section skeleton of a massive *Porites* sp. (**h**) branched, non-scleractinian hydrocoral *Millepora tenera*. Note that all growth forms lack zooxanthellae on the rapidly growing distal branch tips, distal plate margins, and distal edges of septae and trabeculae as shown in Fig. 2.3

accelerated in light due to removal of CO_2 from calcification sites by photosynthetic zooxanthellae. This model requires the zooxanthellae to be located at or near the calcification site, but actually they are located far from the site of calcification (Fig. 2.1). Further, the proposed chemical reactions have not been supported by experimental data. Simkiss (1964) advanced a model based on the removal of phosphate "crystal poisons" from calcification sites by photosynthetic zooxanthellae, but this model also suffers from the requirement that zooxanthellae must be located close to calcification sites. A similar explanation for LEC is that the zooxanthellae act as kidneys to remove the metabolic wastes in the coral animal that can inhibit calcification (Yonge 1968; Crossland and Barnes 1974). Muscatine (1990) suggested that perhaps photosynthesis and calcification are not connected through carbonate chemistry, but rather show a linkage simply because photosynthesis provides energy for calcification. This view was supported by the work of Colombo-Pallotta et al. (2010) who report that calcification in symbiotic corals is not strictly a "light-enhanced" or "dark-repressed" process, but rather, the products of photosynthesis have a critical role in calcification, which should be viewed as a "photosynthesis-driven" process.

Several recent models of coral calcification involve the zooxanthellae in the removal or neutralization of excess protons produced by calcification. McConnaughey and Whelan (1997) proposed that calcification in corals enhances photosynthesis by providing a source of protons that convert seawater HCO_3^- to CO_2 and H_2O, thereby supplying some of the CO_2 used in photosynthesis. Furla et al. (2000a)

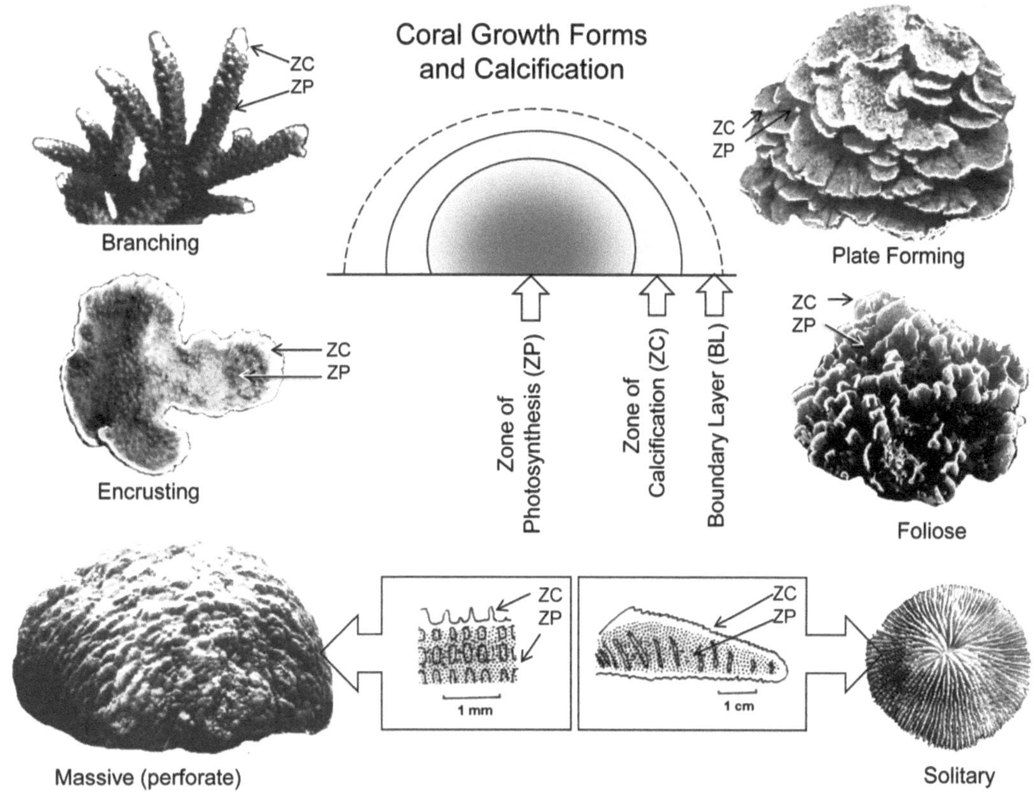

Fig. 2.3 Coral growth forms showing areas of rapid calcification (*ZC*) in relation to areas of photosynthesis (*ZP*) relative to the boundary layer (*BL*). In every case, the ZC is located between the ZP and the BL

determined that the major source of total dissolved inorganic carbon (DIC) used in calcification is from respiration (70–75 % of total $CaCO_3$ deposition), while only 25–30 % originates from the external seawater. The models of Furla et al. (2000a, 2000b) and Allemand et al. (1998) involve various pathways for buffering the H^+ produced during calcification using OH^- produced by photosynthesis (Fig. 2.1a). In contrast, a more recent model (Jokiel 2011a, 2011b, 2013; Jokiel et al. 2014a; Jokiel 2015) is focused on factors controlling dissipation of protons into the water column and uptake of DIC.

During daylight hours the high tissue oxygen tension resulting from photosynthesis will stimulate respiration (Mass et al. 2010). Colombo-Pallotta et al. (2010) found that under normal physiological conditions, a 42 % increase in seawater oxygen concentration promotes a twofold increase in dark-calcification rates relative to controls. Apparently hyperoxia is necessary to maintain a high respiration rate in areas where extremely high calcification is occurring. Colombo-Pallotta et al. (2010) presented a model in which the oxygen and glycerol produced by photosynthesis are translocated to the calicodermal cells, where these materials are used by the mitochondria to generate ATP, which in turn is used to drive calcification. Corals, like other marine animals, are believed to maintain a very low intracellular calcium level when compared to seawater. This implies highly active Ca^{2+}-ATPase with energy supplied from respiration (Al-Horani et al. 2003b). They also contend that Ca^{2+}-ATPase has a dual function: (1). the transport of Ca^{2+} to the site of calcification and (2). the removal of H^+ that increases the aragonite saturation state in the calcifying fluid and facilitates the reaction toward $CaCO_3$ formation. This model does not account for the disposal of the waste product H^+, which ultimately must diffuse into the surrounding bulk water.

2.1.4 Other Models of Coral Calcification

Muscatine (1973) proposed that calcification in corals may be limited by synthesis of skeletal organic matrix produced by the zooxanthellae. Synthesis of organic matrix does appear to be a critical prerequisite for coral calcification, and especially for crystal nucleation, though it is unclear to what extent organic matrix synthesis is likely to limit coral calcification under most circumstances (reviewed by Allemand et al. 2011). The "inhibitory enzyme model" based on the observation that surface seawater is supersaturated with respect to aragonite was developed by Chave (1984). According to this model enzymes prevent mineralization at some locations while allowing mineralization at other specific locations to occur passively by precipitation of aragonite.

2.1.5 Chemistry of Ocean Acidification, Photosynthesis and Calcification

Presentations on ocean acidification inevitably involve three equations (e.g., Royal Society 2005; Kleypas et al. 2006). The first equation describes how increased atmospheric CO_2 caused by anthropogenic burning of fossil fuels dissolves in the oceanic surface waters to form carbonic acid which dissociates into a bicarbonate ion and a proton:

$$CO_2 + H_2O \Leftrightarrow HCO_3^- + H^+ \quad (2.1)$$

The second equation describes the dissociation of a carbonate ion to a bicarbonate ion and another proton:

$$HCO_3^- \Leftrightarrow CO_3^{2-} + H^+ \quad (2.2)$$

The third equation shows the carbonate ion combining with a calcium ion to form calcium carbonate:

$$CO_3^{2-} + Ca^{2+} \Leftrightarrow CaCO_3 \quad (2.3)$$

Changes in seawater pH shift the equilibria among the various forms of dissolved inorganic carbon (DIC) so the distribution of CO_2, HCO_3^- and CO_3^{2-} shifts with pH (Fig. 2.4).

Concentrations of the various forms of inorganic carbon shift with changing pH as follows:

$$CO_2 + H_2O \Leftrightarrow H^+ + HCO_3^- \Leftrightarrow 2H^+ + CO_3^{2-} \quad (2.4)$$

Calcification rate in coral incubation experiments is often determined by measuring change in total alkalinity (A_T) which is defined as the capacity of water to neutralize H^+. Coral calcification lowers A_T through release of protons (Eqs. 2.5, 2.6 and 2.7). In theory, calcification inevitably produces an excess of H^+ and thus reduces total alkalinity (A_T) by two moles for every mole of $CaCO_3$ precipitated (Kinsey 1978; Smith and Kinsey 1978). This relationship has now been verified directly by comparing A_T flux to Ca^{2+} flux in a coral reef flume system (Murillo et al. 2014). Therefore, Eq. 2.3 is misleading if taken out of context. Calcification equations must include two protons on the product side. The correct equations for calcification are as follows:

$$Ca^{2+} + (CO_2 + H_2O) \Leftrightarrow CaCO_3 + 2H^+ \quad (2.5)$$

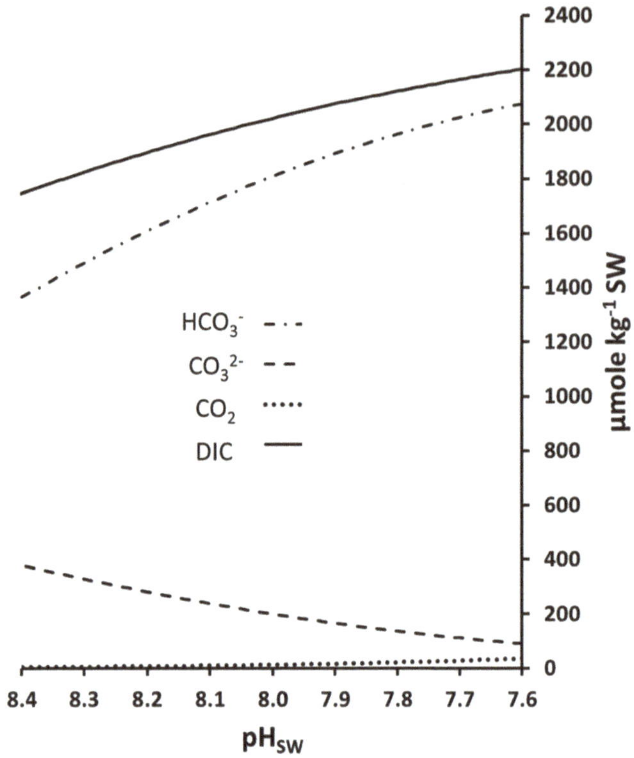

Fig. 2.4 Change in distribution of CO_2, HCO_3^-, CO_3^{2-} and DIC with changes in pH that occur due to coral metabolism and/or increasing ocean acidification (OA). Calculations were performed using CO2SYS (Pierrot et al. 2006) at T = 25 °C, S = 35 ppt, A_T =2300 µmol/kg SW

$$Ca^{2+} + (H^+ + HCO_3^-) \Leftrightarrow CaCO_3 + 2H^+ \quad (2.6)$$

$$Ca^{2+} + (2H^+ + CO_3^{2-}) \Leftrightarrow CaCO_3 + 2H^+ \quad (2.7)$$

Equations 2.5, 2.6 and 2.7 are written in two dimensions with a red arrow showing the relationship among the carbonate species (in parentheses) that shift with the changes in $[H^+]$ described as Eq. 2.4. Dissolution is the reverse of the calcification reaction. Net calcification (G_{net}) is the sum of calcification (positive flux) and dissolution (negative flux). When the equations are written correctly in this manner the importance of protons becomes apparent with two moles of H^+ produced for every mole of $CaCO_3$ precipitated regardless of which form of dissolved inorganic carbon (DIC) is involved.

The following equations describe photosynthetic carbohydrate formation from the various available CO_2 species:

$$(CO_2 + H_2O) \Leftrightarrow CH_2O + O_2 \tag{2.8}$$

$$\updownarrow$$

$$(H^+ + HCO_3^-) \Leftrightarrow CH_2O + O_2 \tag{2.9}$$

$$\updownarrow$$

$$(2H^+ + CO_3^{2-}) \Leftrightarrow CH_2O + O_2 \tag{2.10}$$

The photosynthesis equations are also written in two dimensions with the red arrows showing changes in distribution of species that occurs (Eq. 2.4) with shifts in pH. Note that photosynthesis increases pH (lowers $[H^+]$) while the reverse reaction of respiration decreases pH (increases $[H^+]$). Net photosynthesis (P_{net}) is the sum of photosynthesis (positive flux) and respiration (negative flux). Photosynthesis has a balanced charge (Eqs. 2.8, 2.9 and 2.10), so does not change A_T (Smith and Key 1975).

In sum, photosynthesis and calcification both lower the seawater DIC, while respiration and $CaCO_3$ dissolution raise DIC. Only the precipitation or dissolution of $CaCO_3$ significantly alters A_T. Consequently, changes in A_T can be used to calculate G_{net} and are widely used in this regard. Photosynthesis and respiration can radically alter $[H^+]$ and therefore relative concentration of CO_3^{2-}, HCO_3^- and CO_2. Coral calcification is a biological process that is heavily influenced by the associated processes of photosynthesis and respiration (e.g., changing P_{net}) that modify pH. Protons can be considered a waste product of calcification (Eqs. 2.5, 2.6 and 2.7) and O_2 a waste product of photosynthesis (Eqs. 2.8 and 2.9).

2.1.6 Conceptual Stumbling Blocks

The Calcification Equations The widespread use of Eqs. 2.1, 2.2 and 2.3, with emphasis on Eq. 2.3, fails to communicate the importance of H^+ as a waste product as shown by Eqs. 2.5, 2.6 and 2.7. The importance of Eqs. 2.5, 2.6 and 2.7 cannot be overemphasized – calcification will always result in the production of two moles of H^+ for every mole of $CaCO_3$ precipitated (Kinsey 1978; Smith and Kinsey 1978).
Coral calcification occurs in the space between the innermost tissue layer of the coral (calicodermis) and the skeleton (Fig. 2.1). However, the real physiological questions do not concern the Ω_{arag} at the site of calcification, which is under significant control by the coral animal. Rather, we need to know if the protons produced by calcification in the coral are dissipating out of the organism at a rate sufficient to avoid acidosis of tissues. Measurements of the changes in CO_2-carbonate chemistry of the bulk water do not necessarily relate directly to chemistry of the calcification fluid. Furthermore, measurements made using changes in the chemistry of the bulk water during coral incubations cannot distinguish between whether the supply side or product side of Eqs. 2.5, 2.6 and 2.7 is responsible for the change in G_{net}. The stoichiometry and measured changes in seawater chemistry will be the same for both cases.

The pH Concept An additional conceptual problem traces its roots to the fact that $[H^+]$ is often reported as pH where:

$$pH = -\log[H^+] \tag{2.11}$$

The problem here is that pH represents a double non-linear transformation of $[H^+]$ (i.e. the log of a reciprocal) that disguises the magnitude of change in $[H^+]$. The widespread use of pH rather than $[H^+]$ results from the fact that pH is easily measured with a pH electrode. However, the pH electrode measures activity of H^+ rather than $[H^+]$. Nevertheless, pH has a long history of use and is universally reported in research papers. Physiological systems generally respond to activity of H^+ (rather than concentration), so use of pH is a convenient index of acid-base conditions. Physical processes outside of the organism (such as diffusion through a boundary layer) respond to concentration. The use of pH (-log $[H^+]$) rather than $[H^+]$ clouds the fact that immense changes in $[H^+]$ occur across diffusion barriers that form outside of the tissues. The strength of these gradients increases with increasing OA (Jokiel 2011a).

Until recently the focus on limiting factors for coral calcification rate has been on the reactants (left side of Eqs. 2.5, 2.6 and 2.7), with an emphasis on uptake of specific forms of DIC. Failure to dissipate H^+ from the site of calcification, through the tissues and out through the seawater boundary layer (right side of Eqs. 2.5, 2.6 and 2.7), will cause acidosis and disruption of normal biological processes. An excellent analogy is furnished by the companion process of photosynthesis in reef corals. Photosynthesis produces fixed carbon as a product while calcification produces calcium carbonate as a product. Photosynthesis produces the "waste product" O_2 while calcification produces the "waste product" H^+. Boundary layer thickness can control primary production by limiting efflux of O_2 from the coral. Increased water motion decreases boundary layer thickness, increases O_2 flux rate and increases primary production (Mass et al. 2010). By analogy, boundary layer thickness presumably can control proton efflux and thereby control calcification rate. Corals have evolved a very sophisticated morphology that results in a highly effective means of dealing with the waste products of O_2 and H^+ (Jokiel 2011a, 2011b).

The importance of HCO_3^- uptake to coral metabolism is shown by the abundance of the enzyme carbonic anhydrase (CA) in reef corals. The reaction described in Eq. 2.1 is accelerated by CA which has a reaction rate that is among of the fastest of all enzymes. Coral tissues and zooxanthellae contain large amounts of CA (Graham and Smillie 1976; Weis et al. 1989) which play a major role in controlling transport of CO_2 throughout the coral colony. Al-Horani et al. (2003a) identified CA bound to the membranes of the epidermal cells of the surface body wall. Moya et al. (2008) identified CA in the calicodermis, which controls the precipitation of skeletal material. Wherever the conversion between CO_2 and HCO_3^- is very fast (i.e. such as occurs in the presence of CA) in comparison to the rate of diffusion, a difference in HCO_3^- concentration corresponding to the CO_2 tension difference will be established (Enns 1967).

2.1.7 The Concept of Aragonite Saturation State (Ω_{arag}) in Relation to Ocean Acidification (OA)

Burning of fossil fuels continues to increase the concentration of CO_2 in the atmosphere. When the anthropogenic CO_2 is absorbed by seawater, chemical reactions occur that reduce seawater pH, increase bicarbonate ion (HCO_3^-) and decrease carbonate ion (CO_3^{2-}) concentration (Fig. 2.4) in a process commonly referred to as ocean acidification (OA). These reactions are described in a review by Feely et al. (2009). A doubling of pre-industrial levels of oceanic pCO_2 is predicted to occur at some point within this century (IPCC 2001, 2007), unless we radically limit our burning of fossil fuels. Increased pCO_2 in sea water leads to a decreased aragonite saturation state (Ω_{arag}). Aragonite is the primary mineral form of $CaCO_3$ that is laid down by corals, so the question arose as to how the declining Ω_{arag} would impact living coral populations. Smith and Buddemeier (1992) stated that increased CO_2 would lead to reduced coral calcification rates. Their conclusion was subsequently confirmed by laboratory studies showing that calcification rates of reef-building corals could decline by 20–40 % under twice present day pCO_2 conditions (Gattuso et al. 1999; Langdon et al. 2000; Marubini et al. 2001, 2003; Langdon and Atkinson 2005). These early observations led to a growing concern about the impact of OA on corals and coral reefs (Kleypas et al. 1999a, 1999b; Orr et al. 2005; Hoegh-Guldberg et al. 2007; Carpenter et al. 2008; Veron 2008).

The saturation state concept is widely used by physical chemists in describing seawater carbonate chemistry. The saturation state of aragonite (Ω_{arag}), which is the mineral form of $CaCO_3$ precipitated by reef corals, is of particular interest. The term is defined by the equation:

$$\Omega_{arag} = \frac{[Ca^{2+}][CO_3^{2-}]}{K_{sp}} \quad (2.12)$$

where K_{sp} is the solubility product of aragonite. The $[Ca^{2+}]$ in normal present-day oceanic seawater is essentially constant at 10.3 mmol kg^{-1} SW, normalized to salinity. Likewise, K_{sp} is a constant (at a given temperature, pressure, and salinity), so in shallow oceanic waters Ω_{arag} is directly proportional to $[CO_3^{2-}]$. Intensive work over the past 25 years (reviewed by Feely et al. 2009) has led to a much greater understanding of how combustion of fossil fuels is leading to lower (Ω_{arag}) in the surface waters of the ocean.

2.1.8 Relationship Between Ω_{arag}, the [DIC]:[H^+] Ratio and Coral Calcification (G_{net})

The concept of the [DIC]:[H^+] ratio introduced by Jokiel (2011a) provides a new insight into the controls on coral calcification that has now been supported by observations in other organisms such as coccolithophores (Cyronak et al. 2015) and marine bivalves (Thomsen et al. 2015). A major difficulty with the Ω_{arag} model is failure to explain why G_{net} decreases with increasing OA in the face of increasing [DIC] and increasing $[HCO_3^-]$ (Fig. 2.4). G_{net} increases under higher $[HCO_3^-]$ (Herfort et al. 2008; Marubini et al. 2008; Jury et al. 2010), so G_{net} should increase with increasing OA. Jokiel (2011a) estimated that the increase in G_{net} due to increased $[HCO_3^-]$ caused by a doubling of pCO_2 from pre-industrial levels will only be 3.8 % compared to the predicted decrease in G_{net} of 32 % due to increased $[H^+]$ for a net decrease of 28.2 %. Thus any benefit to skeletal growth caused by higher $[HCO_3^-]$ will be overwhelmed by an order of magnitude greater negative impact due to increased $[H^+]$.

We can demonstrate the relationship between G_{net} and the seawater CO_2-carbonate system parameters of A_T, Ω_{arag}, CO_3^{2-}, HCO_3^-, $CO_{2(aq)}$, H^+ and the [DIC]:[H^+] ratio from a biological perspective. Rather than following the physical chemistry approach of using saturation state we can employ a physiological organism-centered approach based on documented metabolic processes. The major focus must be that any protons generated by the calcification reaction must dissipate out of the coral. Also, there must be uptake of DIC if the coral is to calcify.

The chemistry of the CO_2 -carbonate system is complex, but only two parameters are needed to calculate the distribution of DIC species and Ω_{arag} in seawater at known salinity (S), temperature (T), and pressure (P). The relationship between the major parameters of the system can be demonstrated simply by varying pCO_2 at constant A_T, S, T and P. As pCO_2 increases, pH decreases (i.e. $[H^+]$ increases), [DIC] increases, $[CO_3^{2-}]$ decreases, and Ω_{arag} decreases. The opposite is true for decreasing pCO_2. In other words, $[H^+]$ varies directly with [DIC] under increasing OA and inversely with $[CO_3^{2-}]$. The rules of proportionality (Tourniaire and Pulos 1985) allow us to state this relationship mathematically using proportionality constant (k_1) as follows:

$$[H^+] = \frac{[DIC]}{[CO_3^{2-}]} k_1 \tag{2.13}$$

These terms can be rearranged as follows:

$$[CO_3^{2-}] = \frac{[DIC]}{[H^+]} k_1 \tag{2.14}$$

In oceanic surface water, Ω_{arag} is proportional to $[CO_3^{2-}]$, so we can rewrite the equation with a different proportionality constant (k_2) as:

$$\Omega_{arag} = \frac{[DIC]}{[H^+]} k_2 \tag{2.15}$$

There is a large body of data showing that G_{net} is proportional to Ω_{arag}. Therefore we can rewrite Eq. 2.15 as:

$$G_{net} = \frac{[DIC]}{[H^+]} k_3 \tag{2.16}$$

Equation 2.16 could also be derived from the observation of Schneider and Erez (2006) that G_{net} is directly proportional to [DIC] and inversely proportional to $[H^+]$. The plot of G_{net} versus Ω_{arag} (or G_{net} versus $[CO_3^{2-}]$) should be similar to the plot of G_{net} versus the [DIC]:$[H^+]$ ratio times the appropriate proportionality constant. In other words we need not resort to the Ω_{arag} concept of physical chemistry, but can describe coral calcification based on physiologically relevant parameters.

Bach (2015) used a physical chemistry approach to further investigate these relationships. He rearranged the seawater carbonate system equations to demonstrate the proportional relationship between $[CO_3^{2-}]$ and the $[HCO_3^-]$:$[H^+]$ ratio where $[HCO_3^-]$ is the inorganic carbon substrate and $[H^+]$ functions as a calcification inhibitor as previously defined by Jokiel (2011a). Due to this proportionality rule, he points out that calcification rates will always correlate well with the ratio of $[HCO_3^-]$:$[H^+]$ and equally well to [DIC]: $[H^+]$, $[CO_3^{2-}]$ or Ω_{arag} when T, S, and P are constant. Thus, the correlations between calcification and $[CO_3^{2-}]$ or Ω_{arag} that have previously been reported can be attributed to the combined influence of $[HCO_3^-]$ and $[H^+]$, which provide a more meaningful physiological parameter than Ω_{arag}.

The [DIC]:$[H^+]$ ratio concept is an alternate way of viewing net calcification that can be tested. A high quality data set is available from Langdon et al. (2000), who conducted long term static tests in highly modified sea water chemistries. This work was carried out over a number of years in the 2650 m^3 "ocean" coral reef mesocosm of Biosphere-2 located near Tucson, Arizona. Effects of sea water carbonate chemistry on G_{net} were determined under various sea water chemistries in an assembled community of coral reef organisms consisting of corals, calcifying algae, and other typical reef biota. The investigators manipulated the saturation state of the water by adding various amounts of $NaHCO_3$, Na_2CO_3 and $CaCl_2$. They found that G_{net} was a function of the product of $[Ca^{2+}]$ and $[CO_3^{2-}]$, leading to their conclusion that "saturation state (and not pH, pCO_2, or HCO_3^-) affects coral reef calcification". Data reported for A_T, Ca^{2+}, CO_3^{2-}, HCO_3^-, Ω_{arag}, pH and G_{net} during each of the experimental trials (Appendix Table 2.1) was used to calculate [DIC], $[H^+]$ and the [DIC]:$[H^+]$ ratio. Analysis of these data shows a non-significant relationship between G_{net} and $[Ca^{2+}]$ (Fig. 2.5a), reflecting the superabundance of Ca^{2+} ($\approx$10 mmole kg^{-1}) in relation to CO_3^{2-}($\approx$0.2 mmol kg^{-1}) or DIC ($\approx$2 mmol kg^{-1}). The lack of a relationship between G_{net} and Ca^{2+} supports the observations of Gagnon et al. (2012) that describe exchange of Ca^{2+} and other cations between seawater and the calcifying fluid over the course of a few hours. The mechanism for this type of transport appears to be a voltage-dependent Ca^{2+} channel that accelerates the trans-epithelial transport of Ca^{2+} used for coral calcification (Zoccola et al. 1999), but it has not been shown to transport anions such as CO_3^{2-}. Presumably then, the small differences in $[Ca^{2+}]$ that occur over geologically short timescales are not a major driver of calcification (Fig. 2.5a). In contrast, G_{net} shows a significant correlation with the DIC:H^+ ratio (Fig. 2.5b). In retrospect, the significant relationship between G_{net} and $[CO_3^{2-}]$ or its surrogate Ω_{arag} is due to correlation of Ω_{arag} with the DIC:H^+ ratio (Fig. 2.5c).

2.1.9 Boundary Layers (BL) and Material Exchange Between the Water Column and the Coral

The role of the boundary layer (BL) in controlling material flux in corals and other organisms is one of the keys to understanding calcification in corals. Corals create frictional drag which slows water velocity. Three sub-component layers of the BL have previously been defined and measured (Shashar et al. 1996).

The Diffusion Boundary Layer (DBL) is a quiescent layer of water adjacent to the coral tissue and is important in relation to diffusion-limited processes such as respiration and photosynthesis. Much of the work on boundary layer limitation of material exchange has been focused on this innermost layer. Shapiro et al. (2014) present direct

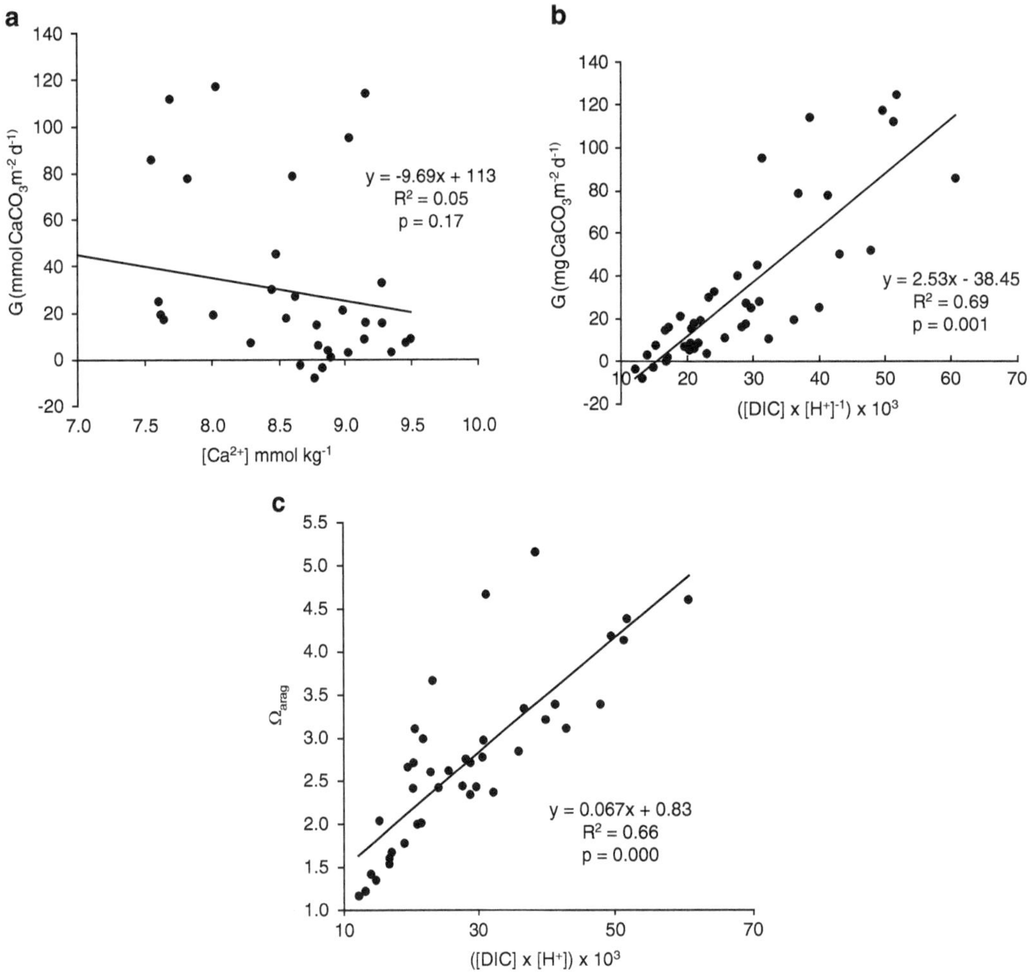

Fig. 2.5 Biosphere-2 data from Langdon et al. (2000) showing: (a.) G_{net} as a function of Ca^{2+}, (b.) G_{net} as a function of the [DIC]:[H^+] ratio and (c.) Ω_{arag} vs. [DIC]:[H^+] ratio (Data are presented in Appendix Table 2.1)

microscopic evidence that corals can at least partially overcome limitation by molecular diffusion in the DBL by developing strong vortical flows driven by motile epidermal cilia covering their entire tissue surface. Ciliary beating produces quasi-steady arrays of counter-rotating vortices that vigorously stir a layer of water extending up to 2 mm from the coral surface, but requires expenditure of energy. Under low ambient flow velocities, these vortices can control the exchange of nutrients and oxygen between the coral and its environment, enhancing mass transfer rates by up to 400 %.

The Momentum Boundary Layer (MBL) controls water movement near the colony and is thicker by an order of magnitude than the DBL. The Benthic Boundary Layer (BBL) incorporates the DBL and MBL to describe the frictional drag of the complex benthic structures on flow near the bottom and controls the exchange of water between the reef and the overlying water column. The BBL studied by Shashar et al. (1996) was more than 1 m thick with a roughness height of 31 cm and a shear velocity of 0.42 cm s^{-1}.

The present discussion of the BL will be focused on the DBL, which produces a thin layer of stagnant seawater adjacent to the coral tissue. This quiescent layer influences the flux of material between the corallum and the water column. The transport of Ca^{2+}, CO_2, CO_3^{2-}, HCO_3^-, O_2, nutrients and H^+ through the BL is limited by the physical processes of diffusion and advection (e.g., Jokiel 1978; Shashar et al. 1993; Lesser et al 1994; Shashar et al. 1996; Kaandorp et al. 2005, 2011). Kühl et al. (1995) found that zooxanthellae photosynthesis in the light resulted in a build-up of O_2 in the photosynthetic tissue of up to 250 % saturation and a tissue pH of up to 8.6 (i.e. 0.7 pH units above the pH value of the overlying seawater). In darkness the O_2 within the coral tissue was depleted by respiration to near anoxic (<2 % air saturation) conditions, with tissue pH of 7.3–7.4. O_2 and pH profiles demonstrated the presence of a 200–300 μ thick BL that separated the coral tissue from the overlying flowing seawater. Two recent models invoke boundary layer controls on coral calcification. One (Kaandorp et al. 2005, 2011) addresses BL limitation of DIC influx and the other (Jokiel 2011a, 2011b) focuses on BL limitation of proton efflux.

Corals experience the highest water motion and thinnest BL at the distal parts of the corallum (Figs. 2.2, and 2.3). Projections of the skeleton and branch tips are covered by thin, colorless tissue which is devoid of zooxanthellae (Fig. 2.2a). These areas are more responsive to changes in water motion than adjacent areas (Jokiel 1978). Increased water flow reduces the thickness of the BL over these structures, increasing local calcification to produce the hoods, papillae, spines, verrucae and other projections that characterize many species of reef corals (Veron 2000). In turbulent water these projections grow outward and increase frictional drag and protect the polyp. In calm, low light environments they remain suppressed (Jokiel 1978).

2.1.10 Material Fluxes

Coral calcification rates based on changes in CO_2-carbonate chemistry describe net flux through the boundary layer that isolates the "black box" of the coral from the water column and do not represent processes at the site of calcification. Distinguishing between calicodermal flux, epidermal flux and gastrodermal flux within the "black box" can be informative. Epidermal flux as defined here refers to the exchange of materials between the epidermis and the external water column (Fig. 2.1b). Rate of epidermal flux is limited by the BL and is characterized by carbonic anhydrase- facilitated transport of bicarbonate HCO_3^- into the coral tissue and efflux of H^+ as described in the Proton Flux Model (Jokiel 2011a, 2011b). Calicodermal flux as defined here includes and emphasizes the exchange of material between coral calicodermal cells and the space between the calicodermis and the $CaCO_3$ accretion site of the skeleton. H^+ must be actively removed from the calcifying space by calicodermal cells if the reaction at the skeleton is to move towards $CaCO_3^{2-}$ precipitation (Eqs. 2.5, 2.6 and 2.7). The mechanism involved appears to be one or more proton pumps (Furla et al. 2000a; Cohen and McConnaughey 2003; Allemand et al. 2004). Allison et al. (2014) used skeletal boron geochemistry to study the DIC chemistry of the fluid used for coral calcification. They showed that corals concentrate DIC in the calcifying fluid at the skeleton calcification site and that bicarbonate makes up a significant amount of the DIC pool used to build the skeleton. Corals actively increase the pH of the calcification fluid to create a diffusion gradient favorable to the transport of molecular CO_2 from the overlying coral tissue into the calcification site. The increased calcification fluid pH and higher [DIC] results in a high aragonite saturation state within the calcifying fluid which is favorable to aragonite precipitation.

However, the waste H^+ being rapidly removed from the calcifying fluid must be dissipated out of the calicodermis and other tissue layers into the water column. Otherwise acidosis will develop in the tissues and block metabolism. The coelenteron fluid can exchange with sea water through the polyp mouths and into the BL. Gagnon et al. (2012) provided evidence for rapid cation exchange (but not anion exchange) between seawater and the calcifying fluid. This mechanism does not alleviate the need to move protons out of the "black box"- through the boundary layer and into the water column. The boundary layer is a physical limitation that is not under biological control.

It appears that metabolic energy is required to transport Ca^{2+} across the calicodermis into the calcifying fluid (between the calicodermis and the skeleton) at a rate sufficient to maintain normal calcification (Al-Horani et al. 2003a, 2003b). Tambutté et al. (1996) concluded that transport of Ca^{2+} across the epidermis and gastrodermis appears to be facilitated by paracellular pathways that connect the calcifying fluid adjacent to the skeleton with the sea water in the BL (Tambutté et al. 2012). The pH in the calcifying space under the calicodermis has been shown to be elevated relative to the polyp surface and to the inside of the coelenteron (Al-Horani et al. 2003a; Ries 2011; Venn et al. 2011). Ca^{2+} is transported over considerable distances within a colony with the direction of transport toward areas of maximum growth and calcification (Taylor 1977). Translocation of metabolic material within the coral has been demonstrated experimentally (Pearse and Muscatine 1971; Taylor 1977; Rinkevich and Loya 1983; Fine et al. 2002). One mechanism for such transport was described by Gladfelter (1983). Polyps of the coral *Acropora cervicornis* are connected in a complex gastrovascular system, which is lined with flagellated cells that can move the gastrovascular fluid at velocities of more than 2 cm min^{-1}. This type of circulation system serves to exchange fluids between the ZP and the ZC.

2.2 The Two-Compartment Proton Flux Model

This model (Jokiel 2011b) considers four major observations not included in earlier models of coral metabolism:

- Boundary-layers control exchange of materials at the tissue-seawater interface, which includes efflux of waste protons as well as influx of dissolved inorganic carbon.
- Zooxanthellae are lacking in rapidly calcifying areas of the coral (Goreau and Goreau 1959; Goreau 1963; Pearse and Muscatine 1971; Crossland and Barnes 1974; Lamberts 1974; Jaubert 1977; Brown et al. 1983; Kajiwara et al. 1997; Marshall and Wright 1998; Fang

et al. 2004; Al-Horani et al. 2005b; Tambutté et al. 2007; Santos et al. 2009).

- Photosynthate (CH_2O) is transported from areas containing zooxanthellae toward areas of rapid calcification that lack zooxanthellae (Pearse and Muscatine 1971; Taylor 1977). Translocation suggests that areas of photosynthesis and areas of rapid calcification are metabolically different and require different chemical environments. A coral colony contains a proximal region of zooxanthellae-rich tissues, termed the zone of rapid photosynthesis (ZP) and a second zone consisting of distal portions of the skeleton (branch tips, outer septal plates, and projecting trabeculae) covered by thin, colorless or lightly pigmented tissues and termed the zone of calcification or ZC (Jokiel 2011b).
- Primary and secondary calcification occurs in corals. Primary calcification in branch tips, septal margins, trabeculae and spines is characterized by rapid outgrowth (extension). This is followed by secondary calcification (accretion) on the sides of branches (Gladfelter 1982, 1983). Skeletal density variations result from differing rates of extension vs. accretion under different conditions of temperature, irradiance and water motion (Barnes and Lough 1993).

2.2.1 Description of the Two-Compartment Proton Flux Model

The model is described using the equations for calcification (Eqs. 2.5, 2.6 and 2.7) and photosynthesis-respiration (Eqs. 2.8, 2.9 and 2.10). The three dimensional hemispherical layered form of the coral tissues is reduced to a two dimensional diagram in Fig. 2.6a. The resulting fluxes and recycling pathways are shown in Fig. 2.6b for protons, Fig 2.6c for carbon and Fig. 2.6d for oxygen. In the ZP, interconversion between HCO_3^- and CO_2 (Eq. 2.2) occurs at an extremely rapid rate due to abundant CA. In the second compartment, or the ZC, both primary calcification and respiration occur but there is no photosynthesis. The major fluxes of H^+, HCO_3^-, O_2 and CH_2O are shown as arrows. In either compartment, as $CaCO_3$ precipitates out of solution, the H^+ must be removed if calcification is to continue. In the ZP, some of the protons produced by the secondary calcification can be used to drive photosynthesis. In the ZC, the excess H^+ is removed via direct flux across the BL. Reducing the description of metabolic reactions to only two sets of equations places the focus on proton flux and eliminates the need to complicate matters by including OH^-. Previous models invoke the use of OH^- derived from photosynthesis as a means of neutralizing the H^+ produced in calcification. However, this approach (Fig. 2.1) requires that OH^- be transported from areas of photosynthesis to the areas of rapid calcification. The focus on pathways of H^+ (Fig. 2.2b) is a very powerful and direct method of balancing material flux and describing the major metabolic processes and pathways in reef coral metabolism.

Placement of the rapidly calcifying areas adjacent to the BL facilitates rapid dissipation of H^+ into the water column from the ZC, and also allows for an efficient method of transporting excess H^+ from secondary calcification sites (ZP) into the water column (Fig 2.6b). The protons produced by secondary calcification are used in the production of photosynthate which is then translocated to the ZC. Respiration of the photosynthate produces ATP energy in the ZC and releases the H^+ into the BL. Thus, translocation of photosynthate serves as a means of transporting both protons and energy from the ZP to the ZC. Furthermore, the major source of carbon (HCO_3^-) used in calcification is derived from the metabolism of photosynthate (Fig. 2.6c), which is consistent with results reported by Furla et al. (2000a, 2000b). Protons being produced by both primary calcification and secondary calcification are concentrated in the ZC, where they can be dissipated into the adjacent water column or into the underlying ZP as needed to maintain maximum metabolic activity. Production, uptake and movement of H^+ within the coral influences localized pH within cells and tissues.

The high oxygen flux required for respiration in the ZC is readily supplied as the by-product from photosynthetic production in the underlying ZP (Fig. 2.6d). Colombo-Pallotta et al. (2010) found that high calcification rate in corals depends on hyperoxic conditions. High oxygen concentration facilitates increased mitochondrial respiration in the ZC which, in turn, generates the large amount of ATP needed to support the rapid deposition of $CaCO_3$. During daylight hours much of the oxygen produced in the ZP is consumed by the high rate of respiration in the overlying ZC. Al-Horani et al. (2003b) found that gross photosynthesis was approximately seven times higher than net photosynthesis, indicating that respiration consumes most of the O_2 produced by the zooxanthellae. The respiration rate in light was approximately 12 times higher than in the dark. The coupling of gross photosynthesis and light respiration produces intense cycling of internal carbon and O_2. Thus hyperoxia is a key feature of reef coral metabolism that is managed very well by the coral under normal conditions through a variety of mechanisms. However, high oxygen tension can lead to oxidative stress and bleaching in corals exposed to abnormally high temperature and high solar irradiance (Lesser 2011).

The skeletal material of the ZC modifies the irradiance in the ZP. Extensive scattering of photons by the skeleton enhances light absorption by symbiotic algae (Enríquez et al. 2005; Marcelino et al. 2013). Coral skeleton can absorb harmful ultraviolet radiation and fluoresce the energy into the visible portion of the spectrum (Reef et al. 2009). Rapid calcification on distal portions of the coral produces conditions in the understory that greatly enhance photosynthetic efficiency (Jokiel and Morrissey 1986). Coral skeletons are efficient at trapping, transporting and redistributing light throughout the colony (Marcelino et al. 2013), so lack of zooxanthellae in the growing tips can also be viewed as an adaptation that allows light to enter the skeleton. As light penetrates a

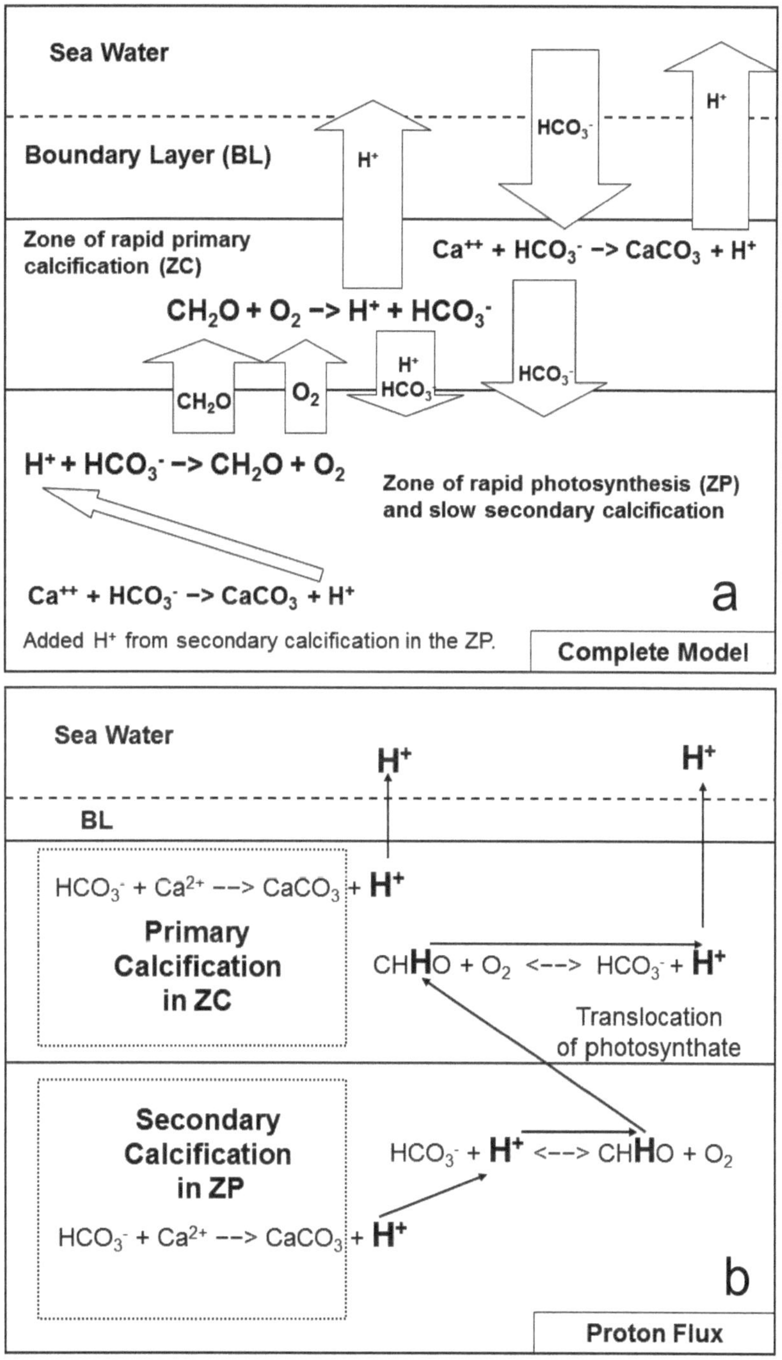

Fig. 2.6 Simplified two compartment proton flux model model: (**a**) spatial arrangement of chemical reactions, (**b**) pathways of protons, (**c**) carbon flux and (**d**) oxygen flux

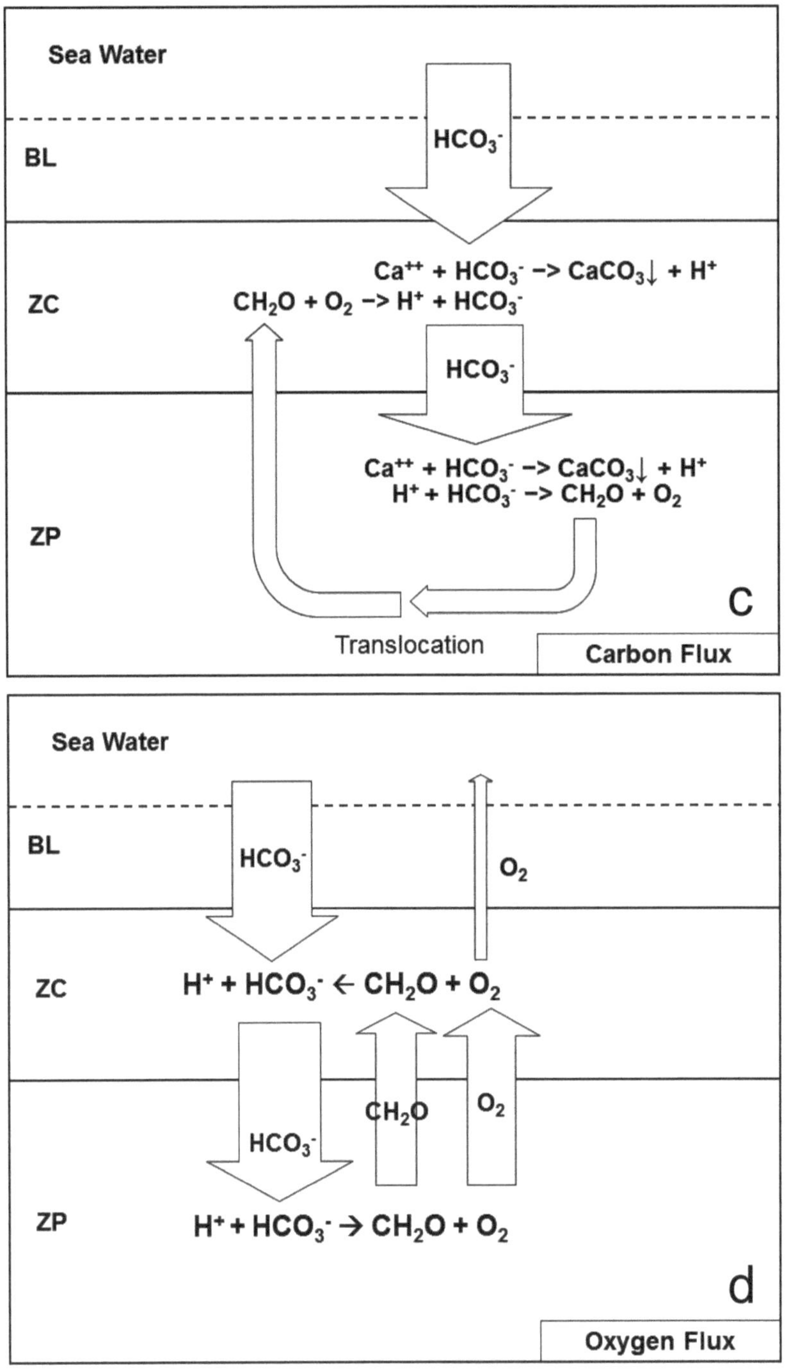

Fig. 2.6 (continued)

skeletal septum in the distal portion of a corallite it scatters and will diffuse into neighboring septa and redistribute throughout the colony. This enables millimeter-size structures to increase amplification as much as twenty-fold by trapping light within coral tissue due to multiple passes. This mechanism of redistribution enhances delivery of light to zooxanthellae and also delivers light to shaded parts of the coral colony.

2.2.2 Application of Model to Other Coral Morphologies

The generalized Proton Flux Model applies over a wide range of coral morphologies (Figs. 2.2 and 2.3) and over broad spatial scales. Branched morphology creates an outer zone of calcifying branch tips exposed to turbulent water where rapid outward growth of the skeleton occurs and where solar irradiation is very high. Rapid photosynthesis occurs largely in the inner quiescent zone of the corallum. In branched colonies the ZC encompasses the outer tips. The morphology of perforate corals replicates the same spatial configuration but at a scale of mm rather than cm. At the upper end of the spatial scale Kajiwara et al. (1997) studied large thickets of the coral *Acropora pulchra* and compared growth of outer white-tipped branches (ZC) to growth of brown-tipped branches (ZP) located deeper in the colony. Zooxanthellae concentration in the white-tipped branches was low compared to the dark-tipped branches. The white-tipped branches showed three times the skeletal weight increase and 14 times the linear extension increase of the brown-tipped branches. The authors concluded that white-tipped branches with a lightly-calcified skeleton expand the area covered by the coral colony while brown-tipped branches develop a heavily-calcified skeleton that strengthens the colony. Fang et al. (2004) showed higher concentrations of ATP in the white tips compared to the brown stalks, providing the ready supply of the energy needed for rapid calcification. At the opposite end of the spatial scale, Al-Horani et al. (2005a, 2005b) employed microprobes and radioisotope techniques to measure the distribution of photosynthesis and calcification across polyps of the coral *Galaxea fascicularis*. The highest rates of photosynthesis occurred in the deeper parts of the calyx (ZP) that contained dense concentrations of zooxanthellae. The exert corallite septae that projected into the water column (ZC) incorporated more ^{45}Ca than the deeper portions (ZP) in both light and dark. Marshall and Wright (1998) report that there are essentially no zooxanthellae in the cell tissues covering the exert septa of *G. fascicularis* where calcium incorporation is highest. Jimenez et al. (2011) found that the BL over the surface of the corals *Platygyra sinensis* and *Leptastrea purpurea* is very thin over protruding skeletal features such as septa and calyx walls. These areas are covered with thin tissue (ZC) that lack zooxanthellae with much thicker tissues containing zooxanthellae located deeper in the calices (ZP). Therefore a wide range of morphologies from single polyps to complex colonial forms fit the general model, often in fractal patterns (Vicsek 1989) at different scales (e.g., a branching perforate coral).

Brahmi et al. (2012) studied the micro- and ultra-structural skeletal growth dynamics of the scleractinian coral *Pocillopora damicornis* and report that the coral is capable of controlling its biomineralization activity with great temporal and spatial precision. They suggested that spatial heterogeneity in coral tissue activity as described by Jokiel (2011b) should be carefully addressed in the development of better biomineralization models for scleractinian corals.

2.3 Ocean Acidification

2.3.1 Attempts to Explain How OA Reduces Coral Calcification

Much of the discussion of OA has been centered on the relationship between coral growth and Ω_{arag} and on the rate that Ω_{arag} will change over time in the surface waters of the sea. The empirical relationship between Ω_{arag} and calcification rate in tropical reef-building corals has been well established (Gattuso et al. 1999; Langdon et al. 2000; Marubini et al. 2001, 2003; Ohde and Hossain 2004; Langdon and Atkinson 2005; Schneider and Erez 2006; Jokiel et al. 2008; Cohen et al. 2009). Temperate corals have a much lower rate of metabolism and skeletal growth shows less of a decrease with decreasing Ω_{arag} (see Fig. 4 I. in Ries et al. 2009; Fig. 4 in Holcomb et al. 2010; Fig. 5 in Rodolfo-Metalpa et al. 2010).

Schneider and Erez (2006) conducted laboratory experiments designed specifically to separate the effects of Ω_{arag}, pH, [CO_3^{2-}], aqueous CO_2, total alkalinity (A_T), and DIC on reef coral calcification. They concluded that calcification (both light and dark) was driven by CO_3^{2-} concentration. However, their data also show similar or higher correlations between calcification and seawater [H^+], [DIC], and A_T. Likewise, Cohen and Holcomb (2009) followed this interpretation and suggested that under conditions of increasing OA corals must expend more energy to remove H^+ from the calcifying fluid between the calicodermis and the skeleton in order to raise the pH of the contained seawater and convert the increasingly plentiful HCO_3^- to CO_3^{2-}. These authors contend that the CO_3^{2-} is then moved into the calcifying fluid between the calicodermis and the skeleton and combines with Ca^{2+} to form the $CaCO_3$ crystals of the skeleton.

Jury et al. (2010) conducted experiments designed to distinguish the effects of Ω_{arag}, pH, [CO_3^{2-}] and [HCO_3^-] on coral calcification by conducting incubations in six regimes of highly modified seawater chemistries. Coral calcification responded strongly and consistently to variation in [HCO_3^-] or DIC, but not to [CO_3^{2-}], Ω_{arag} or pH. Jury et al. (2010) concluded that data from their study showed inconsistencies in the Ω_{arag} model. They suggested that coral calcification in the pH tolerant species *Madracis auretenra* is controlled by [HCO_3^-], but that calcification might be controlled by the combination of seawater [HCO_3^-] and pH in more pH- sensitive species. Experiments designed to test the relative importance of [HCO_3^-] versus [CO_3^{2-}] in coral calcification (de Putron et al. 2010) led to a conclusion opposite to that of Jury et al. (2010) in that calcification showed a better correlation with [CO_3^{2-}] than with [HCO_3^-].

However, calcification in these experiments also correlated with [DIC] and [H^+], consistent with the models proposed by Jury et al. (2010) and by Jokiel (2011a). Schneider and Erez (2006) showed a strong positive relationship between DIC and coral calcification at constant [H^+]. Likewise they showed a strong negative relationship between coral calcification and [H^+] at constant DIC. Comeau et al. (2012) showed that corals and crustose coralline algae uptake HCO_3^- as well as CO_3^{2-}, especially during light-enhanced calcification. Edmunds et al. (2012) studied three species of coral and found that pCO_2 and temperature independently affected calcification, but the response differed among taxa. Massive *Porites* spp. were largely unaffected by the treatments, but branching *Porites rus* grew 50 % faster at 29.3 °C compared with 25.6 °C, and 28 % slower under twice present day levels of pCO_2. Their compilation of results from previous studies revealed a high degree of variation in calcification as a function of pH, [HCO_3^-], and [CO_3^{2-}]. This synthesis supported the hypothesis that coral genera respond in dissimilar ways to pH, [HCO_3^-], and [CO_3^{2-}].

Jokiel (2013) used data on calcification rates of coral and crustose coralline algae from Comeau et al. (2012) to test the Proton Flux Model of calcification. There was a significant correlation between calcification and the ratio of DIC to proton concentration ([DIC] : [H^+] ratio). The ratio is tightly correlated with [CO_3^{2-}] and with Ω_{arag}. Jokiel (2013) noted that correlation does not prove cause and effect, and argued that Ω_{arag} and [CO_3^{2-}] have no basic physiological meaning on coral reefs other than a correlation with the [DIC] : [H^+] ratio. Comeau et al. (2013) responded by describing the type of experiments that are needed to allow further evaluation of the Proton Flux Model in relation to their model. However, they state that their interpretation of the data does not challenge the paradigm that the control of coral calcification is mediated entirely by [CO_3^{2-}]. Subsequent reports (Bach 2015; Cyronak et al. 2015; Jokiel 2015) do not support the [CO_3^{2-}] model.

2.3.2 Shortcomings of the Ω_{arag} Model (i.e., CO_3^{2-} Limitation) in Studies of Coral Calcification

Prior to our awareness of the "OA problem", the disciplines of carbonate physical chemistry and calcification physiology were largely unrelated fields (Roleda et al. 2012). The dominant role that physical chemistry played in the formative years of OA research (i.e. decreasing Ω_{arag} = decreasing [CO_3^{2-}] = decreasing coral calcification) resulted in an incomplete model of how OA will influence the physiology of calcifiers. Thus, two disparate views on calcification chemistry were advanced. The first is the classic biological view that organisms modify local carbonate chemistry of seawater and can use HCO^-or CO_2 for calcification. The second was focused primarily on a physical chemistry view implying that CO_3^{2-} is the main inorganic source of carbon used for calcification. Re-examination of the literature on the metabolic basis of calcification prior to the era of OA research (i.e. 1960–1980) supports the contention that bulk-water CO_3^{2-} is not the substrate for calcification in marine organisms (Roleda et al. 2012), and that other models are more appropriate among the various taxonomic groups.

Control of calcification by [CO_3^{2-}] is an unattractive hypothesis for several reasons. As has been pointed out (McConnaughey and Whelan 1997; Pörtner et al 2005; Wilt 2005; Hofmann and Todgham 2010), CO_3^{2-} is rarely transported across membranes, but rather indirectly passes through tissues via diffusion of CO_2 or through ion exchange transport of HCO_3^- coupled with H^+ transport. Various physiological studies have led to the conclusion that HCO_3^- appears to be the preferred form of inorganic carbon utilized by reef corals (Weis et al. 1989; Furla et al. 2000a, 2000b; Roleda et al. 2012). Bach (2015) used the basic equations that describe the physical chemistry of the sea water carbonate-carbon dioxide system to demonstrate that correlations between calcification and [CO_3^{2-}] or Ω_{arag} can be attributed to the combined influence of [HCO_3^-] and [H^+]. He went on to evaluate whether HCO_3^- or CO_3^{2-} would be the more suitable inorganic carbon substrate for calcification from a physical chemistry point of view. Three lines of analysis led him to the conclusion that HCO_3^- would be favored:

1. *Abundance*. HCO_3^- is the most abundant DIC species in seawater, so it makes sense for an organism to rely on the largest inorganic carbon pool.
2. *Homeostasis*. The hydration time of CO_2 is slow while the hydrolysis of HCO_3^- is fast. Thus CO_3^{2-} transported through the cytosol with a typical pH of 7.2 would quickly turn into HCO_3^- and bind a proton in the cytosol. The resulting HCO_3^- would be transported to the calcification site where the proton would be released back to the cytosol. Hence, the cytosolic pH would remain stable in the case of selective CO_3^{2-} uptake only when CO_3^{2-} uptake and $CaCO_3$ precipitation occur at the same rate. However, both processes probably run out of equilibrium on occasion, especially in a highly variable diurnal environment. In these cases, the utilization of CO_3^{2-} as the inorganic carbon source would constitute a substantial risk for the pH homeostasis. Excess CO_3^{2-} uptake would elevate cytosolic pH while excess $CaCO_3$ precipitation would reduce it. In contrast, a selective uptake of HCO_3^- from seawater would not perturb the cytosolic pH as much under these conditions because HCO_3^- has a relatively low potential to accept or donate H^+ at pH 7.2.

It may therefore be easier for calcifiers to keep cytosolic pH stable at 7.2 by using HCO_3^- as the substrate for calcification.

3. *Stability*. Seawater pH fluctuates substantially in a diurnal and seasonal timescale with HCO_3^- having a dominant and stable concentration over the entire pH range encountered by marine organisms, while $[CO_3^{2-}]$ will show extreme variation. Thus HCO_3^- is a much more reliable inorganic carbon source for calcification.

2.3.3 Increasing Evidence that the Ω_{arag} Model for Coral and Coral Reefs Is Flawed

Venti et al. (2014) summarized their findings as follows: "Using short-term light and dark incubations, we show how the covariance of light and Ω_{arag} can lead to the false conclusion that calcification is more sensitive to Ω_{arag} than it really is." Comeau et al. (2014a) showed further inconsistencies in the Ω_{arag} –calcification relationship. They incubated two coral taxa (*Pocillopora damicornis* and massive *Porites*) and two calcified algae (*Porolithon onkodes* and *Halimeda macroloba*) under 400, 700 and 1000 μatm pCO_2 levels in experiments in Moorea (French Polynesia), Hawaii (USA) and Okinawa (Japan). Environmental conditions differ among the sites. Both corals and *H. macroloba* were insensitive to OA at all three locations, while the effects of OA on *P. onkodes* were location specific. In Moorea and Hawaii, calcification of *P. onkodes* was depressed by high pCO_2, but for specimens in Okinawa, there was no effect of OA. The authors concluded that a linear relationship between calcification and Ω_{arag} for corals is not universal.

Duarte et al. (2013) pointed out that metabolism in inshore waters such as coral reefs results in strong diel to seasonal fluctuations in pH, with characteristic ranges of 0.3 pH units or more on a daily basis. The extreme variability and multiple, complex metabolic controls on pH in coastal waters imply that open ocean conditions cannot be transposed directly to coastal ecosystems. Hence, they contend that ocean acidification from anthropogenic CO_2 is largely an open-ocean syndrome. This concept has been further supported by the work of Cyronak et al. (2014) who showed biogeochemical processes can influence the pCO_2 and pH of coastal ecosystems on diel and seasonal time scales, potentially modifying the long-term predicted effects of increasing atmospheric CO_2. By compiling data from the literature and removing the effects of short-term variability, they showed that the average pCO_2 of coral reefs throughout the globe has increased ~3.5 – fold faster than in the open ocean over the past 20 years. This rapid increase in coastal and reef pCO_2 confounds attempts to predict effects of OA based on oceanic Ω_{arag} (Jury et al. 2013). They constructed a simple model to demonstrate that potential drivers of elevated pCO_2 include additional local anthropogenic disturbances such as increased nutrient and organic matter inputs.

2.3.4 Future Changes in Oceanic Chemistry Due to Human Activity

Caldera and Wickett (2003) found that oceanic absorption of atmospheric CO_2 from fossil fuels may result in larger pH changes over the next several centuries than any inferred from the geological record of the past 300 million years. Pre-industrial, present and future (twice pre-industrial) concentrations of major carbonate system parameters involved in calcification are shown in Table 2.1. Ocean $[Ca^{2+}]$ will not change significantly and is not included in the table. Note that $[CO_3^{2-}]$ decreases while $[HCO_3^-]$ and DIC increase with increasing OA as shown in Fig. 2.4. Carbonate ion concentration decreases from 264 μmol kg^{-1} under pre-industrial levels of atmospheric CO_2 to 170 μmol kg^{-1} under doubled CO_2 conditions, while HCO_3^- increases from 1650 μmol kg^{-1} to 1883 μmol kg^{-1}, and DIC increases from 1922 μmol kg^{-1} to 2059 μmol kg^{-1}. Thus, the majority of the seawater DIC is in the form of HCO_3^-. The majority of the

Table 2.1 Calculated change in carbonate parameters from pre-industrial to twice pre-industrial conditions

	pCO_2 atm	pH	$[H^+]$ (nmol/kg SW)	$[HCO_3^-]$ (μmol/kg SW)	$[CO_3^{2-}]$ (μmol/kg SW)	DIC (μmol/kg SW)	Ω_{arag}
Pre-industrial	280	8.16	6.92	1650	264	1922	4.2
Present	386	8.07	8.51	1742	227	2121	3.6
Twice Pre-industrial	560	7.91	12.3	1883	170	2059	2.7
% Change	+100	−3	+78	+14	−36	+7	−36

Used with permission from Bulletin of Marine Science

Calculated values are based on alkalinity of 2300 μmol/kgSW with T = 25 °C and salinity = 35 ppt using the program CO2SYS (Pierrot et al. 2006)

Estimated pre-industrial saturation state of the tropical ocean in 1880 for pCO_2 is 280 μatm (Kleypas et al. 1999a, 1999b)

host intracellular DIC is also in the form of HCO_3^- (Venn et al. 2009) with very little CO_3^{2-}.

The most dramatic change in the CO_2 system in seawater will be a 78 % increase in $[H^+]$, suggesting that the effect of ocean acidification on coral calcification might directly involve $[H^+]$. According to the Proton Flux Model (Jokiel 2011a, 2011b) the net efflux of H^+ out of the coral and into the water column is influenced by the strength of the diffusion gradient between the coral and the surrounding seawater. This gradient becomes steeper with increasing OA due to increasing $[H^+]$ in the water column, with a consequent decrease in calcification rate. Fick's first law of diffusion links diffusive flux to the concentration field by stating that the flux direction is from areas of high concentration to areas of low concentration with a magnitude that is proportional to the concentration gradient. The efflux of waste protons from the corallum, through the BL and into the water column will occur at a magnitude that is proportional to the concentration gradient. According to this model, increasing the $[H^+]$ in the water column will reduce flux of protons out of the coral. The elimination of H^+ from the coral is just as important as influx or availability of DIC.

2.3.5 Future Regional Changes in Reef Carbonate Production and Dissolution Rates Due to Increasing OA

The most diverse and highly developed reefs occur in areas with very high Ω_{arag} (Kleypas et al. 1999a, 1999b) which is consistent with the hypothesis that $[CO_3^{2-}]$ drives calcification of corals and coral reefs. Thus, it has been assumed that reduction in Ω_{arag} would result in decreased growth (Langdon and Atkinson 2005; Hoegh-Guldberg et al. 2007; Pandolfi et al. 2011). The validity of this assumption was initially challenged by Jokiel (Sect. 2.3.1) who proposed that proton flux expressed as the ratio of substrate to inhibitor [DIC or HCO_3^-]/$[H^+]$ limited calcification rather than $[CO_3^{2-}]$ or Ω_{arag}. The implications of [DIC or HCO_3^-]/$[H^+]$ limited calcification rate on the global distribution of reef calcification has been described by Bach (2015), who showed the absence of a strong latitudinal gradient in $[HCO_3^-]/[H^+]$ in contrast to the strong gradients in $[CO_3^{2-}]$ and Ω_{arag}. The reason for the difference between the two is that temperature has a profound impact on $[CO_3^{2-}]$ and thus Ω_{arag}, but almost no influence on $[HCO_3^-]/[H^+]$. While Ω_{arag} and $[CO_3^{2-}]$ decrease 2–3 fold from the equator towards the poles, $[HCO_3^-]/[H^+]$ is nearly constant. Higher solubility of CO_2 at lower temperature results in an equilibrium shift away from $[CO_3^{2-}]$ towards higher $[CO_2]$ and higher $[HCO_3^-]$. Accordingly, $[CO_3^{2-}]$ declines away from centers of high coral reef development. Ω_{arag} follows the concentration of CO_3^{2-} since $[Ca^{2+}]$ is stable. The slight increase of $[HCO_3^-]$ poleward and eastward of the areas of high coral reef development is balanced by the concomitant increase in $[H^+]$, which explains the stability of $[HCO_3^-]/[H^+]$ over the latitudinal-longitudinal gradient. Thus, the carbonate chemistry conditions controlling calcification on coral reefs will be fairly constant over the globe. Also, the latitudinal pattern of Ω_{arag} and the pattern of $[HCO_3^-]/[H^+]$ are conserved through time in the course of climate change. Likewise, vertical $[CO_3^{2-}]$ and Ω_{arag} gradients in the water column decrease more severely than $[HCO_3^-]/[H^+]$ gradients largely due to the temperature decline, which is strongest in the upper few hundred meters. Lower temperatures negatively affect $[CO_3^{2-}]$ and Ω_{arag} whereas $[HCO_3^-]/[H^+]$ remains unaffected. However, Bach (2015) pointed out that reefs in peripheral areas may be the most severely affected by OA in the future because dissolution is determined by Ω_{arag}. From the carbonate production perspective, however, this is not the case. OA will be equally deleterious in all habitats where $CaCO_3$ formation is controlled by $[HCO_3^-]/[H^+]$. Thus, the relationship between dissolution and calcification undergoes change with increasing OA and increasing global temperature. Rates of secondary calcification, bioerosion, and reef dissolution are important factors in the control of structural complexity and long-term persistence of coral reefs. Silbiger and Donahue (2015) found that secondary reef calcification and dissolution in a coral rubble community responded differently to the combined effect of OA and increased temperature. Calcification had a non-linear response to the combined effect of pCO_2 and temperature: the highest calcification rate occurred slightly above ambient conditions while the lowest calcification rate occurred in the highest temperature–pCO_2 treatment. In contrast, dissolution increased linearly with increasing temperature–pCO_2 . Thus, the coral rubble community switched from net calcification to net dissolution at higher pCO_2 and increased temperature. Jury et al. (2013) reached similar conclusions using a modeling approach.

The values in Table 2.1 are for mean open ocean conditions, which vary little over a diurnal cycle compared to diurnal variations reported for various reefs throughout the world (Table 2.2). In addition to the wide variation between different geographic locations there is considerable variation over small spatial scales at a given location. For example, on a spatial scale of ~ 700 m across a single reef, different magnitudes of pH oscillations have been reported, with open water sites exhibiting less variability than back reef sites (Ohde and van Woesik 1999; Silverman et al. 2007). Processes other than OA, such as changes in nutrient loading from watersheds or change in benthic community structure, can have over-riding effects on long-term pH trends in estuaries and other shallow, nearshore marine environments (Duarte et al. 2013).

Table 2.2 Observed diurnal variation in reported pH (pH_T at 25 °C) for various coral reef habitats throughout the world compared to shallow open ocean values

Location	Habitat	Diurnal Range		Reference
		pH	Δ pH	
Eilat, Israel	Inner moat	8.10–8.30	0.20	Silverman et al. 2007
Shiraho Reef, Okinawa	Inner moat	7.80–8.40	0.60	Suzuki et al. 1995
Rukan-sho, Okinawa	Back reef	7.87–8.52	0.65	Ohde 1995
Shiraho Reef, Okinawa	Inner reef flat	7.90–8.60	0.70	Suzuki et al. 1995
Kiona Beach, O'ahu	Reef flat (sheltered)	7.92–8.13	0.21	Lantz 2011
Makapu'u, O'ahu	Reef flat (open)	7.98–8.10	0.12	Lantz 2011
Heron Island, Australia	Reef flat	8.00–8.40	0.40	Kline et al. 2012
Kāne'ohe Bay, O'ahu	Reef flat	7.90–8.20	0.30	Jokiel et al. 2008
Kāne'ohe Bay, O'ahu	Rreef flat	7.90–8.20	0.30	Martinez et al. 2012
Moloka'i, Hawai'i	Reef flat	7.80–8.40	0.40	Yates and Halley 2006
Moorea, Tahiti	Reef flat	8.02–8.12	0.10	Hofmann et al. 2011
Shiraho Reef, Okinawa	Reef crest	8.00–8.90	0.90	Suzuki et al. 1995
Palmyra, Line Islands	Reef terrace	7.85–8.10	0.25	Hofmann et al. 2011
Palmyra, Line Islands	Fore-reef	7.91–8.03	0.12	Hofmann et al. 2011
Laolao Bay, Saipan	Fore-reef (5 m)	8.16–8.28	0.12	Sean Macduff, unpub.
Laolao Bay, Saipan	Fore-reef (12 m)	8.14–8.24	0.10	Sean Macduff, unpub.
Kingman Reef, Line Is.	Open ocean	8.01–8.03	0.02	Hofmann et al. 2011
North of O'ahu, Hawai'i	Open ocean	8.08–8.12	0.04	Dore et al. 2009
Red Sea	Open ocean	8.20–8.20	0.00	Silverman et al. 2007

2.4 Biological Control or Physical Control of Calcification?

Ries et al. (2010) used the assumption that CO_3^{2-} controls calcification in corals and plotted calcification against Ω_{arag}. They found a curvilinear response which was interpreted to mean that corals exerted strong biological control of the bio-mineralization process. The Two Compartment Proton Flux Model states that coral calcification is limited by the physical process of diffusion across the BL. The Ries et al. (2010) coral calcification data are plotted against $[H^+]$ in Fig. 2.7. If physical control (diffusion of H^+ across the BL) or uptake of $[CO_3^{2-}]$ was the only factor governing calcification, then the relationship between calcification and $[H^+]$ would be linear according to Fick's first law of diffusion, which postulates that the flux of a material goes from regions of high concentration to regions of low concentration with a magnitude that is proportional to the concentration gradient. Rather, the relationship is curvilinear (Fig. 2.7) as we would expect because of the many enzyme-mediated processes involved in photosynthesis, calcification and material transport within the corallum (Fig. 2.6). Therefore a combination of linear and non-linear biological and physical processes including factors such as genetic makeup, biochemical state, temperature, irradiance, nutrient availability and water motion all affect calcification rate. Understanding large-scale spatial variability in coral calcification rates found in nature today, even in a single species, is a complex task (Kuffner et al. 2013), but hopefully establishing baseline datasets will help delineate the most important environmental drivers.

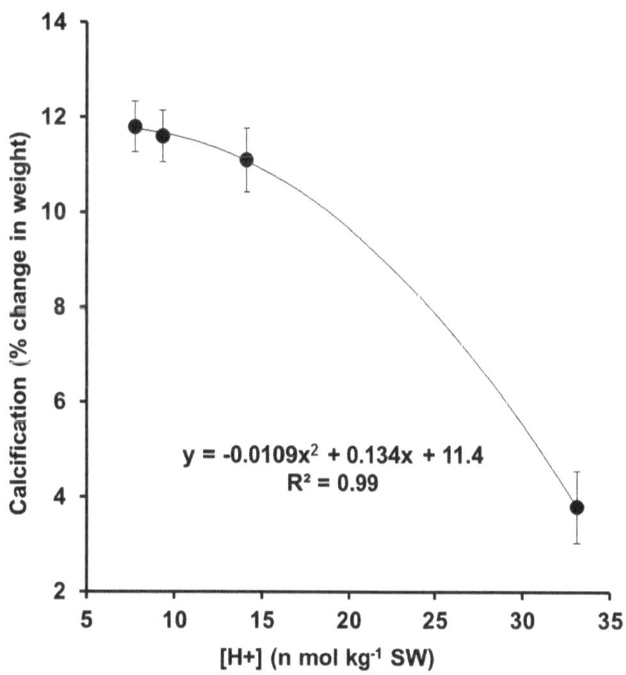

Fig. 2.7 Calcification (as % change over 60 days) plotted against $[H^+]$ rather than Ω_{arag} in the coral *Oculina arbuscula* using Ω_{arag} vs calcification data from Ries et al. (2010)

2.5 Interaction Between Environmental and Biological Factors

2.5.1 Interaction Between OA and Coral-Growth Rate

Marubini et al. (2003) measured calcification in four species of tropical reef corals (*Acropora verweyi*, *Galaxea fascicularis*, *Pavona cactus* and *Turbinaria reniformis*) under 'normal' (280 μmol kg^{-1}) and 'low' (140 μmol kg^{-1}) carbonate-ion concentrations. They report that the calcification rate was affected uniformly across all species tested (13–18 % reduction). An experiment involving the temperate coral *Oculina arbuscula* (Ries et al. 2009) was conducted under similar conditions and provides a useful comparison. Marubini et al. (2003) concluded that a decrease in $[CO_3^{2-}]$ results in a significant reduction in calcification rate for all species tested while Ries et al. (2010) concluded calcification was only minimally impaired in the temperate coral. Plotting these reported calcification rates against $[H^+]$ provides an alternate way of examining the data and provides additional insights (Fig. 2.8, left panel). There was a seven-fold difference in species calcification rate over the range of $[H^+]$ used in the treatments. The corals with higher calcification rate (y-intercept in Fig. 2.8) showed greater calcification reductions (change in slope from −0.81 to −0.11 over the range of equations) in response to increased $[H^+]$. These data are re-plotted in the right panel of Fig. 2.8 to show change in calcification rate for corals grown under normal conditions (pH = 8.06, $[H^+]$ = 8.7 nmol kg^{-1} SW) compared to acidified conditions (pH = 7.75, $[H^+]$ = 17.7 nmol kg^{-1} SW). According to the Proton Flux Model the more rapidly growing tropical corals must dissipate greater quantities of protons through the BL against an increasingly steep $[H^+]$ gradient and will show a stronger reduction in growth. The data suggest that fast-calcifying species are more vulnerable to OA. Comeau et al. (2014b) found that fast calcifiers were more sensitive to ocean acidification than slow calcifiers. The strong linear trend in the graph of metabolic rate (as initial calcification rate) versus change in calcification rate (Fig. 2.8) is consistent with diffusion limitation of material transport at the tissue-water interface.

2.5.2 Temperature and OA

Temperature controls rates of reaction at the site of calcification, in the tissues of the coral and in the water column outside the coral. The combined effects of temperature and OA are influenced by physical chemistry as well as by biochemistry.

Physical Chemistry A model of coral growth based on enhanced kinetics of calcification at higher temperature has been developed (McCulloch et al. 2012). This model

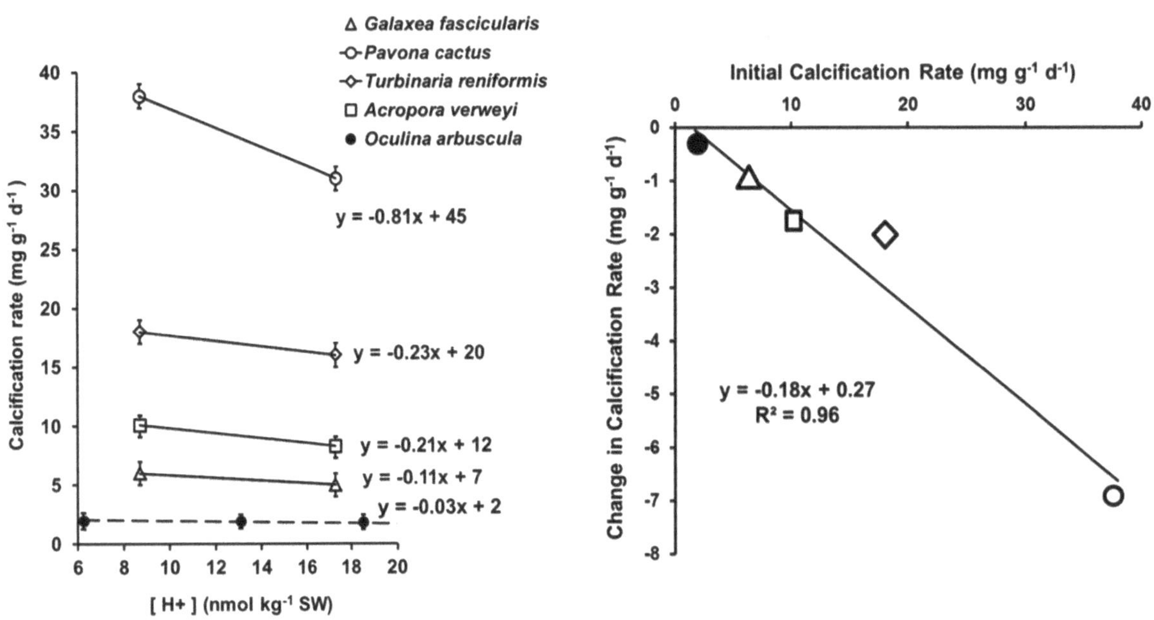

Fig. 2.8 *Left* panel shows calcification rate vs. $[H^+]$ for four species of tropical corals (data from Marubini et al. 2003 as *open symbols*) and for the temperate coral *Oculina arbuscula* (data from Ries et al. 2010 as *solid circles*) with *error bars* as ± SE as re-analyzed by Jokiel (2011a). In the *right* panel the data were re-plotted to show change in calcification rate for corals grown under normal conditions (pH = 8.06) compared to corals grown under acidified conditions (pH = 7.75). The regression is significant with p = 0.004 (Figure from Jokiel (2011a) used with permission from Bulletin of Marine Science)

describes the effect of increased temperature on abiotic processes in the calcifying fluid located in the space between calicodermis and the skeleton and does not consider limiting processes within the coral tissue, processes at the tissue-seawater interface and changes in the boundary layer. The authors concluded that the increase in calcification due to global warming will outweigh the negative effects of declining carbonate ion concentration based solely on physical chemistry considerations in the calcifying fluid. Obviously this conclusion does not fit the preponderance of data showing decrease of coral growth with increasing OA. This model is reminiscent of the earlier model of McNeil et al. (2004) that was based on the assumption that calcification increases linearly with increasing temperature above the present day temperature range. The McNeil model predicted an increase in net coral reef calcification rate of 35 % by the year 2100, a conclusion that runs counter to nearly all experimental observations, which suggest a 15–30 % decrease under these conditions. The McNeil model failed to account for the biological calcification response to temperature (Kleypas et al. 2005). Growth response to temperature is not linear, but declines sharply above peak summer temperature with bleaching and eventual death of corals under future temperature scenarios. McCulloch et al. (2012) ultimately noted that extensive biological experimental and observational data do not support their model, and concluded that the fate of corals will ultimately depend on biochemical adaptation to rapidly changing conditions.

Increasing OA will increase [DIC] in the water column which will enhance influx of the inorganic carbon needed for calcification and photosynthesis. On the other hand, the concomitant increase in [H^+] will reduce efflux of this waste product and thereby slow calcification. By this argument the ratio of [DIC] to [H^+] will correlate with calcification rate. Temperature influences the [DIC] and [H^+] through abiotic carbonate kinetics of seawater (Fig. 2.9). The change in the ratio should have a direct relationship to calcification rate. For example, the shift in the ratio from pre-industrial (280 µatm, 28 °C) to twice pre- industrial (560 µatm, 30 °C) is shown in Fig. 2.9 as a dashed arrow. This is a 33 % reduction in the ratio, which is consistent with the reduction observed in coral calcification under these conditions (Gattuso et al. 1999). As can be seen from the figure, the impact of temperature on the ratio is much less than that of pCO_2.

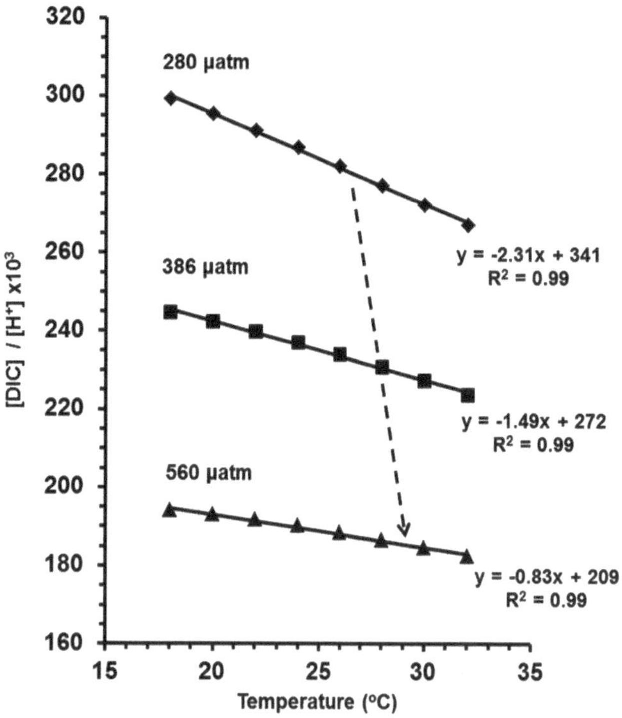

Fig. 2.9 Plot of the ratio of the calcification reactant DIC to the calcification waste product H^+ in the water column relative to temperature under pre-industrial concentrations (280 µatm), 2011 concentrations (386 µatm) and possible future (560 µatm) concentrations of pCO_2 as presented by Jokiel (2011a) (Data used with permission from Bulletin of Marine Science. The *dashed* line shows the 33 % decrease in the ratio from pre-industrial conditions at 280 µatm at 28 °C to twice pre-industrial pCO_2 with an associated greenhouse effect increase of 2 °C)

Biological Response Some corals show a strong biological response to temperature-OA interactions. Reynaud et al. (2003) grew small colonies of the reef coral *Stylophora pistillata* in a matrix of two temperature treatments (25 °C vs. 28 °C) and two pCO_2 treatments (460 µatm vs. 750 µatm) and report no statistical difference between pCO_2 treatments at 25 °C. However, there was a large decline in calcification (approximately 50 %) at 28 °C under acidified conditions. Anlauf et al. (2011) studied the effects of a 1 °C increase in temperature and a 0.20–0.25 unit decrease in pH on the growth of primary polyps in the coral *Porites panamensis*. The growth of polyps was reduced marginally by acidic seawater but the combined effect of higher temperature and lowered pH caused a significant growth reduction of approximately 30 %. A similar 30 % decline at higher temperature – elevated pCO_2 was shown by Edmunds et al. (2012) for the rapidly growing branched coral *Porites rus*, but a slower growing massive *Porites* sp. did not show the effect. The temperature – pCO_2 interaction has been observed in other calcifying reef organisms. Martin and Gattuso (2009) observed the same effect on the coralline alga *Lithophyllum cabiochae*. Algae were maintained in aquaria for one year at ambient or elevated temperature (+3 °C) and at ambient pCO_2 (~400 µatm) or elevated pCO_2 (~700 µatm). During summer the net calcification of the algae decreased by 50 % when both temperature and pCO_2 were elevated while no effect was found under elevated temperature or elevated pCO_2 alone.

A biochemical mechanism can be proposed to explain the temperature-pCO_2 synergism. Coles and Jokiel (1977) showed that coral photosynthesis and respiration increase with increasing temperature, but respiration increases more rapidly. Consequently the ratio of photosynthesis to respiration decreases with increasing temperature to a value of unity near the upper thermal limit, and a mutually beneficial symbiosis cannot be maintained above that level. As temperature increases, the resulting high rates of photosynthesis and respiration require much higher exchange of materials through the boundary layer. Thus corals show greater demand to uptake inorganic carbon and a greater need to dissipate waste H^+ at higher temperature. Perhaps this explains why coral skeletal growth shows a positive correlation with increasing temperature, but only up to an "optimal temperature" (Jokiel and Coles 1977) and then declines rapidly to lethal conditions.

2.5.3 Water Motion and Irradiance

Models of Coral Growth Kaandorp et al. (2005, 2011) used simulation experiments and isotope analyses of coral skeletons to test the hypothesis that water motion and localized external BL gradients of DIC determine gradients of calcification that directly control the morphogenesis of branching, phototrophic corals. Their model is entirely driven by a diffusion-limited BL process and can generate coral growth patterns and morphologies that are virtually indistinguishable from three dimensional images of the actual colonies. This model provides strong support for the contention that water motion increases calcification by breaking down diffusion barriers in the BL. They concluded that inorganic carbon supply on the reactant side of Eqs. 2.5, 2.6 and 2.7 represents the limiting factor for calcification rate. The Kaandorp et al. (2005, 2011) model will show an erroneous increase in coral growth as OA increases, because DIC increases with increasing OA. This problem would be resolved if the model incorporated the influence of H^+ flux; i.e., the model would produce a similar morphological output, but with reduction rather than an increase in coral growth under increasing acidified conditions.

Changes in Growth Form with Increasing Depth From a topological point of view we can treat the myriad coral growth forms (Fig. 2.2) as simple hemispheres, with the hemispherical ZC encapsulating the ZP (Fig. 2.3). The model as presented in Fig. 2.6 represents a cross section at a given point on the hemispherical corallum. Coral reef environments show strong vertical and horizontal gradients of both water motion and irradiance, so these factors are not uniform over the surface of a colony and can influence colony growth form as shown in Fig. 2.10. The colonies of many massive and branching coral species become more flattened and plate-like with increasing depth (e.g., Roos 1967; Graus and Macintyre 1976; Jaubert 1977) in response to submarine irradiance distribution (direction and intensity). Growth along an axis diminishes with decreasing irradiance and with decreasing water motion.

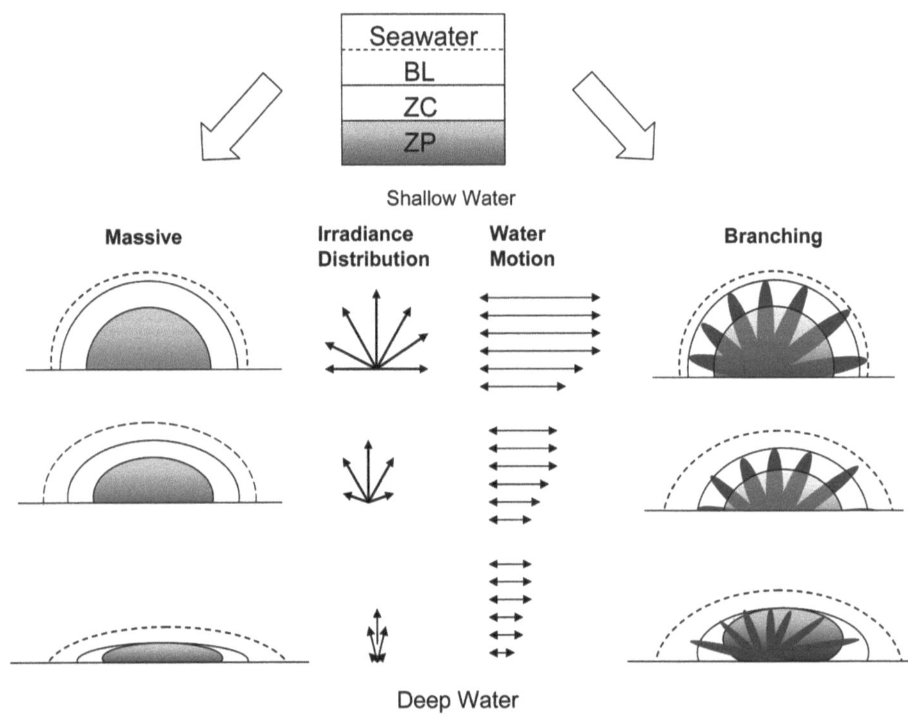

Fig. 2.10 The relationship between the zone of primary calcification (*ZC*), zone of primary photosynthesis (*ZP*) and the boundary layer (*BL*) showing the relative changes in different growth axes due to gradients in the irradiance-water motion field with increasing depth (Figure from Jokiel (2011b) used with permission from Journal of Experimental Marine Biology and Ecology)

2.6 Coral Nutrition

2.6.1 Inorganic Nutrients

Corals grown at elevated levels of inorganic nutrients (nitrogen and phosphorous) increase photosynthetic rate while simultaneously decreasing calcification rate (Hallock and Schlager 1986; Stambler et al. 1991). Corals grown under increased levels of pCO_2 also show a decline in calcification (Gattuso et al. 1999; Schneider and Erez 2006). Corals cultured under a combined high inorganic nutrient – high pCO_2 treatment continue to grow, but do not grow as rapidly as corals growing under ambient conditions (Marubini and Atkinson 1999; Renegar and Riegl 2005). Atkinson et al. (1995) described seawater conditions at the Waikiki Aquarium where corals were growing in high-nutrient (PO_4 ~ 0.6 μM; NO_3 ~ 5 μM; NH_4 ~ 2 μM), high pCO_2 (400–880 μatm) water drawn from a saltwater well. These observations led the authors to conclude that corals can flourish under high nutrient, high pCO_2 conditions, although they did not conduct any comparisons with corals grown at low pCO_2 and low nutrient. This anomalous conclusion persisted until aquarists at the Waikiki Aquarium discovered that adding a flow of low nutrient, low pCO_2 oceanic sea water to the coral display tank greatly improved coral growth compared to corals held in flowing well water (Richard Klobuchar, personal communication, 17 May 2011). Therefore, the situation at the aquarium now appears to be in agreement with results of the various controlled experiments conducted to date (e.g., Marubini and Atkinson 1999; Renegar and Riegl 2005), with the aquarium corals growing at a lower rate under conditions of high nutrient and high pCO_2. However, controlled experiments comparing growth in sea water versus well water at the Waikiki Aquarium are needed.

2.6.2 Organic Nutrient Heterotrophy

It has been suggested that coral sensitivity to OA may depend on energetic status (Cohen and Holcomb 2009) which has recently been shown experimentally (Chauvin et al. 2011; Edmunds 2011). It follows that increasing the nutritional status or energy stores of the coral could potentially ameliorate OA effects on calcification if the coral is able to use these resources to increase ion transport. The ability to do this, however, may be species-specific. For instance, some corals are able to recover faster from bleaching when allowed to feed on plankton, whereas other species are unable to take advantage of the opportunity (Grottoli et al. 2006). These ideas are not in conflict with the Proton Flux Model, since the coral's increased energy supply to proton pumps could result in raising pH in the calcifying space. This in turn would increase the gradient in $[H^+]$.

2.6.3 Organic vs Inorganic Nutrients and Coral Calcification

The success of reef corals in shallow tropical seas stems from the symbiotic association between endocellular zooxanthellae and the host (Muscatine and Porter 1977). This association allows corals and coral reef communities to thrive in spite of the low concentrations of nitrogen (N) and phosphorus (P) in the oligotrophic waters. However, a contradiction was presented by Kinsey and Davies (1979) who showed that increasing the concentration of inorganic N and/or inorganic P reduces, rather than increases, coral growth. In contrast, corals supplied with an increased supply of organic N and organic P in the form of zooplankton increase rather than decrease their skeletal growth rate. Ferrier-Pagès et al. (2003) performed laboratory experiments designed to assess the effect of feeding on the tissue and skeletal growth in the coral *Stylophora pistillata*. Fed colonies exhibited significantly higher levels of protein and chlorophyll per unit surface area than starved colonies. Feeding had a strong effect on tissue growth, increasing it by two to eight times. Calcification rates were also 30 % higher in fed than in starved corals. Thus N and P provided to the coral in the inorganic form results in decreased calcification while increased N and P provided from organic sources result in increased calcification rate.

This paradox can be resolved by incorporating the model of Dubinsky and Jokiel (1994) into the Proton Flux Model. According to the Dubinsky-Jokiel model the zooxanthellae produce CH_2O in great excess of their basic metabolic needs which is translocated to the ZC (Fig. 2.6). However, this energy-rich photosynthate has been described as "junk food" because it lacks the N and P needed to support tissue growth (Falkowski et al. 1984). Under normal conditions the zooxanthellae gain access to inorganic N and P as metabolic waste products from their host in return for fixed carbon. The symbiosis retains and recycles N and P, which in the absence of the algae would have been excreted into the sea and lost. The animal fraction of the symbiosis can obtain organic N and P by feeding on zooplankton, but cannot utilize inorganic N and P directly. In contrast, the zooxanthellae assimilate dissolved inorganic N and P compounds, but cannot assimilate zooplankton. Under normal reef conditions the supply of plankton and inorganic plant nutrients is very limited.

Thus zooxanthellae are normally under nutrient-starved conditions, so given additional inorganic N and P they quickly uptake these substances and put their energy into new plant cell growth. Less photosynthate is translocated from the ZP to the ZC with a consequent decrease in skeletal growth. On the other hand, increased supply of organic N and P in the form of zooplankton gives the advantage to the animal portion of the symbiosis. As the animal digests the zooplankton, it can use the organic N and P for increased cell growth and can also benefit from the energy value of the zooplankton food. The increased energy can be utilized to increase calcification as well as increase animal cell growth. Flow of photosynthate from the ZP is not reduced and skeletal growth in the ZC continues, enhanced by energy derived from the digestion of the zooplankton. The fed coral will produce additional metabolic waste N and P for the zooxanthellae and the plant biomass will eventually increase along with increases in the animal biomass, but at a rate that does not give the advantage to the plant fraction of the symbiosis. Thus, according to the combined model the mechanism responsible for reduced calcification under increased inorganic nutrient concentration is related to diminished fixed-carbon flux from the ZP to the ZC. The cause of reduced growth under increased pCO_2 is reduction of proton efflux out of the tissues and into the water column due to increased $[H^+]$ in the water column. However, the two are additive when inorganic nutrient level and pCO_2 are increased simultaneously (Renegar and Riegl 2005).

2.7 Acclimatization and Adaptation

One of the great unknowns in predicting the effects of ocean acidification on corals and coral reef communities in the future is the ability of coral species to acclimatize and/or adapt to future conditions (Gattuso et al. 1999). There is little evidence thus far that corals can acclimatize to ocean acidification. During the course of a nine-month long mesocosm experiment, corals did not show any decrease in response to the OA treatment (Jokiel et al. 2008). Similarly, nubbins of the coral *Stylophora pisitillata* showed the same sensitivity to OA after either 24 h or 1 year of exposure to reduced pH (Venn et al. 2013). The potential for adaptation, on the other hand, seems more promising (Pandolfi et al. 2011). Coral species and populations have shown differential responses to repeated bleaching events (Guest et al. 2012), offering some hope that enough genetic variability exists in at least some coral populations to allow adaptation to chronic environmental stressors. Evidence for local and regional adaptation also exists, with species showing resilience to naturally occurring extremes in seawater carbonate chemistry (Fagan and Mackenzie 2007; Fabricius et al. 2011) and temperature regimes (Coles 1988; Harriott and Banks 2002).

2.8 Resolving Unexplained Paradoxes with New Insights

Understanding of the importance of proton flux (Jokiel 2011a) and clarification of the spatial relationship between the BL, ZC and ZP (Jokiel 2011b) has resolved a number of apparent contradictions in the coral calcification literature as described below.

2.8.1 Paradox of Decreasing Coral Growth Rate in the Face of Increasing HCO_3^- and Increasing DIC

Marubini et al. (2008) found that doubling the $[HCO_3^-]$ resulted in a coral calcification increase of approximately 27 % at all levels of $[H^+]$ tested. Herfort et al. (2008) also found that coral calcification increased rapidly in response to added HCO_3^-. Since OA increases $[HCO_3^-]$ (Fig. 2.4) we might expect increasing skeletal growth due to higher pCO_2. However, Gattuso et al. (1999) compiled existing information and estimated that coral $CaCO_3$ production decreased by 10 % between 1880 and 1990. They projected an additional 9–30 % (mid estimate: 22 %) reduction from 1990 to 2100 due to OA for a total of 32 % reduction between pre-industrial and twice pre-industrial pCO_2 concentrations. More recent estimates (Marubini et al. 2001; Anthony et al. 2008; Jokiel et al. 2008) support this generalization. The calculated increase in $[HCO_3^-]$ that can be attributed to change from pre-industrial to twice pre-industrial is 14 % (Table 2.1). Using the estimate of Marubini et al. (2008) that a doubling of $[HCO_3^-]$ will result in a 27 % increase in calcification, the expected change in coral growth due to projected increased $[HCO_3^-]$ would only be on the order of +3.8 % compared to the predicted change of -32 % due to increased $[H^+]$. Thus any benefit to skeletal growth due to higher future $[HCO_3^-]$ will be overwhelmed by the order of magnitude greater negative impact due to increased $[H^+]$.

2.8.2 Paradox of Rich Coral Reefs Growing Under Low Ω_{arag} Conditions

Palau Rock Islands Shamberger et al. (2014) reported the existence of highly diverse, coral-dominated reef communities in the Rock Islands of Palau under chronically low pH and low Ω_{arag}. They noted that identification of the biological and environmental factors that enable coral communities to persist under these conditions could provide important insights into the future of coral reefs under anthropogenic acidification. Jokiel (2015) re-analyzed their data from the perspective of the Proton Flux Model and provided the following explanation. The Rock Islands are located in the south lagoon of Palau between Koror and Peleliu. This area is an uplifted ancient coral reef that forms a complex carbonate labyrinth of shallow channels and lagoons containing 250–300 small islands. These uninhabited islands are actually carbonate outcrops. Some of the islands display a mushroom shape with a narrow base created by rapid dissolution and bioerosion by sponges and bivalves and intense grazing by chitons, urchins and fish in the intertidal (Lowenstam 1974; Glynn 1997). The extreme bioerosion and carbonate dissolution in the area coupled with very low rates of seawater exchange produces atypical conditions that provide an important insight into CO_2-carbonate system dynamics.

Understanding the hydrodynamics of the ocean in the vicinity of Palau is critical to our understanding of the processes involved. Golbuu et al. (2012) showed that water circulation in and around the Palau archipelago is very different from circulation around an isolated island or reef, because the mean-water circulation is steered away from and around the archipelago. This deflection generates slower mean currents inside the archipelago compared to accelerated currents surrounding the archipelago. Water exchange is further diminished in shallow waters (depth < 20 m) by the non-linear friction-driven interaction between the tidal currents and the prevailing regional currents. The highest water retention is apparent in the southern lagoon in the Rock Islands area where extensive shallow reef formations occur. The Shamberger et al. (2014) sites at 7–9 km from open ocean waters were located in the shallow flow-restricted bays of the Rock Islands, while sites at 1–3 km were located in the areas of accelerated currents outside of the archipelago that experience rapid flushing with oceanic waters.

The investigators did not make measurements during nighttime when respiration greatly increases pCO_2. Dissolution of carbonates increases A_T (Wisshak et al. 2013). Therefore, additional A_T produced during the night was available to the calcifying organisms during the day due to long residence time of sea water in the lagoon. High rainfall and tidal variation result in submarine groundwater discharge (SGD), which serves as another source of A_T to the calcifying organisms. Rainfall is high and varies between 200 and 450 mm month^{-1} due to Palau's location on the edge of the Western Pacific Warm Pool and the year-long influence of the Intertropical Convergence Zone (Australian Bureau of Meteorology and CSIRO 2011). Cyronak et al. (2013a) measured increased A_T flux due to SGD at Muri Lagoon, Cook Islands, with the daily flux rate of up to 1080 mmol m^{-2} day^{-1}. Dissolution of the complex non-living carbonates supplemented with additional A_T from SGD in the low-flushing reefs of the Rock Islands over the 24 h period would provide considerable A_T buffering of the protons being generated by calcification.

G_{net} was not measured in the study, but a high rate of calcification was implied from the dense standing crop of corals and other calcifying organisms. G_{net} includes dissolution, which occurs at a high rate in such carbonate formations. P_{net} was not measured either, but presumably was very high due to high biomass of corals and other photosynthetic organisms. G_{net} is driven by P_{net}, and not by Ω_{arag} (Jokiel et al. 2014a). Ω_{arag} is actually a dependent variable on G_{net}. Ω_{arag} lags behind G_{net} by several hours during the diurnal cycle (Shamberger et al. 2011; McMahon et al. 2013; Jokiel et al. 2014a). Flux data are lacking, but fortunately the concentration data were taken along a horizontal environmental gradient, which permits calculating flux rates. A crude description of the dynamic benthic processes involved in coral calcification at this site can be made with the available data for $[A_T]$, [DIC] and $[H^+]$. This analysis requires that we accept the assumptions implicit in the experimental design that horizontal mixing is uniform throughout the area of study and that $[A_T]$, [DIC] and $[H^+]$ are not greatly modified by pelagic processes in relation to benthic processes. The environmental gradient between the lagoon and open ocean is shown for $[A_T]$ (Fig. 2.11a), [DIC] (Fig. 2.11b) and $[H^+]$ (Fig. 2.11c). The strongest gradient ($p < 0.001$, Fig. 2.11c) is for net flux of protons out of the lagoon. Net flux of A_T into the lagoon also shows a strong gradient ($p < 0.001$, Fig. 2.11a). Net DIC flux into the lagoon did not show a statistically significant correlation ($p = 0.089$, Fig. 2.11b). This pattern is consistent with detailed observations made by Jokiel et al. (2014a) in mesocosm studies. Dissipation of the protons generated by calcification is a major factor limiting coral growth. Flushing of the Rock Islands with oceanic waters removes H^+ and brings in water with higher $[A_T]$. Water motion can further diminish boundary layers and enhance H^+ and A_T exchange with the benthos (Cyronak et al. 2013b). DIC is abundant in sea water and is not as important as proton flux (Jokiel et al. 2014a) in relation to calcification as reflected in Fig. 2.11b, which shows a weaker relationship than $[A_T]$ or $[H^+]$ with distance from the open ocean.

The situation in the Rock Islands is defined by extremely high rates of carbonate dissolution and restricted water flow. Figure S4 of the Shamberger et al. (2014) report shows a rich coral community at the Rock Islands lacking in macroalgae and turf algae, probably due to intense grazing pressure and low inorganic nutrient supply. A larger macroalgae component would have increased pH during daylight hours without altering A_T. The higher pH in such situations shifts the equilibria toward increased $[CO_3^{2-}]$ and therefore higher Ω_{arag}. In this system, as in other systems, Ω_{arag} simply describes the portion of DIC that is being expressed as CO_3^{2-} under prevailing pH conditions. Such pH conditions can be rapidly modified by algal photosynthesis without changing A_T and thus Ω_{arag} is not very useful as a universal metric related to coral calcification. There is a local correlation

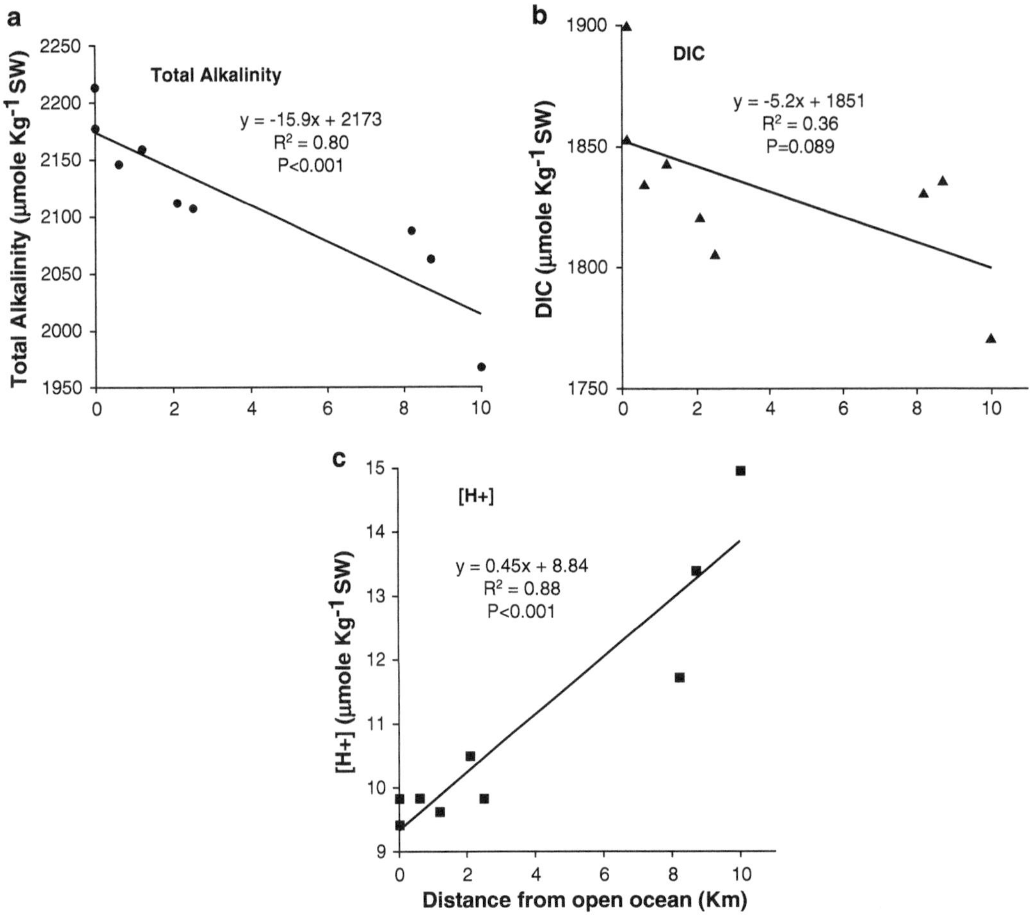

Fig. 2.11 Change in concentration from open ocean to coral lagoon using data from Shamberger et al. (2014) for: (**a**) Total Alkalinity, (**b**) Dissolved Inorganic Carbon, and (**c**) Proton Concentration (Figure from Jokiel (2015) used with permission)

between Ω_{arag} and G_{net}, but Ω_{arag} is a dependent variable on several factors including P_{net} and local dissolution rate of carbonates (e.g., Murillo et al. 2014). The relationship of G_{net} to Ω_{arag} holds within a given system, but varies between systems due to differences in P_{net}, which drives G_{net} (Jokiel 2015). Systems with a higher portion of P_{net} being provided by non-calcifying photosynthetic organisms will have a different relationship to Ω_{arag} than a system dominated by calcifying photosynthetic organisms.

The findings of Murillo et al. (2014) provide additional insight into the situation in the Rock Islands. They conducted flume studies and found that corals isolated from other reef components (carbonate sediment and algae) calcified at a 2:1 ratio of A_T flux to Ca^{2+} flux (ΔA_T:ΔCa^{2+} = 2.0). The same corals incubated in a community that contained carbonate sediment and macroalgae calcified at a lower ratio (ΔA_T:ΔCa^{2+} = 1.6), which indicates the presence of additional sources of alkalinity (i.e. buffering) from carbonate sediments. Carbonate sediments incubated in isolation from the other components buffered the water column, maintaining higher and more stable levels of pH while increasing A_T and DIC.

The A_T of seawater increases with carbonate dissolution in areas such as the Rock Islands that are dominated by carbonate rock and sediment. Corals growing in the presence of such rapidly dissolving carbonates are supplemented with a local source of A_T and live in an environment that is more favorable to calcification compared to environments that lack carbonate deposits. Even so, concentration of A_T on the reefs of the Rock Islands is much lower than offshore (Fig. 2.11a) due to intense calcification which lowers [A_T] by two moles for every mole of $CaCO_3$ precipitated. However, A_T concentration does not tell us anything about A_T flux or its relation to net calcification-dissolution flux (G_{net}). Understanding the dynamics of calcification requires measurement of flux rates. All of the data provided in the Rock Islands study consisted of concentration measurements (pH, [A_T], [DIC], [CO_3^{2-}]). Jokiel et al. (2014a) demonstrated the pitfalls of this approach and pointed out the need to measure flux rates (H^+ flux, DIC flux, A_T flux, and G_{net}) along with net flux of carbon due to photosynthesis-respiration (P_{net}).

In sum, the combination of biological and environmental factors that enable the reef communities in the Rock Islands to persist at chronically low pH and low Ω_{arag} can be identified. First, the extremely high rate of carbonate dissolution increases the A_T available for neutralizing protons. Second, highly restricted hydrodynamic flow maintains the conditions that buffer calcification. Unfortunately, observations under the highly atypical hydrodynamic and geologic conditions at the Rock Islands provide little hope that the global future of coral reefs under anthropogenic acidification can be offset on a broad scale by increased dissolution (Andersson et al. 2003), but perhaps environments such as these can provide refugia on a highly localized scale.

Kāne'ohe Bay, Hawai'i Kāne'ohe Bay, Hawai'i contains well developed coral reef communities that have shown remarkable resilience to various environmental stressors (Hunter and Evans 1995). These rich reefs have developed under high pCO_2 levels. The elevated pCO_2 is due to metabolism of terrigenous organic material transported into the bay by streams. Fagan and Mackenzie (2007) found that pCO_2 was approximately 500 μatm on average in the northern bay while central and southern bay waters had an average pCO_2 of 460 μatm with the entire bay and nearshore reef experiencing levels well above atmospheric pCO_2 (Shamberger et al. 2011). Such levels of pCO_2 are believed to be highly deleterious to coral growth (summarized by Hoegh-Guldberg et al. 2007). One estimate is that when atmospheric partial pressure of CO_2 reaches 560 μatm all coral reefs will cease to grow and start to dissolve (Silverman et al. 2009). So how do we account for the paradox of rich coral reefs of Kāne'ohe Bay growing at levels of pCO_2 that should eliminate them?

The Proton Flux Model provides an explanation. Previous estimates of coral response to ocean acidification have been based on Ω_{arag}. The parameters Ω_{arag} and $[H^+]$ do not always correlate reliably with each other in shallow inshore systems due to the intense metabolic activity of inshore reef communities, input of terrigenous materials, high rates of carbonate dissolution and long residence time of sea water. In a mixed inshore coral-algae community the $[H^+]$ can be very low during daylight hours due to high rates of photosynthesis and low rates of water exchange, while during the night the $[H^+]$ increases dramatically due to respiration of the reef communities. Night $[H^+]$ is of less importance to corals because they do not calcify rapidly in darkness, but it is important during daytime LEC. Use of pH rather than $[H^+]$ can be deceptive. For example, the 0.4 ΔpH range reported for the Moloka'i reef flat in Table 2.2 represents a 5.1 % change in pH, but a 151 % change in $[H^+]$. This strong diurnal signal on shallow coral reefs attenuates with distance offshore and is quite small in oceanic waters. Both Ω_{arag} and $[H^+]$ vary over the diurnal cycle on inshore reefs but are not tightly coupled (Shamberger et al. 2011). Therefore, Kāne'ohe Bay barrier reef daily calcification rates were found to be the same or higher than rates measured on other coral reefs despite the comparatively low Ω_{arag} levels.

2.8.3 Paradox of Rapid LEC in Areas of The Coral Colony That Do Not Contain Photosynthetic Zooxanthellae

This paradox has gone unexplained for half a century. Goreau (1959) noted that "Although the zooxanthellae seem to play an important role in determining calcification rates of reef-building corals, certain, as yet unknown, physiological factors operate to control the basic mineralization process in a manner which bears no obvious relationship to the number of algae present in a given species". More recently, Tambutté et al. (2007) conducted more detailed studies and report that the tissues which calcify at the highest rates do not possess zooxanthellae. The paradox has been resolved by the Two Compartment Proton Flux Model through the realization that calcification and photosynthesis compete for available inorganic carbon and must be spatially separated within the coral (Jokiel 2011b). Reef corals have resolved this conflict by evolving a morphology that places the calcifying sites (ZC) distal to the photosynthetic sites (ZP).

2.9 Alteration of Seawater Chemistry by Corals Over the Diurnal Cycle

Extreme diurnal alteration of pH occurs on shallow coral reefs (Table 2.2). Jokiel et al. (2014a) conducted mesocosm experiments that precisely measured the changes in bulk sea water chemistry and material flux that accompany these pH oscillations over a diurnal cycle. The experiment was conducted in the flow-through mesocosm system at the Hawaii Institute of Marine Biology, Kāne'ohe Bay, O'ahu, Hawai'i. The mesocosm system has been described in detail (Jokiel et al. 2008; Andersson et al. 2009; Jokiel et al. 2014b). Major findings of the Jokiel et al. (2014a) investigation are summarized below.

Calcification over the 24 h period (Fig. 2.12) shows the diurnal pattern related to irradiance, light-enhanced calcification and dark calcification. Values for G_{net} are high due to the large biomass of live coral, high solar irradiance in the shallow mesocosms and absence of sediment or dead carbonate skeleton which could weaken and confound the coral calcification signal through carbonate dissolution. Light saturation of calcification did not occur up to the maximum irradiance which exceeded 1500 μmole photons m^{-2} s^{-1}. This value is many times higher than that supplied by the artificial light typically used in most laboratory studies of coral calcification. There is a drop in calcification to zero around midnight with a dark calcification rate peak at approximately 03:00 h.

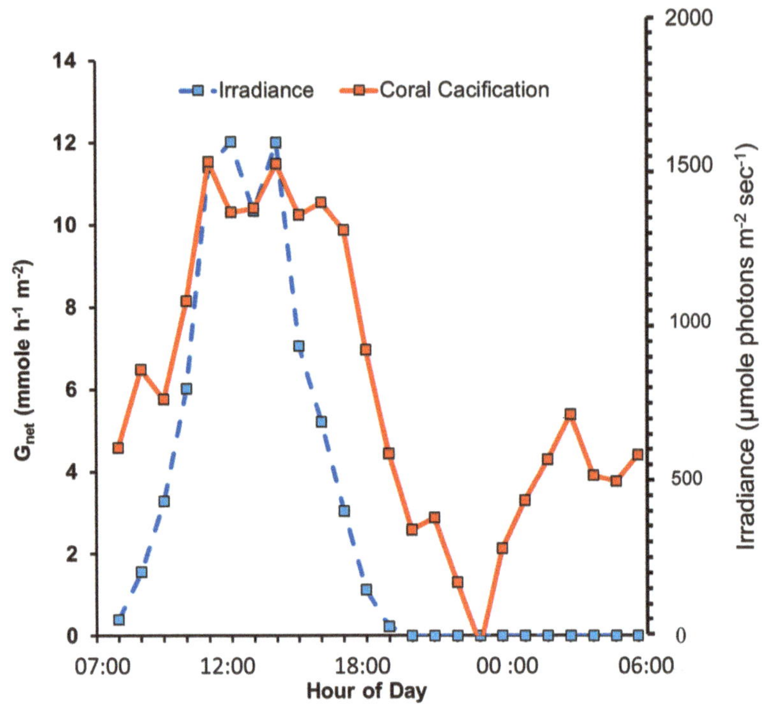

Fig. 2.12 Diurnal net calcification rate (G_{net}) and irradiance for a mesocosm containing corals (Figure from Jokiel et al. (2014a) used with permission)

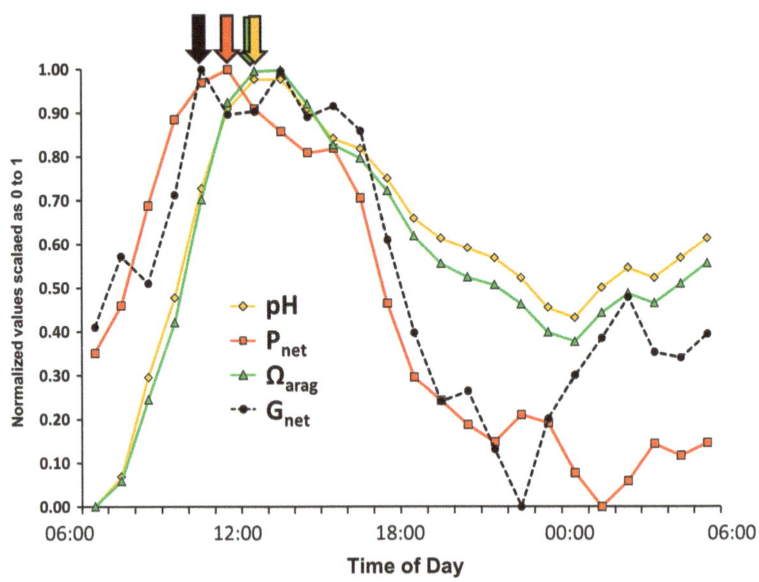

Fig. 2.13 pH, Ω_{arag}, P_{net} and G_{net} values versus time of day with all values normalized to a 0–1 scale. Arrows point to relative maxima for each parameter (Figure and data from Jokiel et al. (2014a) used with permission)

2.9.1 Phase Shifts

Figure 2.13 shows that peak pH and Ω_{arag} lag behind G_{net} throughout the daily cycle by two or more hours. The figure also shows that peak G_{net} follows P_{net} during daylight photosynthetic hours with a reversal during the nighttime hours. Shamberger et al. (2011) previously reported that Ω_{arag} lags behind G_{net} on the reefs of Kāne'ohe Bay. McMahon et al. (2013) reported that peak G_{net} rates occurred 2–3 h before the Ω_{arag} maximum on the Great Barrier Reef. Thus Ω_{arag} (along with closely correlated $[CO_3^{2-}]$, pH and $[DIC]:[H^+]$ ratio) cannot be the primary driver of coral calcification over a diurnal cycle. The use of Ω_{arag} to calculate future changes in G_{net} on a global scale must consider future changes in the other processes that have a great influence on G_{net} on smaller spatial and temporal scales. Figures 2.12 and 2.13 show that diurnal irradiance drives P_{net}, which in turn drives G_{net}. P_{net} and G_{net} alter pH (Eqs. 2.5, 2.6, 2.7, 2.8, 2.9 and 2.10), which controls $[CO_3^{2-}]$ and Ω_{arag} in addition to the other variables based on concentration such as the ratio of [DIC] to $[H^+]$.

2.9.2 Night Calcification

Laboratory studies show that coral calcification continues in darkness, but at a lower rate than observed in light enhanced calcification (Schneider and Erez 2006). Night calcification rates have generally been assumed to be low and constant at night, although this assumption has largely gone untested. Figures 2.12 and 2.13 show decreasing dark calcification following sunset, reaching zero near midnight followed by an increasing rate of dark calcification and an increase in respiration that rises to a peak at 03:00, which is well before dawn. This pattern has occurred consistently in our mesocosm experiments, with the same pattern observed in 30 separate mesocosm runs with different communities under various conditions as well as in flume studies (Murillo et al. 2014). Barnes and Crossland (1980) used time-lapse photography to measure diurnal growth in the staghorn coral *Acropora acuminata* and found that night-time extension rate was similar to or greater than day-time extension. They suggested that, "symbiotic association permits rapid growth because the coral can invest in flimsy scaffolding at night with the certainty that bricks and mortar will be available in the morning". Wooldridge (2013) has proposed a new model for "dark" coral calcification, whereby O_2-limitation of aerobic respiration during the night initiates a homeostatic host response that forms the skeletal organic matrix. The matrix formed at night subsequently allows rapid growth of the aragonite fibers during the "light-enhanced" period of calcification, when abundant energy derived from photosynthesis is available. Perhaps the midnight calcification minimum observed in Figs. 2.12 and 2.13 at 00:00 reflects this period of organic matrix formation that precedes the 03:00 night calcification peak.

Diurnal changes in pH, DO and plankton feeding also have an effect on diurnal calcification in light and darkness. Wijgerde et al. (2012) measured the short-term effects of zooplankton feeding on light and dark calcification rates of the scleractinian coral *Galaxea fascicularis* at oxygen saturation levels ranging from 13 to 280 %. Significant main and interactive effects of oxygen, heterotrophy and light on calcification rates were found. Light and dark calcification rates of unfed corals were affected by hypoxia and hyperoxia. Light calcification rates of fed corals showed highest calcification rates at 150 % saturation. In contrast, dark calcification rates of fed corals were close to zero under all oxygen saturations. The authors concluded that oxygen exerts a strong control over light and dark calcification rates of corals, and proposed that *in situ* calcification rates are highly dynamic. Nevertheless, the inhibitory effect of heterotrophy on dark calcification appeared to be oxygen-independent. They hypothesized that dark calcification is impaired during zooplankton feeding by decrease in pH and aragonite saturation state of the calcifying fluid adjacent to the skeleton resulting from the increased respiration rates.

2.9.3 Diurnal Changes in Concentration of A_T, pH, Ω_{arag} and DO

The variables of A_T, pH, Ω_{arag} DIC, and DO are concentrations while P_{net} and G_{net} are flux rates. Caution must be taken when comparing concentrations to flux rates because flux rate can be high when concentration is high or low, or flux rate can be low when concentration is high or low. Figure 2.13 shows patterns that are difficult to interpret because the figure mixes flux rates with concentrations. Much can be learned by plotting DIC flux and H^+ flux rather than [DIC], [H^+] or pH in relation to P_{net} and G_{net}. DIC flux and H^+ flux are plotted with P_{net} and G_{net} in Fig. 2.14. This figure illustrates the dynamic geochemical and physiological relationships involved in coral metabolism.

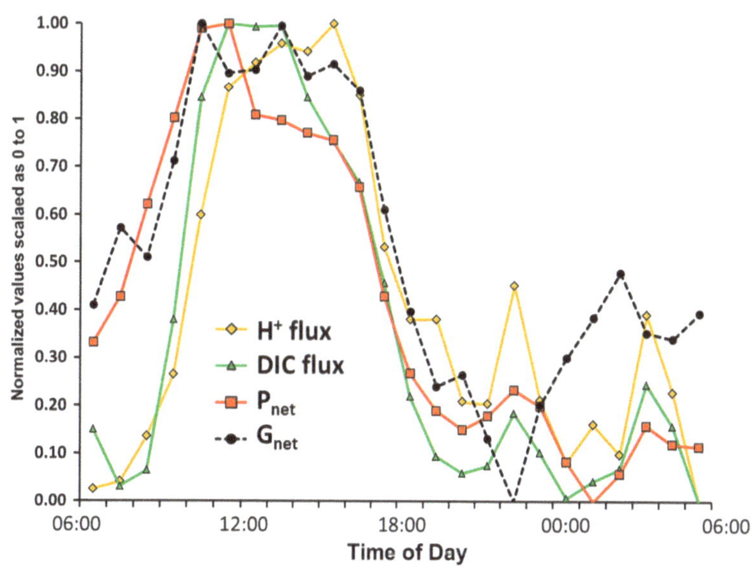

Fig. 2.14 Plot of normalized data for P_{net}, G_{net}, inverse DIC flux and H^+ flux with all values normalized to a 0 to 1 scale (Figure from Jokiel et al. (2014a) used with permission)

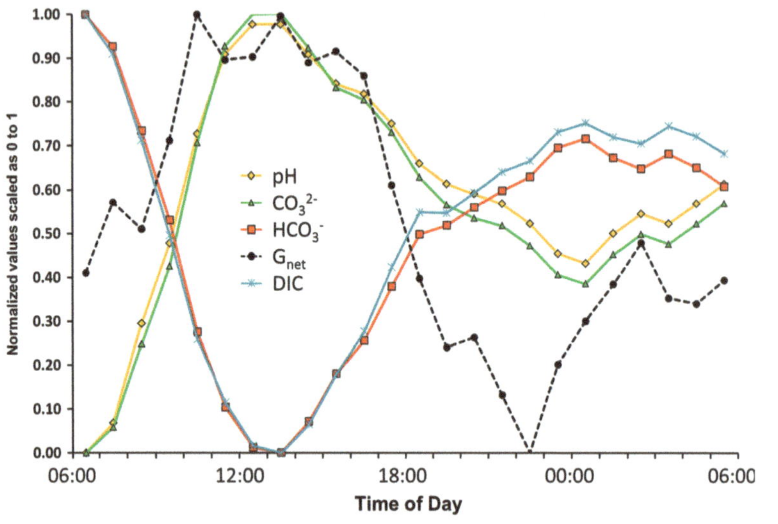

Fig. 2.15 The flux rate of calcification-dissolution (G_{net}) plotted against the concentrations of important variables in the CO_2-carbonate system with all values normalized to a 0–1 scale (Figure from Jokiel et al. (2014a) used with permission)

DIC flux (uptake) in the rapidly calcifying mesocosms increases with increasing P_{net} from 06:00 until mid-day peak P_{net} and then decreases rapidly as P_{net} decreases with decreasing irradiance. Furla et al. (2000a) demonstrated the presence of a DIC pool within coral tissues. The size of this pool was dependent on the lighting conditions, since it increased 39-fold after 3 h of illumination. If we apply this observation to the data shown in Fig. 2.14, it appears that the DIC pool had increased by mid-day, so rate of DIC uptake dropped rapidly as irradiance and photosynthesis declined. However, note that the high dissipation rates of H^+ continued for 2–3 h following the peak rates of P_{net} and G_{net} as the corals rid the backlog of H^+ generated by rapid calcification. Thus the lag of pH behind the peak flux rates of P_{net} and G_{net} represents a disequilibrium resulting from the lag in proton efflux from the corals. The correlation between Ω_{arag} and G_{net} is simply the response of the CO_2-carbonate system to pH as [H^+] shifts the equilibria and redistributes the [CO_3^{2-}] relative to the other DIC components of [HCO_3^-] and [CO_2] (Eqs. 2.4, 2.5, 2.6, 2.7, 2.8, 2.9 and 2.10). Therefore Ω_{arag} closely tracks pH whereas G_{net} tracks P_{net}. Changes in Ω_{arag} are a consequence of changes in both P_{net} and G_{net}. Hence the Ω_{arag} peak and the pH peak lag behind the P_{net} and G_{net} peak (Fig. 2.13) due to lag in proton efflux. This observation demonstrates the importance of understanding the difference between H^+ concentration and H^+ flux. During the night the H^+ flux rate is very responsive to changes in G_{net} due to changes in respiration.

2.10 Back to the Basics

The preceding sections show the importance of using flux rates rather than concentrations when describing a dynamic metabolic system such as a coral or coral reef. Most of the previous research in this area has focused on the relationship between G_{net}, [CO_3^{2-}] (or its surrogate Ω_{arag}), [HCO_3^-], and [H^+] expressed as pH. Plotting these variables in exemplary Figure 2.15 is very informative. A coral must uptake inorganic carbon in order to maintain high rates of photosynthesis and calcification. As a result [DIC] will decrease no matter which carbonate species (HCO_3^-, CO_3^{2-} or CO_2) is taken up by the coral (Eqs. 2.5, 2.6, 2.7, 2.8, 2.9 and 2.10). Thus we see a decline in [DIC] at high rates of G_{net}. [HCO_3^-], which has been identified as the preferred substrate for photosynthesis and calcification (Weis et al. 1989; Goiran et al. 1996; Furla et al. 2000b; Jury et al. 2010; Roleda et al. 2012), drops rapidly as calcification rate increases while closely tracking [DIC] during daylight hours (Fig. 2.15). In contrast, [CO_3^{2-}] lags behind G_{net} and closely tracks pH during the day as shown for Ω_{arag} in Fig. 2.13. If [CO_3^{2-}] (or its surrogate Ω_{arag}) drives calcification, then how do we explain the lag behind G_{net}? And if [CO_3^{2-}] is limiting, how do we explain the fact that [CO_3^{2-}] is increasing rather than decreasing as the coral calcifies rapidly and takes up inorganic carbon? In fact [CO_3^{2-}] increases simply because of the increase in pH caused by rapid photosynthesis that shifts the equilibrium between [HCO_3^-] and [CO_3^{2-}] (Eqs. 2.5, 2.6, 2.7, 2.8, 2.9 and 2.10, Fig. 2.4). Thus, P_{net} is the driver of changes in G_{net} and [CO_3^{2-}]. A basic physiological interpretation of the patterns shown in Fig. 2.15 is that daytime coral metabolism rapidly removes DIC (primarily in the form of HCO_3) while photosynthesis provides the energy that drives G_{net}. Higher pH resulting from rapid photosynthesis pushes the equilibria toward higher [CO_3^{2-}]. This scenario results in a correlation between G_{net} and Ω_{arag}, with both Ω_{arag} and G_{net} as dependent variables on P_{net} along with pH and changes in A_T due to local dissolution. During the night [HCO_3^-], [DIC], [CO_3^{2-}] and pH mirror changes in G_{net}. However, [HCO_3^-] diverges from [DIC], and [CO_3^{2-}] diverges from pH in darkness. The night divergence can be attributed to respiration causing a decrease in pH. The decreasing pH shifts the equilibria so that [CO_3^{2-}] is converted to [HCO_3^-], thereby changing the offset between the points.

2.11 Conclusions

The physical chemist's concept of Ω_{arag} is of critical importance to our understanding of global distribution and changes in the carbonate chemistry of the sea. The vertical and horizontal distribution of Ω_{arag} in the past, present and future will continue to be the subject of extensive research, and the concept of Ω_{arag} as a fundamental driver of abiotic processes, such as the chemical dissolution of carbonates, is indisputable. However, some scientists involved in OA studies have previously adopted Ω_{arag} as the most important independent variable related to coral calcification based on empirical correlation, but without evidence for causation. According to the proton flux hypothesis, coral physiology is responding to $[H^+]$, which shows a correlation with Ω_{arag}. The preoccupation with supply of materials required for calcification (limiting nutrient analogy of N vs. P) with a focus on two interchangeable forms of inorganic carbon (CO_3^{2-} and HCO_3^-) rather than on elimination of waste H^+ prevented a complete understanding of physiological processes. Results must be viewed in the context of reactants and products (Eqs. 2.5, 2.6 and 2.7). Equations describing control of calcification by the ratio of substrate concentration (DIC) to proton concentration (H^+) were derived from a physiological perspective in Sect. 2.1.8. Bach (2015) rearranged the terms in the physical chemistry equations describing the sea water carbonate-carbon dioxide system and demonstrated that calcification is a function of the $[DIC] : [H^+]$ ratio.

Linear regression using Ω_{arag} as the independent variable is a poor descriptor of G_{net} on coral reefs. Much of the existing data on coral calcification was developed in static or low turnover incubation experiments under typical laboratory low-irradiance, artificial-light sources on a 12-h light, 12-h dark cycle (Jokiel et al. 2014b). This regime results in an unrealistic simulation of the actual diurnal cycles that occur on coral reefs. The standard protocol has been to compare linear regressions between or among laboratory treatments. Linear regression provides a very limited description of the actual relationship between the key factors controlling organic and inorganic processes on coral reefs, which are more adequately described by data presentations such as that in Figs. 2.13, 2.14, and 2.15. The linear regression approach does not fully embrace natural diurnal calcification patterns and phase lags because these processes are non-linear (Jokiel et al. 2014a). The linear regression approach can lead to the assumption that Ω_{arag} is the independent variable driving the calcification reaction. Use of Ω_{arag} as an independent variable to compare spatial and temporal variation in G_{net} is known to create difficulties (Shamberger et al. 2011; Falter et al. 2012).

Well-developed reefs occur within a narrow geographic range characterized by open ocean $\Omega_{arag} > 3.3$ (Kleypas et al. 1999a, 1999b), suggesting that high coral Ω_{arag} along with warm shallow waters and high irradiance promotes reef development. Further it has been suggested that reef communities have limited capacity to adapt to lower levels of Ω_{arag} that will occur with future levels of anthropogenic ocean acidification (OA). Recent reports suggest that healthy coral reefs could cease to exist within this time frame as OA continues and oceanic Ω_{arag} decreases (Hoegh-Guldberg et al. 2007; Silverman et al. 2009). However, there are inconsistencies in the relationship (slope and x-intercept) between G_{net} as a function of Ω_{arag} on various reefs throughout the world (Shamberger et al. 2011, 2014; Jokiel 2015). Evenhuis et al. (2015) developed a model of coral reef calcification that embraces the major assumptions that are widely accepted in modeling global coral reef calcification. Jokiel (2015) documented the problems associated with each of the following widely used assumptions: (1) oceanic conditions of Ω_{arag} control (or are at least highly correlated with) G_{net} on coral reefs; (2) calcification rate is driven by bulk water $[CO_3^{2-}]$ expressed as Ω_{arag}; (3) changes in coral calcification rate can be used to estimate future changes in coral reef calcification rate; (4) the impact of OA is additive and not synergistic with other environmental factors such as increased temperature; and (5) predicted Ω_{arag} based on modeled open ocean conditions can be applied to coral reefs. The problems inherent in using these assumptions and the uncertainties and contradictions that result are described in Jokiel (2015) and show the need to re-evaluate basic assumptions.

The physical chemistry concept of Ω_{arag} has no basic physiological meaning in describing G_{net} other than a correlation with the $[DIC]:[H^+]$ ratio (Jokiel 2013; Bach 2015) as well as with other factors such as pH. There is no consistent relationship between Ω_{arag} and G_{net} when comparing reefs throughout the world (Shamberger et al. 2011). Coral reefs are systems in constant disequilibrium with the water column. We must take care not to be led astray in our thinking about the variables that actually drive and control coral and coral reef metabolism and bulk water chemistry. The correlation between G_{net} and other factors is a result of P_{net} driving both G_{net} and Ω_{arag} (McMahon et al. 2013). The observed phenomenon of diurnal hysteresis and diurnal phase lag show the importance of measuring flux rates and emphasizes the challenge in predicting the future effects of OA on coral reefs. The method of using linear extrapolations of Ω_{arag} to determine threshold levels that will shift coral reefs from net calcifying systems to net dissolving states has been questioned (McMahon et al. 2013). Perhaps predicted changes in Ω_{arag} in the open ocean can be used to calculate changes on reefs if we assume that the baseline on the reefs will change in concert with ocean values and that all other processes such as P_{net} and carbonate dissolution will not be influenced by OA. An explanation for the many paradoxes of coral calcification discussed herein has been presented as the "Two Compartment Proton Flux Model of Coral Metabolism" (Jokiel 2011b). This model is focused on localized gradients that influence coral metabolism with a focus on proton flux, carbon pools and translocation of fixed carbon. A major feature of the model is the presence of boundary layers which control local pH gradients and inorganic carbon speciation in addition to proton flux.

2.12 Future Research Directions

The paradigm that G_{net} is controlled by aragonite saturation state Ω_{arag} of bulk seawater on coral reefs is widely embraced in modeling impact of future climate change on coral reefs but is not correct as discussed above. Additional experiments and observations are needed to further examine and resolve entrenched scientific contradictions concerning coral and coral reef carbon metabolism in the face of climate change. Studies are needed on the response of corals and coral reefs to the actual carbonate chemistry of the sea water in contact with the organisms and substrate, rather than relative to changes in the offshore water chemistry. Finely controlled mesocosm investigations focused on measurement of material flux will further test various aspects of the proposed Proton Flux Model. Such experimentation will define the importance of global warming and ocean acidification on G_{net} and P_{net} in a wide range of major coral reef community components under various environmental conditions. A more complete understanding of how boundary layers influence material flux of protons as well as other metabolically important materials is needed. Measurements of changes in the diffusion boundary layer (DBL), momentum boundary layer (MBL) and benthic boundary layer (BBL) as discussed in Sect. 2.1.9 are potentially transformative in the re-evaluation of existing paradigms concerning coral and coral reef metabolism. Such data is vital to the understanding of carbonate dynamics and the ecology of present day reefs and ancient coral reef ecosystems.

The time lag between G_{net} and Ω_{arag} reported previously in field studies (Shamberger et al. 2011; Cyronak et al. 2013b; McMahon et al. 2013) provides evidence that diffusion and advection of materials between the coral and the water column involves time delays. One reason is that corals convert inorganic carbon to organic carbon, translocate the organic carbon to distal calcification sites, store organic carbon as lipid, and can eventually convert stored organic carbon back to inorganic carbon (Jokiel 2011b), creating numerous possible phase lags for metabolic materials. The second reason for the time lag is that rapidly calcifying systems have difficulty dissipating waste protons as shown by continued rapid proton efflux for hours after peak calcification (Fig. 2.14). What other mechanisms can account for the phase lag? Thick boundary layers (BL) resulting from low water motion can slow the exchange of metabolic materials between the coral and the water column. The results of Cyronak et al. (2013b) revealed that stirring had a net stimulatory effect on A_T flux and on the diurnal cycle of hysteresis. Increased attention to the often ignored variable of water-motion regime in experiments could provide insight into results thought previously to be paradoxical. Comeau et al. (2014c) tested effects of water flow on coral reef communities maintained in outdoor flumes under ambient pCO_2 and high pCO_2 (1300 μatm). Net calcification of coral communities, which included sediment communities, was affected by both flow and pCO_2. Calcification correlated positively with flow under both pCO_2 treatments. The effect of flow was less evident for sediments where dissolution exceeded precipitation of calcium carbonate under all flow speeds at high pCO_2. For corals and calcifying algae there was a strong flow effect, particularly at high pCO_2 where positive net calcification was maintained at night in the high flow treatment. These results demonstrate the importance of water flow in modulating the coral reef community response to OA and highlight the need to consider this parameter when assessing the effects of OA on coral reefs.

Studies of reef metabolism on shallow reef flats beginning with the classic work of Odum and Odum (1955) at Enewetak Reef flat were followed by others (Shamberger et al. 2011; Falter et al. 2012) at other locations. All of these studies were based on measurements of diurnal changes in chemistry of sea water within the BBL (see Sect. 2.1.9). Substantial boundary layers also occur over reefs in deeper water. For example, Price et al. (2012) took diurnal metabolic measurements within the BBL for a range of sites from exposed coastal situations to lagoons. They found that ambient variability in pH was substantial and oscillated over a diurnal cycle with diel fluctuations in pH exceeding 0.2. Daily pH maxima were identified as an important control on calcification. Net accretion among sites was positively related to the magnitude and duration of pH above the climatological seasonal low, despite myriad other ecological (e.g., local supply, species interactions, etc.) and physical oceanographic (e.g., temperature, current magnitude and direction, wave strength, latitudinal gradients, etc.) drivers. In general, accretion rates were higher at sites that experienced a greater number of hours at high pH values each day. Where daily pH within the BBL failed to exceed pelagic climatological seasonal lows, net accretion was slower and fleshy, non-calcifying benthic organisms dominated space. Thus, key aspects of coral reef ecosystem structure and function are clearly related to natural diurnal variability in pH, which is driven primarily by photosynthesis and respiration as P_{net}.

We conclude that future progress in understanding of calcification in corals as well as coral reefs will result from a better description of boundary layer processes and from studies of irradiance – water chemistry interactions that occur over a diurnal cycle (e.g., Jokiel et al. 2014a). In addition, studies of interactions among temperature, irradiance, pH and changes in the carbonate-CO_2 seawater system will be very productive.

Acknowledgements This work was supported in part by NOAA Grant "Research in Support of the NWHI Coral Reef Ecosystem Reserve", the EPA Star Grant Program, the Pacific Island Climate Change Cooperative (PICCC), the USGS Cooperative Agreement G13AC00130, and the George Melendez Wright Climate Change Fellowship Program. IBK's involvement was supported by the USGS Coastal and Marine Geology Program. Any use of trade names herein was for descriptive purposes only and does not imply endorsement by the U.S. Government.

Appendix

Appendix Table 2.1 Data reported for A_T, Ca^{2+}, CO_3^{2-}, HCO_3^-, Ω_{arag}, pH and G during each of the experimental trials (Langdon et al. 2000) and calculations for [DIC], [H^+] and the [DIC]:[H^+] ratio

Date		pCO_2	A_T		H^+	Ca^{2+}	CO_3^{2-}	HCO_3^-	CO_2	DIC			G
Start	End	µatm	µeq kg^{-1}	pH	nmol kg^{-1}	mmol kg^{-1}	µmol kg^{-1}	µmol kg^{-1}	µmol kg^{-1}	µmol kg^{-1}	DIC:H^+ $\times 10^{-3}$	Ω_{arag}	mmol $CaCO_3$ m^{-2} day^{-1}
16-Mar-95	21-Mar-95	1394	3384	7.88	131.8	9.27	174.5	2981.3	19.2	3175	24.1	2.43	32.7
21-Mar-95	23-Jun-95	890	2517	7.97	107.2	9.14	146.0	2160.3	11.1	2317	21.6	2.01	8.7
23-Jun-95	11-Jul-95	571	1743	7.97	107.2	8.66	104.1	1471.0	7.7	1583	14.8	1.35	−2.6
11-Jul-95	24-Jul-95	777	1877	7.88	131.8	8.77	93.3	1638.1	10.7	1742	13.2	1.23	−8
25-Jul-95	1-Aug-95	713	2101	7.94	114.8	8.78	118.4	1797.0	10.1	1925	16.8	1.56	14.8
1-Aug-95	19-Aug-95	1160	2120	7.78	166.0	8.83	87.9	1920.5	15.9	2024	12.2	1.17	−3.6
9-Apr-96	17-Apr-95	416	1748	8.09	81.3	8.98	131.7	1404.6	5.4	1542	19.0	1.78	21.1
7-Aug-96	25-Sep-96	688	1834	7.92	120.2	9.34	101.0	1573.9	9.3	1684	14.0	1.42	2.9
4-Mar-97	5-Apr-97	555	1861	8.01	97.7	9.15	121.2	1549.1	7.3	1678	17.2	1.67	16.1
5-Apr-97	5-May-97	443	1633	8.05	89.1	9.02	113.7	1392.8	5.7	1512	17.0	1.54	2.8
5-May-97	6-Jun-97	337	1510	8.11	77.6	8.89	120.2	1184.1	4.4	1309	16.9	1.61	0.8
7-Jun-97	30-Jun-97	458	1941	8.09	81.3	8.80	151.6	1554.2	6.0	1712	21.1	2.01	6.1
3-Jun-97	17-Jul-97	739	2906	8.04	91.2	8.62	209.1	2412.0	10.5	2632	28.9	2.71	27.3
17-Jul-97	9-Sep-97	842	2523	7.96	109.6	8.55	154.8	2136.6	11.4	2303	21.0	1.99	17.9
19-Sep-97	26-Sep-97	773	3249	8.10	79.4	8.60	258.5	2655.5	9.8	2924	36.8	3.34	78.6
26-Sep-97	8-Oct-97	811	3004	8.05	89.1	8.47	218.4	2496.8	10.5	2726	30.6	2.78	45.2
9-Oct-97	12-Oct-97	623	3548	8.20	63.1	8.03	345.1	2772.3	8.0	3125	49.5	4.17	117.2
12-Oct-97	16-Oct-97	710	3347	8.14	72.4	7.82	288.2	2684.9	9.0	2982	41.2	3.39	77.6
23-Oct-97	13-Nov-97	790	2913	8.05	89.1	7.61	212.4	2421.5	10.2	2644	29.7	2.43	24.8
21-Nov-97	26-Nov-97	566	3468	8.23	58.9	7.69	357.5	2651.1	7.1	3016	51.2	4.13	111.9
26-Nov-97	1-Dec-97	717	3287	8.13	74.1	7.60	280.8	2670.5	9.1	2960	39.9	3.21	25.2
1-Dec-97	18-Dec-97	729	3154	8.10	79.4	7.62	258.4	2596.6	9.5	2865	36.1	2.84	19.4
2-Jan-98	12-Jan-98	832	2942	8.03	93.3	7.64	203.8	2473.4	10.9	2688	28.8	2.34	17.5
14-Jan-98	20-Jan-98	526	3644	8.28	52.5	7.55	405.3	2770.8	6.4	3183	60.6	4.6	85.9
20-Jan-98	29-Jan-98	611	3310	8.17	67.6	6.96	297.5	2594.6	8.1	2900	42.9	3.11	50.1
5-Feb-98	19-Feb-98	733	2983	8.08	83.2	6.86	229.1	2449.4	9.6	2688	32.3	2.36	10.6
26-Feb-98	10-Mar-98	548	2728	8.06	87.1	6.84	236.7	2162.9	9.3	2409	27.7	2.44	40
11-Mar-98	15-Mar-98	414	3565	8.23	58.9	6.89	422.1	2615.4	7.3	3045	51.7	4.38	124.7
15-Mar-98	26-Mar-98	548	3278	8.22	60.3	6.90	326.1	2548.1	6.9	2881	47.8	3.39	52
26-Mar-98	16-Apr-98	471	2919	8.08	83.2	6.93	285.7	2269.1	9.4	2564	30.8	2.97	28
16-Apr-98	7-May-98	401	2650	8.09	81.3	6.86	266.6	2016.7	8.2	2292	28.2	2.75	16.3
7-May-98	28-May-98	366	2479	8.08	83.2	6.87	253.4	1867.2	7.9	2129	25.6	2.62	11.1
28-May-98	25-Jun-98	364	2364	8.05	89.1	6.95	231.8	1572.5	8.3	1813	20.3	2.42	5.2
10-Jul-98	22-Jul-98	383	2284	8.02	95.5	9.27	223.3	1737.6	8.7	1970	20.6	3.11	15.7
22-Jul-98	11-Aug-98	397	2194	8.03	93.3	9.49	190.8	1708.1	8.1	1907	20.4	2.72	8.7
11-Aug-98	26-Aug-98	368	2110	8.03	93.3	9.45	187.4	1634.7	7.8	1830	19.6	2.67	7.1
26-Aug-98	3-Sep-98	366	2065	8.11	77.6	8.87	195.0	1578.8	6.0	1780	22.9	2.6	3.8
10-Sep-98	22-Sep-98	457	3227	8.05	89.1	9.03	366.6	2405.3	11.3	2783	31.2	4.66	95
22-Sep-98	1-Oct-98	495	2893	7.96	109.6	8.44	287.8	2242.8	13.1	2544	23.2	3.66	30
1-Oct-98	22-Oct-98	580	2694	7.96	109.6	8.01	236.1	2157.3	12.2	2406	21.9	2.99	19
22-Oct-98	12-Nov-98	821	2561	7.81	154.9	8.29	163.4	2179.5	17.7	2361	15.2	2.04	7.4
2-Feb-99	30-Mar-99	192	2463	8.30	50.1	9.15	375.0	1549.0	4.0	1928	38.5	5.16	114

References

Al-Horani FA, Al-Moghrabi SM, De Beer D (2003a) Microsensor study of photosynthesis and calcification in the scleractinian coral, *Galaxea fascicularis*: active internal carbon cycle. J Exp Mar Biol Ecol 288:1–15 [doi:10.1016/S0022-0981(02)00578-6]

Al-Horani FA, Al-Moghrabi SM, De Beer D (2003b) The mechanism of calcification and its relation to photosynthesis and respiration in the scleractinian coral *Galaxea fascicularis*. Mar Biol 142:419–426 [doi:10.1007/s00227-002-0981-8]

Al-Horani FA, Al-Rousan SA, Manasrah RS, Rasheed MY (2005a) Coral calcification : Use of radioactive isotopes and metabolic inhibitors to study the interactions with photosynthesis and respiration. Chem and Ecol 21(5): 325–335 [doi: 10.1080/02757540500258724]

Al-Horani FA, Ferdelman T, Al-Moghrabi SM, De Beer D (2005b) Spatial distribution of calcification and photosynthesis in the scleractinian coral *Galaxea fascicularis*. Coral Reefs 24:173–180 [doi:10.1007/s00338-004-0461-3]

Allemand D, Ferrier-Pagès C, Furla P, Houlbrèque F, Puverel S, Reynaud S, Tambutté E, Tambutté S, Zaccola D (2004) Biomineralisation in reef-building corals: from molecular mechanisms to environmental control. Comptes Rendus Palevol 3:453–467. [doi:10.1016/j.crpv.2004.07.011]

Allemand D, Furla P, Bénazet-Tambutté S (1998) Mechanisms of carbon acquisition for endosymbiont photosynthesis in Anthozoa. Can J Bot 76:925–941 [doi: 10.1139/b98-086]

Allemand D, Tambutté E, Zoccola D, Tambutté S (2011) Coral calcification, cells to reefs. In: Dubinsky Z, Stambler N (eds), Coral reefs: an ecosystem in transition. New York; Springer Press pp 119–150

Allison N, Cohen I, Finch AA, Erez J, Tudhope AW (2014) Corals concentrate dissolved inorganic carbon to facilitate calcification. Nature Communications 5:5741[doi: 10.1038/ncomms6741]

Allison N, Tudhope AW, Fallick AE (1996) Factors influencing the stable carbon and oxygen isotopic composition of *Porites lutea* coral skeletons from Phuket, South Thailand. Coral Reefs 15:43–57 [10.1007/BF01626076]

Andersson AJ, Kuffner IB, Mackenzie FT, Jokiel PL, Rodgers KS, Tan A (2009) Net loss of $CaCO_3$ from coral reef communities due to human induced seawater acidification. Biogeosci 6, 1811–1823 [doi: 10.5194/bg-6-1811-2009]

Andersson AJ, Mackenzie FT, Ver LM (2003) Solution of shallow-water carbonates: An insignificant buffer against rising atmospheric CO_2. Geology 31:513–516 [doi: 10.1130/0091-7613(2003)031<0513:SOSCAI>2.0.CO;2]

Anlauf H, D'Croz L, O'Dea A (2011) A corrosive concoction: The combined effects of ocean warming and acidification on the early growth of a stony coral are multiplicative. J Exp Mar Biol Ecol 397:13–20 [doi:10.1016/j.jembe.2010.11.009]

Anthony KR, Kline DI, Diaz-Pulido G, Dove S, Hoegh-Guldberg O (2008) Ocean acidification causes bleaching and productivity loss in coral reef builders. Proc Natl Acad Sci 105:17442–17446 [doi: 10.1073/pnas.0804478105]

Atkinson MJ, Carlson B, Crow GL (1995) Coral growth in high-nutrient, low-pH seawater: a case study of corals cultured at the Waikiki Aquarium, Honolulu, Hawaii. Coral Reefs 14:215–223 [10.1007/BF00334344]

Australian Bureau of Meteorology and CSIRO (2011) Climate change in the Pacific: Scientific assessment and new research. Volume 1: Regional Overview (ISBN: 9781921826733 (pbk.) ISBN: 9781921826740 (ebook))

Bach LT (2015) Reconsidering the role of carbonate ion concentration in calcification by marine organisms. Biogeosci Discuss 12:6689–6722. [doi:10.5194/bgd-12-6689-2015]

Barnes DJ, Crossland CJ (1980) Diurnal and seasonal variations in the growth of a staghorn coral measured by time-lapse photography. Limnol Oceanog 35(6): 1113–1117 [doi: 10.4319/lo.1980.25.6.1113]

Barnes DJ, Lough JM (1993) On the nature and causes of density banding in massive coral skeletons. J Exp Mar Biol Ecol 167: 91–108 [doi:10.1016/0022-0981(93)90186-R]

Brahmi C, Kopp C, Domart-Coulon I, Stolarski J, Meibom A (2012) Skeletal growth dynamics linked to trace-element composition in the scleractinian coral *Pocillopora damicornis*. Geochimica et Cosmochimica Acta 99:146–158

Brown BE, Hewit R, Le Tissier MAA (1983) The nature and construction of skeletal spines in *Pocillopora damicornis* (Linnaeus). Coral Reefs 2:81–89 [10.1007/BF02395278]

Caldera K, Wickett ME (2003) Anthropogenic carbon and ocean pH. Nature 425:365 [doi:10.1038/425365a]

Carpenter KE, Abrar M, Aeby GR, Aronson B, Banks S, Bruckner A, Chiriboga A, Cortes J, Delbeek JC, DeVantier L, Edgar GJ, Edwards AJ, Fenner D, Guzmán HM, Hoeksema BW, Hodgson G, Johan O, Licuanan WY, Livingstone SR, Lovell ER, Moore JA, Obura DO, Ochavillo D, Polidoro BA, Precht WF, Quibilan MC, Reboton C, Richards ZT, Rogers AD, Sanciangco J, Sheppard A, Sheppard C, Smith J, Stuart S, Turak E, Veron JEN, Wallace C, Weil E, Wood E (2008) One-third of reef-building corals face elevated extinction risk from climate change and local impacts. Science 321:560–563 [doi: 10.1126/science.1159196]

Chauvin A, Denis V, Cuet P (2011) Is the response of coral calcification to seawater acidification related to nutrient loading? Coral Reefs 30:911–923 [10.1007/s00338-011-0786-7]

Chave KE (1984) Physics and chemistry of biomineralization. Ann Rev Earth Planet Sci 12: 293–305 [doi: 10.1146/annurev.ea.12.050184.001453]

Cohen AL, Holcomb M (2009) Why corals care about ocean acidification. Oceanography 22:118–127 [doi.org/10.5670/oceanog.2009.102]

Cohen AL, McConnaughey TA (2003) Geochemical perspectives on coral mineralization. In: Dove PM, Weiner S, deYoreo JJ (eds) Biomineralization. reviews in mineralogy and geochemistry vol 54, The Mineralogical Society of America, Washington DC, pp 151–187 [doi:10.2113/0540151]

Cohen AL, McCorkle DC, de Putron S, Gaetani GA, Rose KA (2009) Morphological and compositional changes in the skeletons of new coral recruits reared in acidified seawater: Insights into the biomineralization response to ocean acidification. Geochem Geophys Geosyst 10:Q07005 [doi:10.1029/2009GC002411]

Coles SL (1988) Limitations on reef coral development in the Arabian Gulf: temperature or algal competition? Proc 6th Int Coral Reef Symp 3:211–216

Coles SL, Jokiel PL (1977) Effects of temperature on photosynthesis and respiration rates of reef corals. Mar Biol 43:209–216 [doi:10.1007/BF00402313]

Colombo-Pallotta MF, Rodríguez-Román A, Iglesias-Prieto R (2010) Calcification in bleached and unbleached *Montastraea faveolata*: evaluating the role of oxygen and glycerol. Coral Reefs 29:899–907 [doi:10.1007/s00338-010-0638-x]

Comeau S, Carpenter RC, Nojiri Y, Putnam HM, Sakai K, Edmunds PJ. (2014a) Pacific-wide contrast highlights resistance of reef calcifiers to ocean acidification. Proc R Soc B 281: 20141339. [doi:10.1098/rspb.2014.1339]

Comeau S, Carpenter RC, Edmunds PJ (2012) Coral reef calcifiers buffer their response to ocean acidification using both bicarbonate and carbonate. Proc R Soc B 280:20122374 [doi:10.1098/rspb.2012.2374]

Comeau S, Carpenter RC, Edmunds PJ (2013) Response to coral reef calcification: carbonate, bicarbonate and proton flux under conditions of increasing ocean acidification. Proc R Soc B 280:20131153. [doi:10.1098/rspb.2013.1153]

Comeau S, Edmunds PJ, Lantz CA, Carpenter RC (2014c) Water flow modulates the response of coral reef communities to ocean acidification. Sci Reports 4:6681 [doi:10.1038/srep06681]

Comeau S, Edmunds PJ, Spindel NB, Carpenter RC (2014b) Fast coral reef calcifiers are more sensitive to ocean acidification in short-term laboratory incubations. Limnol Oceanogr 59:1081–1091

Connolly SR, Lopez-Yglesias MA, Anthony KRN (2012) Food availability promotes rapid recovery from thermal stress in a scleractinian coral. Coral Reefs 31:951–960. [doi: 10.1007/s00338-012-0925-9]

Crossland CJ, Barnes DJ (1974) The role of metabolic nitrogen in coral calcification. Mar Biol 28:325–332

Cyronak T, Santos IR, Erler DV, Eyre BD (2013a) Groundwater and porewater as major sources of alkalinity to a fringing coral reef lagoon (Muri Lagoon, Cook Islands), Biogeosci 10:2467–2480 [doi:10.5194/bg-10-2467-2013]

Cyronak T, Santos IR, McMahon A, Eyre BD (2013b) Carbon cycling hysteresis in permeable carbonate sands over a diel cycle: Implications for ocean acidification. Limnol Oceanogr 58 (1):131–143 [doi:10.4319/lo.2013.58.1.0131]

Cyronak T, Schulz KG, Santos IR, Eyre BD (2014) Enhanced acidification of global coral reefs driven by regional biogeochemical feedbacks. Geophys Res Lett 41(15): 5538–5546 [doi:10.1002/2014GL060849]

Cyronak T, Schulz KG, Jokiel PL (2015) The Omega myth: what really drives lower calcification rates in an acidifying ocean. ICES J Mar Sci [doi:10.1093/icesjms/fsv075]

de Putron SJ, McCorkle DC, Cohen AL, Dillon AB (2010) The impact of seawater saturation state and bicarbonate ion concentration on calcification by new recruits of two Atlantic corals. Coral Reefs. 30 (2):321–328 [doi:10.1007/s00338-010-0697-z]

Dore JE, Lukas R, Sadler DW, Church MJ, Karl DM (2009) Physical and biogeochemical modulation of ocean acidification in the central North Pacific. Proc Natl Acad Sci USA 106:12235–12240 [doi: 10.1073/pnas.0906044106]

Duarte CM, Hendriks IE, Moore TS, Olsen YS, Steckbauer A, Ramajo L, Carstensen J, Trotter JA, McCulloch M (2013) Is ocean acidification an open-ocean syndrome? Understanding anthropogenic impacts on seawater pH. Estuaries and Coasts 36:221–236 [doi:10.1007/s12237-013-9594-3]

Dubinsky Z, Jokiel PL (1994) Ratio of energy and nutrient fluxes regulates symbiosis between zooxanthellae and corals. Pac Sci 48:313–324

Edmunds PJ (2011) Zooplanktivory ameliorates the effects of ocean acidification on the reef coral *Porites* spp. Limnol Oceanogr 56:2402–2410 [doi:10.4319/lo.2011.56.6.2402]

Edmunds PJ, Brown D, Moriarty V (2012) Interactive effects of ocean acidification and temperature on two scleractinian corals from Moorea, French Polynesia. Global Change Biology. 18:2173–2183 [doi: 10.1111/j.1365-2486.2012.02695.x]

Enns T (1967) Facilitation by carbonic anhydrase of carbon dioxide transport. Science 155:44–47 [doi:10.1126/science.155.3758.44]

Enríquez S, Méndez ER, Iglesias-Prieto R (2005) Multiple scattering on coral skeletons enhances light absorption by symbiotic algae. Limnol Oceanogr 50:1025–1032 [doi: 10.1364/AO.49.005032]

Erez J, Reynaud S, Silverman J, Schneider K, Allemand D (2011) Coral calcification under ocean acidification and global change. In: Dubinsky Z, Stambler N (eds) Coral reefs: an ecosystem in transition. New York, Springer Press, pp 151–176

Evenhuis C, Lenton A, Cantin NE, Lough JM (2015) Modelling coral calcification accounting for the impacts of coral bleaching and ocean acidification. Biogeosci 12:2607–2630 [doi:10.5194/bg-12-2607-2015]

Fabricius KE, Langdon C, Uthicke S, Humphrey C, Noonan S, De'Ath G, Okazaki R, Muehllehner N, Glas MS, Lough JM (2011) Losers and winners in coral reefs acclimatized to elevated carbon dioxide concentrations. Nature Climate Change 1:165–169 [doi:10.1038/nclimate1122]

Fagan KE, Mackenzie FT (2007) Air–sea CO_2 exchange in a subtropical estuarine-coral reef system, Kaneohe Bay, Oahu, Hawaii. Mar Chem 106:174–191 [doi:10.1016/j.marchem.2007.01.016]

Falkowski PG, Dubinsky Z, Muscatine L, Porter J (1984) Light and the bioenergetics of a symbiotic coral. BioScience 34:705–709 [doi:10.2307/1309663]

Falter JL, Lowe RJ, Atkinson MJ, Cuet P (2012) Seasonal coupling and de-coupling of net calcification rates from coral reef metabolism and carbonate chemistry at Ningaloo Reef, Western Australia. J Geophys Res 117:C05003 [doi:10.1029/2011JC007268]

Fang L-s, Chen Y-wJ, Chen C-s (2004) Why does the white tip of stony coral grow so fast without zooxanthellae? Mar Biol 103:359–363 [doi:10.1007/BF00397270]

Feely RA, Doney C, Cooley S (2009) Ocean acidification. Oceanography 22:36–47

Ferrier-Pagès C, Witting, J, Tambtutté E, Sebens KP (2003) Effect of natural zooplankton feeding on the tissue and skeletal growth of the scleractinian coral *Stylophora pistillata*. Coral Reefs 22:229–240 [doi:10.1007/s00338-003-0312-7]

Fine M, Oren U, Loya Y (2002) Bleaching effect on regeneration and resource translocation in the coral *Oculina patagonica*. Mar Ecol Prog Ser 234:119–125 [doi:10.3354/meps234119]

Furla P, Galgani I, Durand I, Allemand D (2000b) Sources and mechanisms of inorganic transport for coral calcification and photosynthesis. J Exp Mar Biol Ecol 203:3445–3457

Furla P, Orsenigo MN, Allemand D (2000a) Involvement of H^+-ATPase and carbonic anhydrase in inorganic carbon absorption for endosymbiont photosynthesis. Am J Physiol 278: R870–R881

Gagnon AC, Adkins JF, Erez J (2012) Seawater transport during coral biomineralization. Earth and Planetary Science Letters 329–330:150–161 [doi:10.1016/j.epsl.2012.03.005]

Galloway SB, Work TM, Bochsler VS, Harley RA, Kramarsky-Winters E, McLaughlin SM, Meteyer CU, Morado JF, Nicholson JH, Parnell PG, Peters EC, Reynolds T, Rotstein DS, Sileo L, Woodley CM (2007) Coral disease and health workshop: Coral histopathology II. NOAA Technical Memorandum NOS NCCOS 56 and NOAA Technical Memorandum CRCP 4. National Oceanic and Atmospheric Administration, Silver Spring, MD 84 pp

Gattuso JP, Allemand D, Frankignoulle M (1999) Photosynthesis and calcification at cellular, organismal and community levels in coral reefs: A review on interactions and control by carbonate chemistry. Am Zool 39:160–183 [doi: 10.1093/icb/39.1.160]

Gladfelter EH (1982) Skeletal development in *Acropora cervicornis*: I. Patterns of calcium carbonate accretion in the axial corallite. Coral Reefs 1:45–51 [doi:10.1007/BF00286539]

Gladfelter EH (1983) Circulation of fluids in the gastrovascular system of the reef coral *Acropora cervicornis*. Biol Bull 165:619–636

Glynn PW (1997) Bioerosion and coral reef growth: a dynamic balance. In: Birkeland C (ed.) Life and death of coral reefs. New York: Chapman and Hall. 68–95

Goiran C, Almoghrabi S, Allemand D, Jaubert J (1996) Inorganic carbon uptake for photosynthesis by the symbiotic coral/dinoflagellate association. 1. Photosynthetic performances of symbionts and dependence on sea water bicarbonate. J Exp Mar Biol Ecol 199:207–225 [doi: 10.1016/0022-0981(95)00201-4]

Golbuu Y, Wolanski E, Idechong JW, Victor S, Isechal AL, Oldiais NW, Idip Jr D, Richmond RH, van Woesik R (2012) Predicting Coral Recruitment in Palau's Complex Reef Archipelago. PLoS ONE 7(11):e50998. [doi:10.1371/journal.pone.0050998]

Goreau TF (1959) The physiology of skeleton formation in corals. I. A method for measuring the rate of calcium deposition by corals under different light conditions. Biol Bull 116:59–75

Goreau TF (1963) Calcium carbonate deposition by coralline algae and corals in relation to their roles as reef-builders. Ann NY Acad Sci 109:127–167 [doi: 10.1111/j.1749-6632.1963.tb13465.x]

Goreau TF, Goreau NI (1959) The physiology of skeleton formation in corals. II. Calcium deposition by hermatypic corals under various conditions in the reef. Biol Bull 117:239–250

Goreau TJ (1977) Coral skeletal chemistry: physiological and environmental regulation of stable isotopes and trace metals in *Montastrea annularis*. Proc R Soc B 196:291–315 [doi: 10.1098/rspb.1977.0042]

Graham D, Smillie RM (1976) Carbonate dehydratase in marine organisms of the Great Barrier Reef. Aust J Plant Physiol 3:113–119 [doi: 10.1071/PP9760113]

Graus RR, Macintyre IG (1976) Light control of growth form in colonial reef corals: computer simulation. Science 193:895–897 [doi: 10.1126/science.193.4256.895]

Grottoli AG, Rodrigues LJ, Palardy JE (2006) Heterotrophic plasticity and resilience in bleached corals. Nature 440:1186–1189 [doi:10.1038/nature04565]

Guest JR, Baird AH, Maynard JA, Muttaqin E, Edwards AJ, Campbell SJ, Yewdall K, Affendi YA, Chou LM (2012) Contrasting patterns of coral bleaching susceptibility in 2010 suggest an adaptive response to thermal stress. PLoS ONE 7(3): e33353 [doi:10.1371/journal.pone.0033353]

Hallock P, Schlager W (1986) Nutrient excess and the demise of coral reefs and carbonate platforms. Palaios 7:389–398 [doi:10.2307/3514476]

Harriott VJ, Banks SA (2002) Latitudinal variation in coral communities in eastern Australia: a qualitative biophysical model of factors regulating coral reefs. Coral Reefs 21:83–94 [doi: 10.1007/s00338-001-0201-x]

Herfort L, Thake B, Taubner I (2008) Bicarbonate stimulation of calcification and photosynthesis in two hermatypic corals. J Phycol 44:91–98 [10.1111/j.1529-8817.2007.00445.x]

Hoegh-Guldberg O, Mumby PJ, Hooten AJ, Steneck RS, Greenfield P, Gomez E, Harvell CD, Sale PF, Edwards AJ, Caldeira K, Knowlton N, Eakin CM, Iglesias-Prieto R, Muthiga N, Bradbury RH, Dubi A, Hatziolos ME (2007) Coral reefs under rapid climate change and ocean acidification. Science 318:1737–1742 [doi: 10.1126/science.1152509]

Hofmann GE, Smith JE, Johnson KS, Send U, Levin LA, Micheli F, Paytan A, Price NN, Peterson B, Takeshita Y, Matson PG, Crook ED, Kroeker KJ, Gambi MC, Rivest EB, Frieder CA, Yu PC, Martz TR (2011) High-frequency dynamics of ocean pH: A multi-ecosystem comparison. PLoS ONE 6(12): e28983 [doi:10.1371/journal.pone.0028983]

Hofmann GE, Todgham AE (2010) Living in the now: physiological mechanisms to tolerate a rapidly changing environment. Annu Rev Physiol 72:127–45 [doi: 10.1146/annurev-physiol-021909-135900]

Holcomb MC, McCorkle DC, Cohen AL (2010) Long-term effects of nutrient and CO_2 enrichment on the temperate coral *Astrangia poculata* (Ellis and Solander, 1786). J Exp Mar Biol Ecol 386:27–33 [doi:10.1016/j.jembe.2010.02.007]

Hönisch B, Ridgwel A, Schmidt DN, Thomas E, Gibbs SJ, Sluijs A, Zeebe R, Kump, L, Martindale RC, Greene SE, KiesslingW, Ries J, Zachos JC, Royer DL, Barker S, Marchitto TM Jr, Moyer R, Pelejero C, Ziveri P, Foster GL, Williams B (2012) The geological record of ocean acidification. Science 335:1058–1063 [doi: 10.1126/science.1208277]

Hunter CL, Evans CW (1995) Coral reefs in Kaneohe Bay, Hawaii: two centuries of western influence and two decades of data. Bull Mar Sci 57:501–515

IPCC (2001) Climate Change 2001: The scientific basis. The contribution of Working Group I to the Third Assessment Report of the Intergovernmental Panel on Climate Change. Houghton JT, Ding Y, Griggs DJ et al. (eds) Cambridge University Press, New York pp 1–881

IPCC (2007) Technical Summary. Climate Change 2007: The physical science basis. Contribution of Working Group I to the Fourth Assessment Report of the Intergovernmental Panel on Climate Change. Solomon S, Qin D, Manning M, Chen Z, Marquis M, Averyt KB, Tignor M, Miller HL (eds). Cambridge University Press, New York pp 19–91

Jaubert J (1977) Light metabolism and growth forms of the hermatypic scleractinian coral *Synaraea convexa* (Verrill) in the lagoon of Moorea (French Polynesia). Proc 3rd Int Coral Reef Symp 1:483–488

Jimenez IM, Kühl M, Larkum AWD, Ralph PJ (2011) Effects of flow and colony morphology on the thermal boundary layer of corals. J R Soc Interface 8:1785–1795 [doi:10.1098/rsif.2011.0144]

Jokiel PL (1978) Effects of water motion on reef corals. J Exp Mar Biol Ecol 35:87–97

Jokiel PL (2011a) Ocean acidification and control of reef coral calcification by boundary layer limitation of proton flux. Bull Mar Sci 87:639–657 [doi.org/10.5343/bms.2010.1107]

Jokiel PL (2011b) The reef coral two compartment proton flux model: A new approach relating tissue-level physiological processes to gross corallum morphology. J Exp Mar Biol Ecol 409:1–12 [doi:10.1016/j.jembe.2011.10.008]

Jokiel PL (2013) Coral reef calcification: carbonate, bicarbonate and proton flux under conditions of increasing ocean acidification. Proc R Soc B 20130031 [doi:10.1098/rspb.2013.0031]

Jokiel PL (2015) Predicting the impact of ocean acidification on coral reefs: evaluating the assumptions involved. ICES J Mar Sci [doi: 10.1093/icesjma/fsv091]

Jokiel PL, Bahr KD, Rodgers KS (2014b) Low-cost, high-flow mesocosm system for simulating ocean acidification with CO_2 gas. Limnol Oceanogr Methods 12:313–322 [doi:10.4319/lom.2014.12.313]

Jokiel PL, Coles SL (1977) Effects of temperature on the mortality and growth of Hawaiian reef corals. Mar Biol 43:201–208

Jokiel PL, Jury CP, Rodgers KS (2014a) Coral-algae metabolism and diurnal changes in the CO_2-carbonate system of bulk sea water. PeerJ 2:e378 [doi:10.7717/peerj.378]

Jokiel PL, Morrissey JI (1986) Influence of size on primary production in the reef coral *Pocillopora damicornis* and the tropical macroalga *Acanthophora spicifera*. Mar Biol 91:15–26 [doi:10.1007/BF00397566]

Jokiel PL, Rodgers KS, Kuffner IB, Andersson AJ, Cox EF, Mackenzie FT (2008) Ocean acidification and calcifying reef organisms: a mesocosm investigation. Coral Reefs 27:473–483 [doi:10.1007/s00338-008-0380-9]

Jury CP, Thomas FI, Atkinson MJ, Toonen RJ (2013) Buffer capacity, ecosystem feedbacks, and seawater chemistry under global change. Water 5:1303–1325

Jury CP, Whitehead RF, Szmant A (2010) Effects of variations in carbonate chemistry on the calcification rates of *Madracis auretenra* (= *Madracis mirabilis* sensu Wells, 1973): bicarbonate concentrations best predict calcification rates. Global Change Biol 16:1632–1644 [doi:10.1111/j.1365-2486.2009.02057.x]

Kaandorp JA, Sloot PMA, Merks RMH, Bak RPM, Vermeij MJA, Maier C (2005) Morphogenesis of the branching reef coral *Madracis mirabilis*. Proc R Soc B 272:127–133 [doi:10.1098/rspb.2004.2934]

Kaandorp, JA, Filatov M, Chindapol N (2011) Simulating and quantifying the environmental influence on coral colony growth form. In: Dubinsky Z, Stambler N (eds) Coral reefs: an ecosystem in transition. New York; Springer Press, pp 177–185

Kajiwara K, Yokoch H, Nagai A, Ueno S (1997) Growth patterns of white-tipped and brown-tipped branches of the reef building coral, *Acropora pulchra* from Amitori Bay, Iriomote Island. Bull Inst Ocean Res Dev, Tokai Univ 18: 1–10

Kawaguti S, Sakumoto D (1948) The effects of light on the calcium deposition of corals. Bull Oceanogr Inst Taiwan 4:65–70
Kinsey DW (1978) Alkalinity changes and coral reef calcification. Limnol Oceanog 23(5):989–991 [doi: 10.4319/lo.1978.23.5.0989]
Kinsey DW, Davies PJ (1979) Effect of elevated nitrogen and phosphorus on coral reef growth. Limnol Oceanogr 24:935–940 [10.4319/lo.1979.24.5.0935]
Kleypas JA, Buddemeier RW, Archer D, Gattuso JP, Langdon C, Opdyke BN (1999b) Geochemical consequences of increased atmospheric carbon dioxide on coral reefs. Science 284:118–120 [doi:10.1126/science.284.5411.118]
Kleypas JA, Buddemeier RW, Eakin CM, Gattuso J-P, Guinotte J, Hoegh-Guldberg O, Iglesias-Prieto R, Jokiel PL, Langdon C, Skirving W, Strong AE (2005) Comment on "Coral reef calcification and climate change: The effect of ocean warming," Geophys. Res. Lett., 32, L08601 [doi:10.1029/2004GL022329]
Kleypas JA, Feely RA, Fabry VJ, Langdon C, Sabine CI, Robbins LL (2006) Impacts of Ocean Acidification on Coral Reefs and Other Marine Calcifiers: A Guide for Future Research, report of a workshop held 18–20 April 2005, St. Petersburg, Fl, sponsored by NSF, NOAA and US Geological Survey. 88 pp
Kleypas JA, McManus JW, Meñez LAB (1999a) Environmental limits to coral reef development: where do we draw the line? Amer Zool 39 (1): 146–159. [doi: 10.1093/icb/39.1.146]
Kline DI, Teneva L, Schneider K, Miard T, Chai A, Marker M, Headley K, Opdyke B, Nash M, Valetich M, Caves JK, Russell BD, Connell SD, Kirkwood BJ, Brewer P, Peltzer E, Silverman J, Caldeira K, Dunbar RB, Koseff JR, Monismith SG, Mitchell BG, Dove S, Hoegh-Guldberg O. (2012) A short-term in situ CO_2 enrichment experiment on Heron Island (GBR). Sci Rep 2:413 [doi:10.1038/srep00413]
Kuffner IB, Hickey TD, Morrison JM (2013) Calcification rates of the massive coral *Siderastrea siderea* and crustose coralline algae along the Florida Keys (USA) outer-reef tract. Coral Reefs 32:987–997 [doi:10.1007/s00338-013-1047-8]
Kühl, M, Cohen Y, Dalsgaard T, Jørgensen BB, Revsbech NP (1995) Microenvironment and photosynthesis of zooxanthellae in scleractinian corals studied with microsensors for O_2, pH and light. Mar Ecol Prog Ser 117:159–172
Lamberts AE (1974) Measurement of alizarin deposited by coral. Proc 2nd Int Coral Reef Symp 2:241–244
Langdon C, Atkinson MJ (2005) Effect of elevated pCO_2 on photosynthesis and calcification of corals and interactions with seasonal change in temperature/irradiance and nutrient enrichment. J Geophys Res [doi:110 C09S07/10.1029/2004JC002576]
Langdon C, Takahashi T, Sweeney C, Chipman D, Goddard J, Marubini F, Aceves H, Barnett H. (2000) Effect of calcium carbonate saturation state on the calcification rate of an experimental coral reef. Global Biogeochem Cycles 14:639–654 [doi: 10.1029/1999GB001195]
Lantz C (2011) Spatiotemporal analysis of the carbonate system on a coral reef, Oahu, Hawaii. MS Thesis. Hawaii Pacific University, Honolulu. 103 pp
Lesser MP (2011) Coral bleaching: causes and mechanisms. In: Dubinsky Z, Stambler N (eds). Coral reefs: an ecosystem in transition. New York, Springer Press. pp 405–419
Lesser MP, Weis VM, Patterson, MR, Jokiel PL (1994) Effects of morphology and water motion on carbon delivery and productivity in the reef coral, *Pocillopora damicornis* (Linnaeus) – diffusion barriers, inorganic carbon limitation, and biochemical plasticity. J Exp Mar Biol Ecol 178:153–179
Lowenstam HS (1974) Impact of life on chemical and physical processes. In: Goldberg ED (ed.), The Sea, vol. 5, Marine chemistry, pp 725–796. Wiley, New York
Marcelino LA, Westneat MW, Stoyneva V, Henss J, Rogers JD, Radosevich A, Turzhitsky V, Siple M, Fang A, Swain TD, Fung J, Backman V (2013) Modulation of light-enhancement to symbiotic algae by light-scattering in corals and evolutionary trends in bleaching. PLoS ONE 8(4): e61492 [doi: 10.1371/journal.pone.0061492]
Marshall AT, Wright A (1998) Coral calcification: autoradiography of a scleractinian coral *Galaxea fascicularis* after incubation in ^{45}Ca and ^{14}C. Coral Reefs 17:37–47 [doi: 10.1007/s003380050092
Martin S, Gattuso J-P (2009) Response of Mediterranean coralline algae to ocean acidification and elevated temperature. Global Change Biol 15:2089–2100 [doi: 10.1111/j.1365-2486.2009.01874.x]
Martinez JA, Smith CM, Richmond RH (2012) Algal mats degrade coral reef physical habitat quality. Estuar Coast Shelf Sci 99:42–49 [doi:10.1016/j.ecss.2011.12.022]
Marubini F, Atkinson MJ (1999) Effects of lowered pH and elevated nitrate on coral calcification. Mar Ecol Prog Ser 188:117–121 [doi:10.3354/meps188117]
Marubini F, Barnett H, Langdon C, Atkinson MJ (2001) Dependence of calcification on light and carbonate ion concentration for the hermatypic coral *Porites compressa*. Mar Ecol Prog Ser 220:153–162
Marubini F, Ferrier-Pages C, Furla P, Allemand D (2008) Coral calcification responds to seawater acidification: a working hypothesis towards a physiological mechanism. Coral Reefs 27:491–499 [doi:10.1007/s00338-008-0375-6]
Marubini F, Ferrier-Pages C, Cuif J (2003) Suppression of skeletal growth in scleractinian corals by decreasing ambient carbonate-ion concentration: a cross-family comparison. Proc R Soc B 270:179–184 [doi:10.1098/rspb.2002.2212]
Mass T, Genin A, Shavit U, Grinstein M, Tchernov D (2010) Flow enhances photosynthesis in marine benthic autotrophs by increasing the efflux of oxygen from the organism to the water. Proc Natl Acad Sci USA 107:2527–2531[doi: 10.1073/pnas.0912348107]
McConnaughey TA, Whelan JF (1997) Calcification generates protons for nutrient and bicarbonate uptake. Earth Sci Rev 42:95–117
McCulloch M, Falter J, Trotter J, Montagna P (2012) Coral resilience to ocean acidification and global warming through pH up-regulation. Nature Climate Change [doi:10.1038/nclimate1473]
McMahon A. Santos IR, Cyronak T, Eyre BD (2013) Hysteresis between coral reef calcification and the seawater aragonite saturation state. Geophys Res Lett 40:4675–4679 [doi:10.1002/grl.50802]
McNeil BI, Matear RJ, Barnes DJ (2004) Coral reef calcification and climate change: The effect of ocean warming, Geophys Res Lett 31, L22309 [doi:10.1029/2004GL021541]
Moya A, Tambutté S, Bertucci A, Tambutté E, Lotto S, Vullo D, Supuran CT, Allemand D, Zoccola D (2008) Carbonic anhydrase in the scleractinian coral *Stylophora pistillata*: characterization, location and role in biomineralization. J Biol Chem 283:25475–25484 [doi: 10.1074/jbc.M804726200]
Murillo LJ, Jokiel PL, Atkinson MJ (2014) Alkalinity to calcium flux ratios for corals and coral reef communities: variances between isolated and community conditions. PeerJ 2:e249 [doi:10.7717/peerj.24]
Muscatine, L (1973) Nutrition in corals. In; Jones OA, Endean R (eds) Biology and geology of coral reefs, 2:271–324. New York: Academic Press
Muscatine L (1990) The role of symbiotic algae in carbon and energy flux in coral reefs. pp 75–87 In: Dubinsky Z (ed) Coral reefs. Elsevier Science Publishers BV, Amsterdam.
Muscatine L, Falkowski PG, Porter J, Dubinsky Z (1984) Fate of photosynthetically fixed carbon in light and shade adapted corals. Proc R Soc Lond B Biol Sci 222:181–202 [doi: 10.1098/rspb.1984.0058]
Muscatine L, Porter J (1977) Reef corals: mutualistic symbioses adapted to nutrient-poor environments. BioScience 27: 454–460 [doi: 10.2307/1297526]
Odum HT, Odum EP (1955) Trophic structure and productivity of a windward coral reef community on Eniwetok Atoll. Ecol Monogr 25:291–320

Ohde S (1995) Calcium carbonate production and carbon dioxide flux on a coral reef, Okinawa. In: Sakai EH, Nozaki Y (eds) Biogeochemical processes and ocean flux in the Western Pacific. Terra Scientific Pub Co, Tokyo

Ohde S, Hossain MMM (2004) Effect of $CaCO_3$ (aragonite) saturation state of seawater on calcification of *Porites* coral. Geochem J 38:613–621

Ohde S, van Woesik R (1999) Carbon dioxide flux and metabolic processes of a coral reef, Okinawa. Bull Mar Sci 65: 559–576

Orr JC, Fabry VJ, Aumont O, Bopp L, Doney SC, Feely RA, Gnanadesikan A, Gruber N, Ishida A, Joos F, Key RM, Lindsay K, Maier-Reimer E, Matear R, Monfray P, Mouchet A, Najjar RG, Plattner G-K, Rodgers KB, Sabine CL, Sarmiento JL, Schlitzer R, Slater RD, Totterdell IJ, Weirig M-F, Yamanaka Y, Yool A (2005) Anthropogenic ocean acidification over the twenty first century and its impact on calcifying organisms. Nature 437:681–686 [doi:10.1038/nature04095]

Pandolfi JM, Connolly RM, Marshall DJ, Cohen AL (2011) Projecting coral reef futures under global warming and ocean acidification. Science 333:418–422 [doi: 10.1126/science.1204794]

Pearse V, Muscatine L (1971) Role of symbiotic algae (zooxanthellae) in coral calcification. Bio Bull (Woods Hole) 141: 350–363.

Pierrot D, Lewis E, Wallace DWR (2006) MS Excel Program Developed for CO_2 System Calculations. ORNL/CDIAC-105. Carbon Dioxide Information Analysis Center, Oak Ridge National Laboratory, US Dept of Energy, Oak Ridge, Tennessee

Pörtner HO, Langenbuch M, Michaelidis B (2005) Synergistic effects of temperature extremes, hypoxia, and increases in CO_2 on marine animals: from Earth history to global change. J Geophys Res 110: C09S10 [doi:10.1029/2004JC002561]

Price NN, Martz TR, Brainard RE, Smith JE (2012) Diel variability in seawater pH relates to calcification and benthic community structure on coral reefs. PLoS ONE 7(8):e43843 [doi:10.1371/journal.pone.0043843]

Reef R, Kaniewska P, Hoegh-Guldberg O (2009) Coral skeletons defend against ultraviolet radiation. PLoS ONE 4(11): e7995 [doi:10.1371/journal.pone.0007995]

Renegar DA, Riegl BM (2005) Effect of nutrient enrichment and elevated CO_2 partial pressure on growth rate of Atlantic scleractinian coral *Acropora cervicornis*. Mar Ecol Prog Ser 293:69–76

Reynaud S, Leclercq N, Romaine-Lioud S, Ferrier-Pages C, Jaubert J, Gattuso J-P (2003) Interacting effects of CO_2 partial pressure and temperature on photosynthesis and calcification in a scleractinian coral. Global Change Biol 9:1660–1668 [doi: 10.1046/j.1365-2486.2003.00678.x]

Ries JB (2011) A physicochemical framework for interpreting the biological calcification response to CO_2-induced ocean acidification. Geochim Cosmochim Acta 75:4053–4064 [doi:10.1016/j.gca.2011.04.025]

Ries JB, Cohen AL, McCorkle DC (2009) Marine calcifiers exhibit mixed responses to CO_2-induced ocean acidification. Geology. 37:1131–1134 [doi: 10.1130/G30210A.1]

Ries JB, Cohen AL, McCorkle DC (2010) A nonlinear calcification response to CO_2-induced ocean acidification by the coral *Oculina arbuscula*. Coral Reefs 29:661–674 [doi: 10.1007/s00338-010-0632-3]

Rinkevich B, Loya Y (1983) Short term fate of photosynthetic products in a hermatypic coral. J Exp Mar Biol Ecol 73:175–184 [doi:10.1016/j.gca.2011.04.025]

Rodolfo-Metalpa R, Martin S, Ferrier-Pages C, Gattuso J-P (2010) Response of the temperate coral *Cladocora caespitosa* to mid- and long-term exposure to pCO_2 and temperature levels projected for the year 2100 AD. Biogeosci 7:289–300

Roleda M, Boyd P, Hurd C (2012) Before ocean acidification: calcifier chemistry lessons. J Phycol 48:840–843 [doi: 10.1111/j.1529-8817.2012.01195.x]

Roos P (1967) Growth and occurrence of the reef coral *Porites astreoides* (Lamarck) on relation to submarine radiance distribution. Dissertation, Drukkerij Elinkwijk, Utrecht, Norway

Royal Society (2005) Ocean acidification due to increasing atmospheric carbon dioxide. Policy document 12/05. ISBN 0 85403 617 2 Download available at www.royalsoc.ac.uk

Santos SR, Toyoshima J, Kin zie III RA (2009) Spatial and temporal dynamics of symbiotic dinoflagellates (*Symbiodinium*: Dinophyta) in the perforate coral *Montipora capitata*. Galaxea, Journal of Coral Reef Studies 11:139–147 [doi:10.3755/galaxea.11.139]

Schneider K, Erez J (2006) The effect of carbonate chemistry on calcification and photosynthesis in the hermatypic coral *Acropora eurystoma*. Limnol Oceanogr 51:1284–1293 [doi:10.4319/lo.2006.51.3.1284

Shamberger KEF, Feely RA, Sabine CL, Atkinson MJ, DeCarlo EH, Mackenzie FT, Drupp PS, Butterfield DA (2011) Calcification and organic production on a Hawaiian coral reef. Marine Chem 127:64–75 [doi:10.1016/j.marchem.2011.08.003]

Shamberger, KEF, Cohen AL, Golbuu Y, McCorkle DC, Lentz SJ, Barkley HC (2014) Diverse coral communities in naturally acidified waters of a Western Pacific Reef. Geophys Res Lett 41 [doi:10.1002/2013GL058489]

Shapiro O, Fernandez VI, Garren M, Guasto JS, Debaillon-Vesque FP, Kramarsky-Winter E, Vardi A, Stocker R. (2014) Vortical ciliary flows actively enhance mass transport in reef corals. Proc Nat Acad Sci 111 (37): 13391–13396 [doi: 10.1073/pnas.1323094111]

Shashar N, Cohen Y, Loya Y (1993) Extreme diel fluctuations of oxygen in diffusive boundary layers surrounding stony corals. Biol Bull 185:455–461[doi: 10.2307/1542485]

Shashar N, Kinane S, Jokiel PL, Patterson MR (1996) Hydromechanical boundary layers over a coral reef. J Exp Mar Biol Ecol 199:17–28 [doi:10.1016/0022-0981(95)00156-5]

Silbiger NJ, Donahue MJ (2015) Secondary calcification and dissolution respond differently to future ocean conditions. Biogeosci 12:567–578 [doi:10.5194/bg-12-567-2015]

Silverman J, Lazar B, Cao L, Caldeira K, Erez J (2009), Coral reefs may start dissolving when atmospheric CO_2 doubles. Geophys Res Lett 36, L05606 [doi:10.1029/2008GL036282]

Silverman J, Lazar B, Erez J (2007) Community metabolism of a coral reef exposed to naturally varying dissolved inorganic nutrient loads. Biogeochem [doi 10.1007/s10533-007-9075-5]

Simkiss K (1964) Phosphates as crystal poisons of calcification. Biol Rev 39:487–504 [doi: 10.1111/j.1469-185X.1964.tb01166.x]

Smith SV, Buddemeier RW (1992) Global change and coral reef ecosystems. Ann Rev Ecol Syst 23:89–118

Smith SV, Key GS (1975) Carbon dioxide and metabolism in marine environments. Limnol Oceanogr 20:493–495 [doi: 10.4319/lo.1975.20.3.0493]

Smith SV, Kinsey DW, 1978. Calcification and organic carbon metabolism as indicated by carbon dioxide. In: Stoddart, DR, Johannes RE (eds) Coral reefs: research methods, pp 469–484

Stambler N, Popper N, Dubinsky Z, Stimson J (1991) Effects of nutrient enrichment and water motion on the coral *Pocillopora damicornis*. Pac Sci 45:299–307

Suzuki A, Nakamori T, Kayanne H (1995) The mechanism of production enhancement in coral reef carbonate systems: model and empirical results. Sediment Geol 99:259–280 [doi: 10.1016/0037-0738(95)00048-D]

Tambutté É, Allemand D, Mueller E, Jaubert J (1996) A compartmental approach to the mechanism of calcification in hermatypic corals. J Exp Biol 199:1029–1041

Tambutté É, Allemand D, Zoccola D, Meibom A, Lotto S, Caminiti N, Tambutté S (2007) Observations of the tissue-skeleton interface in the scleractinian coral *Stylophora pistillata*. Coral Reefs 26:517–529 [doi: 10.1007/s00338-007-0263-5]

Tambutté É, Tambutté S, Segonds N, Zoccola D, Venn A, Erez J, Allemand D (2012) Calcein labelling and electrophysiology: insights on coral tissue permeability and calcification. R Lond B 279:19–27 [doi:10.1098/rspb.2011.0733]

Taylor DL (1977) Intra-clonal transport of organic compounds and calcium in some Atlantic reef corals. Proc 3rd Int Coral Reef Symp pp 431–436

Thomsen J, Haynert K, Wegner KM, Melzner F (2015) Impact of seawater carbonate chemistry on the calcification of marine bivalves. Biogeosci Discuss 12:1543–1571 [doi: 10.5194/bgd-12-1543-2015]

Tourniaire F, Pulos S (1985) Proportional reasoning: A review of the literature. Educ Stud Math 16(2):181–204 [doi: 10.1007/BF02400937]

Vandermeulen JH, Davis ND, Muscatine L (1972) The effects of inhibitors of photosynthesis on zooxanthellae in corals and other marine invertebrates. Mar Biol 16:185–191 [doi:10.1007/BF00346940]

Venn AA, Tambutté É, Holcomb M, Allemand D, Tambutté S (2011) Live tissue imaging shows reef corals elevate pH under their calcifying tissue relative to seawater. PLoS ONE 6(5) e20013 [doi: 10.1371/journal.pone.0020013]

Venn AA, Tambutté É, Holcomb M, Laurent J, Allemand D, Tambutté S (2013) Impact of seawater acidification on pH at the tissue-skeleton interface and calcification in reef corals. Proc Natl Acad Sci 110(5):1634–1639 [doi/10.1073/pnas.1216153110]

Venn AA, Tambutté É, Lotto S, Zoccola D, Allemand D, Tambutté S (2009) Intracellular pH in Symbiotic Cnidarians. Proc Natl Acad Sci USA 106:16574–16579 [doi: 10.1073/pnas.0902894106]

Venti A, Andersson A, Langdon C (2014) Multiple driving factors explain spatial and temporal variability in coral calcification rates on the Bermuda platform. Coral Reefs 33(4):979–997 [doi:10.1007/s00338-014-1191-9]

Veron JEN (2000) Corals of the world. Vol. 1–3. Aust Inst Mar Sci, Townsville MC, Qld, Australia

Veron JEN (2008) Mass extinctions and ocean acidification: biological constraints on geological dilemmas. Coral Reefs 27:459–472 [doi:10.1007/s00338-008-0381-8]

Vicsek T (1989) Fractal growth phenomena. World Scientific, London.

Weis VM, Smith GJ, Muscatine L (1989) A "CO_2 supply" mechanism in zooxanthellate cnidarians: role of carbonic anhydrase. Mar Biol 100:195–202 [doi: 10.1007/BF00391958]

Wijgerde T, Jurriaans S, Hoofd M, Verreth JAJ, Osinga R (2012) Oxygen and heterotrophy affect calcification of the scleractinian coral *Galaxea fascicularis*. PLoS ONE 7(12): e52702. [doi:10.1371/journal.pone.0052702]

Wilt FH (2005) Developmental biology meets materials science: morphogenesis of biomineralized structures. Dev Biol 280:15–25 [doi:10.1016/j.ydbio.2005.01.019]

Wisshak M, Schönberg CHL, Form A, Freiwald A (2013) Effects of ocean acidification and global warming on reef bioerosion—lessons from a clionaid sponge. Aquat Biol 19:111–127 [doi: 10.3354/ab00527]

Wooldridge S (2013) A new conceptual model of coral biomineralisation: hypoxia as the physiological driver of skeletal extension. Biogeosci 10:2867–2884 [doi:10.5194/bg-10-2867-2013]

Yates KK, Halley RB (2006) Carbonate concentration and pCO_2 thresholds for calcification and dissolution on the Molokai reef flat, Hawaii. Biogeosci 3:357–369 [doi:10.5194/bg-3-357-2006]

Yonge CM (1968) Living corals. Proc R Soc Lond B 169:329–344

Zoccola D, Tambutté É, Sénégas-Balas F, Michiels J-F, Failla JP, Jaubert J, Allemand D (1999) Cloning of a calcium channel α1 subunit from the reef-building coral, *Stylophora pistillata*. Gene 227(2):157–167 [doi:10.1016/S0378-1119(98)00602-7]

Photosymbiosis in Past and Present Reefs

3

Jere H. Lipps and George D. Stanley, Jr.

Abstract

Reef organisms are well known for engaging in photosymbiosis in which a heterotrophic protist or animal host partners with one or more kinds of photosynthetic microbes. This relationship provides metabolic advantages in nutrition and rapid calcification, often leading to secretion of massive skeletons in the host. In turn the symbiont receives protection, physical stability in the photic zone and direct access to the sun's energy. On an evolutionary scale, this relationship provided strong selective pressures for producing the algal-host relationship and has occurred multiple times in geological history. Today, different kinds of algae (dinoflagellates, diatoms, chlorophytes, rhodophytes, and cyanobacteria) inhabit various hosts (foraminifera, corals, mollusks) in modern reefs, and multiple phylogenetically separate algae may have also inhabited phylogenetically distinct ancient animals and protists. The modern dinoflagellate photosymbiont *Symbiodinium* occurs in a wide variety of unrelated host organisms from protists to mollusks. Molecular data indicate this genus first evolved either after the end-Cretaceous mass extinction 65 my ago or in the Early Eocene some 55 my ago. Encysted dinoflagellates related to *Symbiodinium* have been traced to the Triassic, and photosymbiosis may have been involved in even earlier reef associations. In all fossils, however, the identity of ancient photosymbionts is difficult to establish because they rarely, if ever, fossilize. Nevertheless, indirect evidence indicates that photosymbiotic ecosystems existed at least as far back as the Cambrian. Inferential lines of evidence, including large colony size, massive skeletons, unusual or complex morphology, the biogeographic distribution of possible hosts and skeletal geochemistry are all consistent with active photosynthesis. In the following pages, we develop the hypothesis that photosymbiosis best explains both the successes and failures of reefs through geologic time. We then review the evidence that suggests photosymbiosis in reef organisms played significant roles through geologic time in both the evolution and extinction of organisms and the reefs they constructed.

Keywords

Photosymbiosis • Reefs • Fossils • Evolution • Extinction • Foraminifera • Corals • Mollusks

J.H. Lipps (✉)
John D. Cooper Archaeological and Paleontological Center, Santa Ana, CA 92701, USA

Museum of Paleontology, University of California, Berkeley, CA 94720, USA
e-mail: jlipps@berkley.edu

G.D. Stanley, Jr.
Department of Geosciences and Paleontology Center, The University of Montana, Missoula, MT 59812, USA
e-mail: george.stanley@mso.umt.edu

D.K. Hubbard et al. (eds.), *Coral Reefs at the Crossroads*,
Coral Reefs of the World 6, DOI 10.1007/978-94-017-7567-0_3

3.1 Introduction

Photosynthesizing organisms have been essential throughout much of geological time for the building of reefs and deposition of carbonate platforms in the shallow, sunlit waters of Earth. In the Archean (3.5 to 2.5 billion years ago), stromatolites were constructed chiefly by photosynthetic cyanobacteria which trapped carbonate in their cells or mucilaginous secretions (Walter 1983; Allwood et al. 2007). Cyanobacteria built and continued to build reef-like structures in Precambrian to modern oceans (Fig. 3.1), initially in the absence of grazers and later on in environments where grazing animals were restricted (Dravis 1983; Dill et al. 1986; Riding 1992 Reid et al. 1995). They have also long been important as encrusters that cement reefs together.

The first photosynthetic eukaryotes arose in the late Paleoproterozoic perhaps as long as 1.63 or more billion years ago (Butterfield 2015), as estimated from molecular evidence (Yoon et al. 2004), morphology (Knoll 2014) and the fossil record (Lipps 2006; Eme et al. 2014). Metazoans, however, did not appear until about 600 million years ago. Once heterotrophic eukaryotes, animals and microbes evolved, another kind of photosynthetic strategy appeared—***photosymbiosis***, the productive association of photosynthesizing unicellular algae or cyanobacteria with heterotrophic microbial eukaryotes and animals. This represented a powerful evolutionary strategy connecting the heterotrophs directly to the sun's energy. Like symbioses in general, photosymbioses occur in both terrestrial and marine organisms (Margulis 1998; Douglas 2010). Among the many biotic relationships that evolved, photosymbiosis was particularly important in marine environments because it produced such profound biological, physical and chemical changes. While no direct evidence for photosymbiotic microbes exists in the fossil record, the process could have evolved among single-celled eukaryotes in the Precambrian even before animals appeared. Huge reefs, 300 m high and 8 km in diameter, were built by microorganisms and animals in the Neoproterozoic (Turner et al. 1993; Wood and Curtis 2014) and similar ones have been constructed ever since the Cambrian (Rowland and Gangloff 1988; Wood 1999; Rowland and Shapiro 2002), primarily by metazoan- or protistan-algal symbioses (Cowen 1983, 1988; Coates and Jackson 1987; Surge et al. 1997).

These associations result in not only reefs, but also in the production of prodigious amounts of carbonate sediment (Hallock and Schlager 1986) on reefs and banks (Lee and Anderson 1991; Hallock 1999; Lee 2006; Langer 2008). This biologic carbonate production annually accounts for very large amounts of reef-related sediments (Fig. 3.2; $\sim 30 \times 10^6$ metric tons of foraminifera alone, Langer 2008) and formation of a variety of carbonate rocks (James 1983). These, in turn, sequester many gigatons of carbon (Langer et al. 1997), thus helping to ameliorate effects of atmospheric CO_2 buildup and global warming now and in the past. When photosymbiosis slows or fails, so does the production of massive amounts of biogenic carbonate.

At many times in the geologic past, entire reef ecosystems collapsed globally in response to environmental changes, and mass extinctions ensued (Fagerstrom 1987; Benton 2003; Erwin 2006; Stanley and Lipps 2011; Clarkson et al. 2015; see also Chap. 8). The breakdown of photosymbioses in today's corals and foraminifera is manifested by bleaching (Stimson et al. 2002; Hallock et al. 2006), and mortality related to bleaching likely

Fig. 3.1 Modern intertidal stromatolites growing at Carbala Point, Shark Bay, Western Australia. Each mushroom-shaped calcareous stromatolite contains chiefly cyanobacteria, although other microbes (foraminifera, diatoms, ciliates, dinoflagellates) and even animals live in and among them now. These are typically 0.5–1.0 + m in height. Microbial mats cover the areas between the stromatolites (Photo by J. H. Lipps 2002)

Fig. 3.2 Aerial view of the northeastern part of Eniwetak Atoll (ocean is to the *left*). Everything in this view except the vegetation on the islets is biogenic carbonate derived either from the complete skeletons or the broken debris of calcifying organisms. Loose sediment, carried from the reef and reef flat in large plumes (*right, center*), eventually ends up on the backreef and lagoon floor (Photo by J. H. Lipps, 1972)

accompanied extinctions of many ancient reef ecosystems. Subsequent diversifications of reef communities following those events may also have been in part due to the reacquisition of symbionts as the environments ameliorated.

Modern reefs face ocean warming and acidification as CO_2 increases in the atmosphere and oceans because of human activities. Degradation of coral-reef ecosystems is already obvious in the increasing incidence of bleaching (Glynn 1996; Douglas 2003; Hallock et al. 2006; van Oppen and Lough 2009), coral disease, ocean acidification (Kleypas et al. 1999; Pelejero et al. 2007), and general human destruction of reef structures (Lipps 2011). For these reasons and others, reef ecosystems appear to be moving toward massive failure (Pandolfi et al. 2005).

We regard reefs as photosynthetically-driven, closely integrated ecosystems much like rain forests on land (Reaka-Kudla 1997). Photosymbiosis is the primary driver of productivity through physiological and morphological adaptations today. In this chapter we develop the hypothesis that photosymbiosis was also integral to reef success and failure through geologic time. We argue that when photosymbiosis succeeded or failed in the past due to environmental perturbation, reefs and related carbonate platforms also succeeded or failed (Chap. 8). In the following pages, we review the occurrence of photosymbiosis on modern and ancient reefs and carbonate-platforms, and its relationship to macroevolutionary processes of diversification, radiation and extinction of reefs and the organisms themselves. While the specifics of this hypothesis must be tested by utilizing an increasingly robust database of taxonomic, paleogeographic, paleoecologic and phylogenetic molecular results, photosymbiosis is an important contributor to reef success today and available evidence indicates that this was also true in ancient reef ecosystems.

3.2 Photosymbioses in Modern, Shallow-Water Carbonate Environments

Photosymbioses by bacteria and single-celled algae living within microbes and larger invertebrates are mostly confined to warm, shallow-water, carbonate settings on reefs and platforms. An assemblage of symbionts living in one host is referred to as a "holobiont", for example "the coral holobiont" (Knowlton and Jackson 2011). For heterotrophic microbes and metazoans, photosymbionts provide added metabolites, nutrients and enhanced calcification. These are particularly advantageous in oligotrophic tropical shallow waters. In kind, the photosymbionts benefit from the stable habitat, protection, and a supply of metabolic wastes, such as CO_2 and nitrogenous compounds, provided by their host (Douglas 2003).

Today, photosymbionts include cyanobacteria, chlorophytes, rhodophytes, dinoflagellates and diatoms hosted by foraminifera (Hansen and Buchardt 1977; Hallock 1999; Lee 2006), radiolaria (Anderson 1983) and ciliates (Lobban et al. 2014) among the microbial forms, plus sponges, cnidarians (including corals), bivalves, tunicates, and possibly bryozoans among larger animals (Fig. 3.3). Fossil invertebrates such as brachiopods, bryozoans, gastropods, and other extinct forms may have hosted photosymbionts in the distant past. That so many different and unrelated lineages of algae and heterotrophs have adopted this cooperative strategy likely indicates a strong

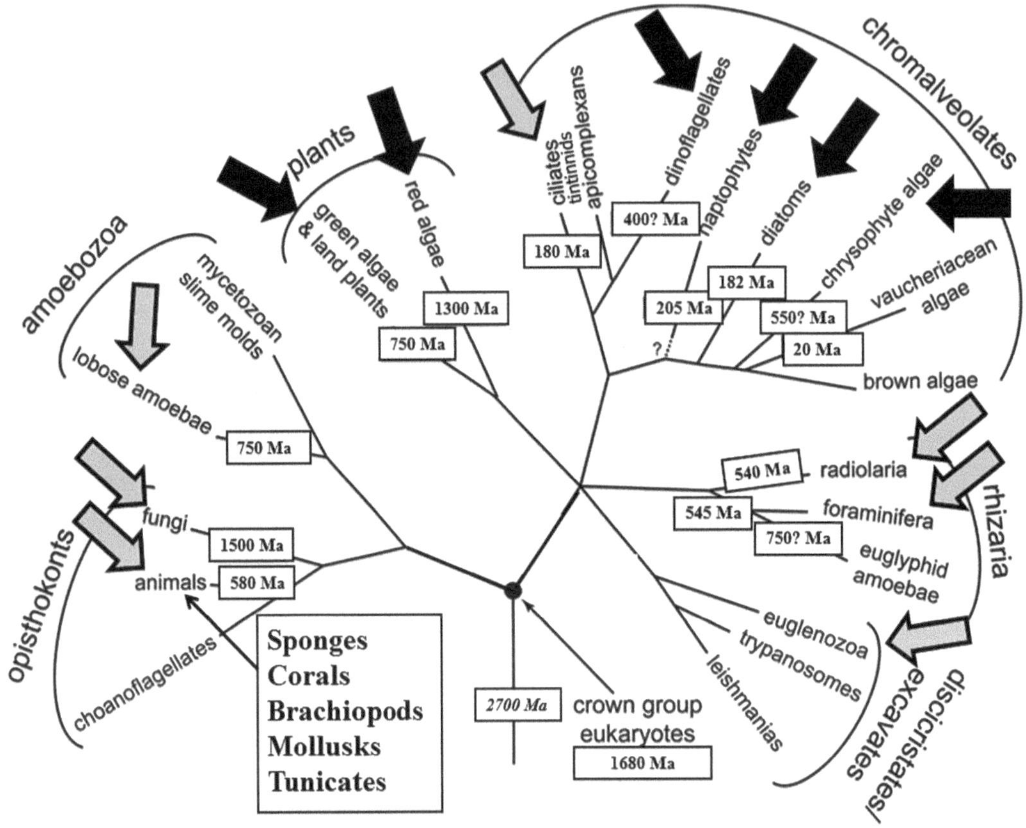

Fig. 3.3 Molecular phylogenetic diagram of the Eukarya showing the polyphyletic distribution of photosymbionts (*gray arrows*) and the common host either eukaryotic single-celled microbes or multicellular animals (*black arrows*). In addition to those taxa named, most other cnidarians, a tunicate, and possibly bryozoans may have hosted photosymbionts now or in the past. Other algae such as cyanobacteria (not shown; dates to 3000+ Ma) and the enigmatic acritarch cysts (not shown; dates to 1600 Ma) may include symbiotic forms as well. The oldest known geologic age based on fossils of each clade is indicated in the box near its root. Molecular or chemical biomarker dates are not included but may indicate earlier origins of most clades although they were not preserved as fossils until much later (Modified from Porter (2004) and Lipps (2006))

selective advantage for photosymbiosis (Baker 2003; Fautin and Buddemeier 2004).

More than one kind of algal symbiont is often found in some hosts. Foraminifera, for example, host *Symbiodinium*, diatoms, rhodophytes, chlorophytes, and cyanophytes, each living alone (Hansen and Buchardt 1977; Hansen and Dalberg 1979; Lee 2006) or as one of multiple symbiont species (Lee 2006) or clades in a single foraminiferan or foraminiferal species (Fay et al. 2009). Sponges too host cyanophytes and dinoflagellates, among other symbionts. Even in modern reef settings, photosymbioses by unrelated symbionts and hosts are quite common (Fig. 3.3).

The most widespread modern photosymbionts are dinoflagellates commonly known as zooxanthellae or, more precisely, by the generic name *Symbiodinium* (Freudenthal 1962). Symbiont-bearing organisms may be called zooxanthellate (z-organisms) and those without symbionts are azooxanthellate (az-organisms). *Symbiodinium* densities measured within coral hosts range from hundreds of thousands to millions per square centimeter (Stimson et al. 2002) and thousands occur in single cells of some larger foraminifera (Fig. 3.4; Fay et al. 2009).

Symbionts may be genetically diverse complexes of closely related forms (Coffroth and Santos 2005). Genomic studies of *Symbiodinium microadriaticum* revealed the presence of a number of different clades (Blank and Trench 1985; Rowan and Powers 1991; LaJeunesse 2002; Fay et al. 2009; LaJeunesse et al. 2010). And, new *Symbiodinium* clades are recognized each year. Currently more than a dozen different genetic clades are known to live in many different hosts, both within and outside of cells. Some of these have been given formal or informal names or letters, and among those clades up to nearly 50 sub-strains also exist (van Oppen et al. 2009; LaJeunesse et al. 2010). The clades of *Symbiodinium* may live in the same host at the same time or in different hosts across many domains of eukaryotes (Fig. 3.3). Different clades may also be found in different parts of a single host. A single foraminiferan, for example, may contain several clades of *Symbiodinium* that live in different parts of its cell (Fay et al. 2009). Other organisms,

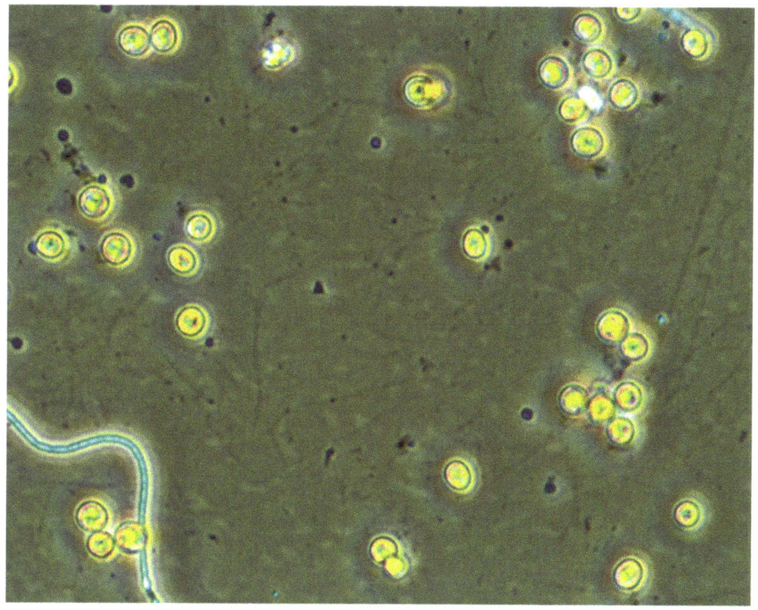

Fig. 3.4 Photosymbionts (*Symbiodinium*) liberated from a single living specimen of the foraminifera *Amphisorus hemprichii*. Thousands of symbiont cells in three different clades are contained in a single cell of these larger foraminifera (Photo courtesy of Scott Fay)

like corals and the giant clam *Tridacna*, also contain *Symbiodinium* in several clades as noted above.

Symbionts live within the cells of the microbial eukaryotes or in special structures in animals (Farmer et al. 2001). In foraminifera, the symbionts live pressed against the interior of the test on the upper (or sunlit) part, and in *Amphistegina* the symbionts occupy cup-shaped depressions that may keep them separated from one another (Lee 2006). Although sponges, like foraminifera, are symbiotic with many algae (Knowlton and Rohwer 2003), only clionid sponges harbor *Symbiodinium* (Hill et al. 2011). Some reef bivalves also maintain *Symbiodinium* in special tubes either in the mantle or in the gills (Farmer et al. 2001; Vermeij 2013).

Photosymbionts produce photosynthates—organic compounds such as glycerol and triglycerides that are translocated within and between cells to supplement the host's nutrient requirements (up to 95 % of that required by the host, Lee 2006). Metabolic CO_2 from the host is utilized by the algal symbionts in photosynthesis. Energy flow and carbon cycling is complex (Fig. 3.5), including the recycling and transport of carbon, and the dynamic energy flux on reefs due to these symbionts (Douglas 2003; Muscatine et al. 2005).

3.2.1 Photosymbiosis in Reef Organisms

Among life strategies both photosymbiotic and non-photosymbiotic organisms exist within taxonomic groups. In corals these are known as zooxanthellate and azooxanthellate (z-corals and az-corals) that today are about equally distributed among species (Cairns 1999, 2007). Zooxanthellate species and their photosynthetic symbionts are restricted to tropical latitudes and shallow depths, whereas azooxanthellate species can inhabit cold and deeper-water environments and expand their geographic distributions far outside the latitudinal ranges of zooxanthellate species (Stanley and Cairns 1988; Kiessling and Kocsis 2015). Molecular data (Barbeitos et al. 2010) suggested that coloniality was the original state of scleractinian corals and that the symbiosis between corals and photosynthetic partners was lost and gained repeatedly during their geologic history. Photosymbiotic organisms normally cannot live without their symbionts, but a few apozooxanthellate species are known to be capable of switching between a symbiotic and a non-symbiotic condition (Stanley and Cairns 1988; Lee 2006, 2011). Others can survive without zooxanthellae but they cannot secrete their carbonate shells as fast. For example, corals and foraminifera from which the symbionts have been removed by herbicides or by growing them in the dark fail to secrete skeletons and eventually die. Why more species are not facultatively zooxanthellate is unclear but it may have an evolutionary and genetic basis.

Oddly, while sunlight is required by photosymbionts, too much of it can kill or damage the host and symbionts due to intense light in very shallow waters. To deal with this, corals make colorful chromoproteins that take up substantial amounts of light (Smith et al. 2013) and foraminifera live in particular light ranges or behaviorally adjust light intensity by moving in or out from under overhangs that shade them (Hohenegger et al. 2000). Oxygen resulting from photosynthesis can also be damaging, and hosts have evolved certain antioxidants as protection (Furla et al. 2005). Carbon dioxide can also be limiting. Wooldridge (2014) coined the

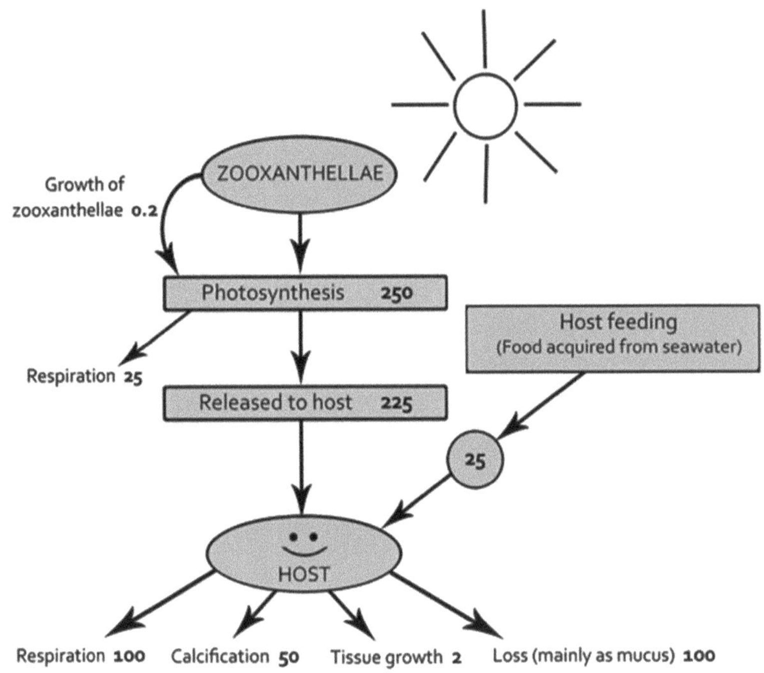

Fig. 3.5 The flow of solar energy (joules) in photosymbiosis: an example from the dinoflagellate *Symbiodinium* to the host coral *Pocillopora*. Only a small amount is retained by the symbiont for growth and maintenance while the majority of it is translocated to the host which receives only a small amount from its feeding. The relative amounts of energy utilized by the host are shown in the *lower part* of the diagram. The host uses just under half to make mucus which is discharged to the environment where it is utilized as food by other organisms (Modified from Cowen 1988). Such diagrams are specific to individuals or species, but photosymbionts greatly enhance the energy flow in similar patterns for other organisms that have been measured

"CO_2 (sink) limitation" model to explain that bleaching at least initially is caused by the host's failure to maintain a sufficient supply of CO_2 which the algal partner needs.

Photosymbionts help calcifying organisms extract calcium (Ca^{2+}) and acquire carbonate ions (CO_3) from an ion-pumping mechanism that brings in Ca^{++} and exports 2 H^+ ions, thus reducing acidity in the calcifying space and resulting in a transformation from CO_2 to CO_3 (Cohen et al. 2001) to facilitate construction of their carbonate skeletons. This has been demonstrated for corals and inferred for other organisms with symbionts (Hallock 1999). The mechanisms of the physio-chemical skeleton formation and the influence of light on the symbionts are not entirely resolved (Goreau and Goreau 1959; Carlon et al. 1996; Goreau et al. 1996; Marshall 1996; Gattuso et al. 1999), but clearly photosymbionts greatly enhance calcification in their skeletonized hosts (Cohen et al. 2001; Hohenegger 2006; Lee 2006; Cohen and Holcomb 2009; Ries et al. 2009; Lee 2011; McConnaughey 2012; see also Chap. 9). Thus carbonate production in these symbiotic organisms is greatest in the upper part of the top 10 m of sea water and falls off to about half that at depths of about 80 m, and then to very low values with depths increasing to 100 m or greater (Hohenegger 2006).

Some z-corals dwelling at greater depths (68–100 m) have adapted to the lower light levels by shifting toward the red end of the spectrum and by skeletal modifications causing the light to pass through their tissues multiple times thus increasing light harvesting efficiency (Kahng et al. 2012). This is directly due to decreasing photosynthetic activity of the symbionts caused by light attenuation with depth in clear tropical waters. The photosymbiotic foraminifera *Cycloclepeus* living at depths of over 100 m also demonstrates similar strategies—they harbor diatom symbionts that function optimally at the light spectra available at those depths and they possess very high surface-to-volume ratios to ensure adequate surface area for photosynthesis (Song et al. 1994).

As mentioned above, skeletal modifications evolved in scleractinian corals to support the dinoflagellate photosymbionts under a selection regime dominated by intense light and also when light is limited. These corals grow as plates and branches to maximize the surface area exposed to light at greater water depths. In addition, calcium carbonate skeletons have evolved to increase irradiance by multiple scattering. The carbonate crystals reflect incoming photons to increase the number of times they pass through the tissue. If it is not absorbed the first time, light bounces off the crystal structure of the calcium carbonate skeleton just underneath the living tissue. Then, it is transmitted back through the coral tissue where the symbionts live. This scattering process provides multiple opportunities for photons to be absorbed by algal pigments, reducing the effects of self-shading and increasing the amount of light absorbed per unit of pigment (Enríquez et al. 2005; Terán et al. 2010; Marcelino et al. 2013). Although not yet studied in detail (Lee 2006), foraminiferal tests of symbiont-bearing taxa also have similar complex internal structures (see the classic work of Carpenter et al. 1862 for detailed drawings of chamberlets, pores, canals, coiling and tubes) that may function to reflect, refract or redirect light within the tests.

Photosynthetic organisms can also live freely in sea water (as plankton, on floating mucus mats, or larger algae) and on a variety of substrates (sediment, rocks, bio-mats) as non-symbiotic forms (Coffroth et al. 2006; Littman et al. 2008; Adams et al. 2009; Pochon et al. 2010; Takabayashi et al. 2012; Sweet 2014). When free-living, they can be dispersed by currents, surge, waves, and even other larger organisms like fish (Castro-Sanguino and Sanchez 2011). Subsequently, their hosts may acquire the symbionts directly, harvesting them from the surrounding environment. They may also be transmitted directly among corals and other metazoans from the parent to offspring and, in some cases, among asexually-dividing foraminifera (Lee 2006).

3.2.2 Photosymbiosis in Hypercalcifiers and Bleaching

On most reefs of the world, z-corals are framework producers and many are ***hypercalcifers***, organisms that can rapidly secrete massive amounts of skeletal calcium carbonate. In the geologic past, Earth's oceans experienced secular shifts in the Mg/Ca ratios driven by changing CO_2 levels (Sandberg 1983) that led to alternating periods that favored or discouraged the precipitation of aragonite versus calcite. As a result, marine organisms were affected by these cycles depending on their preferred skeletal composition. Hypercalcifying organisms such as aragonitic scleractinean corals would be at a disadvantage in a calcite sea cycle, and this relationship may help explain selective patterns of extinctions (Stanley and Hardie 1999). An analysis of this selectivity showed a correspondence between extinctions and hypercalcifying organisms for some extinction events (Kiessling and Simpson 2011). However, the Phanerozoic correspondence for this is far from perfect (Kiessling et al. 2008).

Hypercalcifers today require vigorous water motion and generally prosper in the upper photic zone in optimal temperature ranges of 23–29 °C. While the thresholds change between species, morphologies and location, death will generally occur at prolonged temperatures below 14 °C or above 25 °C. Some photosymbiotic corals are genetically modified to live in warmer water in isolated pools on reefs (Barshis et al. 2013), so the temperature restrictions are not necessarily constant biologically or ecologically. Yet corals, giant clams and foraminifera lose the symbionts on which they depend when temperatures exceed the normal range. In such cases of bleaching, vast numbers of corals turn ghostly white and can die unless the thermal stress is short-lived and the corals can reestablish their photosynthetic relationship (van Oppen and Lough 2009).

Clades of *Symbiodinium* inhabiting hosts vary in their adaptability and response to thermal tolerance (Rowan 2004). Such holosymbionts may explain the survival of some species in bleaching events. Indeed the "Adaptive Bleaching Hypothesis" (Fautin and Buddemeier 2004) posits that, following bleaching, some corals (and presumably other organisms as well) have the ability to reestablish a symbiosis with new clades of symbionts that are better suited to the new post-bleaching environment. This pattern could explain why coral reefs seem so fragile in the short-term when rapid temperature changes can cause widespread mortality but robust in the longer geologic-term as more adaptable species survive and are, therefore, more likely to persist and be preserved.

Symbiont-bearing protists and animals do not live well in areas affected by muddy or terrigenous sediments, an increase in nutrients (Hallock et al. 2006), elevated salinity, pollution, or warming temperatures (Douglas 2003). Corals and other carbonate-producing organisms capable of photosymbiosis are able to prosper in nutrient-deficient environments because of the efficient biochemical cycling of inorganic carbon and nitrogen by zooxanthellae (Hallock 2001). In contrast, low nutrients discourage macroalgae, a primary competitor for space on the reef. Normally, this will favor corals over macroalgae. However, the efficiency with which corals can produce carbonate in low-nutrient waters also makes them susceptible to even small changes.

3.3 Photosymbiosis in Ancient Fossils and Reef Environments

While photosymbiosis very likely occurred in many reefs and reef organisms of the geologic past, photosymbionts are not directly preserved among fossil organisms. As a result, inferring their presence in fossils depends on comparisons with modern animals in general (Cowen 1983, 1988), functional morphology in particular and, rarely, the presence of oxygen or carbon isotopes that are consistent with photosynthesis in the host skeletons (Dreier et al. 2014). Many genera of living scleractinians evolved in the early or middle Cenozoic (Budd 2000; Budd et al. 2011) and some species can be traced back millions of years. In these cases inferences about photosymbionts are more secure than for much older corals (e.g., tabulates and rugose corals). Thus confidence in biological uniformitarianism (i.e., modern biological processes are similar to those of the past) commonly decreases farther back in time. Many Mesozoic and Paleozoic taxa are extinct and many lack extant relatives, posing difficulties for inferring photosymbioses.

Fig. 3.6 Shallow-water Pacific reef corals with a flattened growth form to facilitate the capture of light in the additional area provided to symbionts (Image courtesy of J. Veron)

Although the dominant photosymbiont today is *Symbiodinium*, many reef organisms, including foraminifera, sponges, and even coral, also harbor other kinds of symbionts (Lee 2006, 2011; Ainsworth et al. 2010). These include other types of dinoflagellates that may have existed in earlier geologic time before the evolution of *Symbiodinium*.

The photosymbiotic hypotheses in fossils depend on a variety of indirect criteria. High levels of triaromatic dinosteroids are commonly associated with dinoflagellates in early Cambrian sediments, suggesting that their ancestry may extend to this time (Moldowan et al. 1996). Diagenetically unaltered fossil skeletons of Triassic and Jurassic corals (Stanley and Swart 1995), Paleozoic corals (Zapalski 2014), foraminifera (D'Hondt et al. 1994), and rudistid bivalves (Steuber 1996) have yielded stable isotopes of O and C that have been taken to indicate photosynthesis and hence the likelihood of ancient symbionts. Finally, large, thick and expansive skeletons (Figs. 3.6, 3.7, 3.8, 3.9, 3.10 and 3.11) suggest rapid skeletonization and, therefore, photosynthetic symbionts within the once-living organism. Regular annual bands within skeletons can provide actual linear extension rates and thus useful information about annual growth (Barnes and Lough 1993).

Photosymbiosis in fossil organisms may be inferred from morphologies to capture light along latitudinal or depth gradients (Cowen 1983, 1988; Wood 1999; Stanley and Lipps 2011; Groves et al. 2012). Flattened skeletal shapes and thin tissues spread symbionts over larger living areas within the host resulting in more efficient light capture (Wood 1999). These effects can be seen in flattened corals (Fig. 3.6), the expanded mantle of giant clams (Fig. 3.7) or flattened disc-like foraminifera (Figs. 3.8a and 3.10). Still

Fig. 3.7 The giant clam *Tridacna* at the Palau Mariculture Demonstration Center, Palau. The valves of this clam are huge and massive, a characteristic of the skeletons of animals and protists that host symbionts. The photosymbiont *Symbiodinium* lives in the mantle tissue overlapping the edges of the valves (Photo by J. H. Lipps, 1992)

Fig. 3.8 Large photosymbiotic foraminifera. (**a**) *Marginopora*, harboring the dinoflagellate *Symbiodinium*, ranges in size from 0.5 to 1.5 cm. It lives abundantly on sandy knolls, on algal turfs and *Halimeda* on reefs. (**b**) A living *Alveolinella quoyi* with pseudopodia extended hosts diatom symbionts. It lives on sandy slopes to at least 30 m depth and probably to the base of slopes on the floor of the lagoon, and usually at 40–50 m off reefs near Madang, Port Moresby, Papua New Guinea, and Lizard Island in the Great Barrier Reef, Australia. Smaller specimens may inhabit dead corals and coral rubble wherever it occurs, including just below low-tide level. (**c**) Three dead and cleaned tests of *A. quoyi* exhibit large size (up to 2.5 cm long), numerous long and narrow chambers and extended apertural faces with many large pores. In both *Marginopora* and *Alveolinella*, symbionts are concentrated in upper parts of the interior protoplasm of the tests creating the darker shades on the tests in these black and white images. These larger foraminifera exhibit massive amounts of $CaCO_3$ making up their skeletons relative to non-symbiont bearing benthic foraminifera. (Photos by J. H. Lipps (1986) in Papua New Guinea at Motupore Island (*top left*) near Port Moresby, in the Madang Lagoon (*bottom*), and dead tests of *Alveolinella quoyi* from the Madang Lagoon. See Langer and Lipps (2003) for distributions in the Madang Lagoon)

others like large fusiform foraminifera (Figs. 3.8b, c, 3.9 and 3.11), expand their area through lengthening the skeleton which is occupied by symbionts in the upper part of the protoplasm below the upper test surface (Lipps and Severin 1986). High levels of corallite or modular integration (e.g., interconnection between coral polyps) in colonial photosymbiotic organisms modify their shapes to maximize light and facilitate the transport of photosynthate (Coates and Oliver 1973). On modern reefs, modular organisms modify the colony according to light availability, although the resulting growth form is often a compromise for maximizing light and shedding sediments (Figs. 3.6 and 3.7).

Finally, most photosymbiotic reef organisms tend toward large size (Cowen 1988) at least in comparison to others in their group (Figs. 3.6, 3.7, 3.8, 3.9, 3.10 and 3.11). Increased size of symbiont-bearing taxa is true of most groups from foraminifera through corals to giant clams, although exceptions occur (Fig. 3.12). Therefore, massive skeletons and the large amounts of carbonate rock and sediment have been presumed to be a consequence of high calcification rates (James 1983).

3.4 Important Photosymbiotic Taxa in Ancient Reef Ecosystems

In the previous section, we outlined several lines of evidence that suggest active photosymbioses in the past. The following is a brief overview of specific groups of fossil organisms considered to have been photosymbiotic and the evidence supporting this important relationship.

3.4.1 Foraminifera

In ancient and modern seas, these single-celled eukaryotes are abundant in shallow tropical and semitropical waters, occupying rather specific habitats on the reefs and platforms (Hohenegger et al. 1999; Langer and Lipps 2003; Hohenegger 2006). Some are truly giant and complex protists (Lipps and Severin 1986; Song et al. 1994; Hallock 1999; Lee 2006, 2011; Figs. 3.7, 3.8, 3.9, 3.10 and 3.11). Foraminifera have evolved particular morphologies ranging from the flattened tests in *Marginopora* or *Cycloclepeus* (Song et al. 1994) to large, complicated and elongate forms like *Alveolinella, Praealveolina* and fusulinids (Figs. 3.8, 3.9 and 3.11). Even some smaller foraminifera have morphologies associated with modern symbionts. For example, *Amphistegina* has tiny cups on the interior surface of its test that contain the symbionts (Lee 2006).

Fig. 3.9 Giant Pennsylvanian (Upper Carboniferous) fusulinid foraminifera (*Parafusulina*). Scale = 1 cm. Fusulinids occur in thick, widespread limestone beds in the upper Paleozoic and closely resemble the living symbiont-bearing species *Alveolinella* (Fig. 3.8) in habitat, morphology, size, and internal complexities, all supporting the inference that fusulinids possessed photosymbionts (Photo by J. H. Lipps of University of California Museum of Paleontology specimens)

Fig. 3.10 The large discoidal foraminifera *Nummulites* from the Eocene of Israel. These are very common throughout the ancient Tethys Seaway particularly on carbonate platforms where they occur in thick deposits of limestone. Scale bar = 1 cm (Photo by J. H. Lipps, 1993)

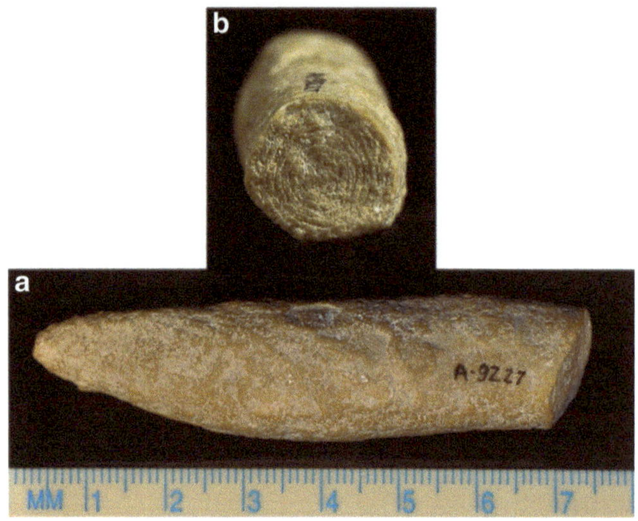

Fig. 3.11 One of the largest calcareous foraminifera known, *Praealveolina* ranges to more than 10 cm in length. These were common in the later Cretaceous, and closely resemble the only large fusiform modern species *Aleveolinella quoyi* as well as the Paleozoic fusulinids, some of which attained even larger sizes. (**a**) This fusiform specimen is broken and about 3 cm are missing on the right end, making its total length and width (near the 4-cm mark) greater than 10 cm and 2.7 cm, respectively. (**b**) Broken end (2.6 cm in diameter) showing the complex inner structures of small chambers divided by partitions. Such complexities are indicative of photosymbionts contained within the test. Photos by Bruce Rubin of University of California Museum of Paleontology specimen A-9227

The first foraminifera that likely had photosymbionts were the mid-Paleozoic fusulinids (Fig. 3.9); they dominated carbonate banks and platforms until the end of that era (Vachard et al. 2010). Evidence for photosymbiosis in these foraminifera includes their relatively large size (for foraminifera), complex internal morphology, ecologic and geographic distributions in tropical environments on reefs and carbonate banks, and overall similarity to modern symbiont-bearing alveolinellids (Lipps and Severin 1986; Severin and Lipps 1989; Lee 2006; Groves et al. 2012).

Large size in fusulinid foraminifera has also been attributed to high atmospheric oxygen levels (~30 %) rather than symbiosis (Payne et al. 2012). When oxygen levels are high a large volume to surface area may allow oxygen to diffuse quickly into the interior of organisms and metabolic rates can be higher, yet foraminifera and other organisms grow to large sizes today when they possess symbionts (Lipps and Severin 1986; Song et al. 1994; Hallock 1999; Lee 2006, 2011; Vermeij 2013). Indeed the largest living fusiform foraminifera *Alveolinella quoyi* (Fig. 3.8b, c), which resembles the large fusulinids (Fig. 3.9), possesses diatom symbionts and can live to depths over 30 m or more. At lengths of 2–3 cm or more, *A. quoyi* achieves large sizes under today's oxygen levels. The volume of the cytoplasm in *A. quoyi* rarely fills more than 45 % of available chamber

Fig. 3.12 *Corculum* (University of Montana Paleontology Center UMIP 14319), a modern photosymbiotic clam, has windows in its shell that allow light to pass to photosymbionts living in the mantle tissue inside the shell. When the clam burrows into sand and the valves are closed, the symbionts are still able to photosynthesize using light that passes through the windows (Photo courtesy of Kallie Moore). Scale bar = 2 cm

volume (mean = 37 %; range = 17–100 %: Severin and Lipps 1989).

The paleobiogeography of fusulinids with smaller individuals in the Polar Regions and larger ones at the equator does not support the oxygen hypothesis for these foraminifera but rather suggests that they possessed symbionts (Zhang and Payne 2012). Fossil alveolinellids (Cretaceous to Neogene) attained very large sizes too (some over 10 cm long; see Fig. 3.11) and all of these lived under a variety of atmospheric oxygen levels. Thus these observations cast "further doubt on the primary role of oxygen as a factor enabling gigantism in photosymbiotic species" (Vermeij 2013). As this variability is inconsistent with a tie to atmospheric O_2 levels, we attribute large size in these symbiont-bearing foraminifera to photosymbiosis rather than oxygen availability.

On modern reefs, larger photosymbiotic foraminifera produce prodigious amounts of calcium carbonate on reefs and carbonate platforms, in some places contributing up to 25 % of the total (avg. ~5 %: Langer et al. 1997; Langer 2008). In the past, vast amounts of carbonate rocks, forming banks and shelves, were similarly produced by large foraminifera that we interpret to have hosted or likely hosted symbionts. Their large size and extended pseudopods would have made them difficult to transport, enhancing their likelihood of in-situ deposition (Severin and Lipps 1989). Paleozoic fusiform fusulinids contributed to thick limestone blankets over many km^2. Mesozoic and Cenozoic alveolinellids, orbitolinids, and others produced thick deposits of carbonate, and the giant, coin-like *Nummulites* (Fig. 3.10), dominated the former Tethyan Seaway in places like the Eocene of Israel. The enormous (up to 10 cm long) elongated foraminifera *Praealveolinella* (Fig. 3.11) appeared in the Cretaceous and surely had symbionts, as its size, carbonate content, distribution, and internal complexity attest.

The phrase "Power of the Pyramids" might be replaced with the "Power of Photosymbiosis" since the huge monoliths of Egypt (Fig. 3.13) are made of nummulitic Eocene limestone blocks (Fig. 3.13 inset). Indeed the "power of photosymbiosis" made the pyramids possible in the first place, since nummulites very likely hosted photosymbionts. Photosymbiotic foraminifera have contributed to the formation of extensive carbonate rocks for ~350 million years of geologic time with exceptions of the post-extinction periods (Chap. 8).

3.4.2 Calcified Sponges

Ancient reefs were also built by a wide variety of calcitic or aragonitic sponges (demosponges, stromatoporoids, chaetetids and other groups). Archaeocyathids were calcitic sponges (Rowland 2001) that dominated reefs during the Early Cambrian (Fig. 3.14). They lived in tropical shallow waters where they produced small mounds, moderate-sized buildups, and even very large complexes, such as the Great Siberian Reef Complex, 200–300 km wide and 1500 km long (Rowland and Hicks 2004).

Archaeocyaths included many species with cup-shaped skeletons that varied in morphology from nearly flat to lobate with flattened edges at the top of the cup to more tubular forms. Individuals ranged in size from a few cm up to 30+ cm and were attached to the substrate with holdfasts. Given their morphologies, shallow water habitats, tropical distribution, and reef-building abilities, archaeocyathids may have possessed symbiotic algae or perhaps cyanobacteria (Cowen 1983; Rowland and Gangloff 1988; Surge et al. 1997; Rowland and Shapiro 2002). However their small size, solitary growth form, low modular integration level and cryptic lifestyles led to the alternative hypothesis that they lacked symbionts and lived in environments with fluctuating nutrients and high input of terrigenous sediments (Wood 1993, 1999; Pratt et al. 2001; Zhuravlev 2001). Both symbiotic and asymbiotic forms may have

Fig. 3.13 "The Power of Photosymbiosis"—The Great Pyramid and Sphinx of Giza, Egypt. The Sphinx consists of several layers of marl and limestone with few nummulites in them as those foraminifera lived on a bank farther away (Gauri et al. 1990). The Great Pyramid was constructed chiefly of local limestone blocks containing abundant *Nummulites* (*inset*). The early historians, Herodotus (Greek, fifth century BCE), Strabo (Greek, second century BCE) and Pliny the Elder (Roman, 23–79 CE), considered stories that the nummulites were lentils dropped by the workmen as they ate which then petrified (Carpenter et al. 1862; Adams 1938; Lipps 1981). Even the earliest of these observers noted that the nummulites occurred widely in the region and hence were not the remains of lentils. The nummulites actually accumulated abundantly in the sediments of the Eocene Tethys Seaway. They likely hosted photosymbionts that made the growth of the large tests possible (Photos by J. H. Lipps, 2007)

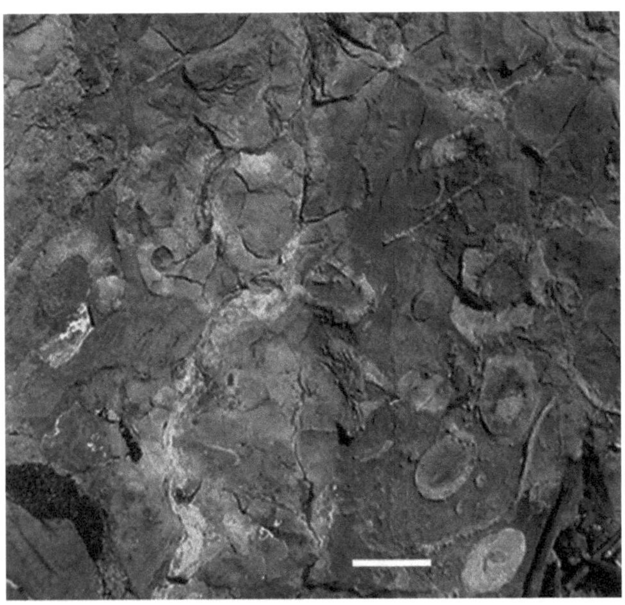

Fig. 3.14 Cross sections of individual archaeocyathan skeletons in the Cambrian Montenegro member of the Poleta formation, White-Inyo Mountains, California. In many places, the archaeocyathans are associated with patch reef and large reef structures; they likely contained photosymbionts. Scale bar = 1 cm (Photo by J. H. Lipps, 1986)

inhabited the same reefs simultaneously, as the same major groups do today (e.g., foraminifera, sponges and corals).

Stromatoporoids were an important group of calcified sponges from Ordovician to Late Devonian time and during the late Mesozoic (Nestor et al. 2010). They were important reef-builders during mid-Paleozoic time, some reaching 10–20 m in diameter. These sponges secreted calcitic skeletons and like corals, appear to have been important in constructing impressive reefs during mid-Paleozoic time (Copper 2002). They lived together with calcareous algae, rugose corals and tabulate corals. Red algae, corals and stromatoporoids formed fringing and barrier reefs of the Silurian and Devonian. These reefs exceeded modern examples in size and volume during a mid-Paleozoic greenhouse time when tropical marine realms reached much higher latitudes than today.

Although stromatoporoids appear to have been reef builders, they yield equivocal evidence of photosymbiosis (Kershaw and Brunton 1999). Among the 5000 different species of living sponges, many harbor photosymbiotic organisms, especially cyanobacteria (Taylor et al. 2007) and some tropical examples show photosymbiotic activity with other organisms in very shallow settings (Steindler

Fig. 3.15 A stromatoporoid pillar in the Silurian reefs of Gotland, showing the platy growths that make up the pillar. Height is approximately 1.5 m (Photo courtesy of Steve Kershaw)

Fig. 3.16 A vertical section cut through a whole stromatoporoid from the mid-Silurian Visby Formation, Gotland, Sweden. In this example, three different species of stromatoporoids grew together and were partly buried by sediment before the next growth, gradually building up the structure. Near the *bottom* a brachiopod lay on the lower growth of stromatoporoid and was then overgrown by the next stromatoporoid layer. This process illustrates how complex reef structures are built by various species (Photo courtesy of Steve Kershaw)

et al. 2002). A radical reinterpretation of stromatoporoids as cyanobacteria rather than metazoans would certainly imply photosynthesis (Kazmierczak 1976) but this interpretation is not widely accepted. Evidence supporting photosymbiosis in Paleozoic stromatoporoids includes growth forms like corals, modular integration and large size (Copper 2002). The limited data suggest that they were slightly slower growing than living corals (Gao and Copper 1997).

Paleoecologically, stromatoporoids in Paleozoic reefs appear to have been limited by nutrients and sediment influx and capable of growing between and over other organisms such as brachiopods, corals and red algae (Fig. 3.15). Some taxa suggest high levels of integration and are interpreted to have lived in shallow, open and sun-lit parts of ancient reefs. This contrasts with living sclerosponges, which are relegated to cryptic and/or deeper water environments.

Whether these organisms harbored photosymbionts is not clear. The large size, platy growth shapes (Fig. 3.16) and integration levels of mid-Paleozoic reef-dwelling stromatoporoids commonly resemble modern photosymbiotic scleractinians (Rosen 2000). Also, stromatoporoids provide evidence for the "thin tissue syndrome" (Wood 1999) and the "solar panel effect" (i.e., flattening with depth to maximize light-gathering capacity). Finally, feeding strategies and paleobiology indicated that at least some stromatoporoids were photoautotrophic (Brunton and Dixon 1994), an idea supported by the co-occurrence of these sponges with large photosymbiotic megalodontid bivalves, which also preferred warm, well-lit marine settings. However other evidence for photosymbiosis among mid-Paleozoic stromatoporoids is equivocal (e.g., are growth bands annual and what was their growth rate relative to corals: Kershaw 1998).

Other sponges which lived during the Permian, Carboniferous and Triassic appear to have been hypercalcifying and capable of building reefs. These include calcified chambered "sphinctozoan" and chaetetid sponges. While some of these reached large size and were primary or secondary reef constructors, they show slow growth rates. Some calcified demosponges such as the Upper Triassic reef-adapted *Stromatomorpha*, are a mimic on Paleozoic stromatoporoids and they may have been photosymbiotic (Senowbari-Daryan and Stanley 2009).

3.4.3 Corals

Like their modern counterparts, corals in the geologic past constructed reefs. Morphological similarities and growth strategies suggest that they also share a photosymbiotic relationship. Their large size and corallite integration argue for rapid skeletonization. Also, what appear to be depth-related changes in colony shape (i.e., flatter colonies at depth: Dodge and Vaišnys 1980; Dustan 1982) and a high degree of corallite integration argue for photosymbiosis. The "edge zone" along the outer corallite wall where tissue extends is similar to modern, zooxanthellate corals. Like foraminifera and bivalves, they also are characterized by thin tissue syndrome (Cowen 1983), displaying large areas of thin tissue for harvesting light. Finally, like some zooxanthellate corals, they also contain abundant radiating features of the skeleton called pennular structures that radiate from the polyp centers and are thought to supplement nutrition in slightly deeper water (Wood 1999; Stanley 2006).

Many ancient corals contain alternating low-density and high-density layers of skeleton (Fig. 3.17a) that in modern corals are annual (Fig. 3.17b). They have been associated with variables of light, temperature, reproduction, nutrients, and other factors affecting the energy budget of coral growth (Buddemeier 1974). Such features in fossils of Permian, Triassic and Jurassic ages allow comparisons with fast-growing living reef species. Some massive Triassic corals contained annual bands almost identical to high and low density bands in living reef corals (Stanley and Helmle 2010). This may indicate ancient photosymbiosis in the Triassic.

Many coral colonies, especially those on modern reefs, have large sizes and this has been used as a proxy for rapid growth and, therefore, photosymbiosis. Some Late Triassic colonies reached 5–10 m in height (Piller 1981; Stanley and Swart 1995), larger than some modern z-coral species. Also, they are similar to "microatolls", which form today as the colonies grow to sea level, their polyps die on top while the colony continues to expand laterally. This distinctive morphology records sea level, and in modern corals occurs in at least 43 (Rosen 1978) colonial or massive species (Scoffin and Stoddart 1978; Smithers and Woodroffe 2000). The oldest-known microatolls from the Late Triassic (Fig. 3.18) have flattened surfaces and a central cavity (Stanley 2005). Since microatolls today are only known in modern zooxanthellate species, their presence in fossil examples has been linked to photosymbiosis.

Cretaceous corals also possess colony shapes and corallite integration that suggest photosymbiosis. Jurassic corals likewise show high integration levels, annual banding and adaptation of colony shapes similar to those of modern reef-building corals (Leinfelder 2001; Barbeitos et al. 2010). Rosen and Turnšek (1989) characterized coral species that survived across the Mesozoic-Cenozoic boundary as either z-coral-like or az-coral-like on the basis of the indirect criteria described above.

A more direct line of evidence comes from isotopic studies of early Mesozoic corals. Thirteen specimens of Triassic scleractinians from reef complexes in Turkey and northern Italy and two specimens from the Jurassic of Poland showed that the isotopic signatures of the Triassic corals from Turkey were more like modern zooxanthellate corals, while the Jurassic samples were similar to azooxanthellate

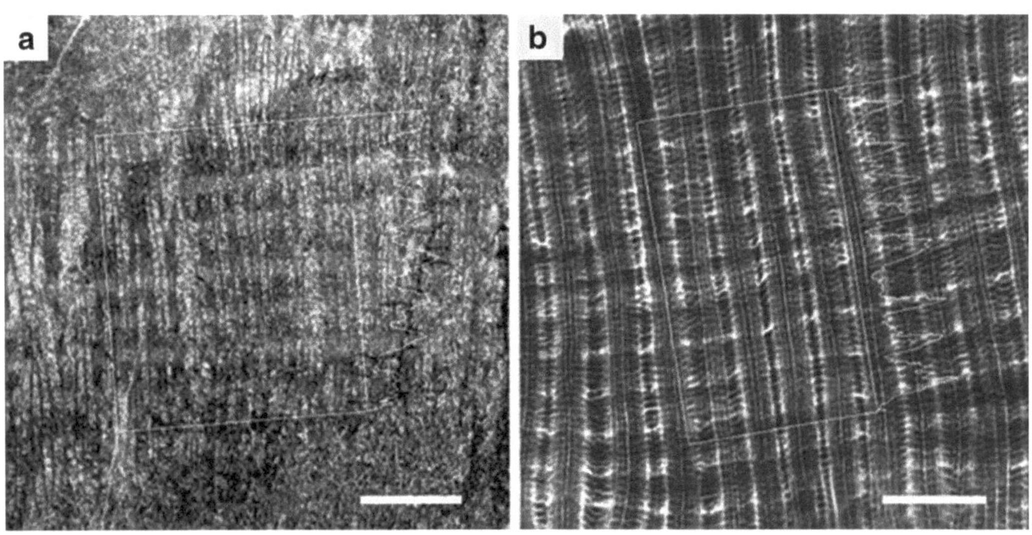

Fig. 3.17 Comparison of growth bands in a Late Triassic coral *Ceriostella* (UMIP 18001) (**a**) with a modern *Montastraea* (**b**) showing high density and low density annual banding. Scale bars = 1 cm (Photos from Stanley and Helmle (2010))

species (Stanley and Swart 1995). Isotopic analyses of skeletons of Late Triassic corals yielded a similar conclusion regarding photosymbiosis (Muscatine et al. 2005).

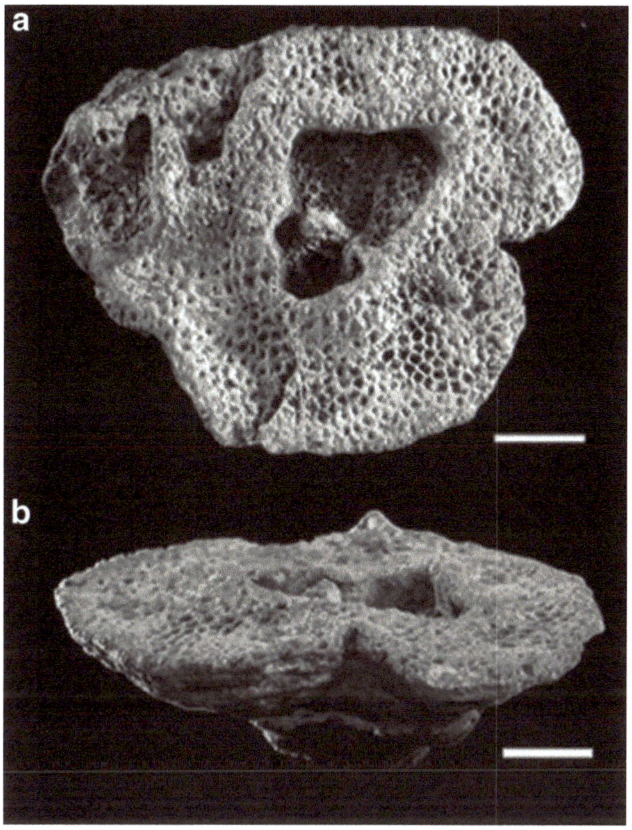

Fig. 3.18 The oldest reported microatolls (University of Montana Paleontology Center UMIP 6813) from the Triassic in Nevada. Like modern microatolls, they formed by growing to sea level to maximize solar radiation, and then spread laterally. Scale bars are 1 cm (Photo modified from Stanley (2005))

Photosymbiosis can be traced back to Paleozoic corals (both Rugosa and Tabulata) which were judged to have been photosymbiotic by paleoecological methods and isotope studies (Zapalski 2014). While the majority of rugose corals were solitary, some colonial species reached large sizes in mid-Paleozoic reefs and like living photosymbiotic species, possessed high levels of corallite integration. Tabulate corals (Fig. 3.19) lived on mid-Paleozoic reefs reached large sizes, some resembling colonies of modern z-corals. Growth rates of many Paleozoic corals (Gao and Copper 1997) also compare favorably to living z-corals and provide evidence for photosymbiosis. However corallum complexity of some Paleozoic species as judged by integration levels, are lower than for scleractinian corals (Coates and Jackson 1987). While these approaches are reasonable, such assessments are really historical hypotheses to be further tested.

3.4.4 Bryozoans

Starting in the late Cambrian (Landing et al. 2015), bryozoans have a detailed fossil record continuing to the present (Taylor and Waeschenbach 2015). Many species over this time are associated with reefs and some grow quite large. The large (up to 7 cm) calcareous bryozoan colonies formerly known as the Trepostomata, for example, may have possessed photosymbionts, and this idea was postulated for Permian examples (Håkansson and Madsen 1991). The large size is consistent with the rapid growth attributed to photosymbiosis. Stable isotope analysis, however, revealed that such Paleozoic bryozoans secreted calcite in isotopic equilibrium with seawater and so did not possess the signature of photosymbiosis (Key et al. 2005). Also, no modern bryozoans are reported to possess photosymbionts. Thus, the role of photosymbiosis in byrozoans remains speculative.

Fig. 3.19 Reconstruction of a Middle Silurian reef illustrating a coral-dominated ecosystem characterized by large colonies of tabulate and rugose corals along with crinoids, bryozoans, brachiopods and other invertebrate taxa. This was the closest approximation in the Paleozoic to scleractinian-dominated reefs (Illustration courtesy of Terry Chase)

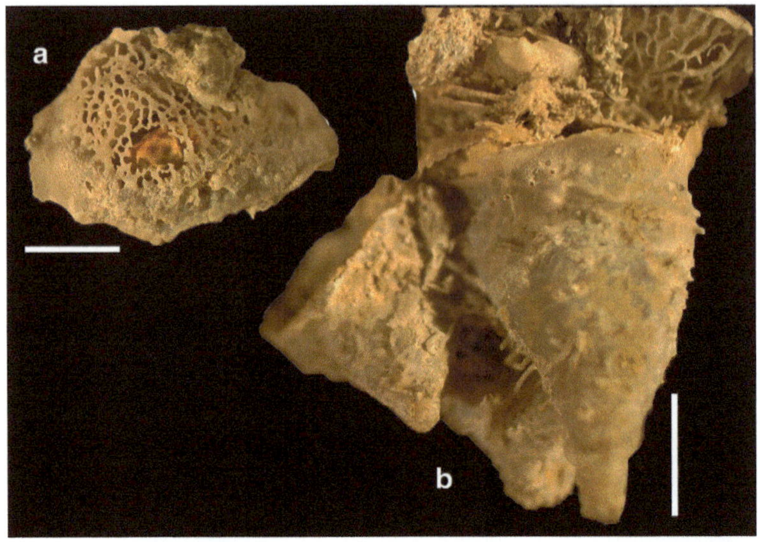

Fig. 3.20 The conical silicified Middle Permian reef brachiopod *Hercosestria* from the Glass Mountains, south-west Texas. (**a**) The sieve-like covering over the ventral valve may have been covered with mantle tissue (UMIP 14291). (**b**) Several individual ventral valves of these brachiopods with their attachment spines and some of the sieve-like covering (UMIP 14292). Scale = 1 cm. University of Montana Paleontology Center (UMIP) (Photo courtesy of Kallie Moore)

3.4.5 Brachiopods

Various Paleozoic productid brachiopods, including Richthofeniacea and Lyttoniacea, may have harbored photosymbionts based on (1) shell adaptations to expose much of the mantle to light, (2) massive calcification, (3) large size, (4) habitat, and (5) paleogeographic distributions in tropical seas of the Late Permian (Cowen 1983, 1988). Among these were the reef-dwelling richthofenid brachiopods (Fig. 3.20) *Hercosia* and *Cyclacantharia* (Grant 1972; Cowen 1983; Fagerstrom 1996; Cowan and Erickson 2010).

Photosymbiosis is a logical hypothesis for the bizarre morphology exhibited by Late Permian lyttoniacean brachiopods which inhabited reef-like buildups (Cowen 1983). However, morphology alone is insufficient to support either a photosymbiosis hypothesis or the possibility that they may have combined photosymbioses with a filter feeding lifestyle.

Some Devonian brachiopods are relative giants, such as *Stringocephalus* in the Givetian, with shells up to 20 cm long; these could also have had algal symbionts (P. Copper, personal communication, 2011). However, unlike the majority of other photosymbiotic organisms that utilize aragonite, brachiopods secreted shells of calcite.

3.4.6 Mollusks

At least 17 independent bivalve groups may have developed photosymbiosis with algae in Earth's history (Vermeij 2013). While evidence is commonly equivocal, these groups display many characters that would promote or result from photosymbiosis. One shelled mollusk may have had photosymbionts in the Eocene to Oligocene, the gastropod *Velates* which was large; it likely had exposed mantle tissue and inhabited well lit habitats (Vermeij 2013). Most photosymbiont-bearing taxa are attached to or buried in the substrate, hence bivalves dominate the shelled mollusks that hosted these symbionts. Many upright, Late Cretaceous rudistid bivalves built mounds and reef-like structures (Kauffman and Johnson 1988) and secreted large, thick shells of both calcite and aragonite (Fig. 3.21). Their fossils, based on calculated growth rates and modifications of the upper shell, strongly infer adaptation to light and photosymbiosis (Vogel 1975). However, some other rudistids may not have possessed photosymbionts (Steuber 2000). While some rudistids display growth rates comparable to the living photosymbiotic *Tridacna* (Fig. 3.7), other living bivalves like *Corculum* (Fig. 3.12) and *Fragum* are small and do not produce large or thick shells, yet they have unique "windows" in their shell, which would have allowed light to reach algal symbionts inside (Watson and Signor 1986; Farmer et al. 2001).

Other bivalves with algal symbionts, like the modern heart cockle *Clinocardium*, do not show any of the characteristics used to infer photosymbiosis (Jones and Jacobs 1992). While size is not always an indicator, it is an obvious characteristic of photosymbiosis for bivalves and other organisms.

Exclusive of rudistids, other giant, reef-dwelling bivalves existed on carbonate platforms through time. Giant clams occurred in the Devonian, Permian, Triassic, and Jurassic. Giant alatoform bivalves in the family Wallowaconchidae (Yancey and Stanley 1999) can be up to a meter in length and occur in reef-related, Upper Triassic carbonate rocks. Not only do these unique bivalves exhibit large size, they also display the "solar panel" effect, secreting a series of

Fig. 3.21 Reconstruction of a cluster of gregarious, upright rudistid bivalves from a Late Cretaceous reef. These large bivalves may have held algal symbionts. Light would have been transmitted through the upper valves, which were reduced to a thin, perforated cap-like morphology (Courtesy of Jose Garcia)

enveloping chambers along the margin of the shell (Fig. 3.22) where presumably photosymbionts were sequestered in the mantle tissue. Like living *Corculum* these bivalves may have hosted symbionts within the chambers and have been able to harvest light transmitted directly through the shell.

3.5 Summary and Conclusions

Photosynthetic algae and microbial eukaryotes and metazoans have formed symbioses throughout much of geologic time (Cowen 1983; Coates and Jackson 1987; Talent 1988; Stanley and Lipps 2011). This cooperative relationship across many unrelated algae and hosts evolved very early on, and has been common and repetitive ever since. Although photosymbionts are not preserved in the fossil record, ample indirect evidence (e.g., isotopes, morphology, size and depth-related changes in host morphologies) indicate that photosynthetic partnerships were strongly selected for the capture of energy directly from sunlight and a reliable nutrient supply for the hosts and a habitat with protection for the symbionts (Cowen 1983, 1988; Hallock 1999; Lee 2006; Vermeij 2013).

Photosymbioses evolved multiple times in geologic time and became quite common, only to go extinct during the extinction events of the past 543 Ma. Most likely these symbioses developed in single-celled hosts and sponges as the algae were harvested as part of the food supply and later sequestered within cells as happens today with the inclusion of both live chloroplasts (Lopez 1979; Cedhagen 1991) and free-living symbionts in the cytoplasm of certain foraminifera (Lee 2006; Fay et al. 2009; Lee 2011). In cases where they occupy those parts of present-day organisms that are irradiated by the sun, symbionts might have been acquired from the environment and transported to specific tissues. Perhaps the same mechanism coevolved in other partners as well. Did hosts acquire symbionts specifically for their own advantage and how complex was the evolution? While modern corals will accept or reject certain algal symbionts, observations of infestations support the hypothesis that forming endosymbiotic associations leading to a "fit" host actually involves a complex series of co-evolutionary steps (Stat et al. 2006).

Although photosymbioses are inferred in the early Paleozoic, they were well-established and widespread by the early Mesozoic. Today's prime photosymbionts are dinoflagellates in the genus *Symbiodinium* belonging to the order Suessiales, which includes closely related symbionts of various planktic organisms (Siano et al. 2010). In their life cycles, dinoflagellates often encyst and these are common as microfossils. The oldest Suessiales cysts that are morphologically similar to those of modern *Symbiodinium* symbionts first appeared in the Late Triassic when scleractinian corals radiated. The fossil record of these cysts tracks those of corals; both groups also experienced an extinction at the end of the Triassic and a recovery in the succeeding Jurassic (Stanley and van de Schootbrugge 2009). Photosymbiosis between scleractinian corals and dinoflagellate symbionts may have occurred in the Triassic (Stanley and Helmle 2010) based on the widespread development of reefs and thick reef-carbonate rocks during the Late Triassic interval (Stanley 1981; Riedel 1991; Kiessling 2010).

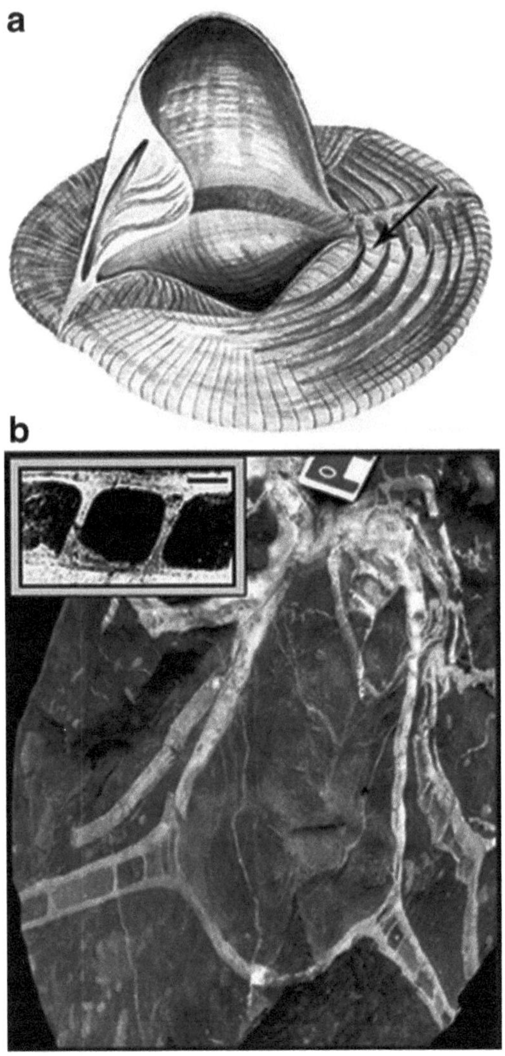

Fig. 3.22 Giant wallowaconchid alatoform clams from the Late Triassic. (**a**) Reconstruction of the shell showing the conch with a cut-away view of the shell that reveals a series of concentric and overlapping chambers (*arrow*) connecting to the central body cavity. The chambers likely housed the photosymbionts inferred to have lived within the chambers of these clams, and the upper surface of the chambers were made of aragonite crystals perhaps oriented as in the modern *Corculum* (Fig. 3.12) to conduct light to the symbionts inside. Courtesy of Jose Garcia. (**b**) Field image of an individual clam cut at an oblique angle, showing the central body cavity and chambered, wing-like extensions (sample UMIP 23530). *Upper left* inset is a thin-section showing details of hollow chambers, now filled with micrite (sample UMIP 24206-F). Scale bar = 5 mm

Molecular-clock data, based on sequences of chloroplasts in modern clades of *Symbiodinium*, place their origin in either the Paleocene or early Eocene, not long after the Cretaceous/Paleogene mass extinction. However, the diversification of the modern lineages of *Symbiodinium* did not occur until the mid-Miocene some 15 million years ago (Pochon et al. 2006; Stat et al. 2006), coincident with the evolution and expansion of the modern coral-reef ecosystems (Perrin 2002).

Because symbiont-host relationships are not monophyletic through geologic time, the coevolution with host species is unclear. Molecular studies of clade D *Symbiodinium* in reef corals, revealed "boom and bust" phases of diversification and extinction over the past 12 million years in response to climate change and the tectonic separation of the Caribbean and Pacific provinces by the emerging Central American Land Bridge (Thornhill et al. 2013). We suggest that in the perspective of deeper time, such "boom and bust" cycles characterized many photosymbionts and their hosts.

Symbionts from the *Symbiodinium* group have established relationships with a wide taxonomic variety of hosts such as sponges, corals and other cnidarians, benthic and planktonic foraminifera, giant clams, among others. Commonly the same clades may be associated with multiple unrelated hosts, e.g., foraminifera, jellyfish, milleporoid hydrozoans, octocorals, nudibranchs, tunicates and bivalves. Different molecular clades of *Symbiodinium* in corals, and perhaps other groups as well, have different ecological preferences for light, temperature, depth and, therefore, hosts (LaJeunesse et al. 2010; Kahng et al. 2012). In foraminifera, different clades may even occupy specific parts of the cell (Fay et al. 2009).

The geologic history of the successes and failures of reefs can be related directly to the acquisition or failure of photosymbioses (Talent 1988). Reefs did well during long periods of stable environmental conditions but became extinct when warming, acidification, and anoxia of the oceans occurred. Following these extinction events, newly acquired photosymbionts fueled the rapid diversification of reef organisms, for example, corals in the Triassic after the great Permian extinctions. Conversely, the loss of photosymbionts could have been a key strategy for surviving extinction crises (Barbeitos et al. 2010).

Based on our review of reef building and the calcifying organisms involved, we confirm our hypothesis that photosymbiosis was integral to the success of both present-day and ancient reefs. We find that the breakdown of the symbiosis most likely was tied to global environmental perturbations that led to mass extinctions. In concert with fluctuations in nutrient, sedimentation and other factors, such breakdowns might explain reef gaps, times of reduced carbonate sedimentation and drastic reductions in reef building observed in the geologic record.

Acknowledgements We are indebted to our numerous colleagues and students. Prof. Dennis Hubbard and Dr. Caroline S. Rogers provided significant editorial assistance that greatly improved our paper. Lipps particularly thanks Drs. Michele Weber, Scott Fay, Richard Cowen, Martin Langer, Ken Severin, Bernard Ormsby and Yan Song for discussion of this problem and assistance in the field and lab. Lipps's research on larger foraminifera was supported by NSF EAR 84–08001. Stanley is indebted to Kallie Moore for photo assistance and thanks go to Jose Garcia for his artwork. Steve Kershaw provided helpful discussions on Paleozoic stromatoporoids.

References

Adams FD (1938) The birth and development of the geological sciences. Dover Pub. New York.

Adams LM, Cumbo VR, Takabayashi M (2009) Exposure to sediment enhances primary acquisition of *Symbiodinium* by asymbiotic coral larvae. Mar Ecol Prog Ser 377:149–156

Ainsworth TD, Thurber RV, Gates RD (2010) The future of coral reefs: a microbial perspective. Trends Ecol Evol 25:233–240

Allwood AC, Walter MR, Burch IW, Kamber BS (2007) 3.43 billion-year-old stromatolite reef from the Pilbara Craton of Western Australia: ecosystem-scale insights to early life on Earth. Precamb Res 158:198–227

Anderson OR (1983) Radiolaria. Springer-Verlag, New York

Baker AC (2003) Flexibility and specificity in coral–algal symbiosis: diversity, ecology and biogeography of *Symbiodinium*. Ann Rev Ecol Evol Syst 34:661–689

Barbeitos MS, Romano SL, Lasker H R (2010) Repeated loss of coloniality and symbiosis in scleractinian corals. Proc Nat Acad Sci 107:11877–11882

Barnes DJ, Lough JM (1993) On the nature and causes of density banding in massive coral skeletons. J Exp Mar Biol Ecol 167:91–108

Barshis DJ, Ladner JT, Oliver TA, Seneca FO, Knowles NT, Palumbi SR (2013) Genomic basis for coral resilience to climate change. Proc Nat Acad Sci 110:1387–1392

Benton MJ (2003) When life nearly died: the greatest mass extinction of all time. Thames & Hudson, London

Blank RJ, Trench RK (1985) *Symbiodinium microadriaticum*: a single species? Proc 5th Int Coral Reef Symp 6:113–117

Brunton, FR, Dixon OA (1994) Siliceous sponge-microbe biotic associations and their recurrence through the Phanerozoic as reef mound constructors. Palaios 9:370–387

Budd AF (2000) Diversity and extinction in the Cenozoic history of Caribbean reefs. Coral Reefs 19:25–35

Budd AF, Klaus JS, Johnson KG (2011) Cenozoic diversification and extinction patterns in Caribbean reef corals: a review. Paleontol Soc Papers 17: 79–94

Buddemeier, RW (1974) Environmental controls over annual and lunar monthly cycles in hermatypic coral calcification. Proc 2nd Int Coral Reef Symp 2:259–267

Butterfield NJ (2015) Early evolution of the Eukaryota. Palaeontol 58: 5–17

Cairns SD (1999) Species richness of recent Scleractinia. Atoll Res Bull 459:1–46

Cairns SD (2007) Deep-water corals: an overview with special reference to diversity and distribution of deep-water scleractinian corals. Bull Mar Sci 31:311–322

Carlon DB, Goreau TJ, Goreau NI, Trench RK, Hayes RL, Marshall AT (1996) Calcification rates in corals. Science 274:117–118

Carpenter WB, Parker WK, Jones TR (1862) Introduction to the study of the Foraminifera. Q J Microsc Sci 2:1–319

Castro-Sanguino C, Sanchez JA (2011) Dispersal of *Symbiodinium* by the stoplight parrotfish *Sparisoma viride*. Biol Letters 299:282–286

Cedhagen T (1991) Retention of chloroplasts and bathymetric distribution in the sublittoral foraminiferan *Nonionella labradorica*. Ophelia 33:17–30

Clarkson MO, Kasemann SA, Wood RA, Lenton TM, Daines SJ, Richoz S, Ohnemueller F, Meixner A, Poulton SW, Tipper ET (2015) Ocean acidification and the Permo-Triassic mass extinction. Science 348:229–232

Coates AG, Jackson JBC (1987) Clonal growth, algal symbiosis, and reef formation by corals. Paleobiol 13:363–378

Coates AG, Oliver WA Jr (1973) Coloniality in zoantharian corals. In: Boardman RS, Cheetham AH, Oliver WA Jr (eds) Animal colonies: development and function through time. Dowden Hutchinson Ross, Stroudsburg, PA, pp 3–27.

Coffroth MA, Lewis CF, Santos SR, Weaver JL (2006) Environmental populations of symbiotic dinoflagellates in the genus *Symbiodinium* can initiate symbioses with reef cnidarians. Current Biol 16:R985–R987

Coffroth MA, Santos SR (2005) Genetic diversity of symbiotic dinoflagellates in the genus *Symbiodinium*. Protist 156:19–34

Cohen AL, Holcomb MC (2009) Why corals care about ocean acidification: uncovering the mechanism. In: Doney SC, Balch WM, Fabry VJ, Feely RA (eds) The future of ocean biogeochemistry in a high CO_2 world. Oceanogr 22:18–127

Cohen AL, Layne GD, Hart SR, Lobel PS (2001) Kinetic control of skeletal Sr/Ca in a symbiotic coral: implications for the paleotemperature proxy. Paleoceanogr 16:20–26

Copper P (2002) Silurian and Devonian reefs: 80 million years of global greenhouse between two ice ages. SEPM Spec Publ 72:181–238

Cowan HL, Erickson JM (2010) *Rafinesquina alternata* as a photosymbiont host: shedding new light on old questions. Geol Soc Amer Abstr Progr 42:287

Cowen R (1983) Algal symbiosis and its recognition in the fossil record. In: Tevesz MJS, McCall PL (eds) Biotic interactions in Recent and fossil benthic communities. Plenum Press, New York, pp 431–478

Cowen R (1988) The role of algal symbiosis in reefs through time. Palaios 3:221–227

D'Hondt S, Zachos JC, Schultz G (1994) Stable isotopic signals and photosymbiosis in late Paleocene planktic foraminifera. Paleobiol 20:391–406

Dill RF, Shinn EA, Jones AT, Kelly K, Steinin RP (1986) Giant subtidal stromatolites forming in normal saline waters. Nature 324:55–58

Dodge RE, Vaišnys JR (1980) Skeletal growth chronologies of Recent and fossil corals. In: Rhoads DC, Lutz RA (eds) Skeletal growth of aquatic organisms: biological records of environmental change. Plenum Press, New York, pp 493–517

Douglas AE (2003) Coral bleaching—how and why? Mar Poll Bull 46:385–392

Douglas AE (2010) The symbiotic habit. Princeton University Press, Princeton

Dravis JJ (1983) Hardened subtidal stromatolites, Bahamas. Science 219:385–386

Dreier A, Loh W, Blumenberg M, Thiel V, Hause-Reitner D, Hoppert M (2014) The isotopic biosignatures of photo- vs. thiotrophic bivalves: are they preserved in fossil shells? Geobiol [DOI: 10.1111/gbi.12093]

Dustan P (1982) Depth-dependent photoadaptation by zooxanthellae of *Montastrea annularis*. Bull Mar Sci 68:253–264

Eme L, Sharpe SC, Brown MW, Roger AJ. (2014) On the age of eukaryotes: evaluating evidence from fossils and molecular clocks. Cold Spring Harb Perspect Biol 6:a016139

Enríquez S, Méndez ER, Iglesias-Prieto R (2005) Multiple scattering on coral skeletons enhances light absorption by symbiotic algae. Limnol Oceanogr 50:1025–1032

Erwin DH (2006) Extinction: how life on earth nearly ended 250 million years ago. Princeton Univ Press, Princeton

Fagerstrom JA (1987) The evolution of reef communities. Wiley, New York

Fagerstrom JA (1996) Paleozoic brachiopod symbioses: testing the limits of modern analogues in paleoecology. Bull Geol Soc Amer 108:1393–1403

Farmer MA, Fitt RK, Trench RK (2001) Morphology of the symbiosis between *Corculum cardissa* (Mollusca: Bivalvia) and *Symbiodinium corculorum* (Dinophyceae). Biol Bull 200:336–343

Fautin DG, Buddemeier RW (2004) Adaptive bleaching: a general phenomenon. Hydrobiol 530/531:459–467

Fay S, Weber M, Lipps JH (2009) The distribution of *Symbiodinium* diversity within individual host foraminifera. Coral Reefs 28:717–726

Freudenthal HD (1962) *Symbiodinium* gen. nov. and *Symbiodinium microadriaticum* sp. nov., a zooxanthella: taxonomy, life cycle, and morphology. J Protozool 9:45–52

Furla P, Allemand D, Shick JM, Ferrier-Pages C, Richier S (2005) The symbiotic anthozoan: a physiological chimera between alga and animal. Integr Compar Biol 45:595–604

Gao JG, Copper P (1997) Growth rates of middle Paleozoic corals and sponges: Early Silurian of Eastern Canada. Proc 8th Int Coral Reef Symp 2:1651–1656

Gattuso J-P, Allemand D, Frankignoulle M (1999) Photosynthesis and calcification at cellular, organismal and community levels in coral reefs: a review on interactions and control by carbonate chemistry. Amer Zool 39:160–183

Gauri KL, Chowdhury AN, Kulsreshtha NP, Punuru AR (1990) Geologic features and durability of limestones at the sphinx. Environ Geol Water Sci 16:57–62

Glynn P (1996) Coral reef bleaching: facts, hypotheses and implications. Glob Change Biol 2:495–509

Goreau TF, Goreau NI (1959) The physiology of skeleton formation in corals. II. Calcium deposition by hermatypic corals under various conditions in the reef. Biol Bull 117:127–167

Goreau TJ, Goreau NI, Trench RK, Hayes RL (1996) Calcification rates in corals. Science 274:117

Grant RE (1972) The lophophore and feeding mechanism of the Productidina (Brachiopoda). J Paleontol 46:213–249

Groves JR, Pike M, Westley K (2012) A test for the possibility of photosymbiosis in extinct fusuline foraminifera: size and shape related to depth of habitat. Palaios 27:739–752

Håkansson E, Madsen L (1991) Symbiosis - a plausible explanation of gigantism in Permian trepostome bryozoans. In: Bigey FP, d'Hondt J-L (eds), Bryozoaires actuales et fossiles. Nantes: Société des Sciences Naturalles de l'Ouest de la France. Memoire hors series 1:151–159

Hallock P (1999) Symbiont-bearing foraminifera. In: Sen Gupta BK (ed), Modern foraminifera. Kluwer Academic Pub, Dordrecht, pp 123–139

Hallock P (2001) Coral reefs, carbonate sediments, nutrients, and global change. In: Stanley GD Jr (ed) The history and sedimentology of ancient reef systems. Kluwer Academic/Plenum Pub, New York, pp 387–427

Hallock P, Schlager W (1986) Nutrient excess and the demise of coral reefs and carbonate platforms. Palaios 1:389–398

Hallock P, Williams DE, Fisher EM, Toler SK (2006) Bleaching in foraminifera with algal symbionts: implications for reef monitoring and risk assessment. Anuario Instit Geociênc 29:108–128

Hansen HJ, Buchardt B (1977) Depth distribution of *Amphistegina* in the Gulf of Elat, Israel. Utrecht Micropaleontol Bull 15:205–239

Hansen, HJ, Dalberg P (1979) Symbiotic algae in milioline foraminifera: CO_2 uptake and shell adaptations. Bull Geol Soc Denmark 28:47–55

Hill M, Allenby A, Ramsby B, Schönberg C, Hill A (2011) *Symbiodinium* diversity among host clionaid sponges from Caribbean and Pacific reefs: evidence of heteroplasmy and putative host-specific symbiont lineages. Mol Phylogen Evol 59:81–8

Hohenegger J (2006) Morphocoenoclines, character combination, and environmental gradients: a case study using symbiont-bearing benthic foraminifera. Paleobiol 32:70–99

Hohenegger J, Yordanova E, Nakano Y, Tatzreiter F (1999) Habitats of larger foraminifera on the upper reef slope of Sesoko Island, Okinawa, Japan. Mar Micropaleontol 36:109–168

Hohenegger J, Yordanova E, Hatta A (2000) Remarks on West Pacific Nummulitidae (Foraminifera). J Foram Res 30:3–28

James NP (1983) Reef environment. In: Scholle PA, Bebout DG, Moore CH (eds) Carbonate depositional environments. Amer Assoc Petrol Geol Mem 33, Tulsa, pp 345–440

Jones DS, Jacobs DK (1992) Photosymbiosis in *Clinocardium nuttalli*: implications for tests of photosymbiosis in fossil molluscs. Palaios 7:86–95

Kahng SE, Hochberg EJ, Apprill A, Wagner D, Luck DG, Perez D, Bidigare, RR (2012) Efficient light harvesting in deep-water zooxanthellate corals. Mar Prog Ser 455:65–77

Kauffman EG, Johnson CC (1988) The morphological and ecological evolution of Middle and Upper Cretaceous reef-building rudistids. Palaios 3:194–216

Kazmierczak J (1976) Cyanophycean nature of stromatoporoids. Nature 264:49–51

Kershaw, S (1998) The applications of stromatoporoid palaeobiology in palaeoenvironment analysis. Palaeontol 41:509–544

Kershaw, S, Brunton FR (1999) Palaeozoic stromatoporoid taphonomy: ecologic and environmental significance. Palaeogeogr Palaeoclim Palaeoecol 149:313–328

Key MM Jr, Jackson PMW, Håkansson E, Patterson WP, Moore DM (2005) Gigantism in Permian trepostomes from Greenland: testing the algal symbiosis hypothesis using δ13 C and δ18 O values. In: Moyano G, Cancino JM, Jackson PMW (eds) Bryozoan studies 2004. Balkema Publ; Leiden, pp 141–151

Kiessling W (2010) Reef expansion during the Triassic: spread of photosymbiosis balancing climatic cooling. Palaeogeogr Palaeoclim Palaeoecol 290:11–19

Kiessling W, Aberhan M, Villier L (2008) Phanerozoic trends in skeletal mineralogy driven by mass extinctions. Nature Geosci 1:527–530

Kiessling W, Kocsis ÁT (2015) Biodiversity dynamics and environmental occupancy of fossil azooxanthellate and zooxanthellate scleractinian corals. Paleobiol 41:402–414

Kiessling W, Simpson C (2011) On the potential for ocean acidification to be a general cause of ancient reef crises. Global Change Biol 17:56–67

Kleypas JA, Buddemeier RW, Archer D, Gattuso JP, Langdon C, Opdyke BN (1999) Geochemical consequences of increased atmospheric carbon dioxide on coral reefs. Science 284:118–120

Knoll AH (2014) Paleobiological perspectives on early eukaryotic evolution. Cold Spring Harb Perspect Biol 6: a016121

Knowlton N, Jackson JBC (2011) Evolutionary diversity and ecological complexity of coral reefs. Paleontol Soc Pap 17:111–120

Knowlton N, Rohwer F (2003) Multispecies microbial mutualisms on coral reefs: the host as a habitat. Amer Nat 162 (4 Suppl):S51–62

LaJeunesse TC (2002) Diversity and community structure of symbiotic dinoflagellates from Caribbean coral reefs. Mar Biol 141:387–400

LaJeunesse TC, Pettay DT, Sampayo EM, Phongsuwan N, Brown B, Obura DO, Hoegh-Guldberg O, Fitt MK (2010) Long-standing environmental conditions, geographic isolation and host–symbiont specificity influence the relative ecological dominance and genetic diversification of coral endosymbionts in the genus *Symbiodinium*. J Biogeogr 37:785–800

Landing E, Antcliffe JB, Brasier MD, English AB (2015) Distinguishing earth's oldest known bryozoan (*Pywackia*, Late Cambrian) from pennatulacean octocorals (Mesozoic–Recent). J Paleontol 89:292–317

Langer MR (2008) Assessing the contribution of foraminiferan protists to global ocean carbonate production. J Eukary Microbiol 55:163–169

Langer MR, Lipps JH (2003) Foraminiferal distribution and diversity, Madang reef and lagoon, Papua New Guinea. Coral Reefs 22:143–154

Langer MR, Silk MT, Lipps JH (1997) Global ocean carbonate and carbon dioxide production: the role of reef foraminifera. J Foram Res 27:271–277
Lee JJ (2006) Algal symbiosis in larger foraminifera. Symbiosis 42:63–75
Lee JJ (2011) Fueled by symbiosis, Foraminifera have evolved to be giant complex protists. In: Seckbach J, Dubinski Z (eds) All flesh is grass: plant-animal interrelationships. Springer, New York, pp 427–452
Lee JJ, Anderson OR (1991) Symbiosis in foraminifera. In: Lee JJ, Anderson OR (eds) Biology of Foraminifera. Academic Press.
Leinfelder RR (2001) Jurassic reef ecosystems. In: Stanley GD Jr (ed) The history and sedimentology of ancient reef systems. Kluwer Academic/Plenum Pub, New York, pp 251–309
Lipps JH (1981) What, if anything, is micropaleontology? Paleobiol 7:167–199
Lipps JH (2006) Major features of protistan evolution: controversies, problems and a few answers. Anuario Inst Geociênc 29:55–80
Lipps JH (2011) Reef restoration—the good and the bad, a paleobiologic perspective. Paleontol Soc Papers 17:139–152
Lipps JH, Severin K (1986) *Alveolinella quoyi*, a living fusiform foraminifera, at Motupore Island, Papua New Guinea. Sci New Guinea 11:126–137
Littman RA, van Oppen MJH, Willis BL (2008) Methods for sampling free-living *Symbiodinium* (zooxanthellae) and their distribution and abundance at Lizard Island (Great Barrier Reef). J Exp Mar Biol Ecol 364:48–53
Lobban CS, Schefter M, Donaldson TJ (2014) Cluster dynamics in *Maristentor dinoferus,* a gregarious benthic ciliate with zooxanthellae and a hypericin-like pigment, in relation to biofilm grazing by the fish *Ctenochaetus striatus*. Symbiosis 63:137–147
Lopez R (1979) Algal chloroplasts in the protoplasm of three species of benthic Foraminifera: taxonomic affinity, viability and persistence. Mar Biol 53:201–211
Marcelino LA, Westneat MW, Stoyneva V, Henss J, Rogers JD, Radosevich A, Turzhitsky V, Siple M, Fang A, Swain TD, Fung J, Backman V (2013) Modulation of light-enhancement to symbiotic algae by light-scattering in corals and evolutionary trends in bleaching. www.PLoS One doi: 10.1371/journal.pone.0061492
Margulis L (1998) Symbiotic planet: a new view of evolution. Basic Books, Amherst, MA
Marshall AT (1996) Calcification in hermatypic and ahermatypic corals. Science 271:637–639
McConnaughey TA (2012) Zooxanthellae that open calcium channels: implications for reef corals. Mar Ecol Prog Ser 460:277–287
Moldowan JM, Dahl J, Jacobson SR, Huizinga BJ, Fago FJ, Shetty R, Watt DS, Peters KE (1996) Chemostratigraphic reconstruction of biofacies: molecular evidence linking cyst-forming dinoflagellates with pre-Triassic ancestors. Geol 24:159–162
Muscatine L, Goiran C, Land L, Jaubert J, Cuif J-P, Allemand D (2005) Stable isotopes ($-^{13}$C and $-^{15}$N) of organic matrix from coral skeleton. Proc Nat Acad Sci 102:1525–1530
Nestor H, Copper P, Stock CW (2010) Late Ordovician and Early Silurian stromatoporoid sponges from Anticosti Island, eastern Canada: crossing the O/S mass extinction boundary. Nat Res Council, Res Press, Ottawa
Pandolfi, JM, Jackson JBC, Baron N, Bradbury RH, Guzman HM, Hughes TP, Kappel CV, Micheli F, Ogden JC, Possingham HP, Sala E (2005) Are US coral reefs on the slippery slope to slime? Science 307:1725–1726
Payne JL, Groves JR, Jost AB, Nguyen T, Moffitt SE, Hill TM, Skotheim JM (2012) Late Paleozoic fusulinoidean gigantism driven by atmospheric hyperoxia. Evol 66:2929–2939
Pelejero C, Calvo E, Mcculloch MT, Marshall JF, Gagan MK, Lough JM, Opdyke BN (2007) Preindustrial to modern interdecadal variability in coral reef pH. Science 309:2204–2207
Perrin C (2002) Tertiary: the emergence of modern reef ecosystems. SEPM Spec Publ 71:587–621
Piller WE (1981) The Steinplatte reef complex, part of an Upper Triassic carbonate platform near Salzburg, Austria. SEPM Spec Publ 30:261–290
Pochon X, Montoya-Burgos JI, Stadelmann, B, Pawlowski JP (2006) Molecular phylogeny, evolutionary rates, and divergence timing of the symbiotic dinoflagellate genus *Symbiodinium*. Mol Phylogen Evol 38:20–30
Pochon X, Stat M, Takabayashi M, Chasqui L, Chauka LJ, Logan DDK, Gates RD (2010) Comparison of endosymbiotic and free-living *Symbiodinium* (Dinophyceae) diversity in a Hawaiian reef environment. J Phycol 46:53–65
Porter SM (2004) The fossil record of early eukaryotic diversification. In: Lipps JH, Waggoner BM (eds) Neoproterozoic—Cambrian biological revolutions. Paleontol Soc Papers 10:35–50.
Pratt BR, Spincer BR, Wood RA, Zhuravlev AYu (2001) Ecology and evolution of Cambrian reefs. In: Zhuravlev AYu, Riding R (eds) The ecology of the Cambrian radiation. Columbia Univ Press, New York, pp 254–274
Reaka-Kudla ML (1997) The global diversity of coral reefs: a comparison with rain forests. In: Reaka-Kudla ML, Wilson DE, Wilson EO (eds) Biodiversity II: Understanding and protecting our biological resources. Joseph Henry Press, Washington, DC, pp 83–108
Reid RP, Macintyre IG, Browne KM, Steneck, RS, Miller T (1995) Modern marine stromatolites in the Exuma Cays, Bahamas: uncommonly common. Facies 33:1–17
Riding R (1992) Temporal variation in calcification in marine cyanobacteria. J Geol Soc London 149:979–989
Riedel P (1991) Korallen in der Trias de Tethys: Stratigraphische Reichweiten, Diversitätsmuster, Entwicklungstrends und Bedeutung als Rifforganismen. Mitteil Gesellsch Geol Bergbaustud Österreich 37:97–118
Ries JB, Cohen AL McCorkle DC (2009) Marine calcifiers exhibit mixed responses to CO_2-induced ocean acidification. Geol 12:1131–1134
Rosen BR (1978) The nature and significance of microatolls, appendix: determination of a collection of coral microatoll specimens from the northern Great Barrier Reef. Proc Trans Royal Soc London B 284:115–122
Rosen BR (2000) Algal symbiosis, and the collapse and recovery of reef communities: Lazarus corals across the K-T boundary. In: Culver SJ, Rawson PA (eds) Biotic response to global change: the last 145 million years. Cambridge Univ Press, Cambridge, pp 164–180
Rosen BR, Turnšek D (1989) Extinction patterns and biogeography of scleractinian corals across the Cretaceous/Tertiary boundary. In: Jell PA, Pickett JW (eds) Fossil Cnidaria 5. Mem Assoc Australas Paleontol 8:355–370
Rowan R. (2004) Coral bleaching—thermal adaptation in reef coral symbionts. Nature 430:742–742
Rowan R, Powers DA (1991) A molecular genetic classification of zooxanthellae and the evolution of animal-algal symbioses. Science 251:1348–1351
Rowland SM (2001) Archaeocyaths—a history of phylogenetic interpretation. J Paleontol 75:1065–107
Rowland SM, Gangloff RA (1988) Structure and paleoecology of Lower Cambrian reefs. Palaios 3:111–135
Rowland SM, Hicks M (2004) The Early Cambrian experiment in reef-building by metazoans. In: Lipps JH, Waggoner BM (eds) Neoproterozoic—Cambrian biological revolutions. Paleontol Soc Papers 10:107–130
Rowland SM, Shapiro SH (2002) Reef patterns and environmental influences in the Cambrian and earliest Ordovician. SEPM Spec Publ 72:95–128

Sandberg PA (1983) An oscillating trend in Phanerozoic nonskeletal carbonate mineralogy. Nature 305:19–22
Scoffin TP, Stoddart DR (1978) The nature and significance of microatolls. Proc Trans Royal Soc London B 284:99–122
Senowbari-Daryan B, Stanley GD Jr (2009) Taxonomic affinities and paleogeography of *Stromatomorpha californica* Smith, a distinctive Upper Triassic reef-adapted demosponge. J Paleontol 83:783–793
Severin KP, Lipps JH (1989) The weight-volume relationship of the test of *Alveolinella quoyi:* implications for the taphonomy of large fusiform foraminifera. Lethaia 22:1–12
Siano AR, Montresorb M, Proberta I, Nota F, DeVargas C (2010) *Pelagodinium* gen. nov. and *P. beii* comb. nov., a dinoflagellate symbiont of planktonic foraminifera. Protist 161:385–399
Smith EG, D'Angelo C, Salih A, Wiedenmann J (2013) Screening by coral green fluorescent protein (GFP)-like chromoproteins supports a role in photoprotection of zooxanthellae. Coral Reefs 32:463–474
Smithers SG, Woodroffe CD (2000) Microatolls as sea-level indicators on a mid-ocean atoll. Mar Geol 168:61–78
Song, Y, Black RG, Lipps JH (1994) Morphologic optimization in the largest living foraminifera: implications from finite element analysis. Paleobiol 20:14–26
Stanley GD Jr (1981) Early history of scleractinian corals and its geological consequences. Geol 9:507–511
Stanley GD Jr (2005) Coral microatolls from the Triassic of Nevada: oldest scleractinian examples. Coral Reefs 24:247
Stanley GD Jr (2006) Photosymbiosis and the evolution of modern coral reefs. Science 312:857–860
Stanley GD, Cairns SD (1988) Constructional azooxanthellate coral communities: an overview with implications for the fossil records. Palaios 5:233–242
Stanley GD Jr, Helmle KP (2010) Middle Triassic coral growth bands and their implication for photosymbiosis. Palaios 25:754–763
Stanley GD Jr, Lipps JH (2011) Photosymbiosis: the driving force for reef success and failure. Paleontol Soc Papers 17:33–60
Stanley GD Jr, Swart PK (1995) Evolution of the coral-zooxanthellae symbiosis during the Triassic: a geochemical approach. Paleobiol 21:179–199
Stanley GD Jr, van de Schootbrugge B (2009) The evolution of the coral-algal symbiosis. In: van Oppen MJH, Lough JM (eds) Coral bleaching: patterns, processes, causes and consequences. Ecol Stud Series 205:7–19
Stanley SM, Hardie, LA (1999) Hypercalcification: paleontology links plate tectonics and geochemistry to sedimentology: GSA Today 9:1–7
Stat M, Carter D, Hoegh-Guldberg, O (2006) The evolutionary history of *Symbiodinium* and scleractinian hosts—symbiosis, diversity, and the effect of climate change. Perspect Plant Ecol Evol Syst 8:23–43
Steindler L, Beer S, Ilan M (2002) Photosymbiosis in intertidal and subtidal tropical sponges. Symbiosis 33:263–273
Steuber T (1996) Stable isotope sclerochronology of rudist bivalves: growth rates and Late Cretaceous seasonality. Geol 24:315–318
Steuber T (2000) Skeletal growth rates of Upper Cretaceous rudist bivalves: implications for carbonate production and organism-environment feedbacks. Geol Soc London Spec Publ 178:21–32
Stimson J, Sakai K, Sembali H (2002) Interspecific comparison of the symbiotic relationship in corals with high and low rates of bleaching induces mortality. Coral Reefs 21:409–421
Surge DM, Savarese M, Dodd JR, Lohmann KC (1997) Carbon isotopic evidence for photosymbiosis in Early Cambrian oceans. Geol 25:503–506
Sweet MJ (2014) *Symbiodinium* diversity within *Acropora muricata* and the surrounding environment. Mar Ecol 35:343–353
Takabayashi M, Adams LM, Pochon X, Gates RD (2012) Genetic diversity of free-living *Symbiodinium* in surface water and sediment of Hawai'i and Florida. Coral Reefs 31:157–167
Talent JA (1988) Organic reef-building: episodes of extinction and symbiosis? Senckenberg Lethaea 69:315–368
Taylor MW, Hill RT, Piel J, Thacker RW, Hentschel U. (2007). Soaking it up: the complex lives of marine sponges and their microbial associates. The ISME J 1:187–190
Taylor PD, Waeschenbach A (2015) Phylogeny and diversification of bryozoans. Palaeontol 58:585–599
Terán E, Méndez ER, Enríquez S, Iglesias-Prieto R (2010) Multiple light scattering and absorption in reef-building corals. Appl Optics 49:5032–5042
Thornhill DJ, Lewis AM, Wham DC, LaJeunesse TC (2013) Host-specialist lineages dominate the adaptive radiation of reef coral endosymbionts. Evol 68:352–367
Turner EC, Narbonne GM, James NP (1993) Neoproterozoic reef microstructure from the Little Dal Group, northwestern Canada. Geol 21:259–262
Vachard D, Pille L, Gaillot J (2010) Palaeozoic foraminifera: systematics, palaeoecology and responses to global changes. Rev Micropaléontol 53:209–254
van Oppen MJ, Baker AC, Coffroth MA, Willis BL (2009) Bleaching resistance and the role of algal endosymbionts. Ecol Stud 205:83–102
van Oppen MJH, Lough JM (2009) Coral bleaching: patterns, processes, causes, and consequences. Springer-Verlag Berlin, Heidelberg
Vermeij GJ (2013) The evolution of molluscan photosymbioses: a critical appraisal. Biol J Linnean Soc 109:497–511
Vogel K (1975) Endosymbiotic algae in rudists. Palaeogeogr Palaeoclim Palaeoecol 17:327–332
Walter MR (1983) Archean stromatolites: evidence of the Earth's earliest benthos. In: Schopf JW (ed) Earth's earliest biosphere. Princeton University Press, Princeton, pp 187–213
Watson ME, Signor PW (1986) How a clam builds windows: shell microstructure in *Corculum* (Bivalvia: Cardiidae). Veliger 28:348–355
Wood R (1993) Nutrients, predation and the history of reef-building. Palaios 8:526–543
Wood R (1999) Reef evolution. Oxford University Press, Oxford
Wood R, Curtis A (2014) Extensive metazoan reefs from the Ediacaran Nama Group, Namibia: the rise of benthic suspension feeding. Geobiol DOI: 10.1111/gbi.12122
Wooldridge SA (2014) Differential thermal bleaching susceptibilities amongst coral taxa: re-posing the role of the host. Coral Reefs 33:15–27
Yancey TE, Stanley, GD Jr (1999) Giant alatoform bivalves in the Upper Triassic of western North America. Palaeontol 42:1–23
Yoon HS, Hackett JD, Ciniglia C, Pinto G, Bhattacharya, D (2004) A molecular timeline for the origin of photosynthetic eukaryotes. Mol Biol Evol 21:809–818
Zapalski MK (2014) Evidence of photosymbiosis in Palaeozoic tabulate corals. Proc R Soc B 281: 20132663. http://dx.doi.org/10.1098/rspb.2013.2663
Zhang Y, Payne JL (2012) Size-frequency distributions along a latitudinal gradient in Middle Permian fusulinoideans. PLoS One 7(6): e38603. doi:10.1371/journal.pone.0038603
Zhuravlev AY (2001) Paleoecology of Cambrian reef ecosystems. In: Stanley GD Jr (ed) The history and sedimentology of ancient reef systems. Kluwer Academic/Plenum Publishers, New York, pp 121–157

4 Bioerosion on Modern Reefs: Impacts and Responses Under Changing Ecological and Environmental Conditions

Chris T. Perry and Alastair R. Harborne

Abstract

Biological erosion (bioerosion) is a key ecological process on coral reefs. It occurs though the grazing activities of specific fish and sea urchin species, and as a result of the colonisation of reef substrates by endolithic species of sponges, bivalves, worms and microorganisms. This activity results either in the direct dissolution of reef (mainly coral) substrate and/or the conversion of this substrate to sediment. As a result, bioerosion plays a key role in defining the structure of the accumulating reef framework, is a key process dictating the balance between rates of carbonate production and erosion, and influences reef-carbonate budget states. This chapter initially explores the key biological agents responsible for reef bioerosion within Holocene reef systems, and the influence that these organisms exert on patterns and styles of reef development. However, in the context of the aims of this book, the most pertinent question is how are reef-bioeroding taxa responding to environmental and ecological change, and how are they interacting with reef substrates under changing conditions. We discuss the current state of knowledge regarding variations in bioerosion rates and the ways in which different bioeroding taxa use space within degrading reef systems. Although much is known about the key taxa that drive reef bioerosion, data on actual bioerosion rates are limited to a few well-cited studies, and information on how these rates vary across spatial and temporal scales is even more limited. Habitat-specific bioerosion budgets for most taxa are also rare. Addressing these knowledge gaps will be critical to predicting future changes in bioeroder abundance and their impacts on changing reef environments.

Keywords

Corals • Bioerosion • Grazing • Parrotfish • Sea urchins • Endolith • Carbonate budgets

4.1 Introduction

The process of *bioerosion* is defined as the corrosion of hard substrates by living agents (terminology of Neumann 1966). In reef environments a wide range of species actively contribute to such biological erosion and they do so either as a function of their feeding strategy (through substrate grazing), or in the process of establishing a protective dwelling space within the reef substrate. Either way the effect of these processes is rather similar: it results in the direct degradation and alteration of both the living structure of

C.T. Perry (✉)
Department of Geography, College of Life and Environmental Sciences, University of Exeter, Exeter, UK
e-mail: c.perry@exeter.ac.uk

A.R. Harborne
Department of Biological Sciences, Florida International University, 3000 Northeast 151 Street, North Miami, FL 33181, USA

Marine Spatial Ecology Laboratory, School of Biological Sciences and ARC Centre of Excellence for Coral Reef Studies, University of Queensland, Goddard Building, Brisbane, QLD 4072, Australia
e-mail: a.harborne@uq.edu.au

D.K. Hubbard et al. (eds.), *Coral Reefs at the Crossroads*,
Coral Reefs of the World 6, DOI 10.1007/978-94-017-7567-0_4

the reef and modification and destruction of the deposited skeletal framework. Through these processes both primary (coral) and secondary (e.g., crustose coralline algal) reef-framework building components are either undermined and weakened, or they are broken down into sediment. Thus this biologically driven process exerts a profound influence on both small spatial-scale and short temporal-scale patterns of reef complexity and reef ecology. Over longer timescales bioerosion also influences important geomorphic facets of reef development, including the compositional structure of a reef (e.g. the internal fabric) and net long-term rates of reef accretion. Where environmental conditions either suppress rates of carbonate production or lead to coral mortality, increased rates of bioerosion relative to production can cause rapid and widespread reef-structural degradation.

These biologically- driven erosional processes are associated with the grazing activities of a range of both fish and echinoid species, as well as the activities of various endolithic (internal substrate dwelling) borers, including species of sponges, bivalves, worms and microorganisms such as cyanobacteria and fungi. In this chapter we outline the key biological agents responsible for bioerosion within Holocene reef systems, we consider their combined influence on reef development, and discuss their spatial distribution within reef environments. A major focus of the chapter, however, is on the role of bioerosion in the changing contemporary reef environment; specifically we ask two key questions: (i) How are bioeroder communities responding to on-going changes in marine environmental and ecological conditions?, and (ii) What are the likely implications of such changes for the balance between reef carbonate production and destruction within near-future reef environments?

4.2 The Reef Bioerosion Process: Key Species and Mechanisms of Bioerosion

The bioerosion of coral reefs is recognized as a major factor influencing reef accretion and development (Hallock and Schlager 1986; Hutchings 1986; Glynn 1997), with the process being driven by two groups of organisms that attack and erode the substrate in different ways; either through the process of boring (a mechanical or chemical process), or through surface grazing. Although the mechanisms of bioerosion differ among species, each contributes to the modification and degradation of reef carbonate material. This may occur either through direct breakdown of reef framework carbonate, by increasing the susceptibility of reef framework to physical damage (e.g. Goreau and Hartman 1963; Scott 1988; Sammarco and Risk 1990), or by directly converting primary skeletal (framework) carbonate to sediment. These sediments may then be either retained in the local reef structure or exported to adjacent onshore environments (lagoons, reef islands) or into deeper water. Bioerosion can thus lead to a reduction in topographic relief with consequent impacts for associated biota (Reaka-Kudla et al. 1996). In addition, bioerosion may affect a reef's structural integrity and increase the potential for wave over-topping as the protective reef crest or rim is eroded away (Sheppard et al. 2005).

Bioeroders can favour either live or dead coral substrate (Scoffin and Bradshaw 2000) and there is some evidence for variable substrate susceptibility, especially to the process of boring (Pang 1973; Perry 1998b). Although poorly quantified this variable susceptibility may result in more rapid degradation of specific coral species or morphologies (Goreau and Hartman 1963; Hubbard et al. 1986) with consequent implications for net reef accretion rates, and for the selective loss or removal of individual coral species from the fossil record. In this first section we discuss the different groups responsible for endolithic boring and for grazing-related bioerosion in reef environments, and outline the key mechanisms they employ in substrate erosion.

4.3 Endolithic Bioerosion

A wide variety of euendolithic organisms (defined as organisms that actively bore into hard substrates: Golubic et al. 1981; Bromley 1994), including species of sponges, bivalves, worms, cyanobacteria, chlorophytes, rhodophytes and fungi, and epilithic organisms (those that grow partly on the surface of hard substrates, such as some sponges) are involved in the process of boring in coral reef environments, a process whose importance to reef development has been recognised since the nineteenth century (Kolliker 1859; Duncan 1876; Duerden 1902; Gardiner 1903; Otter 1937). The borings produced by these endoliths are often well preserved in the fossil record, either as three-dimensional casts or as empty chambers and, because they are often preserved *in situ* relative to their substrate, they have significant potential as palaeo-indicators (Bromley 1978; Perry 1996). The key groups of borers that operate in reef environments include specific sponges, polychaete and sipunculid worms, bivalves, foraminifera, decapods and cirripeds (*macroendolithic borers*), as well as species of cyanobacteria, chlorophytes, rhodophytes and fungi (*microendolithic borers*) (Hutchings 1986). Macroborers are defined as those eroders which produce boreholes with diameters >1 mm and are responsible for two processes of destruction; chemical dissolution of the substrate and mechanical abrasion of the substrate. Microendolithic borers occur widely within carbonate substrates in the marine environment and produce boreholes that range in diameter from ~1 to 100 μm. Bioerosion by these species occurs through a chemical etching process (Golubic et al. 1975). Below we describe the mechanisms and activities of the key reef substrate boring groups.

4.3.1 Sponges

Of all the macroboring groups, sponges have received the greatest attention because, on a reef-wide basis, they are typically the dominant infaunal bioeroders, comprising some 75–90 % of the total macroboring community (in terms of the proportion of substrate infestation, e.g. Goreau and Hartman 1963; MacGeachy and Stearn 1976; Highsmith 1981; Highsmith et al. 1983; Perry 1998b). A wide range of sponge species are believed to be active substrate bioeroders, although the taxonomy of this important group of endoliths is complex and the subject of on-going revision. The following Orders have been reported to include bioeroding species: the Hadromerida (Clionaidae, Spirastrellidae and Alectonidae), the Poecilosclerida (Arcarnidae), the Halichondrida (Halichondriidae) and the Haploschlerida (Phloeodictyidae) (Hutchings 2011). The importance of sponges in reef bioerosion is linked both to their widespread occurrence in reef environments (many are conspicuous on reef substrates; Fig. 4.1a–d) and to the size

Fig. 4.1 Endolithic sponges and their interactions with reef rock. (**a**) Dead brain coral almost completely infested by *C. caribbaea*; (**b**) *Cliona delitrix* invading a colony of *Montastraea cavernosa*; (**c**) Fractured section of dead–in-situ coral framework colonised by *Cliona tenuis* and showing the boring activity extending ~1 cm into the substrate; (**d**) Fractured section of dead–in-situ coral framework colonised by *Siphonodictyon* (*yellow siphons* on substrate surface) and the large (3–4 cm diameter) chambers this sponge borer produces; (**e**) Fractured branched colony showing *orange tissue* of clionid sponge inside the dead coral skeleton; (**f**) Close-up image under scanning electron microscopy showing details of interconnected chambers produced by clionid bioerosion

and extent of the boreholes they excavate. Most bioeroding sponge species produce dense networks of inter-connected chambers that range in size from ~1 to ~10 mm diameter; Fig. 4.1c, e, f), although a few species produce very large (up to 5–6 cm diameter) single chambers (Fig. 4.1d). Different sponge species produce different morphology chambers, but the morphologies of these chambers can also vary between substrate types (probably as a function of skeletal density and corallite form). Despite these limitations, boring chamber morphology has been used to establish an extensive ichnotaxonomy (a taxonomy based on trace fossil size and shape) to describe bioeroding sponges in fossil reefs (see Pleydell and Jones 1988; Perry 1996). Networks of sponge boring chambers can extend to depths of several centimetres below the coral surface, causing the removal of large amounts of framework carbonate. In addition, preferential colonisation of dead basal and cryptic surfaces on corals can lead to significant weakening and undermining, and increase the likelihood of breakage during physical disturbance events.

The boring mechanism employed by sponges involves both a chemical etching process and the mechanical removal of coral skeleton. Specialized etching cells along the advancing sponge tissue front isolate discrete sections of skeleton that are then removed mechanically. These leave behind distinctive scallop-shaped excavations but also results in the production of large volumes of 'sponge chips' (small coral fragments). These chips contribute to the mud-grade fraction of the surrounding sediments after being expelled through the sponge oscules. Studies suggest that only 2–3 % of the coral substrate is actually dissolved during sponge bioerosion, with the remainder being expelled as sediment (Fütterer 1974). Thus the process is important both in terms of coral framework erosion and production of fine-grained carbonate sediments.

Sponge larvae colonise reef substrates from the water column and are capable of colonising both live and dead substrate, although colonization of dead and more cryptic substrates appears more typical. There is little information on whether the type of substrate cover (for example turf as opposed to macroalgal cover) influences sponge settlement, but sponges are not usually early colonising endoliths on dead substrates. This role is usually filled by polychaete worms (Kiene and Hutchings 1994). Once established, however, sponge networks seem to develop rapidly. Rates of boring are poorly known, but estimates from experimental coral substrate deployed at sites in the Indo-Pacific are as follows: Great Barrier Reef: 0.01–0.3 kg $CaCO_3/m^2$-year (Osorno et al. 2005), French Polynesia: 0.1–0.9 kg $CaCO_3/m^2$ -year (Pari et al. 1998), and in the Caribbean rates were 0.05–0.06 kg $CaCO_3/m^2$-year (Bak 1976). How accurately these rates, as derived from experimental substrates, mimic natural sponge bioerosion rates is debatable, but they provide an indication of potential rates of substrate erosion. Rates of sponge bioerosion are probably at their highest in the period immediately following colonisation as the sponge tissue networks advance rapidly through the newly available substrate, but may then decrease. One important implication of this pattern is that whilst newly available dead coral substrate (whatever the cause of mortality) may experience rapid rates of bioerosion over the first few years following colonisation, these rates may not be sustained and may eventually stabilise at a lower rate. It is also clear that bioerosion rate, and the occurrence of different bioeroding species, varies markedly among environments, with factors such as substrate type (especially density), depth, temperature, and nutrient levels also having an influence.

4.3.2 Molluscs

Molluscs (especially some bivalves) are an important agent of coral bioerosion and mainly bore in the basal regions of coral colonies, weakening the structure (MacGeachy and Stearn 1976; Moretzsohn and Tsuchyia 1993). Coral bioeroding species of bivalves belong to six different families: Mytilidae and Lithophaginae, some species of both families also bore into live corals, and Petricolidae, Pholadidae, Clavagellidae and Gastrochaenidae, all of which bore only into dead coral substrate, (Highsmith 1980; Hutchings 2011). Bivalves produce characteristic vase or funnel-shaped boreholes, with a distinct surface aperture (Fig. 4.2a), and which have good potential to be delineated in reef facies (see Evans 1970) and Warme 1975). The trace fossils produced by this group are assigned to the ichnogenus *Gastrochaenolites* (Kelly and Bromley 1984). Other molluscan groups generally account for little coral bioerosion, although they may be locally important, with two species of gastropod, *Leptoconchus* and *Magilopsis*, recognised as live coral borers (Soliman 1969).

Molluscan borers settle from pelagic larval stages and use both mechanical and chemical mechanisms to erode the substrate (Moretzsohn and Tsuchyia 1993). In some cases the ornamented shells are evidence of a mechanical approach, but in most cases, and especially in the case of species boring into live coral, there is strong evidence for a predominantly chemical dissolution process (Kleeman 1990). Neither process produces significant carbonate sediment, although a calcium paste is produced and is used often as a borehole lining. In dead coral, bioerosion proceeds by continued downward penetration as the upper surface is eroded by other means. However, in live corals bivalves essentially backfill their boreholes (Fig. 4.2b), boring and moving upwards to keep pace with the growing coral surface and maintaining their borehole openings between the coral polyps (Soliman 1969; Scott 1988). Rates of boring are

Fig. 4.2 (**a**) Bivalve borehole apertures in reef rock, 4 m depth, Maldives. (**b**) Preserved back-fill structures produced by the boring bivalve *Lithophaga* spp. (trace fossil *Gastrochaenolites torpedo*; Kelly and Bromley 1984) from a Miocene reef deposit in Mallorca. Coin = 20 mm diam. (**c**) Cut section of brain coral showing large *Lithophaga* spp. boring *bottom right* (sample 9 cm wide)

poorly known for molluscs, but estimates from experimental coral substrates deployed at sites in the Indo-Pacific are reported as follows: Great Barrier Reef: 0.01–0.5 kg $CaCO_3/m^2$-year (Osorno et al. 2005) and 0.12–1.2 kg $CaCO_3/m^2$-year (Tribollet and Golubic 2005), the latter being figures for total macroboring but where bivalves were reported as the dominant macroborers; and French Polynesia: 0.01–0.05 kg $CaCO_3/m^2$-year (Pari et al. 1998).

4.3.3 Polychaete and Sipunculan Worms

Several genera of polychaete and sipunculan worms have boring properties and an endolithic life mode. In the polychaetes, species of the following families are known reef bioeroders: the Cirratulidae, Eunicidae, Sabellidae and Spionidae (Hutchings 2011). They occur within both live and dead coral substrates, although preferentially in the latter, and are often early colonizers (Kiene 1989; Hutchings and Peyrot-Clausade 2002). Polychaetes produce elongate, meandering tunnels, the traces of which are assigned to the ichnospecies *Trypanites* (Mägdefrau 1932) or *Maeandropolydora* (Voigt 1965) depending upon their borehole morphology. Species composition among the boring polychaetes appears to change following substrate colonisation, with early, short-lived species giving way to longer-lived species as substrates mature. Mechanisms of boring are not well understood in the polychaetes, but many species clearly use a chemical dissolution process in borehole development, possibly augmented in some cases by mechanical abrasion. Although most boring polychaetes are small and do not typically extract significant amounts of $CaCO_3$ compared with sponges and bivalves (Bak 1976; MacGeachy and Stearn 1976; Highsmith 1981), they may play a critical role in overall bioerosion rates by modifying the substrate and facilitating bioerosion by other agents (Hutchings et al. 1992). However, more significant roles in overall bioerosion rates were noted by Hein and Risk (1975), who reported polychaete erosion as being broadly comparable to that of sponges at shallow water sites in Florida, whilst Klein et al. (1991) reported polychaetes as being responsible for up to 47 % of infaunal bioerosion in fossil and modern corals in the Gulf of Eilat, Red Sea.

Sipunculan worms can also be locally important infaunal borers (Rice and Macintyre 1982) and species from a number of genera are involved: *Aspidosiphon*, *Lithacrosiphon*, *Paraspidosiphon* and *Phascolosoma* (Hutchings 2011). Sipunculans produce elongate, meandering tunnels through reef rock, with the traces of these borers being assigned to the ichnogenus *Trypanites* (Bromley 1994). Mechanisms of boring in the sipunculans are not well understood with both mechanical abrasion and chemical dissolution processes probably involved. Evidence suggests that sipunculans occur most commonly in recently dead coral substrates, and are rare in older well-eroded substrates (Rice and Macintyre 1982). Also locally important in some reef environments are the borings produced by cirripeds, such as the endolithic barnacles (Bromley 1978), the traces of which are assigned to the ichnogenus *Rogerella*.

Rates of substrate boring are not well-known for sipunculan and polychaete worms. Estimates for the two groups combined from experimental coral substrates deployed at sites in the Indo-Pacific are reported as follows: Great Barrier Reef: 0.1–0.3 kg $CaCO_3/m^2$-year (Osorno et al. 2005) and 0.01–0.13 kg $CaCO_3/m^2$-year (Tribollet and Golubic 2005), the latter being figures for total macroboring after 1 year of exposure, but where sipunculans and polychaetes were reported as the dominant bioeroding group; Reunion: 0.05–0.4 kg $CaCO_3/m^2$-year (Chazottes et al. 2002), which again are figures for total macroboring but with polychaetes stated as being the dominant bioeroder; and French Polynesia: 0.01–0.07 kg $CaCO_3/m^2$-year (Pari et al. 1998).

4.3.4 Microbioerosion

Carbonate substrate degradation by euendolithic microorganisms is associated with the activities of photosynthetic cyanobacteria, chlorophytes and rhodophytes, and hetrotrophic fungi and bacteria (Golubic et al. 1975). These microboring organisms produce boreholes that range in diameter from ~1 to 100 μm and occur widely within carbonate substrates in the marine environment (Golubic et al. 1975). Many species occur within relatively narrow ecological niches that are primarily controlled by bathymetric variations in light intensity, and this defines the vertical range over which individual photosynthetic microendoliths occur (Fig. 4.3). Hence they are considered to have good potential as palaeobathymetric and ecological indicators (Radtke and Golubic 2005). Heterotrophic fungi are not restricted in this way and occur to abyssal depths (Golubic et al. 1975). The fossil record clearly indicates that these organisms have played an active role in carbonate substrate degradation over long periods of geological time, having been identified in carbonates of Cambrian (Vogel 1993), Ordovician (Klement and Toomey 1967) and Devonian (Vogel et al. 1987) age.

Microborers penetrate carbonate substrate through biochemical dissolution (Ehrlich 1990) creating intricate networks of tunnels. The morphology of these networks is species-specific (Fig. 4.4) and forms the basis of their ichnotaxonomy (see Golubic et al. 1975; Radtke 1993), but also to an extent their taxonomic identification. Whilst

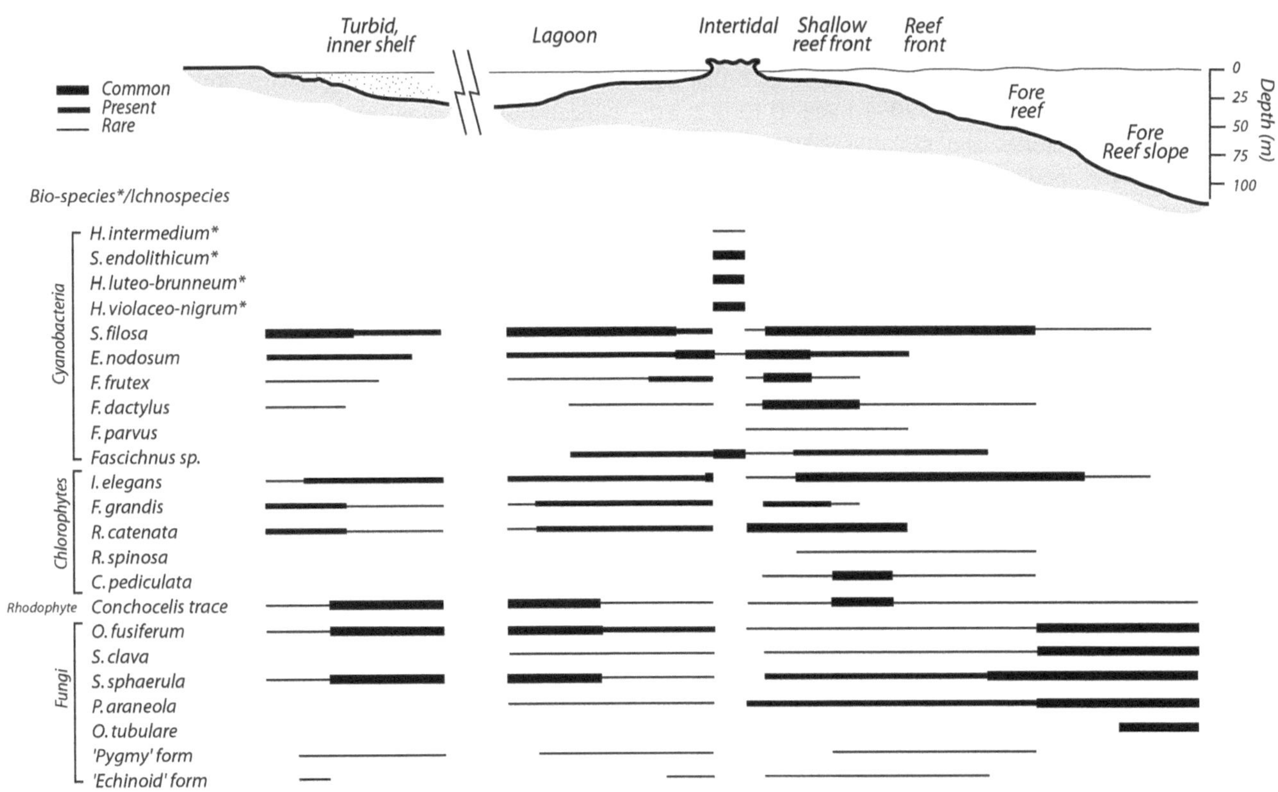

Fig. 4.3 Spatial and bathymetric distribution of microendoliths across a hypothetical reef profile (Reprinted from Perry and Hepburn (2008) Syn-depositional alteration of coral reef framework through bioerosion, encrustation and cementation: taphonomic signatures of reef accretion and reef depositional events. Earth Sci Rev 86:106–144, Copyright 2007, with permission from Elsevier)

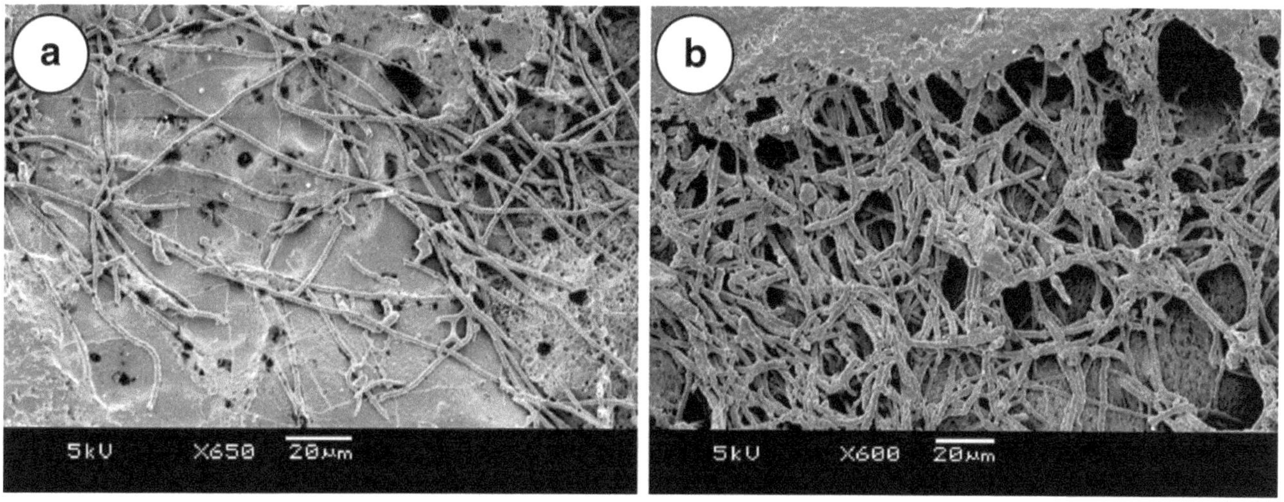

Fig. 4.4 Resin casts within etched coral fragments showing sparse (**a**) and dense (**b**) traces produced by the cyanobacterial endolith *Plectonema terebrans*

microborers have been widely recognised as important agents of reef sediment degradation and alteration, for example through micrite envelope formation (Bathurst 1966) and grain breakdown (Perry 1998a), studies suggest that the process is also quantitatively significant in terms of reef framework degradation (Tribollet and Golubic 2005). Rates of microbioerosion vary temporally and spatially within and between sites, especially in response to light level changes with depth. For example, Vogel et al. (2000), report erosion rates of only 0.001 kg $CaCO_3/m^2$-year at deep water (275 m) reef front sites in the Bahamas, but rates of 0.52 kg/m^2-year at shallow (2 m) leeward sites. Across a range of shallow water sites in Reunion erosion rates of 0.04–0.07 kg $CaCO_3/m^2$-year have been reported (Chazottes et al. 2002), and from sites across an onshore to offshore transect on the Great Barrier Reef Shelf (Australia) rates of between 0.13 and 1.35 kg $CaCO_3/m^2$-year (Tribollet and Golubic 2005). In Reunion highest rates occurred at sites with higher nutrient concentrations, but also varied with epilithic substrate cover, whilst in Australia rates increased with distance offshore. A potential caveat is that these rates are reported from experimental substrates and may reflect rapid initial colonisation and bioerosion, rather than long-term rates. For example, it has been shown that microbioerosion rates often decline in experimental substrates over time (Kobluk and Risk 1977; Tribollet and Golubic 2005), and thus caution is needed in extrapolating these findings to reef substrates in natural systems.

Colonisation by microborers can occur both in live and dead coral substrates, and with some phototrophic species colonising coral substrate very soon after skeletal deposition, growing with the coral skeleton to form a distinct green banding just below the living coral tissue surface (termed an '*Ostreobium* band' after the relevant species of microendolith). However, most dead coral substrates are also often rapidly colonized (seemingly regardless of the cause of mortality) with opportunist phototrophic species being rapidly replaced by low-light tolerant species and by heterotrophic fungi as the substrate becomes overgrown with either turf or macro-/crustose coralline algae. For example, Chazottes et al. (2002) observed that substrates overgrown by turf algae were dominated by the light-loving cyanobacterium *Mastigocoleus testarum*, although rates of microbioerosion were relatively low due to intense grazing pressure. In contrast, substrates which were colonised by the macroalgal species *Lobophora variegate*, and which were only lightly grazed, were dominated by the cyanobacteria *Plectonema terebrans* and *M. testarum*, and exhibited significantly higher rates of bioerosion. At sites with relatively high crustose coralline algal crust cover, the chlorophyte *Ostreobium quekettii* was dominant and rates of microboring were again high.

4.4 External Bioerosion

The process of algal grazing is critical on coral reefs. If allowed to proliferate unchecked, macroalgae can cause partial mortality of corals (Nugues and Bak 2006), total mortality of coral recruits (Box and Mumby 2007), limit coral settlement (Birrell et al. 2008), and trigger coral disease (Nugues et al. 2004). Consequently, key grazing taxa such as parrotfishes and urchins are functionally extremely important in maintaining coral-dominated communities on reefs (Mumby et al. 2006b). The cropping of macroalgae has a further important functional role by simultaneously bioeroding reef structure (Bellwood and Choat 1990). Grazing species may also cause further bioerosion by direct predation on live corals, either as a minor or major component of their diet (Rotjan and Lewis 2006; Hoey and Bellwood 2008). Although less studied than grazing, bioerosion by fishes and urchins is significant because of their high densities and bites rates, and rates of carbonate removal may be higher than for many endolithic taxa. In this section we consider the process of bioerosion by sea urchins, parrotfishes and other fish species, and gastropods. Note that some reviews of bioeroding organisms describe predation of invertebrates protected by shells (e.g. Tribollet and Golubic 2011), but here we focus on the biting of dead carbonate structure and live corals.

4.4.1 Echinoids

Of all the echinoderms that occur on coral reefs, sea urchins are the only group capable of bioerosion that has a significant impact on carbonate budgets (Glynn 1997), a carbonate budget being defined as the balance between the amount of carbonate produced on a reef, less that lost to bioerosion and other chemical dissolution processes. During the process of grazing algae, urchins of genera such as *Diadema* and *Echinometra* (Fig. 4.5) scrape the carbonate substrate with the complex of articulated plates around their mouth (termed the Aristotle's lantern), and excrete $CaCO_3$ particles similar in character to those ingested (Hutchings 1986). Excreted pellets can be categorised into four types (from grazing of sand, coralline algae, calcareous macroalgae such as *Halimeda*, and highly bored carbonate substrate) and, where examined, approximately 43 % of this carbonate has been shown to be reworked sediment i.e., material that was already in sedimentary form, rather than being newly eroded from the reef framework (Hunter 1977). In addition, some urchins graze on small amounts of coral tissue and entire coral recruits, and consequently break down parts of their skeletons (Bak and van Eys 1975; Sammarco 1980; Carpenter 1981; Sammarco 1982). The sum total of this feeding activity means that bioerosion rates for urchins are frequently higher than those of fishes, although the relative importance of bioerosion by fish and urchins will depend on factors such as the reef zone considered and the level of fishing intensity (e.g. Hay 1984; Bak 1994). Urchin bioerosion around the bases of corals also makes the corals particularly susceptible to damage during storms (Bak 1994).

Actual rates of bioerosion vary among urchin species, and increase with increasing body size both within and among species (Bak 1994; Carreiro-Silva and McClanahan 2001). Estimated bioerosion rates range from 0.63 to 1.16 g $CaCO_3$/ind-d for *Diadema antillarum* (Hunter 1977; Scoffin et al. 1980), 0.31–1.8 g $CaCO_3$/ind-d for *Diadema setosum* (Mokady et al. 1996; Ruengsawang and Yeemin 2000; Carreiro-Silva and McClanahan 2001; Muthiga and McClanahan 2007), 0.11–0.14 g $CaCO_3$/ind-d for *Echinometra mathaei* (Russo 1980; Bak 1990; Mokady et al. 1996), and 0.18 g $CaCO_3$/ind-d for *Echinometra viridis* (Griffin et al. 2003). Urchin density is clearly a strong influence on overall bioerosion rate (Bak 1994). In the Caribbean, *D. antillarum* is found in a wide range of reef habitats including rock, coral reef, mangrove roots, seagrass beds, and sand (Randall et al. 1964; Bauer 1980), but is most abundant on patch and continuous reefs from depths of 2–10 m. Densities decrease with increasing depth, and *D. antillarum* is effectively absent deeper than 20–25 m (Ogden and Lobel 1978; Liddell and Ohlhorst 1986; Morrison 1988). Urchins are less common in turbulent water because excessive wave action affects grazing rates

Fig. 4.5 A group of (**a**) *Diadema antillarum* on a Caribbean reef, and (**b**) *Echinometra mathaei* on an Indian Ocean reef. Where urchins are abundant, they are an important bioeroding agent

(Foster 1987), but can be very common in intertidal areas (Muthiga and McClanahan 2007). Juvenile and adult *D. antillarum* densities are also correlated with habitat structural complexity (Bak et al. 1985; Weil et al. 1985; Lewis and Wainwright 1985). Densities of *E. viridis* are much higher in shallow water and decrease rapidly with depth (Griffin et al. 2003), but this species tends to be more abundant in deeper, calmer water than *Echinometra lucunter* (McGehee 1992).

Urchins are less well studied in the Pacific, but *D. setosum* is most common on shallow reefs or in boulder zones and is rare below 10 m (Mokady et al. 1996; Coppard and Campbell 2005; Muthiga and McClanahan 2007). Increasing abundances of *D. setostum* have, however, been linked to increasing sediment sizes and decreasing coral and macrophyte cover (Dumas et al. 2007). *Echinometra mathaei* is a highly adaptable urchin and can live in high wave energy and aerially exposed habitats (Coppard and Campbell 2005). In Fiji, *E. mathaei* was found in all habitats from the forereef to inner seagrass beds, but densities were highest in the mid-back reef zone (Coppard and Campbell 2005). Total echinoid bioerosion on Pacific reefs has been estimated to be 4.5 kg/m^2-year in Moorea (Bak 1990), and 8.3 kg/m^2-year in La Réunion (Peyrot-Clausade et al. 2000).

Calculating the net long-term contribution of urchins to bioerosion is complicated by large reductions or increases in natural populations, primarily because of disease and fishing of their predators. In the Caribbean in the 1970s and early 1980s, there is evidence that the removal of predators, such as triggerfishes, led to *D. antillarum* reaching unnaturally high densities on shallow reefs (Hughes 1994). For example, in Jamaica densities of >70 m^{-2} were recorded on some patch reefs (Sammarco 1980). A subsequent waterborne pathogen decimated *D. antillarum* throughout the region in 1983/1984 (Lessios 1988). This disease event had significant implications for grazing and bioerosion rates in the Caribbean (Carpenter 1988; Mumby et al. 2006b), and is a major factor in the increased areal coverage of macroalgae in the region (Gardner et al. 2003). The lack of recovery of *D. antillarum* has been the subject of much debate (e.g. Lessios 1995; Chiappone et al. 2002). Increased densities have recently been noted on some reefs (e.g. Edmunds and Carpenter 2001), but *Diadema antillarum* remains functionally extinct on many Caribbean reefs, particularly those with high biomasses of predatory fishes (Harborne et al. 2009).

In contrast to the Caribbean scenario, unregulated plagues of urchins are found on some Indian Ocean reefs. Fishing has reduced densities of urchin predators, particularly two triggerfish species, leading to increased urchin populations (McClanahan and Shafir 1990). Consequently, there have been changes in key reef processes, such as increased bioerosion rates and mortality of coral recruits. For example, bioerosion rates as high as 1180 g $CaCO_3$/m^2-year have been reported on unprotected reefs in Kenya with high fishing pressure, compared to 50.3 g $CaCO_3$/m^2-year on protected reefs (Carreiro-Silva and McClanahan 2001). These changes may be reversed inside marine reserves where reduced fishing pressure restores trophic interactions among urchins and their predators (McClanahan and Shafir 1990; McClanahan et al. 1999). Urchins are particularly susceptible to unregulated 'plagues' (i.e., population explosions) because only a few specialist predators can overcome their defensive spines, and food webs containing specialists are more susceptible to perturbations as other species cannot fill the ecological niche if a predator is over-exploited (Pinnegar et al. 2000). Furthermore, urchins have the ability to change body size rapidly with changing resource availability or the varying density of conspecifics, and unchecked population growth may be a consequence of this ability to regulate their size (Levitan 1989).

4.4.2 Parrotfish and Other Fishes

The dental morphology and biting mechanics of parrotfishes means that carbonate substratum is removed during grazing (Bellwood and Choat 1990; Fig. 4.6a). This is particularly true for species categorised as 'excavators'. These species, such as *Sparisoma viride* in the Caribbean and *Cetoscarus bicolor*, *Bolbometopon muricatum*, and *Chlorurus sordidus* in the Pacific, have a short, powerful bite with robust teeth and well-developed musculature, and they focus on convex microhabitats (Bellwood and Choat 1990; Bruggemann et al. 1996). In contrast, scrapers such as *Scarus vetula* in the Caribbean, and *Scarus ghobban* and *Scarus globiceps* in the Pacific, have a wide, but weak, bite and thin teeth, and predominantly bite flat microhabitats (Bellwood and Choat 1990; Bruggemann et al. 1996). A third group of parrotfishes, croppers, remove only algae and do not leave bite scars (Bellwood and Choat 1990). The large numbers of bites taken daily by parrotfishes mean that bioerosion rates can often be high: 0.02–5.4 kg/m^2-year for Caribbean species, and up to 5.6 kg/m^2-year for smaller Pacific species (reviewed by Bellwood 1995b; see also Bruggemann et al. 1996). However, rates of up to 32.3 kg/m^2-year have been reported where populations of large *B. muricatum* are present (Bellwood et al. 2003; Hoey and Bellwood 2008).

While most bioerosion by parrotfish is caused by biting carbonate substrate, additional bioerosion is caused by parrotfishes biting live corals (Rotjan and Lewis 2006; Alwany et al. 2009; Fig. 4.6b). Indeed, live corals are a major component of the diet of the large Pacific species *Bolbometopon muricatum* (Bellwood and Choat 1990), and this feeding can affect coral zonation (Littler et al. 1989). Parrotfishes also interact with other bioeroding taxa. For example, parrotfish grazing facilitates growth and areal

Fig. 4.6 Parrotfish are key grazers of both algae and corals, and different species are recognised as active bioeroders; (**a**) *Scarus forsteni*, and (**b**) *Bolbometopon muricatum* (the bumphead parrotfish), as examples of Indo-Pacific species; and (**c**) *Scarus vetula* (the Queen parrotfish), and initial phase (**d**) and terminal phase (**e**) *Sparisoma viride* (the spotlight parrotfish), as examples of common Caribbean species. Many of these fish leave visible bite scars on coral colonies that represent a significant source of reef bioerosion (Image B courtesy of Valerie and Ron Taylor)

coverage of boring sponges by increasing light levels and clearing areas for settlement (Cebrian 2010). For example, parrotfish corallivory acts synergistically to increase bioerosion rates by the sponge *Cliona tenuis* (Márquez and Zea 2012). All eroded carbonate sediments are defecated by parrotfish, and sediment production rates of 1043 kg/year per individual have been estimated for the Pacific species *Chlorurus gibbus*, and up to 170 kg/m^2-year for Caribbean species, constituting both freshly bioeroded material and reworked sediment (Frydl and Stearn 1978; Bellwood 1996). This sediment is a major source of both interstitial sediment and beach sand (Perry et al. 2015).

While the separation of scrapers and excavators represents a major distinction among parrotfish, species, body size, life phase, and density are also critically important considerations when calculating bioerosion rates. In the Caribbean, bite rates decrease with increasing fish size, and also vary significantly among species and life phases (Mumby 2006), whilst bite sizes also have clear allometric relationships, and increase with fish length (Bruggemann et al. 1996; Mumby 2006). Each species also has clear habitat preferences that influence density and hence bioerosion rates. Species-level habitat preferences are reviewed elsewhere (e.g. Harborne et al. 2006 for Caribbean species), but generally parrotfishes are most common at depths of ~5–10 m depth, and most adult parrotfishes decrease in density below these depths (van Rooij et al. 1996). Consequently bioerosion rates also decrease with increasing depth (Bruggemann et al. 1996).

However, depth also interacts with factors such as proximity to nearby shelter, food resources and biological interactions. For some species, proximity to nursery habitats is also important, and nearby mangrove stands can significantly increase the abundance of species such as *Scarus iseri* and *Scarus guacamaia* (Mumby et al. 2004). The range of data available for Caribbean parrotfish has allowed the production of a bioerosion model which combines fish densities, size structure, scar size, and bite rates to demonstrate rates of approximately 7 kg/m^2-year on shallow reefs which decrease with increasing depth (Bruggemann et al. 1995). Bioerosion rates also varied on smaller scales, and were highest on substrates with high densities of boring algae and low cover of crustose coralline algae (Bruggemann et al. 1995).

In the Pacific, similar relationships to the Caribbean lead to parrotfish sizes, bite rates, bite sizes, and life phases driving analogous patterns of species-specific bioerosion that vary with depth, topography, and reef zone. For example, studies of species distributions on the Great Barrier Reef (GBR) have demonstrated clear differences in bioerosion rates among habitats. A cross-shelf study of *Bolbometopon muricatum*, *Chlorurus microrhinos*, *Chlorurus* spp. and *Scarus* spp. demonstrated significant differences in densities among back reefs, reef flats, reef crests, and reef slopes, but also that these inter-habitat differences varied from inner- to mid- to outer-shelf reefs (Hoey and Bellwood 2008). Significant differences in herbivore assemblages between deep outer-slope zones and shallower habitats on the GBR were also documented by Russ (1984). Further variation in bioerosion rates occur with time of day; with highest parrotfish bite rates recorded in the early afternoon at Red Sea sites, which then declined until dusk (Alwany et al. 2009). In addition, further over-arching patterns of parrotfish density in the Red Sea are apparent at larger spatial scales. These include onshore – offshore gradients, such as those described on the Great Barrier Reef, where inshore reefs have high parrotfish densities of smaller parrotfishes and high rates of sediment reworking, compared to outer reefs where low densities of larger parrotfishes drive high rates of bioerosion and coral predation (Hoey and Bellwood 2008). As for Caribbean species, combining species densities, bite rates, and bite volumes allows the calculation of bioerosion rates, and in the Red Sea the abundant *Scarus ghobban* and *Chlorurus sordidus* were assessed as the most important bioeroders (rates of 41.2 cm^3/h and 7.0 cm^3/h), although individuals of the rarer *Chlorurus gibbus* had the highest rates per fish (47.0 cm^3/h: Alwany et al. 2009).

Bioerosion by parrotfishes has been reduced on many reefs because of over-fishing. Although not a primary target of most fishermen, parrotfish represent an alternative target fishery when the most valuable species, such as grouper and snapper, have been extirpated (Mumby et al. 2012). Furthermore, frequently the largest individuals and species are removed first, and these are the species that have the largest contribution to bioerosion rates. Even in the diverse Indo-Pacific a single large species (*Bolbometopon muricatum*), which is heavily targeted by fisherman, may undertake the majority of fish bioerosion on outer shelf reefs, and there is limited functional redundancy to cope with its loss (Bellwood et al. 2003). The desire to maintain natural bioerosion rates on reefs, and also grazing rates, means that parrotfishes are one of the major fish groups that conservationists aim to conserve through marine reserves. Such reserves can be highly effective and have significantly increased parrotfish densities and biomasses on multiple reefs including those in the Caribbean (Mumby et al. 2006a), Indian Ocean (McClanahan et al. 1999), and the Philippines (Stockwell et al. 2009). Parrotfish-specific fishery regulations have also been introduced in a few countries to try and protect their ecological functions (e.g. Belize, Mumby et al. 2012).

Other grazing fishes on reefs make only limited contributions to carbonate budgets, but surgeonfishes may remove small amounts of carbonate during their grazing of turf algae (Schuhmacher et al. 2008). *Ctenochaetus striatus* can generate relatively high bioerosion rates of up to 70.0g/ m^{-2}-year, but *Acanthurus nigrofuscus* is less important with rates of 2.6 g/m^2-year (Schuhmacher et al. 2008). Perhaps more importantly, the ingestion and defecation of sediment by surgeonfishes means that they do have an important role in sediment transport, potentially moving 18 % of the inorganic sediment on Pacific reef crests (Krone et al. 2011). Damselfishes, which aggressively defend territories on reefs against a range of other species, may have an indirect effect on the small-scale spatial distribution of bioerosion by reducing fish grazing rates inside their territories, leading to parrotfishes preferring to feed on habitat patches without damselfishes (Sammarco et al. 1986). Damselfishes also actively exclude urchins from their territories (Sammarco and Williams 1982). However, the reduction in grazing may lead to increased bioerosion by boring organisms (Risk and Sammarco 1982; Sammarco et al. 1987). Some triggerfishes, filefishes, and puffers also contribute to carbonate budgets by feeding on live corals (Glynn 1997), but data on the rates and processes involved remains limited.

4.4.3 Molluscs – Gastropods/Chitons

Gastropods appear to have a relatively minor role in reef bioerosion (Tribollet and Golubic 2011), but data are limited. Gastropods such as *Urosalpinx cinerea follyensis* can chemically erode carbonate structures using their accessory boring organ and through radula scraping (Carriker 1969). Although not an active bioeroder itself, the feeding activities of the gastropod *Drupella*, which ingest living coral tissue and can cause widespread coral mortality, can also facilitate subsequent bioerosion by urchins and other bioeroders that target dead coral substrate (Lam et al. 2007). Some chitons are reported to use magnetite-capped radula to graze algae, and concurrently scrape away the carbonate structure (Taylor and Way 1976; Fig. 4.7). Bioerosion by the chiton *Acanthopleura gemmata* on the Great Barrier Reef has been estimated as 0.16 kg $CaCO_3$/chiton/year^{-1} (Barbosa et al. 2008). However, bioerosion of actual reef substrata was limited because chitons were most abundant on beachrock, and densities were orders of magnitude greater than on the reef margin (Barbosa et al. 2008). Comparative rates for the Caribbean chiton *Acanthopleura granulata* in a tidal creek were four times lower than the Pacific species *A. gemmata* (Rasmussen and Frankenberg 1990). Rates at Aldabra Atoll were estimated as 40.2 g/m^2-year, but were conducted in a laboratory (Taylor and Way 1976). Additional bioeroders within the phylum include limpets and littorinas (Tribollet and Golubic 2011).

4.5 Spatial Variations in Reef Bioerosion

4.5.1 Regional-Scale Variation

While it is tempting to discuss bioerosion rates on "reefs", these rates clearly vary over a range of spatial scales. The first scale that we consider here is among regions. Published total bioerosion rates are available from multiple reefs in the Caribbean, Eastern Pacific, Pacific, and Indian Ocean (Table 4.1), and demonstrate variability covering two orders of magnitude. Comparing values among regions is difficult because studies are conducted in a range of habitat types, and also use different methodologies. For example, Table 4.1 includes studies using standardised blocks of dead coral, where actual loses of $CaCO_3$ can be measured, and studies that estimate erosion rates by combining taxa-specific erosion rates and measured densities of each organism. Furthermore, carbonate budgets of reefs must be understood in the context of both carbonate destruction and production.

The available data suggest relatively low rates of bioerosion in the Caribbean, typically because of the loss of the urchin *D. antillarum* in the 1980s (Lessios 1988). *D. antillarum* was an important bioeroder on Caribbean reefs, and its disappearance led to very significant changes in net carbonate budgets (Bak et al. 1984). This change is well demonstrated by comparing the pre-mortality rate of >3 kg $CaCO_3$/m^2/year (Bak et al. 1984) and the post-mortality rates of <0.7 kg $CaCO_3$/m^2/year$^-$ (Hubbard et al. 1990; Mallela and Perry 2007). In recent studies in

Fig. 4.7 Chitons, such as this *Acanthopleura granulata* in the Caribbean, have a limited role as reef bioeroders but may be important along rocky shorelines

Table 4.1 Summary of published values of total bioerosion, ordered by region and segregated by habitat type, where results for the entire bioerosion guild have been expressed in kg $CaCO_3/m^2$-year

Region	Location	Habitat	Total bioerosion	Notes	Reference
Caribbean	U.S. Virgin Islands	Forereef	0.65		Hubbard et al. (1990)
Caribbean	Jamaica	Forereef	0.1261, 0.265	Low and medium fluvial disturbances respectively	Mallela and Perry (2007)
Caribbean	Curacao	Forereef	3.3	Estimated	Bak et al. (1984)
Caribbean	Barbados	Fringing reef	5.53		Scoffin et al. (1980)
Eastern Pacific	Uva Island	Back reef	6.37, 6.41	Pre- and post-1983 coral mortality respectively	Eakin (1996)
		Reef flat	4.69, 4.83		
		Forereef	7.43, 8.29		
		Reef base	8.43, 13.64		
Eastern Pacific	Uva Island, Gulf of Chiriqui	Lower seaward slope	~19[1], ~9[1]	Damselfish absent and present respectively, high echinoid density	Glynn (1988)
		Upper seaward slope	~5[1]		
Eastern Pacific	Saboga Island, Gulf of Panama	Lower seaward slope	~17[1]		
Eastern Pacific	Floreana Island, Galapagos	Lower seaward slope	~42[1], ~20[1]	Damselfish absent and present respectively, high echinoid density	
Pacific	Oahu, Hawaii	Nearshore hardgrounds	0.133	Mean from three zones	Harney and Fletcher (2003)
		Landward reef platform	0.13	Mean from four zones	
		Channel margins	0.85	Mean from three zones	
		Seaward reef platform	0.197	Mean from four zones	
		Reef front	0.39	Mean from three zones	
Pacific	Moorea	Barrier reef flat	9.762	Maximum rates	Peyrot-Clausade et al. (1999)
		Fringing reef	5.362		
Pacific	Great Barrier Reef	Slope	0.11, 1.21, 3.43	Llewellyn Reef, One Tree Island, and Wreck Reef respectively, after 2 years	Kiene (1988)
		Flat	1.71, 0.71, 1.67		
		Lagoon	9.11, 1.01, 0.83		
Pacific	Lizard Island, Great Barrier Reef	Lagoon	1.180	1980–1985 data	Kiene and Hutchings (1994)
		Patch reef	1.759		
		Reef flat	0.226		
		Leeward slope	0.996		
		Deep leeward slope	0.430		
		Windward slope	1.914		
Pacific	Moorea	Reef flat	1.39		Pari et al. (2002)
	Tahiti	Reef flat	1.91		
	Tikehau, French Polynesia	Lagoon pinnacles	2.47, 3.02		
	Takapoto, French Polynesia	Lagoon pinnacles	1.80, 1.61		

(continued)

Table 4.1 (continued)

Region	Location	Habitat	Total bioerosion	Notes	Reference
Pacific	Great Barrier Reef	Forereef	0.27, 0.18, 1.09, 1.22, 1.23, 2.19	Snapper Island, Low Isles, Lizard Island, Harrier Reef, Ribbon Reef, and Osprey Reef respectively	Tribollet and Golubic (2005)
Indian Ocean	Reunion Island	Outer flat	9.94	Maximum rates	Peyrot-Clausade et al. (1999)
		Inner flat	2.06		
Indian Ocean	Reunion Island	Outer reef flat	1.72, 2.20, 4.37	High, modest and low nutrient input respectively	Chazottes et al. (2002)
		Inner reef flat	1.81, 2.89, 3.02		
		Back reef	1.56, 3.01, 3.39		

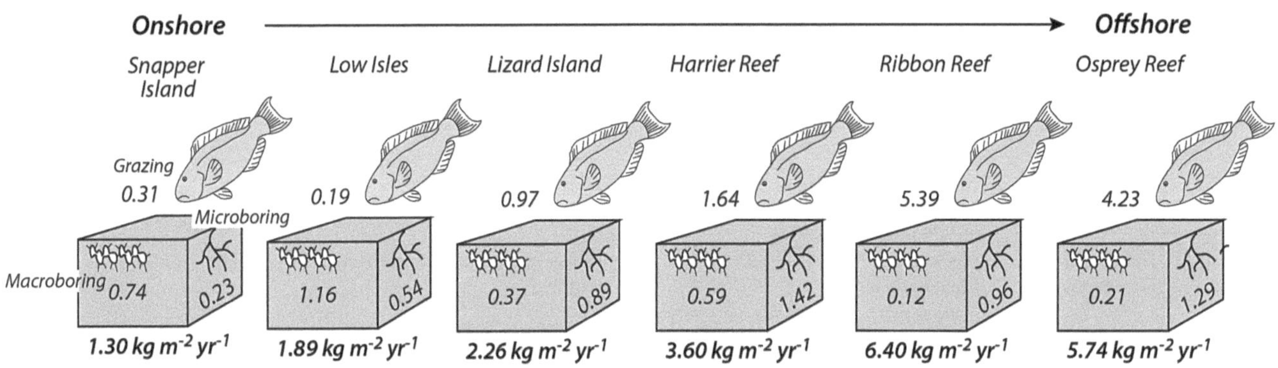

Fig. 4.8 Cross shelf variations in rates of bioerosion by micro- and macroborers, and by grazers, and total rates of bioerosion (kg $CaCO_3$/m^2/year) based on data from experimental blocks deployed for 3 years at sites across the Great Barrier Reef, Australia (Data from Tribollet and Golubic 2005)

Bonaire, urchins were noted to contribute little to overall bioerosion (Perry et al. 2012). In contrast, the high bioerosion rates in the Eastern Pacific appear to be driven by high urchin densities (Eakin 1996). However, such high echinoid densities are not common in the rest of the Pacific, and consequently bioerosion rates are likely to be lower. Few data are available for the Indian Ocean, and appear limited to reef flats. These published rates are high (up to 9.94 kg $CaCO_3$/m^2-year), but may be even higher on reefs where plagues of urchins have been reported after the overfishing of their predators (McClanahan et al. 1999).

4.5.2 Habitat-Scale Variation

In addition to demonstrating inter-regional variation, the review of published bioerosion rates (Table 4.1) also demonstrates variability among reefs in the same region and between habitats (used here in the sense of a reef ecological zone) on the same reefs. That bioerosion rates vary across a seascape, and can be very different among habitats, has been demonstrated in a number of studies (e.g. Bruggemann et al. 1995; Vogel et al. 2000; Tribollet and Golubic 2005; Hoey and Bellwood 2008). Furthermore, bioerosion rates can vary significantly among patches of the same habitat type. For example, external bioeroders have been shown to exhibit very different community structures on inshore reef slopes compared to analogous slopes of offshore reefs (Hoey and Bellwood 2008). Cross-shelf differences in bioerosion rates on the Great Barrier Reef have also been demonstrated in long-term studies of experimental blocks secured at similar depths on inshore, mid-shelf, and offshore sites (Tribollet et al. 2002; Tribollet and Golubic 2005; Fig. 4.8). Understanding the scale and drivers of this inter- and intra-habitat variation in bioerosion is important for assessing their effects on carbonate budgets, especially when predicting rates under future ecological scenarios. Here we use a Caribbean case study to illustrate how bioerosion can vary with reef zonation.

There are insufficient data to describe the pattern of bioerosion rates across a reef profile exclusively from comprehensive bioerosion budgets. Ideally budgets, such as that for a shallow (<10 m) fringing reef in Barbados that estimated gross $CaCO_3$ destruction rates of 123 × 10^{6}g/

year (Scoffin et al. 1980), would be available for multiple examples of each major Caribbean habitat type. In the absence of such data, we follow the example of Harborne et al. (2006) who described patterns of bioerosion across reefs by using better-understood gradients of species densities or areal coverage. These patterns are illustrated across a typical Caribbean reef profile using densities or areal cover from the literature (Fig. 4.9), and described below.

It is relatively well established that *D. antillarum* was, prior to 1983, most abundant on patch reefs, reef crests and shallow *Orbicella* spp.-dominated reefs (Fig. 4.9). Although mortality of *D. antillarum* reduced densities dramatically, they are still most abundant in these habitats and remain a significant source of bioerosion on some shallow reefs (e.g. Jamaica, Perry 1999). Furthermore, where full or partial urchin recovery has occurred in the region, it is re-establishing high densities on shallow reefs (Edmunds and Carpenter 2001). Densities are a little lower on back reefs, and the lack of topographical complexity on gorgonian-dominated forereefs leads to low densities in this habitat (Bauer 1980). Urchin densities are also low or absent on escarpments (Liddell and Ohlhorst 1987; Morrison 1988). There are fewer data for *Echinometra viridis* than for *D. antillarum*, but the majority of studies highlight the high densities on patch reefs and the absence from fore-reef habitats (e.g. Chiappone et al. 2002; Griffin et al. 2003). *E. viridis* was not affected by the pathogen that affected *D. antillarum* (Lessios 1988), and current patterns are apparently similar to historical distributions (Chiappone et al. 2002). The major bioeroding parrotfish, *Sparisoma viride*, are most abundant on shallow reefs (e.g. Nagelkerken et al. 2000). *Sparisoma viride* is less abundant on gorgonian-dominated forereefs because of low topographic complexity (Mumby and Wabnitz 2002). Densities of *Sparisoma viride* are lowest on deep escarpments (van Rooij et al. 1998). The percentage cover of boring sponges tends to decrease with increasing depth on fore-reefs (Liddell and Ohlhorst 1987), but may also be abundant in back reef habitats (Perry 1998b). Systematic abundance data across reef profiles are not available for the other bioeroding taxa, and the likely levels of bioerosion in Fig 4.9 are estimated through a combination of the literature cited in Sect. 4.2 and the authors' knowledge.

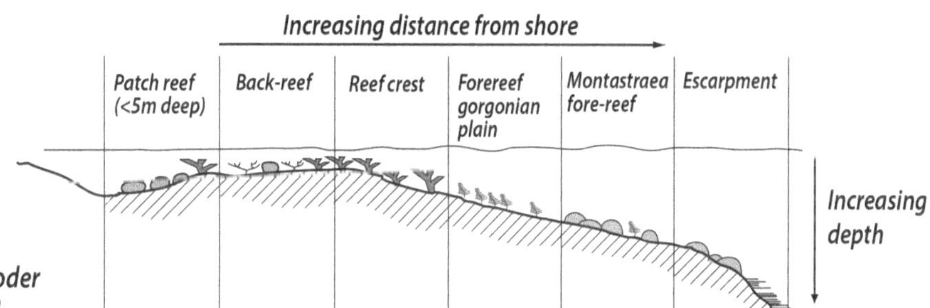

Bioeroder group	Patch reef (<5m deep)	Back-reef	Reef crest	Forereef gorgonian plain	Montastraea fore-reef	Escarpment
Diadema antillarum (m^{-2}) [1]	6.7 - 72.8	0.1 - 3.9	2.4 - 25.8	0.5 - 4.5	3.2 - 18.0	0.0
Echinometra viridis (m^{-2}) [2]	0.0 - 62.6	0.0 - 2.7	No data	0.0 - 0.002	0.0	No data
Sparisoma viride (100 m^{-2}) [1]	1.0 - 16.7	0.8 - 35.0	0.6 - 21.0	0.3 - 1.8	1.1 - 9.4	0.0 - 0.6
Endolithic sponges (% cover) [3]	No data	1.3	No data	No data	20.0	0.7
Molluscs	Medium	Medium	Low	Low	Low	Low
Polychaete and sipunculan worms	Medium	Medium	Low	Low	Low	Low
Phototrophic microborers	High	High	High	High	Medium	Low
Heterotrophic microborers	Low	Low	Low	Low	Medium	High

[1] from Harborne et al. (2006) and Liddell and Ohlhorst (1987)
[2] from Brown-Saracino et al. (2007); Sammarco (1982); Griffin et al. (2003); Chiappone et al. (2002); Lessios (1995); McClanahan (1999); McPherson (1969) and McGhee (1992)
[3] from Liddell and Ohlhost (1987)

Fig. 4.9 Patterns of bioerosion rates for major bioeroding taxa across a schematic Caribbean reef profile. Rate ranges for urchins, fishes and sponges based on published density studies. *Diadema antillarum* densities represent those prior to the major 1983 mortality event. Pattern of bioerosion rates for all other taxa (ranked 'low' to 'high') are estimated from published literature and authors' personal observations

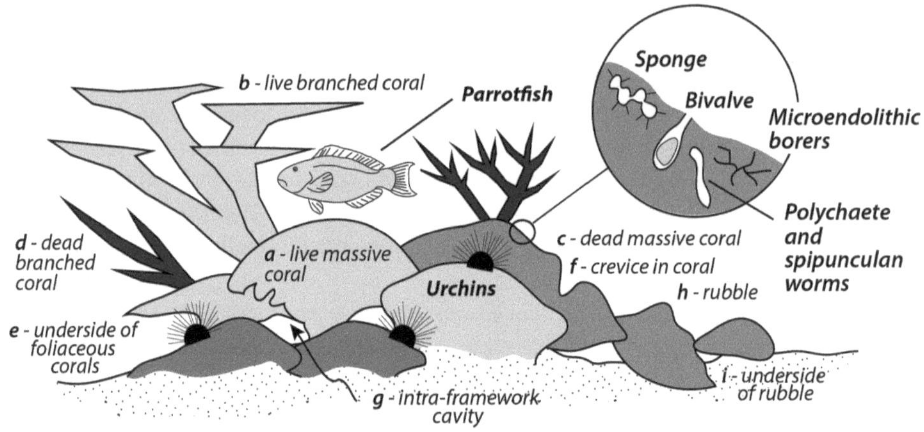

Micro-habitat type

	a	b	c	d	e	f	g	h	i
Parrotfish	✓	✓	✓	✓	×	×	×	✓	×
Urchins	✓ i	✓ i	✓	✓	✓	✓	✓	✓	×
Endolithic sponges	×	×	✓	✓	✓	✓	✓	✓	✓
Bivalves	✓ ii	×	✓	✓	✓	✓	✓	✓	✓
Worms	×	×	✓	✓	✓	✓	✓	✓	✓
Microborers	✓ iii	×	✓	✓	✓	✓	✓	✓	✓

Bold ticks denote preferred substrates
Grey ticks denote minor influence (see superscripts for additional comments)

i - grazing on coral recruits
ii - some species of Lithophaga bore into live corals
iii - Ostreobium bands form in live massive corals

Fig. 4.10 Schematic diagram showing the range of coral architecturally controlled micro-habitats that can exist on a hypothetical reef, and an indication of how different groups of bioeroders utilise these different micro-habitats (Micro-habitat diagram adapted from Kobluk 1988)

4.5.3 Intra-Habitat Variation

The bioerosion processes discussed above operate primarily on reef substrates that are exposed, to varying degrees, on or close to the active surface of a reef. In other words bioerosion occurs on areas of living coral and dead coral framework, but also on more protected (cryptic) framework surfaces. As a function of this within-habitat heterogenity, and given the different life modes of different reef bioeroders, a marked spatial partitioning occurs in the dominant bioeroders that will colonise and/or erode different parts of the potentially available reef substrate (e.g. Scoffin and Garrett 1974). For example, fish primarily graze on open, exposed live and dead substrates, whilst urchins and most macroendolithic borers can attack both exposed and more protected substrates (Fig. 4.10). The smallest and most cryptic surfaces can only be reached and infested by macroborers. Microendolithic bioeroders are pervasive on most substrates. Thus the species involved in reef bioerosion, and the rates at which different substrates and surfaces within an individual reef habitat will be eroded, may be influenced as much by substrate type and orientation as by whether it is living or dead.

Research on the interaction between bioeroders and substrate type has been most extensive for endolithic organisms and provides some evidence that bioerosion differs with substrate density and morphology. In experimental studies, Musso (1993) observed differential breakdown rates as a function of endolithic bioerosion in colonies of *Acropora* with different skeletal morphologies, the highest rates occurring in forms with a table-like morphology. Goreau and Hartman (1963), working in Jamaica, also documented greater impacts of endolithic sponge bioerosion on plating corals in deeper fore-reef environments. They attributed these greater impacts not to any higher rate of boring *per se*, but rather to corals having thinner skeletons, slower rates of growth and larger dead surface areas on their undersides and bases available for endolithic colonization. A similar conclusion was reached by MacGeachy (1977) in studies of endolithic bioerosion in Barbados. Other studies, such as those of Highsmith (1981) and Highsmith et al. (1983) have inferred a strong influence of skeletal density on bioerosion. These studies suggest a positive correlation between endolithic bioerosion rates and substrate skeletal density, but further work on such patterns are urgently needed. There is also evidence that substrate density influences the impacts of grazing, especially by parrotfish.

Generally, 'excavators' have a greater impact on substrate erosion due to their jaw structures and bite rates, but Ong and Holland (2010) suggest that 'scrapers' may have a much greater influence than 'excavators' on erosion rates where substrate densities are lower.

Rates of endolithic bioerosion by a particular taxon are likely to be controlled by several interacting factors: coral growth morphology, the rate of coral growth, the rate of endolith boring and the abundance of those endoliths in the environment (MacGeachy 1977). Thus, at reef-wide scales, factors other than the actual rate of boring become important. For example, because rates of coral growth and framework carbonate production decline with depth, the role of substrate boring may become relatively more important in deeper reef front settings, regardless of whether actual rates of boring are different from those in shallower environments (see Perry 1998b and Hubbard 2009 for discussions of these issues). The impact of this production-erosion balance shift may also be exacerbated by the prevalence of coral substrates whose morphologies, with large dead basal surface areas, favour extensive endolith recruitment.

4.6 The Role of Bioerosion in Reef Structural Development

As evident from previous sections, a wide range of reef-related species play an active role in the biological erosion of reef substrates through the processes of boring and grazing. These processes are an integral aspect of the functioning of 'healthy' coral reef ecosystems and, as the fossil record clearly shows, bioerosion has been an important aspect of coral reef development over long-periods of geological time. Indeed, broadly comparable assemblages of endolithic macroborers, based on preserved trace fossils, have been identified in reef deposits of Miocene age in Mallorca (Perry 1996), of Oligocene-Miocene age in Grand Cayman (Pleydell and Jones 1988) and Puerto Rico (Edinger and Risk 1994), and from the Upper Jurassic of Southern England (Fursich et al. 1994). This influence on reef development is exerted in a number of ways, but most especially (i) by influencing coral preservation potential and by modifying the composition and small-scale heterogeneity of reef substrates; and (ii) through the production and release of the detrital 'products' of bioerosion, which can represent a major source of reef sediment (Perry et al. 2015). Bioerosion rates also influence the balance between carbonate production and erosion in reef environments, and the impacts on reef accretion are further discussed in Sect. 4.8.

Cumulatively, the different processes of reef bioerosion continually interact with the accumulating reef framework and consequently play an important role in determining the breakdown and the selective removal of coral material from reef environments. This has two main implications: (i) it can influence the nature of reef framework development, including the ratio of *in situ* to detrital coral framework (Perry 1999); and (ii) as a consequence, the abundance of coral material entering the fossil record (Pandolfi and Greenstein 1997; Greenstein and Pandolfi 2003). Evidence of extensive coral substrate alteration as a result of bioerosion has been widely documented in numerous studies, including in the classic early descriptive works on the interactions between reef framework producers and modifiers (the processes of taphonomic alteration, see Scoffin 1992) at sites in Bermuda (Ginsburg and Schroeder 1973; Scoffin and Garrett 1974). Cut sections of coral rock, from these and other sites, clearly show the effects of multiple phases of bioerosion and subsequent sediment and cement infilling on the developing fabric of reef rock. The species and processes involved vary spatially across individual reefs (the ecological implications of which are discussed below) and thus the impact of bioerosion on reef framework preservation varies both spatially and temporally across reefs. Studies of shallow sub-surface coral fabrics across a depth gradient at sites in Jamaica (Perry 1999) clearly show the effects of spatially varying substrate modification processes, including endolithic bioerosion, on coral preservation.

Core records through Holocene reef sequences also show that bioerosion (especially evident through preserved macroendolith traces) has been an important long-term influence on the fabric of reef structures throughout their individual accretionary histories. Examples of this come from St. Croix (Hubbard et al. 1986; Hubbard et al. 2005), and Galeta Point, Panama (Macintyre and Glynn 1976). The effect of this bioerosion is, in many cases, to contribute to the generation and accumulation of coral rubble-dominated fabrics within the internal structure of reefs (Hubbard et al. 1990). Indeed, where the process of bioerosion is intensive, either as a function of high bioerosion rates or slow carbonate production rates, multiple cycles of boring can profoundly modify reef framework fabrics, often leading to near-complete obliteration of primary coral skeletal structures (Macintyre and Glynn 1976). Figure 4.11 shows the effects of different bioerosion processes on the alteration and partial replacement of primary coral substrates with increasing age and, in this case, submergence.

A further consequence of reef bioerosion from a reef framework development perspective is a potential biasing (or taphonomic filtering) of the fossil record, whereby preferential or substrate specific targeting of substrates (as a function of coral species or morphologies) may remove particular coral species from the preservable coral assemblage. The extent of this 'filtering', and the processes responsible, vary among reef environments (Goreau and Hartman 1963; Hubbard et al. 1986; Pandolfi and Michin 1995; Pandolfi and Greenstein 1997). However, given sufficient

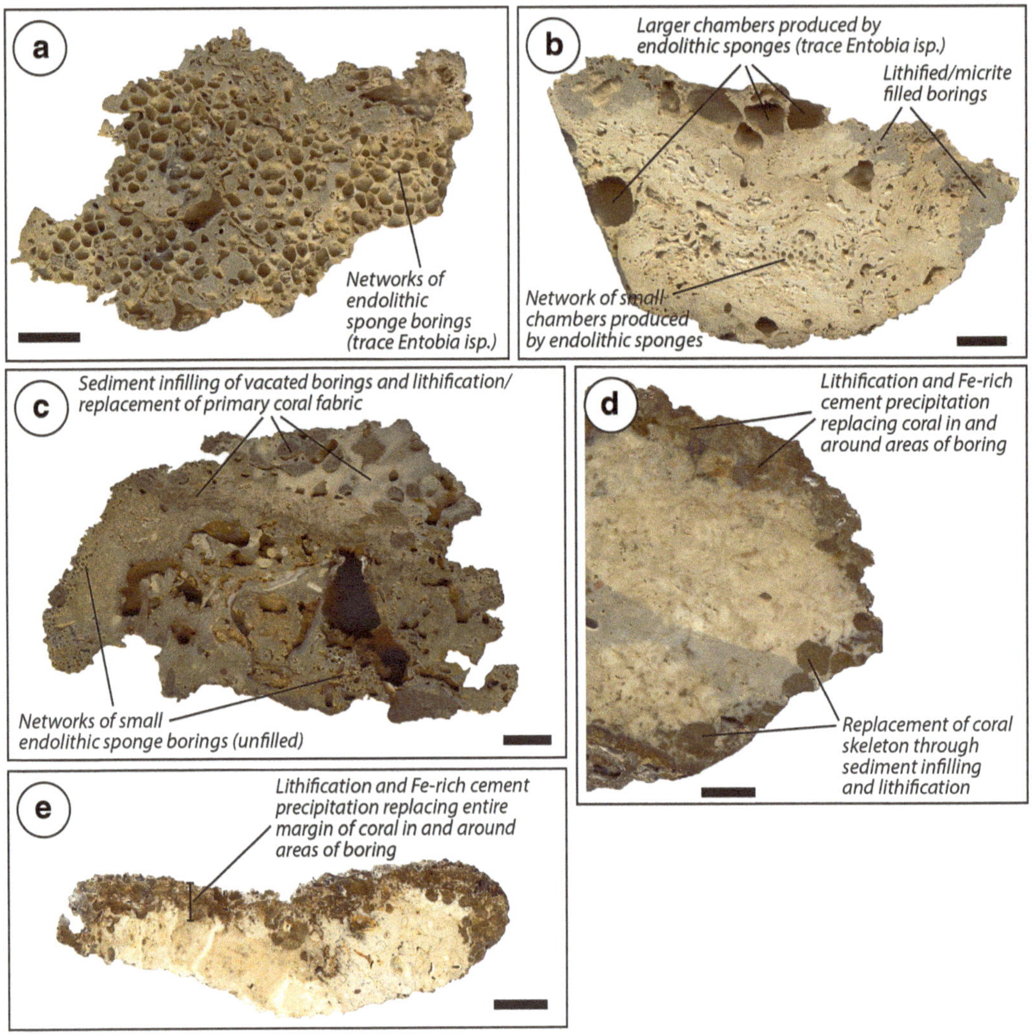

Fig. 4.11 Cut sections through coral samples from the submerged reefs off Hawaii showing the effects of endolithic boring on reef rock fabrics. (**a**, **b**) Empty vacated sponge borings that have removed varying amounts of primary coral substrate; (**c**) Bioeroded coral sample with partial secondary (sediment) infilling borings around the margin; (**d**, **e**) Replacement of primary reef rock through sediment infilling/ lithification within vacated borings. All scale bars = 1 cm (Images courtesy of Charlotte Humphrey)

knowledge of the spatial and/or bathymetric distribution of the species that produce preserved bioerosional traces (usually endoliths) these can actually be used to aid palaeoenvironmental and palaeobathymetric interpretations (Blanchon and Perry 2004; Perry and Hepburn 2008).

Evidence of very rapid and intensive impacts of bioerosion on reef frameworks are most prevalent at sites with especially high densities of bioeroding urchins and parrotfish which can result in rapid framework breakdown. Reaka-Kudla et al. (1996), for example, documented the effects of intense bioerosion by the urchin *Eucidaris thouarsii* at shallow sites in the Galapagos, where rates of bioerosion were sufficient to rapidly change the topographic complexity of these reefs. Similarly in Moorea, urchin erosion removed more carbonate than was being produced, and was thus responsible for net reef structural breakdown (Bak 1990). High densities of parrotfish have also been shown to ingest a significant proportion of the carbonate produced on reefs. For example, in Hawaii, Ong and Holland (2010) report that parrotfish are responsible for removing as much as 60 % of estimated shallow fore-reef carbonate production.

These various processes of bioerosion thus act as important agents of reef structural change over short temporal and small spatial scales, both by breaking down coral skeletons, but also through the generation of numerous boreholes and cavities in the accumulating reef structure. The former can influence the topographic complexity of reefs, especially where larger stands of branching or massive corals are undermined by high levels of boring or grazing, leading to a flattening of the reef surface. The latter, in contrast, will influence the small scale heterogeneity of the reef structure. Such localised framework modification will create a wide range of microhabitats on the reef surface, essentially small and diverse ecological niches, that harbour a diverse range of reef-associated cryptic species (termed cryptofauna: Kobluk 1988). Such interactions between skeletal breakdown and cavity boring on cryptofauna have been well demonstrated by a study on two blenny species (Clarke 1996). Holes created by borers in dead *Acropora palmata* stands are occupied by both *Acanthemblemaria spinosa* and *A. aspera*, but subsequent collapse of the *A. palmata* stands favoured *A. aspera* populations. Although *A. spinosa* is competitively superior, its metabolic requirements necessitate it living higher in the water column, on intact *A. palmata* colonies, where planktonic food densities are higher.

The by-products of many bioerosional processes also feedback into the overall process of reef development because they represent a source of both coral rubble and reef sediment. Sediment is a detrital product of several bioerosional processes produced across a range of size fractions in different reef environments (Fig. 4.12). Estimates of the volume of sediment and rubble in reef cores from St. Croix averaged 58 % (Hubbard et al. 1990). In their own right, boring sponges can produce very large quantities of mud-grade (<63 μm) coral 'chips' during the excavation of their chamber networks, and these 'chips' are expelled into the adjacent environment. Volumetric contribution estimates such as these are rare, but Fütterer (1974) suggested the process may contribute to as much as 30 % of the volume of lagoon sediments at Fanning Island in the Pacific. Both parrotfish and urchins also produce sediment-grade carbonate as a by-product of their grazing activities. Parrotfish, for example, excrete large amounts of fine coral sand, milled from the material they ingest. This material may be either randomly dispersed across reefs or, for some species, excreted within specific reef settings. Where the latter occurs there is a net export of grazing- derived carbonate sediment off-reef (Bellwood 1995a). Urchins also excrete large amounts of faecal pellets comprised of aggregates of smaller skeletal fragments (mostly of coral and coralline algae). Post-excretion degradation of the organic binding of these pellets releases smaller fragments into the surrounding sediments. Estimates by Hunter (1977) suggest, in fact, that *D. antillarum* was probably the major fine-grained sediment producer on the fringing reefs of Barbados in the 1970s, estimated to be producing 9.7 kg $CaCO_3/m^2$-year. Because of spatial variations in parrotfish and urchin abundance across individual reefs the importance of these sediment- producing processes will exhibit a high degree of spatial heterogeneity.

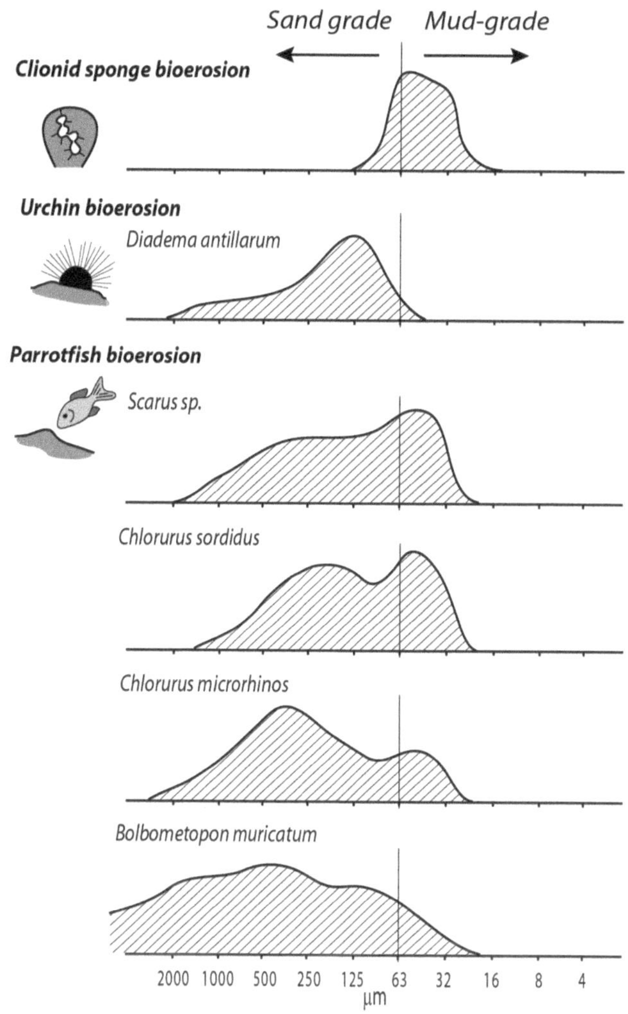

Fig. 4.12 Size-fraction distributions of sediments produced as a by-product of reef bioerosion. Note the variable range of size fractions produced by different species: thus spatial variations in the abundance and diversity of bioeroders across individual reefs are likely to result in differing impacts on sediment production (rates and size fractions). Clionid data from: Fütterer (1974); *Diadema* data from: Hunter (1977); Parrotfish data from: Hoey and Bellwood (2008). Further work is needed to fully understand the sedimentary significance of these processes across individual reefs. NB. Bivalves, worms and microendoliths produce little or no measurable sediment

4.7 Impacts of Ecological and Environmental Change: Ecological Feedbacks and the Changing Role of Bioerosion in Contemporary Reef Systems

Bioeroding taxa clearly exert varying degrees of influence on dead and living components of benthic communities within different reef habitats. However, changes to benthic communities may in turn affect the bioeroding taxa, setting up complex feedbacks that shape rates of both carbonate production and destruction. These links between bioeroders and the nature of benthic communities, are further complicated by a range of natural and anthropogenic stressors. In this section we examine how natural and anthropogenic changes to reef environments can affect the process of bioerosion.

4.7.1 Impacts of Eutrophication

The impacts of increasing eutrophication on bioerosion have probably been the most widely studied. Elevated nutrient concentrations promote higher primary productivity and thus may favour and enhance the activities of filter-feeders such as endolithic bivalves and sponges (Fabricius 2005). Indeed, several studies have reported higher rates of sponge bioerosion in areas of elevated nutrient concentrations: Rose and Risk (1985) at sites in Grand Cayman, Ward-Paige et al. (2005) on the Florida Reef Tract, and Holmes et al. (2000) at sites in Indonesia. However, whilst the influence of higher nutrient levels is likely to be partly a function of higher productivity by these filter-feeding taxa, modified benthic community composition (and especially increased algal biomass) can also influence how grazers interact with the substrate, and thus with the associated endolithic bioeroding communities.

In Tahiti, Pari et al. (2002) showed that at the most eutrophic sites studied, urchin grazing was the dominant bioerosional force, whilst at low nutrient sites endolithic sponge bioerosion was dominant. Much of the difference between sites was attributed to different benthic algal communities, with high macroalgal cover at the most eutrophic sites associated with high rates of urchin erosion. These same erosional processes were also reported to limit coral recruitment and to have shifted the reefs into states of net erosion. Similarly, different benthic substrate types and ecological states were reported to influence bioerosion rates at sites around Reunion which had different nutrient enrichment regimes (Chazottes et al. 2002). At low nutrient sites the reef communities were characterised by high live coral cover, high rugosity, high fish and urchin densities, and turf algal was the primary coloniser of dead coral substrate. At these sites grazing and macrobioerosion rates were 'high' relative to other sites and microbioerosion rates 'low'. Amongst the microbioeroder community the cyanobacterium *Mastigocoleus testarum* was dominant in turf algal covered habitats where light level reaching the substrate were higher. At the nutrient-impacted sites live coral cover was low, fish and urchin densities were low, and dead substrate was mainly colonised by macroalgae, especially *Lobophora variegata*. The greater shading effect of this alga favoured lower light specialist microborers such as *Plectonema terebrans* and *M. testarum*. At these sites grazing and macroboring were 'low' relative to other sites and microbioerosion rates were relatively 'high'.

Understanding the interactions between eutrophication and bioerosion necessitates considering the feedbacks between macroalgal cover and grazing rates. While eutrophication might be expected to increase macroalgal cover, and consequently grazing and bioerosion, grazing may in turn reduce macroalgal cover. The relative importance of nutrients (bottom-up control) versus grazing (top-down control) has been heavily debated in the literature (e.g. Diaz-Pulido and McCook 2003). While this debate is beyond the scope of this review, it is clear that under particular circumstances nutrification can have minimal impacts on macroalgal cover. For example, using a simulation-based model for Caribbean forereefs after the functional loss of *Diadema antillarum*, Mumby et al. (2006b) demonstrated that when hurricane frequencies and coral mortality from disease are low, parrotfishes can compensate for nutrification impacts. In contrast, when coral cover is low and macroalgal cover high, nutrification can be important, and coral cover will only increase over time when nutrification is absent. However, it is important to remember that while parrotfish may be able to maintain the ability of reefs to resist shifting from coral-dominated to macroalgal-dominated states, nutrification may increase bioerosion rates without changing coral cover. If we consider two reefs with similar coral cover, but with one having higher nutrient levels that drive increased macroalgal growth rates, bioeroder responses may be rather different. While parrotfishes may be able to increase grazing rates (by increasing their density or bite rates) on the higher nutrient reef and maintain similar macroalgal cover to the lower nutrient reef, bioerosion will be higher on the higher nutrient reef. Furthermore, as previously described, the increased nutrients may drive higher bioerosion rates by boring organisms such as sponges.

In addition to short-term changes in bioerosion rates by external grazers, increased macroalgal cover may have longer-term impacts because of changes in microhabitat availability caused by structural complexity changes. Ecological feedbacks are characteristic of reefs, and increased macroalgal cover will reduce coral recruitment and consequently coral cover, which will in turn reduce structural complexity. Consequently fish recruitment may be reduced because of limited shelter, and this may lead to reduced adult populations, lower grazing rates, and higher macroalgal cover (Mumby and Steneck 2008). Structural complexity of Caribbean reef has been further reduced by loss of *Acropora* corals from hurricanes and disease (Aronson and Precht 2001), colonies of which are among the most complex on these reefs. A lack of structural complexity, leading to high predation rates on juveniles, may also be a contributing factor explaining why densities of *Diadema antillarum* have not recovered in many parts of the Caribbean. It is clear that macroalgal cover is increasing on many reefs (Gardner et al. 2003; Bruno and Selig 2007) and that many reefs are less complex than they were (Alvarez-Filip et al. 2009). However, the relative importance of reduced coral recruitment and changing bioerosion rates on this process is poorly understood. Given the changing complexity of many reefs and the fact that different guilds of bioeroders are strongly partitioned across different intra-site architectural scales (see Fig. 4.9 above), it is reasonable to hypothesise that as habitat complexity changes, so too will the types and abundance of bioeroders. Furthermore, because topographic complexity will change in different ways and at different rates between habitats the nature and rate of bioeroder community change is likely to be highly non-uniform. At least initially, bioerosion rates may increase with increased grazing, and then decrease with decreasing structural complexity and recruitment limitation.

Nutrient-related changes in bioerosion rates may also affect reef sediment production. In Reunion, Chazottes et al. (2008) identified a shift from coral to coralline algal dominance in the sediment, a change they attribute to a shift in benthic algal communities. They also observe a shift from predominantly coarse- to fine-grained sands within higher nutrient sites, a change attributed to increased sponge borer activity. A long timescale study (37 years) in the Florida Keys also documented marked transitions in sediment composition on the reefs, with a doubling of molluscan material at some sites and a tripling of coral sand, the latter being attributed to an increase in sponge bioerosion under higher eutrophic conditions (Lidz and Hallock 2000). Where algal overgrowth of dead coral substrates has occurred, but bioerosion by fish and urchins was reported to have not increased, *Halimeda* plates and molluscan fragments have become far more abundant in the sediment. At sites with higher nutrient levels, however, increases in endolithic sponge and bivalve bioerosion in the widely available dead substrate have increased fine-grained coral sediment production.

4.7.2 Impacts of Sedimentation

Often linked to eutrophication (but also often operating as a singular 'environmental impact') are increased sedimentation rates. Several studies have focused on examining cross-shelf gradients in bioerosion, mostly working on the Great Barrier Reef from high sedimentation (and high turbidity) inshore settings, to 'clear water' offshore environments. In these studies evidence has been presented to suggest that, perhaps contrary to expectations, inshore sites have been characterised by lower rates of bioerosion compared to offshore sites, and that whilst the former has been associated mostly with sponge boring (but low grazing), the latter are generally dominated by grazers and endoliths. The conclusions from these experimental studies on inshore sites are clearly supported in core records from numerous inshore reefs in the region which are generally characterised by exceptionally good coral preservation (Perry and Smithers 2006; Perry et al. 2008a). Inshore corals often show no signs of internal bioerosion, a pattern attributed to high sedimentation rates and thus low residence time of dead coral substrates at the substrate-water interface. These critical interactions between the type and abundance of bioeroders, the availability of substrates (depending on sediment burial rate), and any additional stimulating effect on bioerosion from elevated nutrient levels are further discussed by Risk and Edinger (2011).

4.7.3 Impacts of Climatic Change

In addition to spatial and temporal variations in bioerosion that can be attributed to differences in water quality regimes, the direct and indirect impacts of climate change on reef bioerosion is an increasingly relevant issue. Such impacts may arise either from associated changes in the wider reef ecosystem, such as reduced live coral cover and/or changed structural complexity, or actual changes in rates of bioerosion and their impacts. Several studies, for example, have documented very high rates of bioerosion at sites where large-scale coral mortality has occurred (as a result of sea-surface temperature (SST) related coral bleaching events and/or coral disease). Examples include the widespread colonisation of dead *A. palmata* substrates in the Caribbean by endolithic sponges (López-Victoria and Zea 2004) and bivalves (*Lithophaga* sp.) in the Maldives (Kleeman 2008). Only a few studies have examined the actual response of endolithic borers to higher SSTs or reduced aragonite saturation, but those that have report differing responses in different groups of bioeroders. For example, Tribollet et al. (2006) found no evidence for an increase in microendolith metabolic activity under higher pCO_2 conditions. In contrast, Wisshak et al. (2012) indicated the likelihood of elevated rates of endolithic sponge bioerosion under such conditions, with lower pH conditions suggested to enhance the rate of substrate dissolution by sponges at reduced metabolic cost. The most significant increases occur when pCO_2 reach levels projected for well into the future (beyond 2100), but clearly point to a potential impact on future reef carbonate budgets. More recently, Barkley et al. (2015) have reported elevated levels of *Lithophaga* sp. bioerosion under naturally elevated pH conditions in Palau that mimic those predicted for the region by 2100. Bleaching-induced coral mortality can also increase bioerosion rates by external grazers, as densities of this trophic group typically increase after bleaching events (Pratchett et al. 2008). These higher abundances may be driven by herbivores (specifically fish) exploiting increased algal availability on coral skeletons or the loss of competitively dominant coral-specialist species. Additional direct impacts of climate change related to environmental parameter shifts on rates of bioerosion remain unclear, although one might reasonably speculate that any impacts on coral skeletal densities, as well as on the amount of dead coral substrate, will influence the abundance of bioeroders and thus their cumulative impact on net reef accretion potential.

4.7.4 Other Ecological and Environmental Impacts

Coral-bleaching events are not the only causes of large-scale mortality on reefs, and disturbances such as corallivory by outbreaks of *Acanthaster planci* (crown-of-thorns starfish: Fabricius et al. 2010), disease (Aronson and Precht 2001), or hurricanes (Woodley et al. 1981) can also generate large areas of dead (or partially dead) substrate for potential bioeroder exploitation. Frequently, research considering the responses of reefs to these disturbances focuses on immediate changes to fish and benthic community structure and subsequent recovery. Changes to bioerosion rates are less well studied, but some effects can be inferred from available data. *A. planci* outbreaks on Indo-Pacific reefs have been recognised since the 1960s and can have major impacts on coral cover at regional scales (Fabricius et al. 2010). The biggest effect of this loss of coral cover tends to be on obligate coral feeders, such as chaetodontids (Williams 1986). However, the loss of coral tissue typically leads to increased cover of turf algae that are favoured by grazers. Despite this increase in food, the density and biomass of herbivores did not differ between a set of *A. planci* affected reefs and non-affected control reefs (Hart et al. 1996), although it is possible that bioerosion rates did change because of increased bite rates (but see Hart and Klumpp 1996). Greater effects will be observed if reef structural complexity is lost through biological and physical erosion of dead coral skeletons, in which case dramatic changes to fish communities have been documented (Sano et al. 1987). Loss of structural complexity is likely because of the increased substratum open for settlement by endolithic bioeroders, but in the absence of other major stresses both benthic and fish communities may recover to pre-mortality states (Sano 2000). However, further chronic disturbances or low stocks of grazing fishes can lead to large increases in the amount of macroalgal cover that can persist over long timescales (e.g. Mumby et al. 2006b).

Like *A. planci* outbreaks, coral disease tends to leave intact skeleton and reef complexity after killing the tissue, and subsequent changes in benthic and fish communities and bioerosion rates are likely to be similar to those described above. A more direct effect of disease on bioerosion rates was the mass mortality of the urchin *D. antillarum* in the Caribbean (Lessios 1988). The loss of >90 % of urchins throughout the region had profound effects on the ecology of the region, including a reduction in bioerosion rates (Bak et al. 1984). Another example of direct impacts on bioerosion rates, as opposed to indirect effects such as coral mortality opening areas of substratum for endolithic borers, is harvesting of grazing fishes. Parrotfishes are increasingly a target for fishermen who have extirpated more valuable species such as groupers (Mumby et al. 2012). The impacts of exploiting parrotfishes, and other bioeroding fishes, on bioerosion rates will depend on the species targeted (excavators versus scrapers), and the sizes of captured fish, since bite volumes scale with body size (Mumby 2006). For example, targeted fishing of the large humphead parrotfish *Bolbometopon muricatum* means bioerosion by parrotfishes varies from almost none on exploited reefs to virtually balancing levels of calcification (Bellwood et al. 2003). Changes in both the abundance and size of parrotfishes and of urchins would have further impacts on the amount of sediment being produced, and hence reef framework development. Exploitation of parrotfish can also increase macroalgal cover on reefs (Mumby et al. 2006a; Hughes et al. 2007) and, as in the case for eutrophication, subsequently drive further changes to bioerosion rates because of altered abundances of boring endoliths. Fishing may also target urchin predators, leading to plague populations of urchins on reefs and dramatic increases in bioerosion rates (McClanahan and Shafir 1990).

4.8 Quantifying the Role of Bioerosion: Carbonate Budgets and the Changing Face of Reef Accretion

The combined effects of coral substrate bioerosion, by both grazers and borers, will influence various aspects of contemporary reef structure and habitat complexity, as well as longer-term reef framework development. Particularly important in this respect, and a factor likely to become increasingly important as reef ecosystems continue to decline, is the influence that bioerosion exerts on net rates of reef accretion (Perry et al. 2008b; Hubbard 2009; Perry et al. 2012). This occurs as a function of the role bioerosion plays in influencing both how much and in what form carbonate is preserved on a reef and, in essence, relates to the balance between the amount of carbonate produced less that lost or removed through bioerosion. This concept is defined by the biological carbonate budget approach to conceptualizing and quantifying reef geomorphic functioning i.e., carbonate production and reef growth potential (Chave et al. 1972), the balance between production and erosion representing the net accumulation rate of $CaCO_3$. These budget measures can be considered a quantitative indicator of the functional state (from a geomorphic perspective) of a reef. Inherent within this concept, and as an integral aspect of reef carbonate budget estimates, is the role of biological erosion as a major framework erosional process (Perry et al. 2012, 2014).

Most significantly, changes in the abundance or activity of bioeroders relative to the rates of carbonate production can lead to rapid transitions in net reef accretion rates. For example, short-term ecological changes can drive episodic fluctuations or more permanent transitions in the abundance of individual carbonate producers and destroyers (see examples in Sect. 4.7), some of which may have an especially significant impact on reef carbonate budgets and thus on reef structural integrity. The interactions between these production and erosion processes (and the effects of rapid changes in both) are well illustrated in a number of classic case studies that document transitions to states of short-term net reef erosion. For example, at Uva Island, off the Pacific coast of Panama, net carbonate production rates declined as a result of: (1) an ~50 % loss of live coral cover caused by the warming of surface waters during the 1982/1983 El Niño, and (2) an associated increase in the densities of grazing urchins. The impact of this changed production-erosion balance was to shift the reef $CaCO_3$ budget from a positive (0.34 kg/m^2-year) to a negative state (−0.19 kg/m^2-year: Eakin 1996). However, marked variations in the magnitude of the reported shifts in the net budget figures (and in reef framework degradation i.e., the geomorphic response) occurred in different reef environments. For example, reef flat environments exhibited little net change, whilst erosion rates increased markedly at seaward fore-reef sites. Similarly, rapid increases in urchin bioeroder abundances at reef sites around the Galapagos were reported to have caused widespread coral framework degradation (Reaka-Kudla et al. 1996). At nutrient influenced sites in Indonesia, Edinger et al. (2000) showed the effects of high rates of sponge bioerosion on reef carbonate budgets: sponges within the most eutrophic environments caused the reefs to enter net erosional states. In Curacao, Bak et al. (1984) concluded that post-mortality carbonate dynamics had been significantly altered by the mass mortality of *D. antillarum*. Before mass mortality, there was a slight negative carbonate budget (production of 3.1 kg $CaCO_3$/m^2-year < bioerosion 3.3 kg $CaCO_3$ m^2-year) on shallow forereefs (5–10 m), but changes in urchin abundance and the consequent reduction of bioerosion (to 0.4 kg $CaCO_3$/m^2-year) meant there could be a shift to states of net carbonate accumulation. Such studies clearly demonstrate the potential impacts of medium-term (decadal timescale) changes in reef carbonate production as driven by environmental disturbance, and provide insight into the types of geomorphic changes that might be anticipated on reefs that have undergone, or may undergo, shifts in ecological states and in the actual abundance or relative importance of reef bioeroders.

A conceptual approach to considering the impacts of these relative changes in reef carbonate producers and eroders was proposed by Perry et al. (2008b). This conceptual model recognises that the relative importance of carbonate producing and erosional processes are transitional over time and space within individual reefs, and that these can reflect wider ecological changes in the reef environment. Resultant budgetary states (net productional, net erosional and neutral i.e., a state of budgetary stasis) are considered within a ternary space, within which different budgetary pathways can be envisaged that reflect different ecological conditions and disturbances. For example, many (unimpacted) reefs will exist within the positive net budget domain of the ternary space (Fig. 4.13). However, within this domain, where rates of production exceed bioerosion, different reefs will vary in terms of the relative importance of primary (coral) and secondary (calcareous encruster) carbonate production (e.g., points A and A^1), and in the net rate of production (points A and A^2). Reefs may also move temporally between these states in response to small-scale environmental perturbations. However, where major or prolonged ecological disturbances occur, more fundamental changes in reef production status can arise. These may result in a shift from a production-dominated to a bioerosion-dominated state, shifts that are analogous to those described by Eakin (1996) from Panama, or Edinger et al. (2000) from Indonesia. Such conceptual approaches thus have considerable merit as a framework within which to consider the wider implications of modified bioerosional regimes, and to link these response scenarios to a wide range of ecological and environmental changes.

As evident from earlier discussions in this chapter, an additional axis of change might be added to such a diagram to allow considerations of budgetary state changes with depth. This is perhaps an especially important concept given the strong spatial (and bathymetric) variations that occur in the types of bioeroding taxa (and their actual and relative rates of erosion) across such environmental gradients. These ideas have also been discussed, from a reef accretion perspective, by Hubbard (2009), and have relevance here because of the different impacts of ecological and structural change that are occurring in different (depth-related) reef habitats. A topical question that can thus be asked in this respect is: How does loss of habitat complexity affect the abundance and diversity of bioeroders active

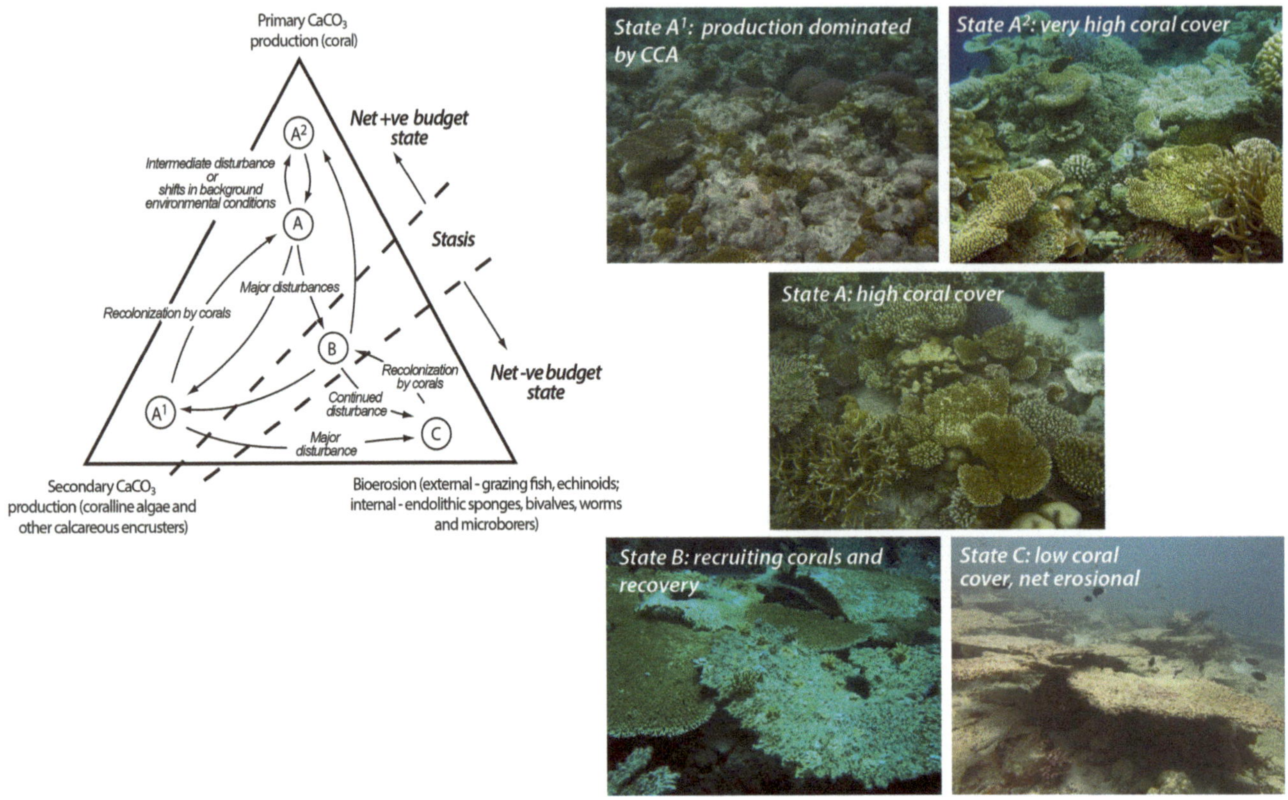

Fig. 4.13 Schematic diagram illustrating the temporal dynamics of reef framework carbonate production in relation to the budget states approach of Perry et al. (2008b). Relative changes in the importance of bioerosion can drive reefs into states of stasis or net erosion, but also back again if conditions ameliorate. Photos are indicative of the different states *A* to *C* and do not relate to any individual reef (Adapted from Perry et al. (2008b))

within individual reef environments? Do reefs along a trajectory of ecological decline and topographic 'flattening' reach a point where rates of bioerosion decline, and the system effectively approaches a state closer to budgetary stasis (*sensu* Perry et al. 2008b; see Fig. 4.13) rather than being net erosional. And how does this threshold vary with the physical environment, such as with increasing depth? For example, given documented links between habitat structural complexity and echinoid and fish densities (see Sect. 4.2), as well as substrate availability for endolithic bioeroders, it is reasonable to speculate that as complexity declines so too will the populations of bioeroders these habitats support, and that these responses will vary between habitats.

The responses and feedbacks between the structural complexity of individual habitats and the bioeroder communities that they support are likely to be very habitat specific, and to be influenced by the nature of the extrinsic drivers of habitat change. These ideas are illustrated through two Caribbean examples: one for a shallow *A. palmata* dominated habitat (Fig. 4.14), and the other an *Orbicella* spp. dominated fore-reef (Fig. 4.15), where the immediate capacity for recovery is assumed to be limited. For the *Acropora* habitat we discuss two contrasting scenarios caused by different disturbance events. In the first (Fig. 4.14), white band disease effectively kills the corals tissue but leaves the skeletons standing intact. Thus, although live coral cover has declined, the structural complexity of the habitat is relatively unaffected in the short term. Under such conditions, and with an increase in turf algae growing on dead coral substrate, numbers of grazing parrotfish and urchins may actually increase, driving an increase in the rates of substrate bioerosion, whilst coral carbonate production rates have declined significantly. Associated with this phase of habitat change, steady increases in the abundance of shallow water endolithic species might be anticipated (see López-Victoria and Zea 2004) as they progressively colonise newly available substrates. However, over time, as the standing *Acropora* framework starts to degrade, the abundance of endoliths may continue to increase, but progressive loss of architectural complexity may drive a reduction in the grazing fish and urchin populations that can be supported (and which are susceptible to architectural decline). The net effect of this is likely to be a reduction in the overall rate of bioerosion, but also a shift in the main groups of bioeroders operating within the habitat. The projected budgetary consequences of these changes are shown in Fig. 4.14, and are likely to see the habitat moving into a state of net erosion.

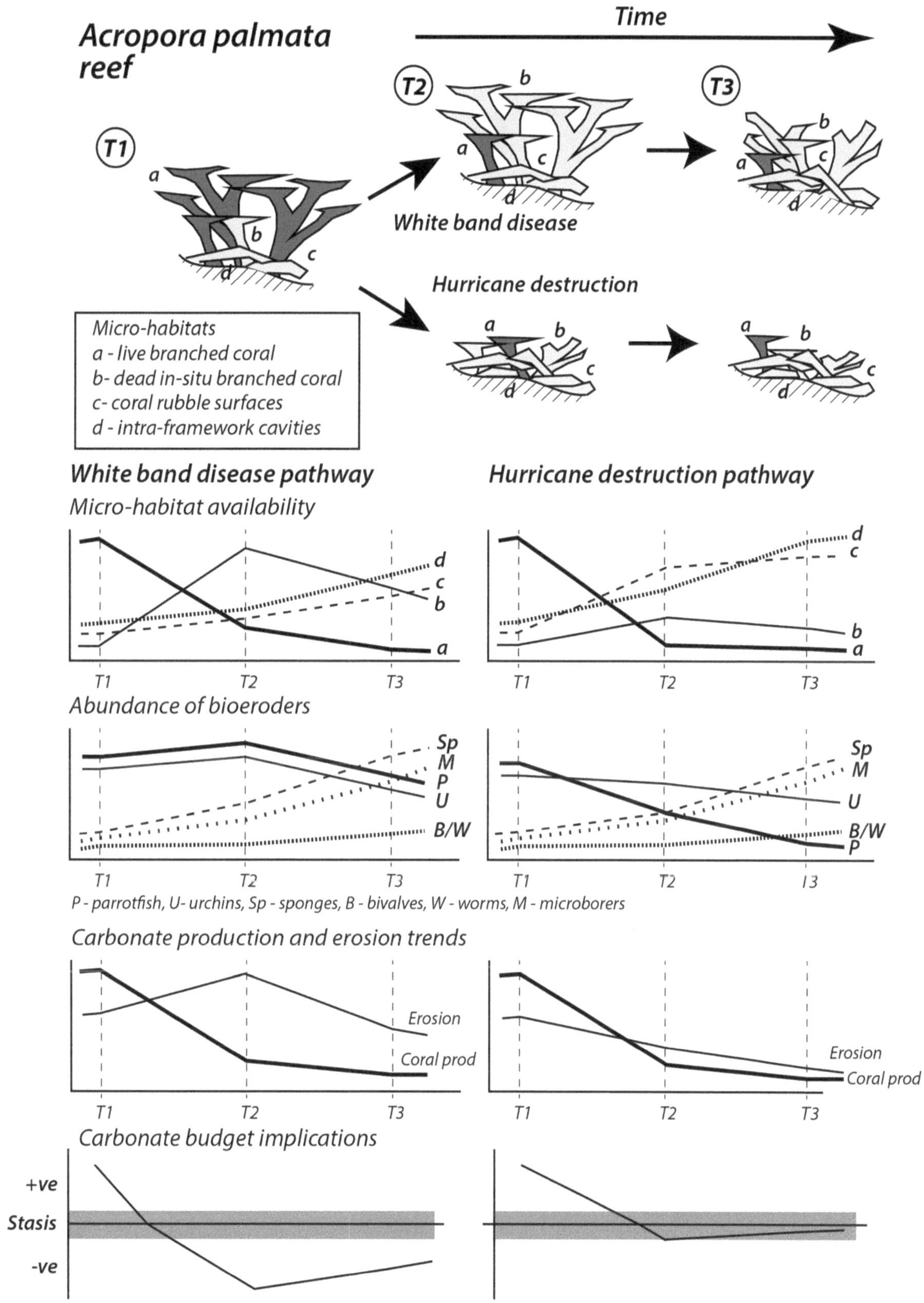

Fig. 4.14 Conceptual diagram showing the impacts of different extrinsic disturbances (hurricane damage and *white* band disease) on a branched coral-dominated (e.g. *Acropora palmata*) reef in the Caribbean

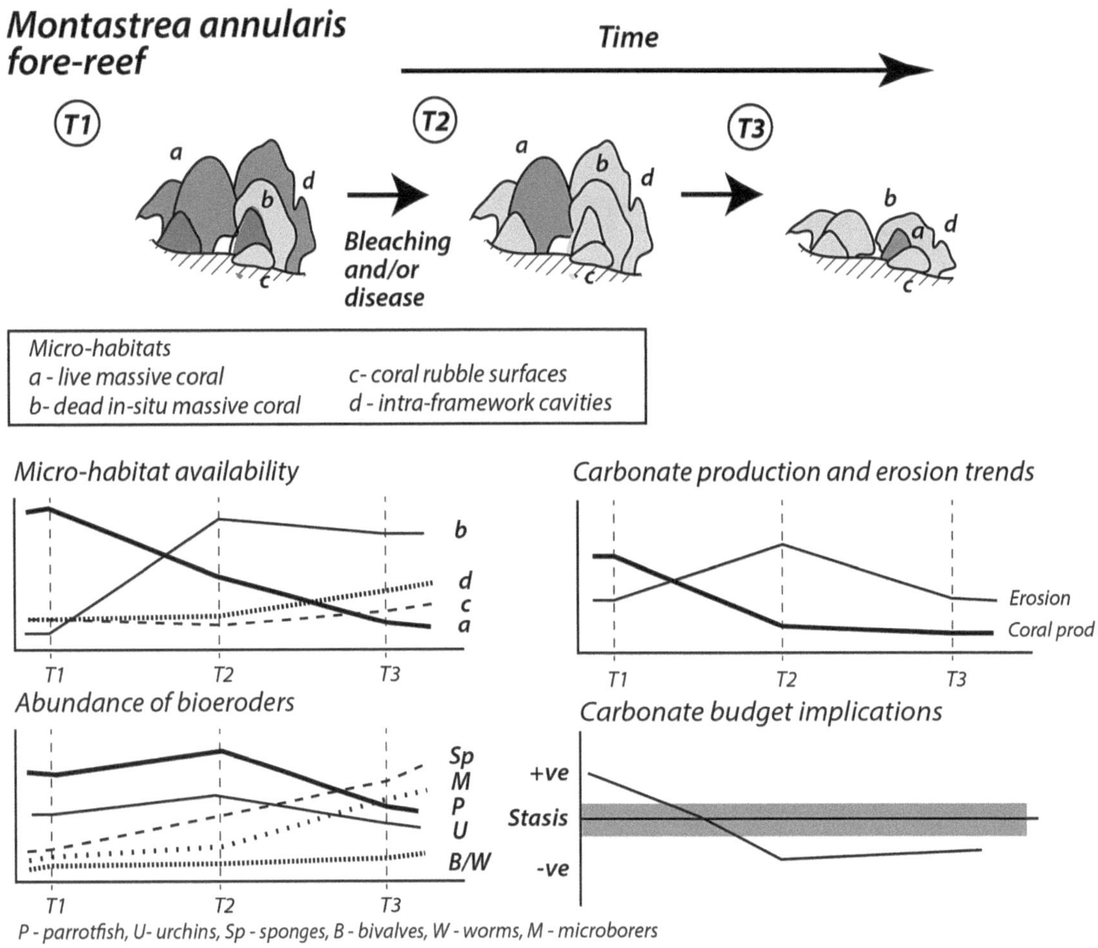

Fig. 4.15 Conceptual diagram showing the impacts of extrinsic disturbances (bleaching and/or disease) on a massive coral-dominated (*Orbicella* spp.) reef. Projected impacts on complexity, micro-habitat availability, the abundance of different bioeroder groups, and projected budgetary implications of these changes are shown. See text for an explanation of these changes

A different set of responses can, however, be envisaged where the same *Acropora*-dominated habitat is impacted by a hurricane event (Fig. 4.14). Here much more rapid (immediate) loss of structural complexity occurs, leaving a habitat with less *in situ* branched coral habitat, but significantly more rubble and intra-framework habitat. Such rapid changes are likely to impact the bioeroding community in a different way. Firstly, one would anticipate a relatively rapid decline in grazing fish numbers as the high structural relief of the reef is lost. Urchin populations, conversely, may remain stable as the abundance of newly available rubble habitat is more favourable to this group. Over time, progressive breakdown of the remnant *Acropora* framework will further limit suitable habitat for grazing fish and, eventually, urchins. However, as in the white band disease-induced scenario, the abundance of endolithic bioeroders is also likely to increase over time, colonising the newly available dead coral substrate. The impacts of these changes are likely to be: (i) a relatively rapid decline in coral carbonate production rates; (ii) a steady (albeit lagged, but less so for fish) decline in erosion rates; and (iii) ultimately a budgetary transition from states of high to low net carbonate production.

In contrast, in habitats dominated by massive corals, where rates of structural change (even following mass-mortality events) may be much slower, much more gradual changes in budgetary states are likely. For example, in *Orbicella* spp. fore-reef habitats (Fig. 4.15) extrinsic disturbances such as bleaching and/or coral disease-induced mortality events, will typically cause limited immediate architectural loss, the key transition being a marked decline in live coral cover, but a rapid increase in dead coral substrate availability. The impact of this is likely to be an initial increase in parrotfish and urchin bioerosion as these groups immediately exploit the newly available substrate, although rates of urchin bioerosion start from a lower point than they would in more favourable shallow water *Acropora*-dominated habitats. Rates of endolithic bioerosion are also likely to increase as new substrate is exploited. Over time slow degradation of the massive framework structure by bioerosion occurs, but leads to far less rubble generation that in *Acropora* habitats. The net impact of these changes, however, is likely to be a budgetary transition from net carbonate production states to states of net erosion. Thus different production-erosion pathways and relationships probably exist in different habitat types.

Inherent within the examples given here is the idea that, over time, different interactions and feedbacks are likely to develop between the coral architectural components of a given reef habitat that are modified, as a result of ecological or environmental change, and the bioeroding communities that degrade those same coral substrates. Thus, for example, high rates of bioerosion may drive rapid degradation of newly available (dead) coral substrate but, over time, these same processes may then also destroy much of the coral architectural complexity that provides the habitat space for those same bioeroders (an example of a negative feedback). Further interactions may also occur that complicate the relatively simple concepts outlined in Figs. 4.14 and 4.15. These may include: (i) the effects of bioeroder preference for specific substrates within a habitat; (ii) species specific (within a group) utilisation of sub-habitat types within a habitat; (iii) feedbacks between the types of algal cover and substrate bioerosion (e.g. Chazottes et al. 2002); and (iv) grazing facilitating some recovery of coral cover. Relatively limited datasets exist to parameterise these relationships at present, but they are important for understanding the implications of environmental and ecological change on the balance between reef carbonate production and bioerosion, and thus for the future budgetary state of reefs.

4.9 Summary and Key Research Gaps

Bioerosion is a major ecological process on reefs that defines the very structure of their framework. However, perhaps because of the taxonomic or logistical difficulties of research, bioerosion is less well studied than many other major processes such as primary productivity and grazing. In this chapter we have demonstrated that the key taxa are well established, but that data on actual bioerosion rates are limited to a few well-cited studies. More detailed information on how these rates vary across spatial and temporal scales is even more limited, and habitat-specific bioerosion budgets for all taxa are rare. This lack of data is not merely an academic problem. Before we have even established a conceptual framework of how bioerosion budgets might vary from region to region, reef to reef, and from habitat to habitat, profound changes in reef ecology are altering these baseline values. Coral mortality, decreasing water quality, and exploitation and diseases of external grazers affect bioerosion in many ways, including disruption of complex ecological feedbacks that have been described, but for which quantitative supporting data are scarce. It is clear that the balance between bioerosion and carbonate production is critical for the long-term health of reefs, but neither half of this equation is well described. Consequently, trying to initiate management initiatives towards a goal of accreting, complex reefs that are resilient to disturbance and to provide a full range of goods and services to human populations is hugely problematic. We thus conclude this chapter by outlining some of the major research that will aid our understanding of bioerosion on modern coral reefs.

In terms of understanding the functioning of the bioerosion process the following are perhaps the most critical questions:

1. How do rates of bioerosion by different species within the same functional group (e.g., sponges) vary, and how do these rates vary on different reef substrate types? This is important for understanding the impacts of different species on substrate bioerosion rates, and the consequences of changing abundances of individual species as reef ecosystems change (Perry et al. 2014);
2. To what extent does the architectural complexity of a habitat and its substrate composition influence the types, abundance and rates of substrate bioerosion by different bioeroder groups? This is important to allow us to project the effects of changing habitat complexity on rates of bioerosion, and thus on rates and styles of architectural complexity decline; and
3. How and why do bioerosion rates of individual taxa vary across and among reefs, and how do these rates integrate across the entire community to generate spatially variable bioerosion budgets? Establishing a much greater range of bioerosion rates at habitat-scales is critical for understanding the interactions with carbonate production, and their combined influences on reef framework growth (Kennedy et al. 2013; Perry et al. 2013).

In terms of understanding the ecological dynamics of bioerosion and the feedbacks between ecological and environmental change, and bioerosion, the following are among the key questions we need to ask:

1. What are the factors limiting the recovery of *Diadema antillarum* in the Caribbean, and is it possible to assist any recovery? Thirty years after the loss of this keystone species, recovery has been modest and is limited geographically. Aiding the re-establishment of this species will have profound effects on the bioerosion, grazing, and resilience of Caribbean reefs (see Miller et al. 2009);
2. How are the various threats to reef health affecting bioeroding taxa, their bioerosion rates, and ecological feedbacks? For example, following a mass coral

bleaching event, understanding the rate of loss of reef complexity compared to reef recovery will help establish whether parrotfish populations will be limited by loss of refuges before new coral growth re-establishes habitat rugosity (see Jackson et al. 2014);

3. How will changing bioerosion rates on different coral colonies interact with increasingly powerful hurricanes to alter the impacts of major storms? Research suggests that global climate change will cause hurricanes to become more powerful, and perhaps more frequent (Knutson et al. 2010), and that bioerosion can make coral colonies more susceptible to toppling;
4. How will decreased aragonite saturation impact on reef carbonate budgets? Ocean acidification has been inferred as a future impact on coral calcification rates and skeletal density (Hoegh-Guldberg et al. 2007), both of which would have the potential to reduce reef accretion and increase bioerosion rates (e.g. Wisshak et al. 2012). How these changes may be manifested in the responses of different substrate types is very poorly understood; and;
5. How will increased sea-surface temperatures, as a more immediate global stress to reefs, affect bioerosion rates? Global climate change has the potential to increase the metabolic rates of bioeroding taxa and the productivity of macroalgae, but whether these changes will impact bioerosion rates is not clear.

References

Alvarez-Filip L, Dulvy NK, Gill JA, Côté IM, Watkinson AR (2009) Flattening of Caribbean coral reefs: region-wide declines in architectural complexity. Proc R Soc Lond B 276:3019–3025

Alwany MA, Thaler E, Stachowitsch M (2009) Parrotfish bioerosion on Egyptian Red Sea reefs. J Exp Mar Biol Ecol 371:170–176

Aronson RB, Precht WF (2001) White-band disease and the changing face of Caribbean coral reefs. Hydrobiol 460:25–38

Bak RPM (1976) The growth of coral colonies and the importance of crustose coralline algae and burrowing sponges in relation with carbonate accumulation. Neth J Sea Res 10:28–337

Bak RPM (1990) Patterns of echinoid bioerosion in two Pacific coral reef lagoons. Mar Ecol Prog Ser 66:267–272

Bak RPM (1994) Sea urchin bioerosion on coral reefs: place in the carbonate budget and relevant variables. Coral Reefs 13:99–103

Bak RPM, Carpay MJE, de Ruyter van Steveninck ED (1984) Densities of the sea urchin *Diadema antillarum* before and after mass mortalities on the coral reefs of Curaçao. Mar Ecol Prog Ser 17:105–108

Bak RPM, Carpay MEJ, de Ruyter van Steveninck ED (1985) Densities of the sea urchin *Diadema antillarum* before and after mass mortalities on the coral reefs of Curaçao. Mar Ecol-Prog Ser 17:105–108

Bak RPM, van Eys G (1975) Predation of the sea urchin *Diadema antillarum* Philippi on living coral. Oecologia 20: 111–115

Barbosa SS, Byrne M, Kelaher BP (2008) Bioerosion caused by foraging of the tropical chiton *Acanthopleura gemmata* at One Tree Reef, southern Great Barrier Reef. Coral Reefs 27:635–639

Barkley HC, Cohen AL, Golbuu Y, Starczak VR, deCarlo TM, Shamberger KEF (2015) Changes in coral reef communities across a natural gradient in seawater pH. Sci Adv e1500328

Bathurst RGC (1966) Boring algae, micrite envelopes, and lithification of molluscan biosparites. J Geol 5:15–32

Bauer JC (1980) Observations on geographical variations in population density of the echinoid *Diadema antillarum* within the western North Atlantic. Bull Mar Sci 30:509–515

Bellwood DR (1995a) Carbonate transport and within-reef patterns of bioerosion and sediment release by parrotfishes (family Scaridae) on the Great Barrier Reef. Mar Ecol Prog Ser 117:127–136

Bellwood DR (1995b) Direct estimate of bioerosion by two parrotfish species, *Chlorurus gibbus* and *C. sordidus*, on the Great Barrier Reef, Australia. Mar Biol 121:419–429

Bellwood DR (1996) Production and reworking of sediment by parrotfishes (family Scaridae) on the Great Barrier Reef, Australia. Mar Biol 125:795–800

Bellwood DR, Choat JH (1990) A functional analysis of grazing in parrotfishes (family Scaridae): the ecological implications. Env Biol Fishes 28:189–214

Bellwood DR, Hoey AS, Choat JH (2003) Limited functional redundancy in high diversity systems: resilience and ecosystem function on coral reefs. Ecol Lett 6:281–285

Birrell CL, McCook LJ, Willis BL, Diaz-Pulido GA (2008) Effects of benthic algae on the replenishment of corals and the implications for the resilience of coral reefs. Oceanogr Mar Biol: Ann Review 46:5–63

Blanchon P, Perry CT (2004) Taphonomic differentiation of *Acropora palmata* facies in cores from Campeche Bank reefs, Gulf of México. Sedimentol 51:5–76

Box SJ, Mumby PJ (2007) Effect of macroalgal competition on growth and survival of juvenile Caribbean corals. Mar Ecol Prog Ser 342:139–149

Bromley RG (1978) Bioerosion of the Bermudan reefs. Palaeogeogeog, Palaeoclim, Palaeoecol 23:169–197

Bromley RG (1994) The palaeoecology of bioerosion. In: Donovan SK (ed) The Palaeobiology of trace fossils, John Wiley & Sons, Chichester

Bruggemann JH, van Kessel AM, van Rooij JM, Breeman AM (1995) A parrotfish bioerosion model: implications of fish size, feeding mode and habitat use for the destruction of reef substrates. In. Bruggemann JH (ed) Parrotfish grazing on coral reefs: a trophic novelty. PhD thesis, University of Groningen

Bruggemann JH, van Kessel AM, van Rooij JM, Breeman AM (1996) Bioerosion and sediment ingestion by the Caribbean parrotfish *Scarus vetula* and *Sparisoma viride*: implications of fish size, feeding mode and habitat use. Mar Ecol Prog Ser 134:59–71

Bruno JF, Selig ER (2007) Regional decline of coral cover in the Indo-Pacific: timing, extent, and subregional comparisons. PLoS One 2, doi:10.1371/journal.pone.0000711

Carpenter RC (1981) Grazing by *Diadema antillarum* Philippi and its effects on the benthic algal community. J Mar Res 39:749-765

Carpenter RC (1988) Mass mortality of a Caribbean sea urchin: immediate effects on community metabolism and other herbivores. Proc Natl Acad Sci 85:511–514

Carreiro-Silva M, McClanahan TR (2001) Echinoid bioerosion and herbivory on Kenyan coral reefs: the role of protection from fishing. J Exp Mar Biol Ecol 262:133–153.

Carriker MR (1969) Excavation of boreholes by the gastropod, *Urosalpinx*: an analysis by light and scanning electron microscopy. Am Zool 9:917–933

Cebrian E (2010) Grazing on coral reefs facilitates growth of the excavating sponge *Cliona orientalis* (Clionaidae, Hadromerida). Mar Ecol Prog Series 31:533–538
Chave KE, Smith SV, Roy KJ (1972) Carbonate production by coral reefs. Mar Geol 12:123–140
Chazottes V, Le Campion-Alsumard T, Peyrot-Clausade M, Cuet P (2002) The effects of eutrophication-related alterations to coral reef communities on agents and rates of bioerosion (Reunion Island, Indian Ocean). Coral Reefs 21:375–390
Chazottes V, Reijmer JJ, Cordier E (2008) Sediment characteristics in reef areas influenced by eutrophication-related alterations of benthic communities and bioerosion processes. Mar Geol 250:114–127
Chiappone M, Swanson DW, Miller SL, Smith SG (2002) Large-scale surveys on the Florida Reef Tract indicate poor recovery of the long-spined sea urchin *Diadema antillarum*. Coral Reefs 21:155–159
Clarke RD (1996) Population shifts in two competing fish species on a degrading coral reef. Mar Ecol Prog Ser 137:51–58
Coppard SE, Campbell AC (2005) Distribution and abundance of regular sea urchins on two coral reefs in Fiji. Micron 37:249–269
Diaz-Pulido G, McCook LJ (2003) Relative roles of herbivory and nutrients in the recruitment of coral-reef seaweeds. Ecol 84:2026–2033
Duerden J (1902) Boring algae as agents in disintegration of corals. Bull Am Mus Nat Hist 16:323–332
Dumas P, Kulbicki M, Chifflet S, Fichez R, Ferraris J (2007) Environmental factors influencing urchin spatial distributions on disturbed coral reefs (New Caledonia, South Pacific). J Exp Mar Biol Ecol 344:88–100
Duncan P (1876) On some thallophytes parasitic within recent *Madreporaria*. Proc Royal Soc London 24:101–107
Eakin CM (1996) Where have all the carbonates gone? A model comparison of calcium carbonate budgets before and after the 1982-1983 El Niño at Uva Island in the eastern Pacific. Coral Reefs 15:109–119
Edinger EN, Limmon GV, Jompa J, Widjatmoko W, Heikoop JM, Risk MJ (2000) Normal coral growth rates on dying reefs: are coral growth rates good indicators of reef health? Mar Poll Bull 40:404–425
Edinger EN, Risk MJ (1994) Oligocene-Miocene extinction and geographic restriction of Caribbean coral reefs: role of temperature, turbidity and nutrients. Palaios 9:576–598
Edmunds PJ, Carpenter RC (2001) Recovery of *Diadema antillarum* reduces macroalgal cover and increases abundance of juvenile corals on a Caribbean reef. Proc Natl Acad Sci 98:5067–5071
Ehrlich HL (1990) Microbial formation and degradation of carbonates. In: Ehrlich HL (ed) Geomicrobiology. Dekker, New York
Evans JW (1970) Palaeontological implications of a biological study of rock-boring clams (Family Pholadidae). In: Crimes TP, Harper JC (eds) Trace fossils. Seel, Liverpool.
Fabricius KE (2005) Effects of terrestrial runoff and the ecology of coral reefs: review and synthesis. Mar Poll Bull 50:125–146
Fabricius KE, Okaji K, De'ath G. (2010) Three lines of evidence to link outbreaks of the crown-of-thorns seastar *Acanthaster planci* to the release of larval food limitation. Coral Reefs 29:593–605
Foster SA (1987) The relative impacts of grazing by Caribbean coral reef fishes and *Diadema*: effects of habitat and surge. J Exp Mar Biol Ecol 105:1–20
Frydl P, Stearn CW (1978) Rate of bioerosion by parrotfish in Barbados reef environments. J Sed Petrol 48:1149–1157
Fursich FT, Palmer TJ, Goodyear KL (1994) Growth and disintegration of bivalve-dominated patch reefs in the upper Jurassic of southern England. Palaeont 37:131–171
Fütterer DK (1974) Significance of boring sponge *Cliona* for the origin of fine-grained material of carbonate sediments. J Sed Petrol 44:79–80
Gardiner J (1903) The Maldive and Laccadive groups, with notes on other coral formations in the Indian Ocean. The Fauna of the Maldive and Laccadive Archipelagoes I: 333–341
Gardner TA, Côté IM, Gill JA, Grant A, Watkinson AR (2003) Long-term region-wide declines in Caribbean corals. Science 301:958–960
Ginsburg RN, Schroeder JH (1973) Growth and submarine fossilization of algal cup reefs, Bermuda. Sedimentol 20:575–614
Glynn PW (1988) El Nino warming, coral mortality and reef framework destruction by echinoid bioerosion in the Eastern Pacific. Galaxea 7:129–160
Glynn PW (1997) Bioerosion and coral-reef growth: a dynamic balance. In: Birkeland C (ed) Life and death of coral reefs. Chapman & Hall, New York 69–98
Golubic S, Friedmann I, Schneider J (1981) The lithobiontic ecological niche, with special reference to microorganisms. J Sed Petrol 51:475–478
Golubic S, Perkins RD, Lukas KJ (1975) Boring microorganisms and microborings in carbonate substrates. In: Frey RW (ed) The study of trace fossils. Springer-Verlag, Berlin
Goreau TF, Hartman WD (1963) Boring sponges as controlling factors in the formation and maintenance of coral reefs. Am Assoc Adv Sci Pubs 75:25–54
Greenstein BJ, Pandolfi JM (2003) Taphonomic alteration of reef corals: Effects of reef environment and coral growth form II: The Florida Keys. Palaios 18:495–509
Griffin SP, García RP, Weil E (2003) Bioerosion in coral reef communities in southwest Puerto Rico by the sea urchin *Echinometra viridis*. Mar Biol 143: 79–84
Hallock P, Schlager W (1986) Nutrient excess and the demise of coral reefs and carbonate platforms. Palaios 1:389–398
Harborne AR, Mumby PJ, Micheli F, Perry CT, Dahlgren CP, Holmes KE, Brumbaugh DR (2006) The functional value of Caribbean coral reef, seagrass and mangrove habitats to ecosystem processes. Adv Mar Biol 50:57–189
Harborne AR, Renaud PG, Tyler EHM, Mumby PJ (2009) Reduced density of the herbivorous urchin *Diadema antillarum* inside a Caribbean marine reserve linked to increased predation pressure by fishes. Coral Reefs 28:783–791
Harney JN, Fletcher CH (2003) A budget of carbonate framework and sediment production, Kailua Bay, Oahu, Hawaii. J Sed Res 73:856–868
Hart AM, Klumpp DW (1996) Response of herbivorous fishes to crown-of-thorns starfish *Acanthaster planci* outbreaks. I. Substratum analysis and feeding ecology of *Acanthurus nigrofuscus* and *Scarus frenatus*. Mar Ecol Prog Ser 132:11–19
Hart AM, Klumpp DW, Russ GR (1996) Response of herbivorous fishes to crown-of-thorns starfish *Acanthaster planci* outbreaks. 11. Density and biomass of selected species of herbivorous fish and fish-habitat correlations. Mar Ecol Prog Ser 132:21–30
Hay ME (1984) Patterns of fish and urchin grazing on Caribbean coral reefs: are previous results typical? Ecol 65:446–454
Hein FJ, Risk MJ (1975) Bioerosion of coral heads: inner patch reefs, Florida reef tract. Bull Mar Sci 25:133–138
Highsmith RC (1980) Geographic patterns of coral bioerosion: a productivity hypothesis. J Exp Mar Biol Ecol 46:177–196
Highsmith RC (1981) Coral bioerosion: damage relative to skeletal density. Am Natural 117:193–198
Highsmith RC, Lueptow RL, Schonberg SC (1983) Growth and bioerosion of three massive corals on the Belize Barrier Reef. Mar Ecol Prog Ser 13:261–271
Hoegh-Guldberg O, Mumby PJ, Hooten AJ, Steneck RS, Greenfield P, Gomez E, Harvell CD, Sale PF, Edwards AJ, Caldeira K, Knowlton N, Eakin CM, Iglesias-Prieto R, Muthiga N, Bradbury RH, Dubi A, Hatziolos ME (2007) Coral reefs under rapid climate change and ocean acidification. Science 318:1737–1742

Hoey AS, Bellwood DR (2008) Cross-shelf variation in the role of parrotfishes on the Great Barrier Reef. Coral Reefs 27:37–47

Holmes KE, Edinger EN, Hariyadi, Limmon GV, Risk MJ (2000) Bioerosion of live massive corals and branching coral rubble on Indonesian coral reefs. Mar Poll Bull 40:606–617. Hubbard DK (2009) Depth-related and species-related patterns of Holocene reef accretion in the Caribbean and western Atlantic: a critical assessment of existing models. In: Swart PK Eberli GP, MacKenzie JA (eds) Perspectives in carbonate geology: a tribute to the career of Robert Nathan Ginsburg. IAS Special Publication No. 41: 1–18.

Hubbard DK (2009) Depth and species-related patterns of Holocene reef accretion in the Caribbean and western Atlantic: a critical assessment of existing models. Spec Publ Int Assoc Sedimentol 40:1–18

Hubbard DK, Burke RB, Gill IP (1986) Styles of reef accretion along a steep, shelf–edge reef, St. Croix, U.S. Virgin Islands. J Sed Petrol 56:848–861

Hubbard DK, Miller AI, Scaturo D (1990) Production and cycling of calcium carbonate in a shelf-edge reef system (St. Croix, U.S. Virgin Islands): applications to the nature of reef systems in the fossil record. J Sed Petrol 60:335–360

Hubbard DK, Zankl H, Heerden IV, Gill IP (2005) Holocene reef development along the northeastern St. Croix shelf, Buck Island, U.S. Virgin Islands. J Sed Res 75:9–113

Hughes TP (1994) Catastrophes, phase shifts, and large-scale degradation of a Caribbean coral reef. Science 265:15471551.

Hughes TP, Rodrigues MJ, Bellwood DR, Ceccarelli D, Hoegh-Guldberg O, McCook L, Moltschaniwskyj N, Pratchett MS, Steneck RS, Willis B (2007) Phase shifts, herbivory, and the resilience of coral reefs to climate change. Current Biol 17:360365.

Hunter IG (1977) Sediment production by *Diadema antillarum* on a Barbados fringing reef. Proc 3rd Int Coral Reef Symp. Miami, FL 1:105–109

Hutchings PA (1986) Biological destruction of coral reefs - A review. Coral Reefs 4:239-252

Hutchings PA (2011) Bioerosion. In: Hopley D (ed) Encyclopedia of modern coral reefs: structure, form and process. Springer Dordrecht, The Netherlands, 139–156

Hutchings PA, Kiene WE, Cunningham RB, Donnelly C (1992) Spatial and temporal patterns of non-colonial boring organisms (polychaetes, sipunculans and bivalves molluscs) in *Porites* at Lizard Island, Great Barrier Reef. Coral Reefs 11:23–31

Hutchings PA, Peyrot-Clausade, M. (2002) The distribution and abundance of boring species of polychaetes and sipunculans in coral substrates in French Polynesia. J Exp Mar Biol Ecol 269:101–121

Jackson JBC, Donovan MK, Cramer KL, Lam VV (eds) (2014) Status and Trends of Caribbean Coral Reefs: 1970-2012. Global Coral Reef Monitoring Network, IUCN, Gland, Switzerland.

Kelly SRA, Bromley RG (1984) Ichnological nomenclature of clavate borings. Palaeontol 27:793–807

Kennedy EV, Perry CT, Halloran PR, Fine M, Carricart-Ganivet JP, Iglesias-Prieto R, Form A, Wisshak M, Schönberg CHL, Mumby PJ. (2013) Avoiding coral reef functional collapse requires combined local and global action. Current Biol 23:912–918

Kiene WE (1988) A model of bioerosion on the Great Barrier Reef. Proc 6th Int Coral Reef Symp 3:449–454

Kiene WE (1989) Biological destruction on the Great Barrier Reef. PhD thesis, Australian National University, Canberra.

Kiene WE, Hutchings PA (1994) Bioerosion experiments at Lizard Island, Great Barrier Reef. Coral Reefs 13:91–98

Kleeman K (1990) Boring and growth in chemically boring bivalves from the Caribbean, Eastern Pacific and Australia's Great Barrier Reef. Senckenbergiana maritima 21:101–154

Kleeman K (2008) *Parapholas quadrizonata* (Spengler, 1792) dominating dead-coral boring from the Maldives, Indian Ocean. In: Wisshak M, Tapanila L (eds) Current developments in bioerosion. Springer-Verlag, Berlin Heidelberg

Klein R, Mokady O, Loya Y (1991) Bioerosion in ancient and contemporary corals of the genus *Porites*: patterns and palaeoenvironmental implications. Mar Ecol Prog Ser 77:245-251

Klement K, Toomey D (1967) Role of the blue-green alga *Girvanella* in skeletal grain destruction and lime-mud formation in the Lower Ordovician of West Texas. J Sed Petrol 37:1045–1051

Knutson TR, McBride JL, Chan J, Emanuel K, Holland G, Landsea C, Held I, Kossin JP, Srivastava AK, Sugi M (2010) Tropical cyclones and climate change. Nature Geosci 3:157–163

Kobluk DR (1988) Cryptic faunas in reefs: ecology and geologic importance. Palaios 3:379–390

Kobluk DR, Risk MJ (1977) Rate and nature of infestation of carbonate substrate by a boring alga. J Exp Mar Biol Ecol 27:107–115

Kolliker A (1859) On the frequent occurrence of vegetable parasites in the hard structure of animals. Proc R Soc Lond 10:95–99

Krone R, Paster M, Schuhmacher H (2011) Effect of the surgeonfish *Ctenochaetus striatus* (Acanthuridae) on the processes of sediment transport and deposition on a coral reef in the Red Sea. Facies 57:215–221

Lam K, Shin PK, Hodgson P (2007) Severe bioerosion caused by an outbreak of corallivorous *Drupella* and *Diadema* at Hoi Ha Wan Marine Park, Hong Kong. Coral Reefs 26:893

Lessios HA (1988) Mass mortality of *Diadema antillarum* in the Caribbean: what have we learned? Ann Rev Ecol Syst 19:371–393

Lessios HA (1995) *Diadema antillarum* 10 years after mass mortality: still rare, despite help from a competitor. Proc R Soc Lond B 259:331–337

Levitan DR (1989) Density-dependent size regulation in *Diadema antillarum*: effects on fecundity and survivorship. Ecol 70:1414–1424

Lewis SM, Wainwright PC (1985) Herbivore abundance and grazing intensity on a Caribbean coral reef. J Exp Mar Biol Ecol 87:215–228

Liddell WD, Ohlhorst SL (1986) Changes in benthic community composition following the mass mortality of *Diadema* at Jamaica. J Exp Mar Biol Ecol 95: 271–278

Liddell WD, Ohlhorst SL (1987) Patterns of reef community structure, north Jamaica. Bull Mar Sci 40:311–329

Lidz BH, Hallock P (2000) Sedimentary petrology of a declining reef ecosystem, Florida reef tract (U.S.A). J Coastal Res 16:675–697

Littler MM, Taylor PR, Littler DS (1989) Complex interactions in the control of coral zonation on a Caribbean reef flat. Oecol 80:331–340

López-Victoria M, Zea S. (2004) Storm-mediated coral colonization by an excavating Caribbean sponge. Climate Res 26:25–256

MacGeachy JK (1977) Factors controlling sponge boring on Barbados reef corals. Proc 3rd Int Coral Reef Symp 2:477–483

MacGeachy JK, Stearn CW (1976) Boring by macro-organisms in the coral *Montastrea annularis* on Barbados reefs. Hydrobiol 61:715–745

Macintyre IG, Glynn PW (1976) Evolution of a modern Caribbean fringing reef, Galeta Point, Panama. Am Assoc Petrol Geol Bull 60:1054–1072

Mägdefrau K (1932) Über einige Bohrgänge aus dem unteren Muschelkalk von Jena. Paläontologia Z. 14:513–523

Mallela J, Perry CT (2007) Calcium carbonate budgets for two coral reefs affected by different terrestrial runoff regimes, Rio Bueno, Jamaica. Coral Reefs 26:129–145

Márquez JC, Zea S (2012) Parrotfish mediation in coral mortality and bioerosion by the encrusting, excavating sponge *Cliona tenuis*. Mar Ecol 33: 417–426

McClanahan TR, Muthiga NA, Kamukuru AT, Machano H, Kiambo RW (1999) The effects of marine parks and fishing on coral reefs of northern Tanzania. Biol Cons 89:161–182

McClanahan TR, Shafir SH (1990) Causes and consequences of sea urchin abundance and diversity in Kenyan coral reef lagoons. Oecol 83:362–370

McGehee MA (1992) Distribution and abundance of two species of *Echinometra* (Echinoidea) on coral reefs near Puerto Rico. Carib J Sci 28:173–183

Miller MW, Kramer, KL William SM, Johnston L, Szmant AM (2009) Assessment of current rates of *Diadema antillarum* larval settlement. Coral Reefs 28:511–515

Mokady O, Lazar B, Loya Y (1996) Echinoid bioerosion as a major structuring force of Red Sea coral reefs. Bio Bull 90:367–372

Moretzsohn F, Tsuchyia M (1993) Preliminary survey of the coral boring Bivalvia fauna of Okinawa, southern Japan. Proc of the 7th Int Coral Reef Symp 1: 404–412

Morrison D (1988) Comparing fish and urchin grazing in shallow and deeper coral reef algal communities. Ecol 69:1367–1382

Mumby PJ (2006) The impact of exploiting grazers (Scaridae) on the dynamics of Caribbean coral reefs. Ecol Apps 16:747–769

Mumby PJ, Dahlgren CP, Harborne AR, Kappel CV, Micheli F, Brumbaugh DR, Holmes KE, Mendes JM, Broad K, Sanchirico JN, Buch K, Box S, Stoffle RW, Gill A B (2006a) Fishing, trophic cascades, and the process of grazing on coral reefs. Science 311:98–101

Mumby PJ, Edwards AJ, Arias-González JE, Lindeman KC, Blackwell PG, Gall A, Gorczynska MI, Harborne AR, Pescod CL, Renken H, Wabnitz C, Llewellyn G (2004) Mangroves enhance the biomass of coral reef fish communities in the Caribbean. Nature 427:533–536

Mumby PJ, Hedley JD, Zychaluk K, Harborne AR, Blackwell PG (2006b) Revisiting the catastrophic die-off of the urchin *Diadema antillarum* on Caribbean coral reefs: fresh insights on resilience from a simulation model. Ecol Mod 196:131–148

Mumby PJ, Steneck RS (2008) Coral reef management and conservation in light of rapidly evolving ecological paradigms. Trends Ecol Evol 23:555–563

Mumby PJ, Steneck RS, Edwards AJ, Ferrari R, Coleman R, Harborne AR, Gibson JP (2012) Fishing down a Caribbean food web relaxes trophic cascades. Mar Ecol Prog Ser 445:13–24

Mumby PJ, Wabnitz C (2002) Spatial patterns of aggression, territory size, and harem size in five sympatric Caribbean parrotfish species. Env Biol Fish 63:265–279

Musso B (1993) Rates of skeletal degradation following death in three species of *Acropora*. Proc 7th Int Coral Reef Symp 1:413–418

Muthiga NA, McClanahan TR (2007) Ecology of *Diadema*. Edible sea urchins: In: Lawrence JM (ed) Biology and ecology. Elsevier Science B.V., Amsterdam

Nagelkerken I, van der Velde G, Gorissen MW, Meijer GJ, van't Hof T, den Hartog C (2000) Importance of mangroves, seagrass beds and the shallow coral reef as a nursery for important coral reef fishes, using a visual census technique. Est Coast Shelf Sci 51:31–44

Neumann AC (1966) Observations on coastal erosion in Bermuda and measurements of the boring rate of the sponge, *Cliona lampa*. Limnol Oceanogr 11:92–108

Nugues MM, Bak RPM (2006) Differential competitive abilities between Caribbean coral species and a brown alga: a year of experiments and a long-term perspective. Mar Ecol Prog Ser 315:75–86

Nugues MM, Smith GW, van Hooidonk RJ, Seabra MI, Bak RPM (2004) Algal contact as a trigger for coral disease. Ecol Lett 7:919–923

Ogden JC, Lobel PS (1978) The role of herbivorous fishes and urchins in coral reef communities. Env Biol Fish 3:49–63

Ong L, Holland KN (2010) Bioerosion of coral reefs by two Hawaiian parrotfishes: species, size differences and fishery implications. Mar Biol 157:1313–1323

Osorno A, Peyrot-Clausade M, Hutchings PA (2005) Patterns and rates of erosion in dead *Porites* across the Great Barrier Reef (Australia) after 2 years and 4 years of exposure. Coral Reefs 24:292–303

Otter GW (1937) Rock-destroying organisms in relation to coral reefs. Scientific Reports of the Great Barrier Reef Expedition 1928–1929 1:323–352

Pandolfi JM, Greenstein BJ (1997) Taphonomic alteration of reef corals: effects of reef environment and coral growth form. I. The Great Barrier Reef. Palaios 12:27–42

Pandolfi JM, Michin PR (1995) A comparison of taxonomic diversity between reef coral life and death assemblages in Madang Lagoon, Papua, New Guinea. Palaeogeog, Palaeoclim, Palaeoecol 119:321–341

Pang RK (1973) The systematics of some Jamaican excavating sponges (Porifera). Postilla 161: 75pp

Pari N, Peyrot-Clausade M, Le Campion-Alsumard T, Hutchings P, Chazottes V, Golubic S, Le Campion J, Fontaine MF (1998) Bioerosion of experimental substrates on high islands and on atoll lagoons (French Polynesia) after two years of exposure. Mar Ecol Prog Ser 166:119–130

Pari N, Peyrot-Clausade M, Hutchings PA (2002) Bioerosion of experimental substrates on high islands and atoll lagoons (French Polynesia) during 5 years of exposure. J Exp Mar Biol Ecol 276:109–127

Perry CT (1996) Distribution and abundance of macroborers in an upper Miocene reef system, Mallorca, Spain: implications for reef development and framework destruction. Palaios 11:40–56

Perry CT (1998a) Grain susceptibility to the effects of microboring: implications for the preservation of skeletal carbonates. Sedimentol 45:39–51

Perry CT (1998b) Macroborers within coral framework at Discovery Bay, north Jamaica: species distribution and abundance, and effects on coral preservation. Coral Reefs 17:277–287

Perry CT (1999) Reef framework preservation in four contrasting modern reef environments, Discovery Bay, Jamaica. J Coastal Res 15:796–812

Perry CT, Edinger EN, Kench PS, Mumby PJ, Murphy G, Steneck RS, Smithers SG (2012) Estimating rates of biologically driven coral reef framework production and erosion: a new census-based carbonate budget methodology and applications to the reefs of Bonaire. Coral Reefs 31:853–868

Perry CT, Hepburn LJ (2008) Syn-depositional alteration of coral reef framework through bioerosion, encrustation and cementation: taphonomic signatures of reef accretion and reef depositional events. Earth Sci Rev 86:106–144

Perry CT, Kench PS, O'Leary MJ, Morgan KM, Januchowski-Hartley F (2015) Linking reef ecology to island-building: Parrotfish identified as major producers of island-building sediment in the Maldives. Geology 43:503–506

Perry CT, Murphy GN, Kench PS, Edinger EN, Smithers SG, Steneck RS, Mumby PJ (2014) Changing dynamics of Caribbean reef carbonate budgets: emergence of reef bioeroders as critical controls on present and future reef growth potential. Proc Royal Soc B 281: 2014–2018

Perry CT, Murphy GN, Kench PS, Smithers SG, Edinger EN, Steneck RS, Mumby PJ (2013) Caribbean-wide decline in carbonate production threatens coral reef growth. Nature Communications. doi: http://dx.doi.org/10.1038/ncomms2409

Perry CT, Smithers SG (2006) Taphonomic signatures of turbid-zone reef development: examples from Paluma Shoals and Lugger Shoal, inshore central Great Barrier Reef, Australia. Palaeog, Palaeoclim, Palaeoecol 242:1–20

Perry CT, Smithers SG, Palmer SE, Larcombe P, Johnson KG (2008a) A 1200 year paleoecological record of coral community development from the terrigenous inner-shelf of the Great Barrier Reef. Geology 36:691–694

Perry CT, Spencer T, Kench PS (2008b) Carbonate budgets and reef production states: a geomorphic perspective on the ecological phase-shift concept. Coral Reefs 27:853–866

Peyrot-Clausade M, Chabanet P, Conand C, Fontaine MF, Letourneur Y, Harmelin-Vivien M (2000) Sea urchin and fish bioerosion on La Réunion and Moorea reefs. Bull Mar Sci 66:477–485
Peyrot-Clausade M, Chazottes V, Pari N (1999) Bioerosion in the carbonate budget of two Indo-Pacific reefs: La Réunion (Indian Ocean) and Mooréa (Pacific Ocean). Bull Geol Soc Denmark 45:151–155
Pinnegar JK, Polunin NVC, Francour P, Badalamenti F, Chemello R, Harmelin-Vivien ML, Hereu B, Milazzo M, Zabala M, D'Anna G, Pipitone C (2000) Trophic cascades in benthic marine ecosystems: lessons for fisheries and protected-area management. Env Cons 27:179–200
Pleydell SM, Jones B (1988) Borings of various faunal elements in the Oligocene-Miocene Bluff Formation of Grand Cayman, British West Indies. J Paleontol 62:348–367
Pratchett MS, Munday PL, Wilson SK, Graham NAJ, Cinner JE, Bellwood DR, Jones GP, Polunin NVC, McClanahan TR (2008) Effects of climate-induced coral bleaching on coral-reef fishes - Ecological and economic consequences. Oceanogr Mar Biol: Ann Rev 46:251–296
Radtke G (1993) The distribution of microborings in molluscan shells from recent reef environments at Lee Stocking Island, Bahamas. Facies 29:81–92
Radtke G, Golubic S (2005) Microborings in mollusk shells, Bay of Safaga, Egypt: morphometry and ichnology. Facies 51:118–134
Randall JE, Schroeder RE, Starck WA (1964) Notes on the biology of the echinoid *Diadema antillarum*. Carib J Sci 4:421–433
Rasmussen KA, Frankenberg EW (1990) Intertidal bioerosion by the chiton *Acanthopleura granulata*; San Salvador, Bahamas. Bull Mar Sci 47:680–695
Reaka-Kudla ML, Feingold JS, Glynn W (1996) Experimental studies of rapid bioerosion of coral reefs in the Galápagos Islands. Coral Reefs 15:101–107
Rice ME, Macintyre IG (1982) Distribution of Sipuncula in the coral reef community, Carrie Bow Cay, Belize. In: Rützler K, Macintyre IG (eds) The Atlantic Barrier Reef Ecosystem at Carrie Bow Cay, Belize I: Structure and Communities. Smithsonian Institution Press, Washington 311–320
Risk MJ, Edinger EN (2011) Impacts of sediment on coral reefs. In: Hopley D (ed) Encyclopedia of modern coral reefs. Springer, Netherlands 575–586.
Risk MJ, Sammarco PW (1982). Bioerosion of corals and the influence of damselfish territoriality: a preliminary study. Oecologia 52:376–380
Rose CS, Risk MJ (1985) Increase in *Cliona delitrix* infestation of *Montastrea cavernosa* heads on an organically polluted protion of the Grand Cayman fringing reef. Mar Ecol-Pub Staz Zool Napolii, 6:345–363
Rotjan RD, Lewis SM (2006) Parrotfish abundance and selective corallivory on a Belizean coral reef. J Exp Mar Biol Ecol 335:292–301.
Ruengsawang N, Yeemin T (2000) Bioerosion caused by grazing activities on coral communities in the Gulf of Thailand. Proceedings Ninth International Coral Reef Symposium, Bali. 1:289–294
Russ G (1984) Distribution and abundance of herbivorous grazing fishes in the central Great Barrier Reef. II. Patterns of zonation of mid-shelf and outershelf reefs. Mar Ecol Prog Ser 20:35–44
Russo AR (1980) Bioerosion by two rock boring echinoids (*Echinometra mathaei* and *Echinostrephus aciculatus*) on Enewetak Atoll, Marshall Islands. J Mar Res 38:99–110
Sammarco PW (1980) *Diadema* and its relationship to coral spat mortality: grazing, competition, and biological disturbance. J Exp Mar Biol Ecol 45:245–272
Sammarco PW (1982) Echinoid grazing as a structuring force in coral communities: whole reef manipulations. J Exp Mar Biol Ecol 61:31–55
Sammarco PW, Carleton JH, Risk MJ (1986) Effects of grazing and damselfish territoriality on bioerosion of dead corals: direct effects. J Exp Mar Biol Ecol 98:1–19
Sammarco PW, Risk MJ (1990) Large-scale patterns in internal bioerosion of *Porites*: cross continental shelf trends on the Great Barrier Reef. Mar Ecol Prog Ser 59:145–156
Sammarco PW, Risk MJ, Rose C (1987) Effects of grazing and damselfish territoriality on internal bioerosion of dead corals: indirect effects. J Exp Mar Biol Ecol 112:185–199
Sammarco PW, Williams AH (1982) Damselfish territoriality: influence on *Diadema* distribution and implications for coral community structure. Mar Ecol Prog Ser 8:53–59
Sano M (2000) Stability of reef fish assemblages: responses to coral recovery after catastrophic predation by *Acanthaster planci*. Mar Ecol Prog Ser 198:121–130
Sano M, Shimizu M, Nose Y (1987) Long-term effects of destruction of hermatypic corals by *Acanthaster planci* infestation on reef fish communities at Iriomote Island, Japan. Mar Ecol Prog Ser 37:191–199
Schuhmacher H, Krone R, van Treeck P (2008) Underestimated eroder among reef fishes - experimental comparison between *Ctenochaetus striatus* and *Acanthurus nigrofuscus* (Acanthuridae). Proc 11th Int Coral Reef Symp, Ft. Lauderdale, Florida. 1:337–340
Scoffin T (1992) Taphonomy of coral reefs: a review. Coral Reefs 11:57–77
Scoffin TP, Bradshaw C (2000) The taphonomic significance of endoliths in dead-versus live-coral skeletons. Palaios 15:248–254
Scoffin TP, Garrett P (1974) Processes in the formation and preservation of the internal structure in Bermuda patch reefs. Proc 2nd Int Coral Reef Symp 2: 429–448
Scoffin TP, Stearn CW, Boucher D, Frydl P, Hawkins CM, Hunter IG, MacGeachy JK (1980) Calcium carbonate budget of a fringing reef on the west coast of Barbados. Part II - Erosion, sediments and internal structure. Bull Mar Sci 30: 475–508
Scott PJB (1988) Distribution, habitat and morphology of the Caribbean coral- and rock-boring bivalve, *Lithophaga bisulcata* (d'Orbigny) (Mytilidae: Lithophaginae). J Molluscan Studies 54:83–95
Sheppard C, Dixon DJ, Gourlay M, Sheppard A, Payet R (2005) Coral mortality increases wave energy reaching shores protected by reef flats: examples from the Seychelles. Est Coast Shelf Sci 64:223–234
Soliman G (1969) Ecological aspects of some coral-boring gastropods and bivalves of the northwestern Red Sea. Am Zool 9:887–894
Stockwell B, Jadloc CRL, Abesamis RA, Alcala AC, Russ GR (2009) Trophic and benthic responses to no-take marine reserve protection in the Philippines. Mar Ecol Prog Ser 389:1–15
Taylor JD, Way K (1976) Erosive activities of chitons at Aldabra Atoll. J Sed Pet 46:974–977
Tribollet A, Atkinson MJ, Langdon C (2006) Effects of elevated pCO_2 on epilithic and endolithic metabolism of reef carbonates. Glob Change Biol 12:2200–2220
Tribollet A, Decherf G, Hutchings PA, Peyrot-Clausade M (2002) Large-scale spatial variability in bioerosion of experimental coral substrates on the Great Barrier Reef (Australia): importance of microborers. Coral Reefs 21:424–432
Tribollet A, Golubic S (2005) Cross-shelf differences in the pattern and pace of bioerosion of experimental carbonate substrates exposed for 3 years on the northern Great Barrier Reef, Australia. Coral Reefs 24:422–434
Tribollet A, Golubic S (2011) Reef bioerosion: agents and processes. In: Dubinsky Z, Stambler N (eds) Coral reefs: an ecosystem in transition. Springer, Dordrecht, Netherlands pages?
van Rooij JM, de Jong E, Vaandrager F, Videler JJ (1996) Resource and habitat sharing by the stoplight parrotfish, *Sparisoma viride*, a Caribbean reef herbivore. Env Biol Fishes 47:81–91
van Rooij JM, Videler JJ, Bruggemann JH (1998) High biomass and production but low energy transfer efficiency of Caribbean

parrotfish: implications for trophic models of coral reefs. J Fish Biol 53(Suppl. A):154–178

Vogel K (1993) Bioerosion in fossil reefs. Facies 28:109–114

Vogel K, Gektidis M, Golubic S, Kiene W, Radtke G (2000) Experimental studies on microbial bioerosion at Lee Stocking Island, Bahamas and One Tree Island, Great Barrier Reef, Australia; implications for paleoecological reconstructions. Lethaia 33:190–204

Vogel K, Golubic S, Brett CE (1987) Endolithic associations and their relation to facies distribution in the Middle Devonian of New York State, USA. Lethaia 20:26–29

Voigt E (1965) Uber parasitische Polychaeten in Kreide-Austernsowie einige andere in Muschelschalen bohrende Würmer. Paläontol Zeit 39:193211

Ward-Paige CA, Risk MJ, Sherwood OA, Jaap WC (2005) Clionid sponge surveys on the Florida Reef Tract suggest land-based nutrient inputs. Mar Poll Bull 51: 570579

Warme JE (1975) Borings as trace fossils, and the processes of marine bioerosion. In: Frey RW (ed) The Study of Trace Fossils. Springer, Berlin. 181–229

Weil E, Torres JL, Ashton M (1985) Population characteristics of the sea urchin *Diadema antillarum* in La Parguera, Puerto Rico, 17 years after the mass mortality event. Rev Biol Trop 53:219–231

Williams DM (1986) Temporal variation in the structure of reef slope fish communities (central Great Barrier Reef): short-term effects of *Acanthaster planci* infestation. Mar Ecol Prog Ser 28:157–164

Wisshak M, Schönberg CHL, Form A, Freiwald A (2012) Ocean acidification accelerates reefs bioerosion. PLOS One 7:e45124

Woodley JD, Chornesky EA, Clifford PA, Jackson JBC, Kaufman LS, Knowlton N, Lang JC, Pearson MP, Porter JW, Rooney MC, Rylaarsdam KW, Tunnicliffe VJ, Wahle CM, Wulff JL, Curtis ASG, Dallmeyer MD, Jupp BP, Koehl MAR, Neigel J, Sides EM (1981) Hurricane Allen's impact on Jamaican coral reefs. Science, 214:749–755

Sponge Contributions to the Geology and Biology of Reefs: Past, Present, and Future

5

Janie Wulff

Abstract

Histories of sponges and reefs have been intertwined from the beginning. Paleozoic and Mesozoic sponges generated solid building blocks, and constructed reefs in collaboration with microbes and other encrusting organisms. During the Cenozoic, sponges on reefs have assumed various accessory geological roles, including adhering living corals to the reef frame, protecting solid biogenic carbonate from bioeroders, generating sediment and weakening corals by eroding solid substrate, and consolidating loose rubble to facilitate coral recruitment and reef recovery after physical disturbance. These many influences of sponges on substratum stability, and on coral survival and recruitment, blur distinctions between geological vs. biological roles.

Biological roles of sponges on modern reefs include highly efficient filtering of bacteria-sized plankton from the water column, harboring of hundreds of species of animal and plant symbionts, influencing seawater chemistry in conjunction with their diverse microbial symbionts, and serving as food for charismatic megafauna. Sponges may have been playing these roles for hundreds of millions of years, but the meager fossil record of soft-bodied sponges impedes historical analysis.

Sponges are masters of intrigue. They play roles that cannot be observed directly and then vanish without a trace, thereby thwarting understanding of their roles in the absence of carefully controlled manipulative experiments and time-series observations. Sponges are more heterogeneous than corals in their ecological requirements and vulnerabilities. Serious misinterpretations have resulted from over-generalizing from a few conspicuous species to the thousands of coral-reef sponge species, representing over twenty orders in three classes, and a great variety of body plans and relationships to corals and solid carbonate substrata.

Dynamics of living sponges are difficult to document because most sponges heal after partial mortality and vanish quickly after death. Thus observations of localized increases or overgrowths of corals by a few unusual sponge species have led to recent assertions that sponges are in the process of overwhelming coral reefs. However, a consistent pattern of high mortality in the few long-term census studies done on full assemblages suggests that, perhaps for the first time in their long history, sponges may actually be unable to keep up with changes in the sea. Diminished sponge populations could have profound consequences, many of them negative, for corals and coral reefs.

J. Wulff (✉)
Department of Biological Science, Florida State University,
Tallahassee, FL 32303, USA

Smithsonian Tropical Research Institute, Panama
e-mail: wulff@bio.fsu.edu

D.K. Hubbard et al. (eds.), *Coral Reefs at the Crossroads*,
Coral Reefs of the World 6, DOI 10.1007/978-94-017-7567-0_5

Keywords
Sponges • Reef-building • Reef restoration and repair • Water column filtering • Positive ecosystem roles of sponges on coral reefs

5.1 Introduction: Sponges and Reefs Have Been Linked from the Beginning

Sponges are daunting creatures, diverse and difficult to identify. Their growth forms are challenging to quantify, and they impede post-mortem analysis by vanishing quickly without a trace. Sponges are also entrancingly beautiful, expressing an unsurpassed diversity of color and form. They are masters of wound healing, regeneration, and mutually beneficial associations. Geological and biological roles of sponges on reefs are inextricably inter-tangled by the many strong influences that sponges have had, and continue to have, on the stability of solid biogenic substrata and the viability of the organisms producing these substrata. Paleozoic and Mesozoic sponges built primary reef framework blocks with their dense skeletons of calcium carbonate or densely interlocking silica spicules (e.g., Hartman 1977; Wood 1990). Most modern sponges play various accessory roles, many of them required for reef building and maintenance, and played only by sponges. These roles include: (1) fortifying the framework with dense solid carbonate; (2) breaking down solid substrate into silt-sized chips and eroding and weakening the skeletons of framework-builders; (3) aiding reef repair by facilitating consolidation of loose rubble and stabilizing it until carbonate secreting organisms can bind it permanently; (4) improving survival of living corals by "gluing" them to the reef frame if their bases are eroded, and protecting their skeletons from excavators; (5) harboring hundreds of symbiont species (microbes, plants, animals) for all or part of their life cycles; (6) maintaining water clarity and possibly also minimizing water-borne pathogens by efficiently filtering and digesting picoplankton; (7) serving as food for mobile organisms like angelfishes, hawksbill turtles, and nudibranchs; and, (8) in collaboration with their microbial symbionts, influencing seawater concentrations of dissolved inorganic and organic components (reviews in Rützler 1978; Diaz and Rützler 2001; Wulff 2001; Rützler 2004; Wulff 2006e; Bell 2008; Rützler 2012; Wulff 2012). In addition, aesthetic considerations are not trivial in a world in which conservation can be motivated by recreational enthusiasm, and sponges are star contributors of intriguing colors and shapes on reefs.

5.2 The Nature of Sponges

The structure of sponges is more homogeneous and simple than that of other multicellular heterotrophs (e.g., Simpson 1984). Most modern sponges have relatively soft bodies, with living tissue throughout their three-dimensional forms. The living tissue is pervaded by a supporting skeletal meshwork, as well as a system of canals through which the sponges pump water, from which they very efficiently remove picoplankton and in some cases dissolved organic material. Informal homogeneous construction, along with a high degree of cellular totipotency, allows sponges to heal wounds extremely rapidly, attach to substrata with any portion of their bodies, and accommodate intimate associations with symbionts of every group of organisms without mortally disrupting their own function.

Versatility and lability, in both ecological and evolutionary senses, have contributed to the astonishing persistence and diverse functional roles of the Porifera. Sponges of four different body plans, each with a unique set of relationships with corals and reef substrata, have figured prominently throughout the history of reefs:

(1) *free-standing, epibenthic*: of every possible growth form, from thin crusts to giant baskets, clusters of tall tubes, and bushes of erect branches. Their skeletons may be fine meshworks of spongin fibers or spicules or, usually, both. The skeletons entirely pervade the body, which is relatively soft and flexible when spongin dominates the skeletal composition and rigid when there is a higher proportion of spicules. The majority of the sponge biomass on many current coral reefs represents this group of sponges, and in the following account this is the group I refer to if no further specification than 'sponges' is given.
(2) *cryptic*: inhabiting crevices and other cryptic spaces within the reef framework. Most have the same set of skeletal properties as the free-standing sponges, and some species have no skeletons at all, or skeletons of calcareous spicules. These occur in either of two growth forms: thinly encrusting the walls of crevices or else entirely filling small internal spaces in the reef framework. Some members of this sciophilic (shade-loving)

community also live on exposed surfaces, but many are confined to cryptic spaces and evidently never achieve large sizes (van Soest 2009).

(3) *excavating*: boring into solid carbonate substrata, and either living entirely within their burrows, or in some cases also forming thin or thick crusts on the substratum surface.

(4) *hypercalcified* or *coralline*: tissue confined to the surface of extremely dense, solid and massive carbonate skeletons, often with silica spicules as well. On modern reefs these sponges are also sciophilic.

Curiously, each of these four relationships to reef substrata is expressed by extant species in from 3 to all 21 of the currently recognized orders of marine demosponges, suggesting that this range of possible roles on reefs has ancient roots in this clade. Ecological interactions, ecosystem roles, and vulnerabilities to environmental challenges differ substantially among these four types of sponges, and much confusion relating to sponges on reefs has been caused by generalizing from work on a few species representing one of these four sponge types to all "sponges".

5.3 Species Diversity of Sponges on Present-Day Reefs

Of the 8553 described sponge species at the time of the most recent review of global sponge diversity (van Soest et al. 2012), about 42 % inhabit realms with coral reefs (i.e., Western, Central, and Eastern Indo-Pacific, Tropical Eastern Pacific, Tropical Atlantic). The Demospongiae, with 83 % of the extant species, are by far the most speciose of the four classes of sponges; and the proportional representation of Demospongiae among reef-associated species is even greater. Diversity at high taxonomic levels (e.g., more than twenty orders, representing three of the four classes of the Phylum Porifera: Demospongiae, Homoscleromorpha, and Calcarea) is reflected in a variation in geological and biological roles of sponges far exceeding that of the relatively homogeneous reef-building corals, most of which are in the single anthozoan order Scleractinia.

Sponges are also more heterogeneous than corals with respect to abiotic factors and ecological interactions that cause them to thrive or perish. Abiotic factors such as chronic rough water, periodic storm-associated wave action, temperature extremes, freshwater, UV light, sunlight in photosynthetically useful wavelengths, water column nutrients and resulting picoplankton abundance and composition, sediment, competition with other sponges, and opportunistic spongivory have all been demonstrated to influence habitat, depth, or latitudinal distribution and abundance of sponges (detailed review in Wulff 2012). Most striking is how differently sponges respond to factors that influence distribution and abundance. One sponge's nightmare can be another's paradise.

Species diversity of sponges on present day coral reefs exhibits a similar pattern all over the world. When many sites representing a range of depths and local circumstances are sampled within a local area, 100–300 sponge species are typically reported regardless of the ocean basin, e.g.: 157 species in 102 stations in Jamaica (Lehnert and Fischer 1999), 300 species in 417 stations in the Bahamas (Reed and Pomponi 1997), 96 species at 42 stations on three remote southeastern Caribbean atolls (Zea 2001), 261 species at 103 stations at Ningaloo Reef (Schönberg and Fromont 2012), 150 species at 43 stations in the Dampier Archipelago NW Australia (Fromont et al. 2006), 226 species in 22 stations in SE Queensland (Hooper and Kennedy 2002; Hooper et al. 2002), 150 species at 37 stations in the Spermonde Archipelago, Indonesia (Cleary and de Voogd 2007), and 148 species at 30 stations in the Thousand Islands, Indonesia (de Voogd and Cleary 2008). Another consistent world-wide pattern within local areas is extreme heterogeneity from station to station with respect to which subsets of the regional species pool are represented (e.g., Zea 1994; Hooper and Kennedy 2002; Hooper et al. 2002).

Realm-wide faunas include up to ten or more times as many species of sponges as of corals. Moreover, species diversity is relatively similar among tropical ocean basins rather than dramatically lower in the Tropical Atlantic as it is for corals. Van Soest et al. (2012) listed 945, 975, and 1325 described sponge species in the tropical Western Atlantic, Indian Ocean, and Coral Sea/NE Australia, respectively. Reported diversity patterns still reflect the amount of taxonomically focused study, and new sponge species are being described at a rate of 35–87 per year (van Soest et al. 2012). Biogeographic comparisons of current and past diversity and the detailed tracings of taxonomic patterns through time that have been so informative for corals (e.g., Collins et al. 1996; Budd 2000; Schwartz et al. 2012) are impeded for sponges, with the exception of the coralline sponges, by their extremely poor fossil record. Sponge skeletons disarticulate quickly after death, and silica spicules are subject to dissolution in seawater (Hartman 1977; Rützler and Macintyre 1978; Hartman et al. 1980).

5.4 Geological Roles of Sponges: Reef Frame-Building and Fortifying

Ancient groups of reef builders have led humans in a merry chase with respect to their relationships with extant taxa. In her 1990 review of the history of the study of reef-building sponges, Wood (1990) aptly referred to them as "ancient waifs". Fossils designated as archaeocyaths, sphinctozoans, stromatoporoids, and chaetetids suggested tantalizing possibilities to generations of paleontologists. For example, between 1826 and 1970, stromatoporoids were placed with Anthozoa, Bryozoa, Hydrozoa, Cyanobacteria, tabulate corals, Foraminifera, Cephalopoda, Plantae, as well as Porifera (see table in Debrenne 1999, after Wood 1987). Discovery of living sponges with dense calcium carbonate skeletons, some with an initially surprising combination of solid carbonate with silica spicules and collagenous fibers, finally allowed definitive assignment of many of these creatures previously known only as fossils to the sponges (Hartman and Goreau 1970; Hartman and Goreau 1975; Vacelet 1970). Canal systems of the living coralline sponges were strongly reminiscent of traces on the surfaces of the skeletons of fossil stromatoporoids. As soon as Hartman and Goreau (1970) had proposed the shift of stromatoporoids to the Porifera, based on their analysis of *Ceratoporella* from Jamaica (Fig. 5.1) features that had not been previously observed or accorded importance were observed in other fossils. Focused searches for sponge characteristics in fossil material revealed siliceous spicules in some fossils in which they had been assumed to be absent, and astrorhizae were noted to be surface features of fossil chaetetids, providing another link to living examples (review by Wood 1990). The revelation that siliceous spicules in the living relicts can dissolve as they grow helped to further link living examples to fossils lacking spicules (e.g., Stock 2001).

Hartman and Goreau's (1970) elegant discussion of the challenges and joys of relating unusual living organisms to fossils, as well as to other living groups, engaged their readers with questions revolving around what constitutes reliable evidence of clade membership rather than grade (i.e., groups defined by evolutionary relationship vs. groups defined by observable structural similarity). This issue became an important focus of researchers who discovered additional living species and availed themselves of the exciting possibility of learning about ancient groups by detailed study of living representatives. Accumulated details of their biology, skeleton formation, larvae, soft tissue, and spicules have revealed that chaetetid, stromatoporoid, and sphinctozoan are indeed grades rather than clades (e.g., Vacelet et al. 2010; West et al. 2010); and living coralline sponges represent at least five orders of Demospongiae that are represented by non-coralline sponges on modern coral reefs: Clionaida, Merliida, Agelasida, Haplosclerida, and Dictyoceratida; as well as the Class Calcarea, which is far less represented on modern reefs. Curiously sponges of the same grade (i.e., stromatoporoid, sphinctozoan, chaetetid) can be separated by live tissue characteristics into different higher taxa; and skeletons that are readily grouped together as the same grade may exhibit very different microstructure (Willenz and Hartman 1989; Vacelet et al. 2010). Delving into skeletal structure at very fine scales has demonstrated shared pathways in skeletal formation among sponges with different microstructure, a further surprise (e.g., Gilis

Fig. 5.1 Living hypercalcified sponges. All photos were taken by, and contributed to this paper by courtesy of Philippe Willenz: (**a**) *Ceratoporella nicholsoni* (Hickson) a large healthy specimen on a cave wall, Pear Tree Bottom, Jamaica, (**b**) the same specimen of *C. nicholsoni* as in photo a, 3 years later; note the virtual lack of growth that is typical of these extraordinarily slow-growing animals, and also the damage on the top; (**c**) a broken specimen of *C. nicholsoni*, showing the extreme density of the basal calcareous skeleton and the very thin layer of living tissue; (**d**) entrance to the Pear Tree Bottom cave, within which live the few species of hypercalcified sponges that are the surviving remnants of a diverse set of species that thrived on open surfaces and built reefs prior to the Cenozoic

et al. 2013). Hypercalcified sponges known only from fossils represent additional orders of demosponges, but the lack of matching between grade and clade requires that caution be used in assignment to higher taxa, and a classification based on observable morphological characters must remain in practice for fossil taxa (West et al. 2010). Diversity of living hypercalcified sponges is a small remnant of those that built reefs in Paleozoic and Mesozoic oceans.

5.4.1 Archaeocyatha

Archaeocyaths were the earliest reef-building sponges. These Lower Cambrian builders of sturdy carbonate skeletons have been grouped with cnidarians, algae, sponges, vascular plants, and foraminiferans at various times, assigned their own phylum or kingdom, and finally in the early 1990s grouped back where they had been placed in the 1860s and again in the 1930s – among the sponges (Rowland 2001). Similarities in skeleton formation between the living *Vaceletia* and the extinct Archaeocyatha help to link them to the demosponges, as do morphological evidence of filter feeding, crypt cells, and style of asexual propagation (Debrenne and Zhuravlev 1994; Debrenne 1999; Reitner et al. 2001). Although the solitary cone shapes of earlier Archaeocyathans were not conducive to formation of solid reef frameworks, later forms were more integrated (Wood et al. 1992). Reef-formation may nevertheless have depended on the collaboration of associated calcimicrobes (calcium secreting micro-organisms) with the archaeocyaths (Debrenne 2007; Kiessling 2009).

This central theme of the need for collaboration between primary framework builders and various groups of accessory reef-binders for successful reef building has persisted ever since this ancient example. Just as for modern reefs, environmental factors, including water movement and depositional setting, as well as temperature, determined where archaeocyath-calcareous depositing cyanobacterial associations resulted in resistant reefs (Gandin and Debrenne 2010). Environmental requirements must be satisfied for both the organisms contributing solid building blocks and those binding the blocks into a reef.

5.4.2 Hypercalcified Sponges

Following the archaeocyath extinction 500 MYA, sponges of stromatoporoid, sphinctozoan, and chaetetid grades built reefs at various times throughout the Paleozoic and Mesozoic, generally in conjunction with microalgae and metazoan taxa capable of growing in encrusting forms (Hartman et al. 1980; Wood 1995, 2011). Hypercalcified sponges suffered substantial extinctions at the end of both the Devonian and the Triassic (e.g., Kiessling et al. 2007). Extracting clues from ancient sponge reefs that might advise us on the long-term prospects for modern reefs becomes more complex the more we learn. Among the factors that must be considered are Mg/Ca concentrations in seawater (e.g., Stanley and Hardie 1998), as well as temperature, nutrients, sediment, and interactions of all of these factors with each other and with a variety of biotic agents (e.g., Wood 1993; Kiessling 2009; Wood 2011; Chaps. 8 and 9).

Correlations of paleoenvironmental conditions with reef development must be interpreted cautiously. For example, Middle Carboniferous reef mound building by chaetetids is known from low light, low sediment habitats, similar to the situations in which present day coralline sponges are found (West and Kershaw 1991). This could be interpreted as evidence that these were always the favored habitats of coralline sponges, or that during this time period they were forced to such sites, or that such sites were simply where preservation and/or subsequent discovery were more likely. In a comprehensive evaluation of taphonomic issues, Wood (2011) gives many examples of how to avoid misinterpretations by focusing on detailed mechanisms and processes of reef-building, and understanding form-function relationships. Historic roles of solid-skeletoned organisms can be problematic enough when all we have is a snapshot in rock. The likelihood of misinterpretation is exacerbated by the possibility that soft-bodied sponges have played roles in reef construction, maintenance, and repair that leave no traces in the finished reef frame.

Extremely slow growth rates of living coralline sponges (e.g., 0.18–0.23 mm/year for *Ceratoporella*: Willenz and Hartman 1985, 1999; 0.05–0.1 mm/year for *Acanthochaetetes*: Reitner and Gautret 1996; 0.236 mm/year for *Astrosclera*: Wörheide 1998) lend credence to the idea that competition from rapidly growing scleractinians may have played a role in restricting reef-building sponges to caves and other cryptic habitats. Changes in reefs that coincided with the blossoming of scleractinian zooxanthellate corals in the middle Jurassic included the creation of caves and other cryptic spaces by the combined foliaceous, branching, and plate-like morphologies of rapidly growing corals needing to collect sunlight (Jackson et al. 1971). These cryptic spaces provided a new habitat in which sediment and competition from organisms that are fueled by sunlight are minimized (Jackson et al. 1971). Although species diversity may now be relatively low, coralline sponges continue to be key fortifiers of the reef frame (Fig. 5.1), working from inside by depositing skeletons that are at least twice as dense and with compressive strength several times as great as those of scleractinian corals (Schumacher and Plewka 1981; Willenz and Hartman 1999; Vacelet et al. 2010). Individual *Ceratoporella nicholsoni* Hickson can be a meter in diameter and populations can be dense, with as many as 5–12 individuals of greater than 10 cm in diameter per m^2 (Lang et al. 1975). Large individuals must be thousands of years old, suggesting a strategy that has favored resistance to physical damage over efficient recovery from damage (Vacelet et al. 2010). The disadvantage to this strategy, i.e., less efficient recovery, is increasingly apparent on modern reefs that are beset by multiple, larger, and more chronic disturbances (e.g., Wulff 2006b).

5.4.3 Reef-Building Sponges with Siliceous Skeletons: Lithistids and Hexactinellids

Hypercalcified sponges were not the only reef-builders; sponges with hard dense skeletons made of elaborate silica spicules called desmas also built reefs (e.g., Hartman et al. 1980). Ordovician reefs containing large proportions of these lithistid sponges, as well as stromatoporoids, depended on crust-producing microbes, and sometimes encrusting bryozoans, to bind the sponges together and fill gaps between them, thus helping to hold the framework together (Adachi et al. 2011). These sponges may have also served as baffles encouraging deposition of sediment, and subsequent lithification. In the Mesozoic, lithistid sponges contributed especially to Jurassic reefs of the Tethys Sea, but lithistids diminished dramatically in the Cretaceous and early Tertiary and became largely confined to deep water (Maldonado et al. 1999). Experimental support for the idea that these reef-builders diminished near the Cretaceous-Tertiary boundary due to depletion of silicon in shallow water as diatoms proliferated comes from studies of recent sponges grown in silica-enhanced seawater. When Maldonado et al. (1999) grew the common Mediterranean encrusting species *Crambe crambe* in silica-enriched water, it augmented its typical spicule assortment with elaborate spicules similar to those found in fossil deposits. Conversely, the high abundance of lithistids in the Jurassic may have been promoted by higher dissolved silica levels due to volcanic activity in the Triassic (Maldonado et al. 1999). Another possible contributor to the demise of reefs built by siliceous sponges is the extreme post-Jurassic decline of calcimicrobes that both the lithistids and the fused-silica-spicule sponges in the Class Hexactinellida required as collaborators in reef-building (Brunton and Dixon 1994).

Differences among the reef-building sponges in skeletal materials can have far-reaching ramifications for reef accretion. One important difference between lithistid and hypercalcified reef building sponges is the resistance of the lithistids' silica skeletons to boring organisms. On a geological time scale, the same Triassic volcanism that may have boosted silica for lithistid sponges may have also altered seawater chemistry to the detriment of hypercalcified reef-building sponges, which suffered substantial extinction at the end of the Triassic (e.g., Kiessling et al. 2007; Kiessling 2009; Pandolfi and Kiessling 2014).

5.5 Geological Roles of Sponges: Promoting Reef-Frame Integrity, Increasing Coral Survival, and Facilitating Repair

Geological roles of sponges in building and maintaining reefs shifted profoundly in the Tertiary, after over 490 million years of primary framework building (Wood 1990). Currently, sponges serve chiefly as binders, consolidators, eroders, reinforcers, and protectors of solid carbonate (Table 5.1). Soft-bodied sponges may also have played these accessory roles during the Paleozoic and Mesozoic, but (aside from excavations that are readily attributable to boring sponges) it is hard to know how we would recognize such roles of soft-bodied sponges in the fossil record, given that these sponge roles are so difficult to perceive on modern reefs. Even where sponges have been experimentally demonstrated to significantly affect the success of reef building, their contributions are far from obvious by observation alone.

5.5.1 Increasing Coral Survival by Adhering Living Corals to the Reef and Protecting Exposed Skeletons Against Eroders

Goreau and Hartman (1963, 1966) observed that sponges could adhere living corals securely to the reef frame even after their basal attachments were eroded by excavating organisms, and suggested that association with sponges could therefore increase coral survival. In addition to compensating for the erosion of the bases of the corals by gluing them to the reef, sponge cover of coral skeletons where they lack living coral tissue can simultaneously block further invasion by eroders (Fig. 5.2). Wulff and Buss (1979) confirmed these benefits of sponges to corals by mapping and measuring all of the corals on eight fore-reef patch reefs in the San Blas Islands, Panama, and then removing sponges from half of the patch reefs. Only 6 months later, 40 % of the corals, representing 46 % of the percent cover of living coral tissue, had fallen off the reefs from which sponges were removed, in striking comparison with losses of only 4 % of the coral colonies (3 % of the surface area of live coral tissue) from the control reefs. Thus the observably negative role of boring sponge species can be countered by the positive roles of adhesive and protective coating by epibenthic and cryptic sponges. These results illustrate how easy it is to misinterpret the net effect of an interaction of a sponge and

Table 5.1 Evidence for beneficial geological roles of sponges on modern coral reefs

Authors	Process	Evidence type	Experimental results
Goreau and Hartman (1966)	Semi-cryptic sponge species adhering living corals to reefs and protecting exposed solid carbonate from eroders	Skeletons of corals molded around sponge oscules, stable associations for long time periods	
Lang et al. (1975)	Framework reinforcement by coralline sponges within cryptic spaces	Dense populations in caves of sponge species with solid carbonate skeletons 2× density of corals	
Wulff and Buss (1979)	Living sponges of many species adhering living corals to reefs and preventing access by boring sponges to exposed skeleton parts	Removed sponges from patch reefs after mapping and measuring corals on control and experimental reefs, re-measuring after 6 months	40 % of live corals lost from reefs from which sponges were removed, while only 4 % of corals were lost from reefs with sponges
Wulff (1984)	Sponges binding coral rubble into stable piles until coralline algae can grow between rubble pieces to permanently consolidate them, providing surfaces for coral recruits	Experimentally compared fates of coral rubble in piles with and without sponges in Caribbean; compared survival of small corals on stabilized vs. loose rubble; compared rubble consolidation on E Pacific reefs with and without sponges	Rubble piles w sponges were bound into stable piles in 5 mo, corals had recruited by 10 mo; no consolidation of rubble or coral recruitment in piles without sponges; survival of small corals on stable rubble 13 % in 4 years, 1 % on loose rubble; rubble in E Pacific only consolidated where sponges were present
Biggs (2013)	Sponges binding coral rubble into stable piles, corals recruiting onto stabilized rubble	Experimentally compared fates of rubble in piles seeded with sponges and without sponges, as well as rubble made of cement and bound together with cement	Significantly more rubble piles with sponges retained rubble and height; significantly more corals, and of more species, recruited to rubble bound with sponges vs. rubble bound with cement

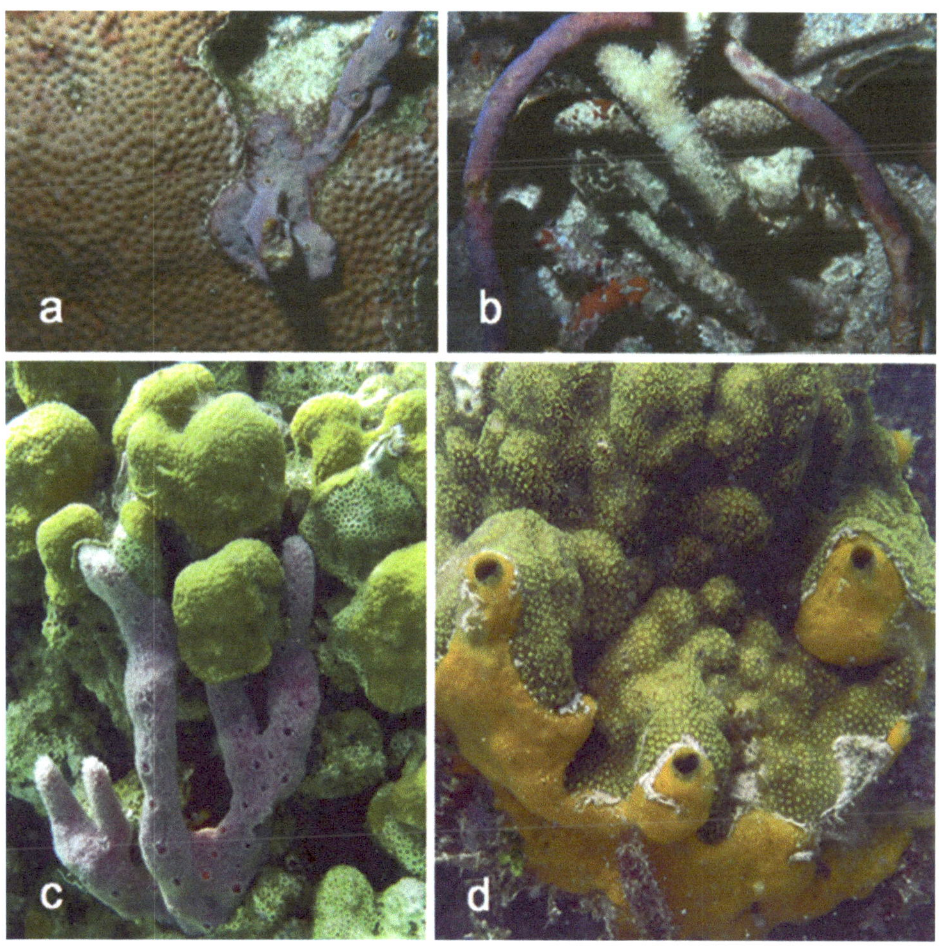

Fig. 5.2 Sponges adhering corals to the reef and protecting exposed solid carbonate from excavators. (**a**) base of a branching sponge, *Aplysina cauliformis* (Carter) from which the erect portion was broken during Hurricane Allen, 1980, Jamaica; the wound healed quickly and regrowth could already be seen after a few weeks; (**b**) a branch of *A. cauliformis*, broken during Hurricane Allen, and caught in a pile of coral rubble also generated by the storm. Within a few days the sponge had attached to several pieces of rubble, binding them together; (**c**) the branching sponge *Niphates erecta* Duchassaing and Michelotti covering bare coral (*Orbicella annularis* (Ellis and Sollander)) skeleton and adhering to live portions of the colony; (**d**) the semi-cryptic sponge *Mycale laevis* (Carter) protecting bared coral (*Porites astreoides* Lamarck) skeleton, gluing the coral to the reef, and even providing an increasing substratum over which the coral can grow (Goreau and Hartman 1966)

coral. Even when the sponge is actually saving the coral's life, it may appear to be engaged in aggressive overgrowth. Time-series observations are essential for determining if a sponge is advancing over the coral. Wulff and Buss (1979) framed their report of this mutually beneficial association in terms of carbonate balance in order to underscore how the net effect of sponge-coral interactions on reef building and maintenance may not be surmised correctly by simple observation.

5.5.2 Rubble Stabilization: A Key Step in Reef Recovery After Physical Damage

Rubble generated by storm waves and other disturbances can be inhospitable to coral recruits, as they are jostled about by chronic water movement and foraging animals. Sponges can solve this instability problem in two ways: (1) sponges living in cryptic spaces under the reef surface can grow upwards into rubble piled upon them, and (2) epibenthic sponges that have been broken by storms can be incorporated into rubble piles as errant fragments (Fig. 5.2). In both cases, its homogeneous 3-dimensional body allows a single sponge to quickly attach (within 2–5 days) to several pieces of rubble, holding them steady against each other until carbonate-secreting organisms, especially encrusting coralline algae, can bind them together permanently (Wulff 1984; Biggs 2013). Without rapid binding by sponges, slower-growing carbonate-secreting binders could not grow from one piece of rubble to the next. The sponges are the "fingers holding the pieces together while the superglue sets" (thank you to D. Hubbard for this analogy).

Experimental exploration of each step of this process on a shallow Caribbean coral reef in Panama revealed that rubble piles with sponges remained elevated above the reef surface, became bound together by encrusting coralline algae within 5 months, and became colonized by coral recruits within 10 months. Rubble piles without sponges remained loose and increasingly collapsed, although each individual piece of rubble became encrusted with coralline algae (Wulff 1984). Small corals on stabilized rubble survived significantly better than small corals on loose rubble (for respectively stabilized and unstabilized rubble: 66 % undamaged vs. 35 % undamaged after 4 months, and 13 % survival vs. 1 % survival after 4 year).

In the tropical eastern Pacific, rubble on the tops of reefs in the Gulf of Chiriqui, Panama, was stabilized as cryptic sponges grew up through the reef frame to bind it, but the absence of exposed sponges on the reefs resulted in aprons of rubble at their bases, each rubble piece thickly coated with many layers of coralline algae (Glynn 1974; Wulff 1997c). The lack of a mechanism for stabilizing rubble against the challenges of episodic storms and chronic disturbance agents such as large foraging triggerfish and sea urchins prevents these rhodoliths from being incorporated into solid reef framework: pieces of loose rubble do not remain still next to each other long enough for encrusting coralline algae to grow from one piece of rubble to another, welding them into a stable structure. A similar dearth of epibenthic sponges in the Galapagos may contribute to extensive rhodolith piles, in which individual pieces of rubble resulting from massive coral mortality during the 1982/83 ENSO event have become encrusted by coralline algae so that they are up to 15 cm in diameter (Halfar and Riegl 2013). No recovery has occurred, and what was once incipient coral framework has remained as a rubble bed with no signs of recovery for over 20 years.

Discrepancies between growth of individual corals and reef accretion remind us of the diversity and complexity of the processes that contribute to reef building, maintenance and repair (e.g., Hubbard 1985, 1988; Hubbard et al. 1998). It is possible that some of these discrepancies may be explained at least partially by differences in abundances of epibenthic and cryptic sponges that are capable of mediating the cycling of loose rubble back to stable substrata suitable for coral recruitment and growth. The coincidence of coral reefs and hurricanes in shallow tropical water suggests that the cycle of rubble generation, consolidation, and recruitment of corals has long been a normal part of scleractinian reef-building. Sponges have likely played key roles in ensuring that it is actually a cycle instead of a one-way path from living corals to rubble. The only other organisms capable of rapidly adhering to multiple rubble pieces are fleshy algae, but their need for light causes them to overgrow the stabilized substrata, impeding coral recruitment.

Goreau and Hartman (1966) and Hartman (1977) pointed out that sponge binding could also aid reef growth by preventing piles of rubble from cascading down steep slopes, sweeping everything in their path into talus piles at the bases of fore-reef walls. Soft-bodied, non-excavating sponges, i.e., the vast majority of sponges, that participate in gluing living corals to the reef frame and mediating rubble consolidation, vanish shortly after they perish because their skeletal frameworks of protein fibers and spicules fall apart (Wulff 2006c, 2008a). Thus they do not appear in fossil reefs; and even on modern reefs, these roles are invisible unless observed in action or explored by experiments that explicitly compare dynamics with and without sponges.

5.5.3 Improving Reef Restoration by Harnessing the Ability of Sponges to Bind Rubble

Expanding on experiments demonstrating sponge-mediated rubble consolidation (Wulff 1984; Biggs 2013) experimentally showed how sponges can be sustainably used to restore damaged reef sites. Erect branching sponges, the growth form most likely to become broken and included in rubble piles under natural circumstances (e.g., Wulff 2006b; Fig. 5.2), can be harvested sustainably because the branches from which fragments are cut grow back quickly. Once the sponge fragments are inserted into rubble piles, each fragment rapidly reattaches to several pieces of rubble. Using sponges to bind loose rubble into stable structures on which coral recruits are more likely to survive is not only less expensive and more attractive than artificial cements, but sponge-mediated binding is autocatalytic, as the sponge fragments grow and multiply, continually adding to their binding power. Moreover, framework-building coral species recruit significantly more to coral rubble bound with sponges than to cement bound structures, adding another reason to use living sponges in reef restoration (Biggs 2013).

5.6 Geological Roles of Sponges: Bioerosion

Bioeroding sponges have provided mysteries aplenty, and in spite of publications throughout the nineteenth century declaring them to indeed be sponges and also active borers rather than inhabitants of holes made by other creatures, it was not until nearly the twentieth century that these were accepted as facts (see Schönberg 2008 for a comprehensive history). The exact mechanism of boring, in particular the relative importance of chemical dissolution and mechanical removal of chips, is still an active area of research (review by Schönberg 2008). Although the ability to excavate and otherwise whittle down solid biogenic carbonate may seem as astonishing a feat for sponges as generating dense solid carbonate skeletons, the ability to excavate is currently represented in five orders of the class Demospongiae, suggesting the possibility of an ancient origin within the sponges.

Reefs may have been re-shaped by sponges from the start. Excavations that could have been made by sponges have been found in Cambrian archaeocyath reefs and middle Ordovician hard substrata (Kobluk 1981). However, although bioerosion by a variety of macro-organisms was common in Paleozoic tropical biogenic carbonate, the radiation of the group currently responsible for the majority of excavations in reefs, the clionaid sponges (Order Clionaida), was a Mesozoic phenomenon (Tapanila et al. 2004). Cenozoic boring in reef substrata is dominated by sponges (e.g., Perry 2000), and on currently accreting reefs sponges accomplish up to 90 % of the macroscopic boring (e.g., Goreau and Hartman 1963; Perry 1998; Rützler 2012). Although the great majority of sponge species are not capable of excavating corals, and the biomass of excavating species is relatively small, their influence can be dramatic (excellent reviews dispersed over the last few decades include Goreau and Hartman 1963; Hartman 1977; Wilkinson 1983; Rützler 2002; Schönberg 2008).

Abundance of boring sponges and the rate at which they break down solid carbonate varies widely. This variation has been recognized as a possible source of clues about environments for fossil reefs, and the value of sponge borings has been discussed for paleo-reconstruction, e.g., for bathymetric patterns (Bromley 1970; Bromley and d'Alessandro 1984, 1990; see also Chap. 4). Schönberg and Tapanila (2006) matched the morphology of bioerosion by the modern *Siphonodictyon paratypicum* to the fossil bioerosion trace *Entobia devonica* for paleoecological interpretation with respect to bathymetry and sedimentation, and their findings largely matched earlier ones, i.e. that *Siphonodictyon* spp. typically occur in shallow, low energy environments (Reitner and Keupp 1991). Evidence from the late Oligocene suggested that bioeroding sponge distributions were influenced by salinity gradients, just as they are today (Lawrence 1969). In Ordovician-Silurian reefs built by tabulate corals and stromatoporoids, a pattern of higher proportion of specimens bored in off-reef facies could have been caused by the greater competition for space on reefs, which may have diminished the success of boring (Tapanila et al. 2004). A similar pattern in the amount of boring was found in Pleistocene deposits of north Jamaica, where the percent of framework removed by borers was greater in back-reef/lagoonal settings than on the fore-reef. Sponges were responsible for most of the excavations on fore-reefs. Overall 64.7 % of framework carbonate was removed by sponges, and only 8.2 % by bivalves and 25.8 % by a variety of worms (Perry 2000). Pleydell and Jones (1988) reported similar rates for Grand Cayman Oligocene-Miocene bioerosion.

The clearest environmental correlate of sponge bioerosion has been eutrophication, and increased sponge bioerosion with nutrient increases has been detected in different settings and geological times (e.g., Hallock 1988; Edinger and Risk 1997; see also Chap. 4). On modern reefs, excavating sponges have been demonstrated to increase in abundance with increased water column nutrients (e.g., Rose and Risk 1985; Holmes 1997; Ward-Paige et al. 2005; Alcolado 2007). The relationship with nutrients depends on the species, and is not monotonic. Even boring sponges cannot cope with extremely high nutrient levels, and the toxic effects of the resulting eutrophication (Rützler 2002). Chaves-Fonnegra et al. (2007) found increased abundance of *Cliona delitrix* Pang as they evaluated sites closer and closer to a sewage outfall on San Andrés, Colombia, but this species declined to zero at the site closest to the outfall. As the authors pointed out, negative influence on sponges of the high nutrients at the outfall could have been confounded with increased sediment, a frequent covariant. Negative effects of sediment may also explain increased importance of boring by bivalves and worms relative to sponges within the bay at Discovery Bay, Jamaica, in spite of increased food for sponges in the water column (Macdonald and Perry 2003).

Advance of boring sponges into coral skeletons can be influenced by characteristics of the interacting species and the idiosyncrasies of immediate context, including angle of encounter, coral growth form or species, sponge species (Rützler 2002; Schönberg 2002, 2003; López-Victoria et al. 2006), and even parrotfish bites at the sponge-coral interface (Marquez and Zea 2012). The excavating sponge *Siphonodictyon coralliphagum* Rützler produces mucus that kills coral tissue, allowing this sponge to penetrate coral tissue and possibly also settle on live coral as a larva (Rützler 2004, 2012); and other boring sponges can undermine polyps in order to make their way into the skeleton (e.g., Chaves-Fonnegra and Zea 2011). The strong preference of the voraciously excavating species *Cliona delitrix* for massive corals might even, over time, change the composition of the coral community to favor species of foliose and branching corals (Chaves-Fonnegra and Zea 2011).

Rate of spread by excavating sponges is not only enhanced by factors that spur on the sponges, but also by the more sheltered habit of the sponges (Schönberg and Wisshak 2012) and factors such as temperature that stress the corals enough to hinder their ability to fend off the sponges (Rützler 2002). Spread of boring sponges can be slowed or halted by prior encrustation of solid carbonate or by overgrowth of coralline or other macroalgae; and recruitment and excavation can be prevented by cover of other sponges (e.g., López-Victoria et al. 2006; Chaves-Fonnegra and Zea 2011; González-Rivero et al. 2012; pers. observ. Figs. 5.2 and 5.3).

Some excavating clionaids harbor symbiotic zooxanthellae, but this symbiosis does not tend to break down under abnormally high temperatures as readily as in scleractinians. When 84–87 % of the corals on Orpheus Reef, GBR, bleached in 1998, the boring sponge *Cliona orientalis* Thiele retained its zooxanthellae (Schönberg and Wilkinson 2001), an advantage that may be conferred by the sponge's ability to move the zooxanthellae deeper into its tissue during stressful events (Schönberg and Suwa 2007) and by their relatively heat-resistant G-clade zooxanthellae (Schönberg and Loh 2005; Schönberg et al. 2008). The abundance of zooxanthellate sponges significantly increased after the 1998 bleaching, which was interpreted to be a result of their ability to survive, remain healthy, and recruit where corals died (Schönberg and Ortiz 2009). It may appear obvious to ascribe aggressive behavior of boring sponges, as well as their consistent increases with water column nutrients, to benefits from symbiotic zooxanthellae (e.g., Fang et al. 2014), but some species of rapidly advancing excavators, e.g., *Pione lampa* (de Laubenfels), *S. coralliphagum*, and *C. delitrix* in the Caribbean, do not have photosynthetic symbionts (Rützler 2002).

As part of an overall carbonate budget for five sites in Bonaire, Perry et al. (2012) calculated that loss rates related

Fig. 5.3 Sponges boring, overgrowing, and protecting coral. (**a**, **b**) the boring sponge *Cliona caribbaea* Carter being overgrown and killed by the encrusting sponge *Chondrilla caribensis*; (**c**) the boring sponge *C. tenuis* Zea and Weil steadily diminishing chances of survival for a coral, in the absence of epibenthic or semi-cryptic sponges; (**d**) the readily storm-broken sponge *Svenzea zeai* (Alvarez et al.) temporarily overgrowing coral (*Diploria*)

to boring sponges ranged from 0.002 to 0.07 kg/$CaCO_3$/m^2-year, which is smaller than losses to parrotfishes (0.95–2.75 kg/$CaCO_3$/m^2-year) at the same sites (for context, $CaCO_3$ production by corals ranged from 0.20 to 12.07 kg/$CaCO_3$/m^2-year). This relationship was different in Jamaica, where fish bioerosion was only 8–20 % of the internal macro-bioerosion, which was dominated by sponges (Mallela and Perry 2007). Water quality and maturity of the community may influence the relative losses to endolith and grazer bioerosion, with bioerosion by sponges often being more important on reefs where they have had time to establish and where nutrient concentrations are higher (Carreiro-Silva and McClanahan 2012). As all of these reports point out, variation is great, even among sites near each other. Perry et al. (2012) remark on the difficulties with interpreting differences when comparing data collected in different ways [they cite 0.2 kg/$CaCO_3$/m^2-year in Barbados from Scoffin et al. (1980) and 0.02–1.04 kg/$CaCO_3$/m^2-year in French Polynesia from Pari et al. (2002)]. Environmental conditions play a key role, and recently ocean acidification was recognized as a strong driver of sponge bioerosion (Wisshak et al. 2012; Fang et al. 2013; Wisshak et al. 2013; Fang et al. 2014; Wisshak et al. 2014; Stubler et al. 2014), while it simultaneously suppresses coral calcification (Jokiel et al., Chap. 2 this volume).

The amount of solid carbonate eroded into sediment may not be the most important measure of the influence of boring sponges on reef building and maintenance. Although at many sites parrotfishes may reduce more substrate mass to sediment than do boring sponges, the result may reduce coral survival and reef growth far less. Parrotfishes scrape only from the surface, whereas sponges can erode the bases of corals, causing them to topple from the reef frame and perish in the sediment (e.g., Goreau and Hartman 1963; Wulff and Buss 1979). Thus with only a small amount of sponge erosion, entire living coral colonies may be lost (Fig. 5.3). Preventing this aspect of coral death and the loss of large chunks of solid carbonate may be one of the most important roles of epibenthic, semi-cryptic, and cryptic sponges on coral reefs (Wulff and Buss 1979, Fig. 5.2), especially if boring sponges are increasing in abundance, and if they increase their activity as climate change progresses.

5.7 Biological Roles of Sponges: Overgrowth of Living and Dead Coral

Some sponge species have been demonstrated to kill coral tissue by allelochemicals, and a few species have been demonstrated to aggressively overgrow living corals at some sites (recent review in Wulff 2012, pp. 308–312). Still the combined number of species that have been shown to be able to kill corals, or to kill them conspicuously in at least some places is fewer than 0.4 % of the sponge species that have been described in biogeographic realms with coral reefs. Other sponges may kill a small patch of coral tissue to allow the sponge to adhere to the underlying skeleton, but this can bind the corals securely onto the reef frame (Wulff and Buss 1979). As with most examples of mutualism, there is a price to pay for the benefits. In this case, even several cm^2 of tissue is a very small price for a tenfold gain in the entire colony's survival rate. Further expansion of branching, semi-cryptic, or massive sponge species over living coral has been reported only rarely. Time-series observations of interactions that had initially appeared to be overgrowths on reefs in Colombia showed that most sponges did not actually progress over living coral. Only 16 of the 95 sponge species present overgrew coral at all, and only three of these overgrew coral in more than 10 % of contacts (Aerts and van Soest 1997; Aerts 2000).

Cases in which field observations have demonstrated overgrowth of live corals over time usually fall into three categories: (1) sponges that are alien to the reefs on which they are overgrowing corals, e.g., *Mycale grandis* Gray, an Indonesian and Australian native, in Kaneohe Bay, Hawaii (Coles and Bolick 2007), and *Chalinula nematifera* (de Laubenfels), an Indo-Pacific native, in the Mexican Pacific (Ávila and Carballo 2008); (2) thinly encrusting sponges that are densely inhabited by cyanobacteria, e.g., *Terpios hoshinota* Rützler and Muzik, in the Pacific (Rützler and Muzik 1993), and *Chondrilla caribensis* Rützler et al. (Vicente 1990) in the Caribbean; and (3) cases in which the corals are particularly stressed (Wulff 2012).

At a particular moment and site, an aggressive sponge species can devastate corals. For example, sponges of an aggressive species may infest half a locale's corals (Benzoni et al. 2008), cover half the substratum (Reimer et al. 2010), or spread over coral tissue at rates of nearly a mm a day (Bryan 1973). In none of these cases, however, has the aggressive sponge species caused continually increasing devastation. Rather, there is a consistent, curious pattern of infestations being found only at some sites, and being ephemeral at any particular site. The most dramatic example of a sponge that can overgrow corals is the cyanobacteria-packed thinly encrusting sponge *T. hoshinota*. Since it was first reported in Guam (Bryan 1973), it has been found at far ranging sites across the Pacific including Okinawa, Taiwan, Indonesia and Lizard Island, Australia, but has vanished from some sites where it was once common (e.g., review in Wulff 2012, pp. 309–310; de Voogd et al. 2013). Coral-threatening *Mycale grandis* in Hawaii has recently diminished in abundance (pers. observ.); and *Chalinula nematifera* was only found at two of 150 sites in the Mexican Pacific, and has not increased (Ávila and Carballo 2008). Although *Chondrilla caribensis* quickly covered dead coral skeletons at a central lagoon site on the Belize Barrier Reef where the coral *Agaricia tenuifolia* had suffered catastrophic mortality due to bleaching (Aronson et al. 2002), at other nearby sites it is extremely rare on coral reefs (Wulff 2012, pp. 310–312). Aronson et al. (2002) pointed out that *Chondrilla* did not overgrow living coral, but only recruited and grew after coral death. Although the sponge cover would prevent recruitment of corals, this cover also protects solid carbonate from being reduced to sediment by boring organisms. *Chondrilla* has been observed to overgrow *Cliona*-infested coral skeletons in Belize, putting the boring sponge out of business on the spot (Fig. 5.3).

The role of stress in spurring overgrowth of living corals by sponges is uncertain, perhaps because stress has been defined in a variety of ways. But just as coral health can influence the advance of boring sponges, coral health can influence overgrowth. Time series observations of *T. hoshinota* in Okinawa revealed a pattern of devastation to live corals at sites where development had increased turbidity of coastal waters (Rützler and Muzik 1993), and recent experiments have shown that circumstances allowing *T. hoshinota* to overgrow corals depend on relative health of both the corals and sponges at a particular site (Wang et al. 2012). The threat of a thinly encrusting *Clathria* species that was killing recently transplanted massive reef corals, *Porites lutea* Milne Edwards and Haime, at a Gulf of Aden site diminished as the infested corals recovered from the stress of being transplanted (Seguin et al. 2008). Although Aerts and van Soest (1997) found that overgrowth of corals by sponges was not more likely on reefs deemed stressed (evaluated by higher sedimentation rate and poorer water column visibility), they did discover that the thinly encrusting *Clathria (Thalysias) venosa* (Alcolado), which one-time observations suggested could be a threat, only overgrew living coral if the coral was first experimentally damaged (Aerts 2000).

5.8 Biological Roles of Sponges: Water-Column Influences

As sponges pump water through their internal canals, their uniquely fine-scale filter system (the collars of the choanocytes) captures picoplankton that pass through the coarser filters of other filter-feeding taxa. Reiswig (1971) demonstrated that sponges of three Caribbean species representing different orders could capture an astonishing 96.4 % of the bacteria in the water column. Reiswig's classic, and still unsurpassed, studies (1971, 1973, 1974) relating sponge feeding, respiration, abundance, and population dynamics, allowed him to estimate that the sponges of the fore-reef at Discovery Bay, Jamaica, could filter the entire water column above them every 24 h. Technological advances have made it possible to add further details. Now we also know that sponges can use dissolved organic matter, as well as filter a suite of minute particles, including procholorophytes, picoeukaryotes, cyanobacteria, and heterotrophic bacteria. The efficiency with which they capture each of these components of the picoplankton or absorb dissolved organic matter is influenced by their species, shape, size, densities of microbial symbionts, and internal morphology, as well as by what is available (e.g., Strimaitis 2012 and reviews in Rützler 2004, 2012; Wulff 2012).

5.8.1 Maintaining Water Clarity

Losses of sponges have unfortunately corroborated Reiswig's (1974) estimate of the great importance of sponge filtering. After Hurricane Allen in Jamaica (Woodley et al. 1981), pulverized organisms and the bacteria devouring them kept the water column murky as long as sponges that had survived the storm remained shut down. When the sponges resumed their pumping, the water cleared rapidly (pers. observ.). Florida Bay, from which water flows out to the reefs of the Florida Keys, has been devastated by many cyanobacteria blooms since 1982, when the first blooms killed up to 90 % of the sponges (Butler et al. 1995). Peterson et al. (2006) combined estimates of sponge biomass and filtering abilities to figure the cost to the water column of sponge loss, and concluded that reduced filtration of the water column resulting from heavy sponge mortality during the first bloom can entirely explain the subsequent blooms. This conclusion raises the disturbing possibility that the recent loss of 71 % of sponge biomass from a shallow reef in the central Belize Barrier Reef during an extended phytoplankton bloom (Wulff 2013) could allow future incipient phytoplankton blooms to billow forth because badly depleted sponge populations can no longer nip them in the bud.

Zooxanthellae or cyanobacteria contribute in various degrees to the nutrition of some sponge species, with zooxanthellae largely confined to excavating species of the order Clionaida (Rützler 1990; López-Victoria and Zea 2005; Hill et al. 2011; but also see Garson et al. 1999; Scalera-Liaci et al. 1999, for interesting exceptions). Sponges that harbor photosynthetic symbionts are not as consistently reliant on them as are scleractinian corals, and complete shading of photosymbiont-bearing sponges can result in diminished growth, or loss of biomass, or no apparent negative effects at all (e.g., Erwin and Thacker 2007; Freeman and Thacker 2011). The coral-killing *Terpios hoshinota* is capable of escaping from shading by extending fine threads until they reach sunlit substrata, where they resume growth as a continuous sheet (Soong et al. 2009). Variation in the importance of the photosymbionts may depend in part on symbiont identity (e.g., Thacker 2005;

Erwin and Thacker 2007) and also on the ability of the sponge host species to switch between different modes of acquiring food (review in Wulff 2012, pp. 301–303).

5.8.2 Influences on Dissolved Organic and Inorganic Water-Column Components

Sponges can have profound effects on dissolved water-column components, especially carbon, nitrogen, and silicon (review by Maldonado et al. 2012). In collaboration with their microbial symbionts, some sponges can soak up and make use of dissolved organic material (Reiswig 1981; de Goeij et al. 2008; Weisz et al. 2008). Some species of sponges that inhabit cryptic spaces within the reef frame, may acquire a significant portion of their nutrition from dissolved organic matter (DOM) generated by corals or coralline algae rather than relying on sparse picoplankton (van Duyl et al. 2011). Recent reports have suggested that this could be a major force in nutrient cycling on coral reefs (de Goeij et al. 2013), with sponges and their symbionts transforming DOM into sponge biomass, and extremely rapid cell cycling resulting in the shedding of cells that serve as food to organisms incapable of using dissolved organic matter directly. Scaling-up processes identified for a few species at the level of cubic centimeters, to entire communities and the vastly larger water column above a coral reef must be done cautiously, as reefs vary widely in biomass of both cryptic and epibenthic sponges (e.g., Wilkinson 1987; Wilkinson and Cheshire 1990). For example, a conclusion that cryptic sponges account for orders of magnitude more biomass than epibenthic sponges was based on endoscopic observations of a Red Sea reef (Richter et al. 2001) where epibenthic biomass estimates were extremely small (0.8–1.2 % cover, no volume measurements given). On reefs where epibenthic sponges are more evident (e.g., in most Caribbean locations), the assumption that biomass of cryptic sponges is greater is less likely to be applicable.

While it is not yet clear how important these processes are in overall nutrient cycling on coral reefs in general, sponge-mediated nutrient cycling within the reef frame is an intriguing reminder of the possibility that there are other surprising sponge-mediated processes on coral reefs that we have not yet even imagined. This is underscored by the recent discovery that the diet of two species of Caribbean excavating sponges can be mainly dissolved organic carbon (Mueller et al. 2014).

Sponges, in collaboration with their symbiotic microbes, can also greatly influence nitrogen cycling on coral reefs. Transformations in which sponge microbes participate include nitrification, nitrogen fixation, denitrification, and anaerobic ammonium oxidation (e.g., Corredor et al. 1988; Webster and Taylor 2012). This is a rapidly growing area of sponge and microbial research, as new techniques are developed and the potential importance to coral reef ecosystems becomes more apparent (Maldonado et al. 2012).

5.9 Biological Roles of Sponges: Providing Shelter and Food

5.9.1 Animal and Plant Guests of Sponges

How the hundreds of species of echinoderms, worms, molluscs, arthropods, fishes and multicellular algae hosted by sponges, within and on the surfaces of their bodies, might influence coral reef functioning is not clear, but these species certainly bolster biodiversity substantially (Cerrano et al. 2006; Wulff 2006e). Sponges and their symbionts offer opportunities to study community and population ecology in patchy habitats in which the patches (i.e., individual sponges) can be readily manipulated. Among the surprising and fascinating results of studies on sponge inquilines is the discovery of eusocial shrimps in sponges (Duffy 1996). Some symbionts use their hosts only as a shelter or breeding site, but others also consume their host (Wulff 2006e; Schönberg and Wisshak 2012). For obligate symbionts, the loss of their host sponges can be a disaster, leading to a cascade of local extinctions with potentially grave consequences. After a couple of major sponge mortality events on the Belize Barrier Reef (Wulff 2013) in which a total of 74 % of the sponge biomass was lost, eusocial shrimps became extremely difficult to find (J.E. Duffy, pers. comm.), and other inquilines vanished or died when their sponge host died (Fig. 5.4). Economically important spiny lobster populations were negatively influenced when 71 % of the sponges that provided shelter for their juveniles perished in a dense cyanobacterial bloom (Butler et al. 1995). We have barely begun to explore this aspect of coral-reef sponge interactions.

5.9.2 Consumers of Sponges

Although most epibenthic sponges are well defended against consumption by most of the large mobile predators with which they share habitat, a few spongivores depend on sponges. Angelfishes tend to consume most of the sponge species that they encounter in a "smörgåsbord" fashion, by taking only small amounts of any particular sponge (a mean of 2.8 bites in Wulff's 1994 study of unmanipulated angelfishes) before moving on to another sponge that is generally of a different species (Randall and Hartman 1968; Wulff 1994, 2006e, 2012 pp. 313–315; Fig. 5.4).

Fig. 5.4 Interactions of coral reef sponges. (**a**) a recently dead *Callyspongia vaginalis* (Lamarck) with dying symbiotic zoanthids, and a symbiotic goby that was gone the following day; (**b**) *Verongula rigida* (Esper) with one bite removed by an angelfish just before the photo was taken, and a wound healing where bites had been taken 2 days earlier; (**c**) the easily confounded congeners *Tedania ignis* (Duchassaing and Michelotti) and *T. klausi* (Wulff) which differ from each other with respect to vulnerability to starfish predation, disease, and temperature and salinity extremes; (**d**) *Aplysina fulva* (Pallas) suffering (skeleton exposed where tissue died a few days earlier, and black or white signs of necrosis where tissue has died more recently; the ochre yellow portions are still alive) in the midst of a dense cyanobacteria bloom in which 71 % of the biomass of the sponge fauna was killed (Wulff 2013)

Epibenthic sponges on coral reefs are not severely limited by routine spongivory of this type, because they readily regenerate where they have been bitten, and because only a small amount is ever eaten at one time.

Although angelfishes disproportionately feed on some species (Wulff 1994), they spread their feeding over most of the sponge community. Randall and Hartman (1968) found a total of 70 sponge species in gut contents of four species of angelfishes, Hourigan et al. (1989) observed that three species of angelfishes consumed 23 sponge species, Padilla Verdín et al. (2010) found 24 sponge species in gut contents of two species of angelfishes, and Wulff (1994) observed that angelfishes of three species consumed 64 sponge species on a coral reef in Panama, including 36 of the 39 species in a fully censused 16 m^2 plot.

The other large dedicated spongivores on coral reefs are hawksbill turtles, which can devour large quantities of sponge tissue; but they only eat a handful of species in three orders of demosponges (Meylan 1988, 1990; van Dam and Diez 1997; León and Bjorndal 2002). The presence on coral reefs of charismatic mobile species, such as angelfishes, trunkfishes, and hawksbill turtles that routinely consume sponges may depend on diverse, thriving sponge assemblages. Curiously routine spongivory may have less dramatic effects on prey sponge species than opportunistic spongivory (Wulff 2006e).

Opportunistic spongivory can be an important trophic pathway on coral reefs. The conspicuous large Caribbean starfish *Oreaster reticulatis* may depend on occasional consumption of coral reef sponges that wash off reefs into adjacent seagrass meadows where the starfish reside. *Oreaster* (adults and juveniles) maintained in tanks on their usual diet of microalgae fared poorly relative to those fed on coral reef sponges (Scheibling 1979); and populations of *Oreaster* inhabiting seagrass meadows into which sponges were more frequently washed by storms included a significantly higher proportion of large individuals (Wulff 1995). *Oreaster* reject sponge species that inhabit seagrass, but readily eat many of the coral reef species that are only available to *Oreaster* if they are washed off the reef into the seagrass (Wulff 1995). Opportunistic spongivory by herbivorous parrotfishes may also exert control on habitat distribution of sponge species, by preventing some of the species that are typically confined to cryptic spaces within the reef frame and in rubble piles from growing out of their hiding places, as some of these species appear to be relatively undefended against predators. Herbivorous parrotfishes, *Sparisoma aurofrenatum* (Cuvier and Valenciennes) *S. viride* (Bonnaterre) and *S. chrysopterum* (Bloch and Schneider) and to a lesser extent *Scarus iserti* Bloch battled each other over the opportunity to consume normally cryptic sponges that were exposed when researchers broke open their hiding places within the reef framework or rubble piles (Wulff 1997b). The possibility that sponges constitute an important supplement to their possibly nitrogen-limited diet is suggested by their battles for the sponges, and also by the alacrity with which they responded to exposure of cryptic sponges, veering from their paths and charging straight to the sponges as soon as they were exposed. Similar behavior was observed in the eastern Pacific, where the angelfish *Holacanthus passer* Valenciennes usually feeds on plankton in the water column above the reef, but responds immediately to the exposure of cryptic sponges when the reef is cracked apart, plummeting to the seafloor and engaging in battles with other fishes (including the parrotfish *Scarus ghobban* Forsskål and the Moorish idol, *Zanclus cornutus* (Linneaus) that are also attracted to the exposed sponges (Wulff 1997c).

5.10 Future of Sponges on Coral Reefs: Assessing and Ascribing Causes to Increases and Decreases

Literature concerning coral reef sponge abundance and dynamics, and interactions of corals with sponges, includes some striking discrepancies. Demonstrated dramatic declines of sponges contrast with assertions that sponges are increasing unchecked; reports of experimentally demonstrated extreme benefit to corals by associations with sponges contrast with assertions that sponges constitute one of the chief enemies of corals and reefs. Roots of these discrepancies are embedded in: (1) application of inappropriate methods for assessing and monitoring sponges; (2) lumping together as "sponges" a highly heterogeneous group of animals with a wide range of responses to changing conditions and influences on corals and coral reefs, rather than distinguishing sponge species; and (3) a tendency to generalize from studies on single conspicuous, and often unusual, species to entire regional faunas of many hundreds of species. In the hope of clarifying the pitfalls involved, each of these problems is discussed in detail below.

5.10.1 Inappropriate Methods for Assessing and Monitoring Sponges Yield Data That Are Difficult to Interpret

As sessile animals that can be large and exhibit a diversity of growth forms, sponges superficially appear to be amenable to the same field assessment and monitoring methods that work well for corals. However the tissue in most sponges is not a thin layer over the surface of a solid skeleton, as it is for corals, but fully three-dimensional. Ecosystem roles of sponges, therefore, scale with their volume rather than the surface area of live tissue, and their abundance in the context of population dynamics and vulnerability to local extinction must also be measured by volume (detailed discussions in Rützler 1978; Wulff 2001; Rützler 2004; Wulff 2012). Video transects, that adequately record corals that are oriented to sunlight, fail for sponges that live on vertical surfaces, under corals, and within crevices or embedded in the substratum (eg., Abdo et al. 2004). Point counts, and other percent-cover measures, do not reflect the abundance of most sponges. An encrusting sponge 2 cm in diameter and 1-mm thick, a spherical sponge with 2-cm diameter, and a tube sponge 2 cm across (with a 0.5-cm diameter cavity) and height of 8 cm, all have the same percent cover (i.e., 3.14 cm ^{2}) on a planar projection, but the sphere has 13 times the volume and the tube has 75 times the volume of the encrusting sponge. Number of sponge individuals is rarely informative, as the size of an individual can range over many orders of magnitude (e.g., McMurray et al. 2010; Schönberg and Fromont 2012). A barrel sponge, 1 m in diameter and 1 m tall has 2.5 million times the volume of the encrusting sponge 2 cm across; thus it might filter 2.5 million times as much picoplankton from the water column and provide 2.5 million times as many bites of food to spongivores! Moreover, a physical disturbance or pathogen infection can quickly increase the numbers of individuals by fragmentation, while simultaneously diminishing biomass.

Explicit comparison of sponges of different growth forms on a shallow reef in Caribbean Panama (Wulff 2001) revealed that sponges in the four growth form categories of erect branching, massive, thickly encrusting, and encrusting each contributed about 25 % of the total percent cover, but with respect to volume, the erect branching sponges were 63 % of the total, and the encrusting sponges were a trivial 1.8 %. The one-quarter of the volume that constituted massive sponges was contributed by only 8 % of the individuals.

Sponges differ from corals in another way that diminishes usefulness of data acquired by methods often used for studying corals: most sponges vanish shortly after their death because the skeletal scaffolding (made of protein fibers or protein and spicules) that supports their living tissue deteriorates quickly when not embedded in tissue (e.g., Wulff 2006c, 2008a). Thus mortality cannot be documented, and is likely to go unnoticed unless individual sponges were monitored before a mortality event. Substantial biomass can also be lost from sponges due to various agents of partial mortality (disease, predators, storms, e.g., Wulff 2006a, 2006b, 2006c, 2008a, 2013). Extreme regeneration ability of many sponges renders partial mortality quickly invisible unless pre-mortality monitoring includes measurements of total volume of each sponge (Wulff 2010, 2013).

Growth of sponges over exposed coral skeletons from which the tissue was previously lost can readily be confused with aggression against living corals in snap-shot observations. Time-series observations of the boundary between live sponge and live coral are the only way in which the two very different processes of aggression against living coral and protection of coral skeletons from excavators (by covering exposed portions) can be distinguished (e.g., Aerts 2000).

In summary, data on sponge population and community dynamics and interactions with corals can be readily misinterpreted, unless the data are gained by the somewhat arduous process of monitoring volume changes of individual sponges over time (Wulff 2001, 2006e, 2012, 2013 pp. 276–281). One-time observations tend to under-estimate sponge mortality and over-estimate negative effects on corals perpetrated by sponges.

5.10.2 Lumping Together Sponges of Diverse Talents, Vulnerabilities, and Relationships with Corals

Different sponge species, even closely related ones, can have dramatically different relationships with corals and reefs (Hartman 1977; Rützler 1978; Wulff 2001; Rützler 2004; Wulff 2006e; Rützler 2012; Wulff 2012; Wulff 2013). Abundance of sponges on present day coral reefs, and the determinants of abundance, must be considered separately for each of the four types of sponges (i.e., epibenthic, cryptic, boring, hypercalcified). Identification to species is key, because sponges that look similar can play very different roles. For example two Caribbean species, *Iotrochota birotulata* and *Desmapsamma anchorata*, are both erect brancing forms in the Order Poecilosclerida, but exhibit growth and mortality rates that differ by an order of magnitude (Wulff 2008b). *Iotrochota* forms mutually beneficial associations with branching sponges of other species that increase growth rates and survival of participating individuals, while *Desmapsamma* behaves as a parasite on other sponges (Wulff 1997a, 2008b), and also overgrows gorgonians (e.g., McLean and Yoshioka 2008). Species that are of the same genus can react very differently to environmental changes. Four Caribbean species of *Ircinia* responded to a mass mortality caused by dense phytoplankton in contrasting ways: two species that grow as clusters of mounds lost much biomass but began to recover within months, while two other species (one grows as large spheres and one as thick-walled vases) were entirely eliminated by the phytoplankton bloom (Wulff 2013). Likewise, two conspicuous Caribbean *Tedania* species were long thought to be a single species because of their similar appearances and spicule complements, but one is immune to starfish predation and can therefore inhabit seagrass meadows from which the other is barred by *Oreaster* readily consuming it, and the two species differ in susceptibility to disease and to extreme environmental conditions as well (Wulff 2006d, Fig. 5.4). Sponge taxonomy is unquestionably challenging, but failing to distinguish sponge species in field surveys makes no more sense than combining data on parrotfishes and snappers as "fishes", or *Acropora* and *Porites* as "corals".

5.10.3 Are "Sponges" Overwhelming Coral Reefs?

Assertions that some coral reefs may be turning into sponge reefs and that sponges are increasingly overwhelming corals have been presented recently from two different viewpoints. Bell et al. (2013) support their assertion by pointing out that Mesozoic reefs of siliceous sponges provide historical precedents for reefs dominated by sponges, that reports on sponge disease are less prevalent than reports on coral disease, and that sponges can be abundant in high nutrient and turbidity settings such as lagoons. In contrast, Pawlik (2011) and Loh and Pawlik (2014) assert that palatable sponges that can outcompete corals are increasing in response to loss of spongivorous fishes by overfishing, a scenario that parallels increases in fleshy algae after herbivores have been overfished. I will discuss whether or not sponges are actually increasing at all in a later section, and for the moment only discuss two of these possible influences on future abundance of coral reef sponges: disease and water column nutrients. For the former it is clear that there are inadequate data for any conclusions, and for the latter there may be sufficient reports from a variety of reef sites to allow some tentative conclusions.

Diseases of corals have caused huge declines (e.g., Miller et al. 2009), and seem likely to continue to be devastating; but fewer reports of sponge diseases than of coral diseases may not necessarily reflect fewer losses of sponges to disease (Rützler 2004, 2012; Webster 2007) because sponge disease is very likely to be underestimated and under-reported. Sponge disease is virtually impossible to document unless it is caught in progress. Sponges that have died from disease tend to deteriorate quickly and vanish (e.g., Cowart et al. 2006; Wulff 2006c), while the skeletons of dead corals remain to proclaim for years afterwards that they existed before a disaster killed them, even if the exact disaster cannot be determined from the skeletons. Evidence for partial mortality due to disease is readily seen months later for corals, in the form of denuded skeleton; but partial mortality in sponges is entirely effaced within days or weeks as the denuded skeleton deteriorates and the sponge generates a new surface. Monitoring programs that have tracked coral disease for decades tend to not include sponges at all. Even in cases where sponge disease prevalence is reported, an inverse relationship between the speed at which a pathogen can entirely kill a particular sponge, and the probability that the diseased sponge will be observed before it disintegrates and vanishes argues against accurate evaluation of sponge disease in one-time field surveys (Wulff 2006c). Monitoring disease in sponges will have to be done differently (e.g., at greater frequency, and at sites in which every sponge has previously been mapped and measured with respect to volume) than for corals if we are to learn how important sponge disease really is, or is not.

Water column nutrient concentrations, and the consequent productivity and availability of picoplankton (heterotrophic bacteria, cyanobacteria, prochlorophytes, and pico eukaryotes), are factors that appear to influence the abundance of sponges in general on coral reefs. Greater sponge biomass has been related to greater availability of sponge food both within and between regions. Wilkinson and Cheshire (1990) measured much greater sponge biomass on nearshore parts of the Great Barrier Reef, where water column production is high, than on oceanic reefs in the highly oligotrophic waters of the outer Great Barrier Reef. Taking this comparison a step further, making an explicit comparison between oceans, Wilkinson (1987) measured 7.9–570 g of living sponges per m^2 on the Great Barrier Reef vs. 367–2458 g of living sponges per m^2 on Caribbean coral reefs, which are characterized by greater water column productivity. Transplant experiments have demonstrated higher growth rates with higher water-column picoplankton concentrations between depths on the same reef (sponges of a tube-shaped species grew faster in deep relative to shallow water, Lesser 2006; Trussell et al. 2006). Reef sponges transplanted onto mangrove prop roots, where picoplankton densities were much higher than on the reef, grew 2–3 times as fast as they grew on the coral reef where they normally live (Wulff 2005). The relationship between sponge biomass and growth rates with picoplankton is not monotonic however, and the down-side of dense phytoplankton is that blooms have caused the most striking mass mortalities of sponges that have ever been documented by comparisons of census data from both before and after a mortality event (Butler et al. 1995; Stevely et al. 2011; Wulff 2013).

5.10.4 Data on Sponge Increases and Decreases

More to the point perhaps, than examining scenarios that might explain proliferation of epibenthic sponges over coral reefs, is determining whether or not there are data that demonstrate this proliferation. Although they have opposing ideas about what might cause proliferation of sponges, Pawlik (2011), Bell et al. (2013) and Loh and Pawlik (2014) cite a similar set of papers to support assertions of a phase shift to sponge dominance on coral reefs, including Aronson et al. (2002), Maliao et al. (2008), Norström et al. (2009), McMurray et al. (2010) and Colvard and Edmunds (2011). Because second-hand citations can result in plausible scenarios becoming established facts, it may be useful to examine the data in this set of papers, as not all of the authors claimed that their data demonstrate a general increase in epibenthic sponges. Aronson et al. (2002) discovered that the encrusting sponge *Chondrilla caribensis* increased from 15 to 43 % cover at an unusual site where the corals had previously suffered catastrophic mortality. This sponge species is virtually absent from other reefs nearby, as well as from most other Caribbean reefs in which full fauna surveys have been made (Wulff 2012 pp. 310–312). It would be interesting to know why it became so abundant so quickly at this site. McMurray et al. (2010) also monitored a single species, the barrel sponge *Xestospongia muta*, and acknowledged that although numbers of individuals increased at their two sites, total percent cover and volume did not. Mortality of large individuals, which constituted the bulk of the biomass of the populations, and which are susceptible to a fast-moving disease and to hurricane damage, could abruptly diminish abundance. Colvard and Edmunds (2011) monitored sites in the US Virgin Islands for 14 years, with a primary focus on corals. They documented a slight increase in numbers of individuals ($0.17/m^2$ in 1992, $0.21/m^2$ in 2006) of three sponge species with either erect branching or thinly encrusting forms. Because sponges of these growth forms are readily fragmented by disease or storms into more but smaller individuals by partial mortality, it is possible for an increase

Table 5.2 References commonly used to support assertions that epibenthic sponges in general are overwhelming coral reefs, even in cases in which this was not the conclusion of the original authors

Authors	Locations	Sponge species	Data	Comments
Aronson et al. (2002)	Belize Barrier Reef, Channel Cay	*Chondrilla caribensis*	% cover increased from 15 to 43 %	This single encrusting species covered skeletons of coral that had suffered catastrophic mortality
McMurray et al. (2010)	Florida Keys, Key Largo	*Xestospongia muta*	# individuals increased by 33 % and 46 % at the two sites, but no increases in volume or % cover	Vulnerability of very large individuals to disease and hurricanes could decrease the population dramatically and quickly
Maliao et al. (2008)	Florida Keys, one site	sponges in general	% cover did not change, but was 2.2 % in 1996 and 2.2 % in 2000	Claims of proliferation of sponges at the expense of corals, i.e., a "phase shift" are not supported by the data
Colvard and Edmunds (2011)	US Virgin Islands	*Amphimedon compressa*, *Aplysina cauliformis*, *Spirastrella coccinea*	# individuals increased from 0.17/m^2 in 1992 to 0.21/m^2 in 2006	All three sponge species readily fragment, thus # individuals can increase while biomass decreases
Norström et al. (2009)	Caribbean, in general	sponges in general	Literature review only	Rely on Aronson et al. (2002) as the sole non-boring sponge example
Bell et al. (2013)	SE Sulawesi, Indonesia	*Lamellodysidea herbacea*	60–80 indiv/m^2 to 100–120 indiv/m^2	Sites with high sedimentation and turbidity; increases in # could be due to fragmentation, not biomass increases

in numbers to be linked with decrease in biomass. Maliao et al. (2008) refer to "proliferation of macroalgae and sponges" and include a figure showing how a "phase-shift" has occurred, illustrated by a pair of drawings in which there is apparently three times the amount of sponge mass in the post-phase-shift drawing. This is a puzzling conclusion, given that the data they present are 2.2 % cover of sponges at the start of the study, and 2.2 % at the end, indicating not only very low abundance, but also no sponge increase. Norström et al. (2009) compile data from the publications listed above, and also include studies of boring sponges, but no independent data. In addition to these Caribbean studies, Bell et al. (2013) include a study in which numbers of sponge individuals, most of the species *Lamellodysidea herbacea*, increased from 60–80 per m^2 to 100–120 per m^2 at sites in southeast Sulawesi with high sedimentation and turbidity. Without volume or percent cover information it cannot be determined if this represents an increase in sponges or merely fragmentation into more but smaller individuals of the sponges present earlier. Bell et al. (2013) were careful to make the point that, although dramatic increases of *Terpios hoshinota* have been well documented, these have not been stable. Excavating sponges are a very different story, and increases have been well documented (see Sect. 5.6 and references cited therein). However, even though an impression may be given that sponges are relatively more abundant in places where corals have decreased, with the exception of a few unusual species (refer to Sect. 5.7, and Wulff 2012, pp. 308–312), data have not yet been published to support the assertion that epibenthic sponges in general are proliferating over corals and coral reefs (Table 5.2).

5.10.5 Sponge Dynamics Documented by Full Censuses in Time Series

Mapping, identifying, and measuring the volume of every sponge within a permanently marked plot, again and again at regular intervals, is not fashionable, and often not feasible. This arduous process has only been accomplished at a few sites (see below), but is required if we really want to know whether sponges are increasing or decreasing on coral reefs. As discussed earlier, dead sponges tend to quickly fall apart and vanish so that there is no record that they existed unless they had been previously mapped. Signs of partial mortality are effaced quickly by regeneration, and thus repeated volume measurements are the sole way to know about non-fatal biomass losses.

Three census sites in the Florida Keys have revealed extreme losses of sponges. Stevely et al. (2011) reported losses over just 2 years, at Marathon Key and Long Key of 93 % and 88 % by volume (respectively 69 % and 45 % by number of individuals). The cause was cyanobacteria blooms, which also caused prior losses of 90 % of the sponges in Florida Bay (Butler et al. 1995). At a third site in the Florida Keys, Biggs and Strimaitis (pers. comm.) documented losses of 30 % by volume on a reef influenced by an extended cold snap.

The other two sites where all sponges of all species have been measured in time series were both chosen to represent especially healthy reefs with little human interference. In the San Blas Islands, Panama, in the course of 14 years, 41 % of the volume was lost, and 44 % of the species were lost from a 16 m^2 plot (Wulff 2006a). Although the plot was small, it included 1395 individuals representing 39 species at the

Table 5.3 Documented sponge dynamics on coral reefs (i.e., individual sponges of all or most of the species present were followed in time series)

Authors	Location	Data	Losses	Causes
Wulff (2006a)	Kuna Yala, Panama	14 years, five full censuses, all sponge species	Losses: 41 % of volume, 44 % of species	Disease primarily, same species lost on nearby reefs
Stevely et al. (2011)	Marathon, Fla. Keys, USA	2 years, full censuses, most abundant sponge species	Losses: 93 % of volume, 69 % of individuals	Cyanobacteria blooms
Stevely et al. (2011)	Long Key, Fla. Keys, USA	2 years, full censuses, most abundant sponge species	Losses: 88 % of volume, 45 % of individuals	Cyanobacteria blooms
Wulff (2013)	Belize Barrier Reef	6 years, six full censuses, all sponge species	Losses: 74 % of volume, 44 % of individuals	Cyanobacteria bloom chiefly, and an unknown event 3 years earlier
Biggs and Strimaitis, pers comm	Florida Keys	4 years, five full censuses, all sponge species	Losses: 30 % of volume	Cold snap

start, and loss of species could not be explained by simple stochastic loss of rare species from a small plot. Rare species were not disproportionately represented among those that vanished, and the same species that disappeared from the plot were also missing from other reefs in Kuna Yala that were being followed more qualitatively. Disease was observed in many of the species that vanished, but it is not known that this was the cause of all losses. On the Belize Barrier Reef, two mortality events occurred during 6 years of annual censusing, with the second having a more dramatic effect and a clear cause: an extended dense phytoplankton bloom (Wulff 2013). A total of 74 % of the volume and 44 % of the individuals were lost.

These are not many studies on which to base generalizations, but it should be noted that enormous losses have been documented in every case in which individuals of all or most species in an assemblage have been followed over time, even when sites were chosen as especially favorable for corals and sponges. It would seem prudent to at least reevaluate our assumptions about the overgrowth of coral reefs by sponges and to encourage more studies that are adequate to resolve this issue (Table 5.3).

5.11 Summary: What Would Happen to Coral Reefs if Sponges Were Entirely Deleted?

If we plucked all sponges out of present day coral reefs, the changes would be dramatic and varied, given the very different roles played by epibenthic, cryptic, excavating, and hypercalcifying sponges. Possible positive changes include: (a) living corals would no longer be threatened by the few aggressive sponge species and (b) excavations into coral skeletons would decrease unless boring bivalves increased in response to absence of boring sponges. Possible negative changes include: (a) the water column could become clogged with prokaryotic and other picoplankton growing in response to nutrients, with the loss of the only biological filters efficient enough at capturing picoplankton to keep it in check; (b) living corals would lose adhesive to bolster their grips on the reef frame when their bases are eroded, and would be more likely to fall to their deaths in the sediment; (c) reef repair might cease, and accretion rates could diminish wherever coral rubble remains unstabilized and, therefore unsuitable for successful coral recruitment; (d) the loss of sponge protection on portions of coral skeletons that are not covered with tissue would allow greater access to boring organisms that remain, such as some bivalves and worms; (e) hundreds of species of invertebrates, fishes, and microbes that are obligate symbionts of sponges would lose their habitat, possibly resulting in extinction; (f) obligate spongivores, many of which are attractive mobile fauna, such as angelfishes, hawksbill turtles, and dorid nudibranchs, would lose their prey, and (g) reef frames would be weakened by the loss of reinforcement contributed by skeletons of hypercalcifying sponges that are twice as dense as those of scleractinian corals. Many of these situations have already been documented by controlled experiments or time-series observations. Substantial losses of epibenthic and semicryptic sponge species have been documented by all of the few studies in which coral-reef sponge assemblages have been censused in time-series.

We have insufficient data for confident prediction about whether or not sponges will increase or decrease, but two consistent patterns appear to be emerging, both of them related to nutrient levels. Water column nutrient increases may cause increases in sponges in general, and appear to especially spur on boring sponges and the few species that can overgrow living corals. Simultaneously corals may be more susceptible to both boring and overgrowth when they are stressed by water column issues, such as increased sedimentation and diminished light, that are frequently concomitant with higher nutrient levels. Epibenthic and semicryptic sponges (i.e., the only organisms capable of binding coral rubble after physical disturbance and mitigating losses of live corals due to boring sponges) appear to be highly vulnerable to phytoplankton blooms that are caused by especially large increases in water column nutrients. There may be a fine line between increases and complete loss; but it seems all too possible that losses of sponges may accelerate, and that coral reefs deprived of the many positive roles that sponges play will suffer.

Dedication and Acknowledgements This paper is dedicated to Willard D. Hartman. In his many thoughtful papers, and in his teaching and informal interactions as a beloved professor and researcher at Yale University and the Discovery Bay Marine Laboratory in Jamaica, as well as on expeditions and at conferences, Willard modeled the value of integrating processes spanning temporal scales from evolutionary to seconds, and spatial scales from world-wide to those revealed by high-quality electron microscopy. He had the breadth and depth of understanding of invertebrate form, function, ecosystem roles, and evolutionary history that allowed him to quickly connect a living sponge sporting a dense carbonate skeleton to fossils weighted with a motley history of presumed relations. Work of his students and colleagues continues to be inspired by Willard's kindness, wit, elegance of expression, and his vision of geological and biological aspects of sponges on coral reefs as inseparable.

I am very grateful to Philippe Willenz, Egbert Leigh, Christine Schönberg, and Caroline Rogers for contributing significantly to this chapter with many thoughtful comments. In particular the section on boring sponges was enhanced, and Pacific examples bolstered, by Christine's energetic application of her expertise. Philippe's generous contributions of his beautiful photos of, and expertise on, the hypercalcified sponges were essential. Along with Willard Hartman, Klaus Rützler, Michele Sarà, and Clive Wilkinson fired my inspiration for many long-term studies of sponges; and my funding sources for those long-term sponge studies have been primarily the NSF, Smithsonian Tropical Research Institute, and the Caribbean Coral Reef Ecosystem program of the US National Museum of Natural History, SI.

References

Abdo D, Burgess S, Coleman G. Osborne K (2004) Surveys of benthic reef communities using underwater video. Long-term monitoring of the Great Barrier Reef. Standard operational procedure 2. 3rd revised edition. Australian Institute of Marine Science, Townsville, 62 p.

Adachi N, Ezaki Y, Liu J (2011) Early Ordovician shift in reef construction from microbial to metazoan reefs. Palaios 26:106–114

Aerts LAM (2000) Dynamics behind stand-off interactions in three reef sponge species and the coral *Montastrea cavernosa*. Mar Ecol 21:191–204

Aerts LAM, van Soest RWM (1997) Quantification of sponge/coral interactions in a physically stressed reef community, NE Colombia. Mar Ecol Prog Ser 148:125–134

Alcolado PM (2007) Reading the code of coral reef sponge community composition and structure for environmental bio-monitoring: some experiences from Cuba. Porifera Research: Biodiversity, Innovation and Sustainability. Série Livros 28. Museu Nacional, Rio de Janiero. Pp. 3–10

Aronson RB, Precht WF, Toscano MA, Koltes KH (2002) The 1998 bleaching event and its aftermath on a coral reef in Belize. Mar Biol 141:435–447

Ávila E, Carballo JL (2008) A preliminary assessment of the invasiveness of the Indo-Pacific sponge *Chalinula nematifera* on coral communities from the tropical Eastern Pacific. Biol Invas 11:257–264

Bell JJ (2008) The functional roles of marine sponges. Est Coastal Shelf Sci 79:341–353

Bell JJ, Davy SK, Jones T, Taylor MW, Webster NS (2013) Could some coral reefs become sponge reefs as our climate changes? Global Change Biology 19:2613–2624

Benzoni F, Calcinai B, Eisinger M, Klaus R (2008) Coral disease mimic: sponge attacks *Porites lutea* in Yemen. Coral Reefs 27:695

Biggs B (2013) Harnessing natural recovery processes to improve restoration outcomes: An experimental assessment of sponge-mediated coral reef restoration. Plos One 8:e64945

Bromley RG (1970) Borings as trace fossils and *Entobia cretacea* Portlock, as an example. In: Crimes TP, Harper JH (eds) Trace fossils. Geol J Special Issue 3:49–90

Bromley RG, d'Alessandro A (1984) The ichnogenus *Entobia* from the Miocene, Pliocene and Pleistocene of southern Italy. Riv It Paleont Strat 90:227–296

Bromley RG, d'Alessandro A (1990) Comparative analysis of bioerosion in deep and shallow water, Pliocene to recent, Mediterranean Sea. Ichnos 1:43–49

Brunton FR, Dixon OA (1994) Siliceous sponge-microbe biotic associations and their recurrence through the Phanerozoic as reef mound constructors. Palaios 9: 370–387

Bryan PG (1973) Growth rate, toxicity and distribution of the encrusting sponge *Terpios* sp.(Hadromerida: Suberitidae) in Guam, Mariana Islands. Micronesica 9:237–242

Budd AF (2000) Diversity and extinction in the Cenozoic history of Caribbean reefs. Coral Reefs 19:25–35

Butler MJ, Hunt JH, Herrnkind WF, Childress MJ, Bertelsen R, Sharp W, Matthews T, Field JM, Marshall HG (1995) Cascading disturbances in Florida Bay, USA: cyanobacteria blooms, sponge mortality and implications for juvenile spiny lobsters *Panulirus argus*. Mar Ecol Prog Ser 129:119–125

Carreiro-Silva M, McClanahan TR (2012) Macrobioerosion of dead branching *Porites*, 4 and 6 years after coral mass mortality. Mar Ecol Prog Ser 458:103–122

Cerrano C, Calcinai B, Pinca S, Bavestrello G (2006) Reef sponges as hosts of biodiversity: cases from North Sulawesi. Proc 10th Int Coral Reef Symp 1:208–213

Chaves-Fonnegra A, Zea S (2011) Coral colonization by the encrusting excavating Caribbean sponge *Cliona delitrix*. Mar Ecol 32:162–173

Chaves-Fonnegra A, Zea S, Gómez ML (2007) Abundance of the excavating sponge *Cliona delitrix* in relation to sewage discharge at San Andrés Island, SW Caribbean, Colombia. Bol Invest Mar Cost 36:63–7

Cleary DFR, de Voogd NJ (2007) Environmental associations of sponges in the Spermonde Archipelago, Indonesia. J Marine Biol Assoc UK 87:1669–1676

Coles SL, Bolick H (2007) Invasive introduced sponge *Mycale grandis* overgrows reef corals in Kaneohe Bay, Oahu, Hawaii. Coral Reefs 26:911

Collins LS, Budd AF, Coates AG (1996) Earliest evolution associated with closure of the Tropical American Seaway. Proc Natl Acad Sci 93:6069–607

Colvard NB, Edmunds PJ (2011) Decadal-scale changes in abundance of non-scleractinian invertebrates on a Caribbean coral reef. J Exp Mar Biol Ecol 397:153–160

Corredor JE, Wilkinson CR, Vicente VP, Morell JM, Otero E (1988) Nitrate release by Caribbean reef sponges. Limnol Oceanogr 33:114–120

Cowart JD, Henkel TP, McMurray SE, Pawlik JR (2006) Sponge orange band (SOB): a pathogenic-like condition of the giant barrel sponge, *Xestospongia muta*. Coral Reefs 25:513

de Goeij JM, van den Berg H, van Oostveen MM, Epping EHG, van Duyl FC (2008) Major bulk dissolved organic carbon (DOC) removal by encrusting coral reef cavity sponges. Mar Ecol Prog Ser 357:139-151 de Goeij JM, Oevelen D van, Vermeigh JJA, Osinga R, Middelbury JJ, de Goeij AFPM, Admiraal W (2013) Surviving in a marine desert: the sponge loop retains resources within coral reefs. Science 342:108-110

de Voogd NJ, Cleary DFR (2008) An analysis of sponge distribution and diversity at three taxonomic levels in the Thousand Islands/ Jakarta Bay reef complex, West Java, Indonesia. Mar Ecol 29:205–215

de Voogd NJ, Cleary DFR, Dekker F (2013) The coral-killing sponge *Terpios hoshinota* invades Indonesia. Coral Reefs 32:755

Debrenne F (1999) The past of sponges – sponges of the past. Memoirs of the Queensland Museum 44:9–21

Debrenne F (2007) Lower Cambrian archaeocyathan bioconstructions. Comptes Rendus Palevol 6:5–19

Debrenne F, Zhuravlev AY (1994) Archaeocyathan affinities: How deep can we go into the systematic affiliation of an extinct group. In: van Soest RWM, van Kempen TMG, Braekman J-C (eds) Sponges in time and space: biology, chemistry, paleontology. A.A. Balkema, Rotterdam, pp 3-12

Diaz C, Rützler K (2001) Sponges: An essential component of Caribbean coral reefs. Bull Mar Sci 69:535–546

Duffy JE (1996) Eusociality in a coral-reef shrimp. Nature 381:512–514

Edinger EN, Risk MJ (1997) Sponge borehole size as relative measure of bioerosion and paleoproductivity. Lethaia 29:275–286

Erwin PM, Thacker RW (2007) Incidence and identity of photosynthetic symbionts in Caribbean coral reef sponge assemblages. J Mar Biol Ass UK 87:1683–169

Fang JHK, Athayde MAM, Schönberg CHL, Kline DI, Hoegh-Guldberg O, Dove S (2013) Sponge biomass and bioerosion rates under ocean warming and acidification. Global Change Biol 19:3581–3591

Fang JHK, Schönberg CHL, Athayde MAM, Kline DI, Hoegh-Guldberg O, Dove S (2014) Effects of ocean warming and acidification on the energy budget of an excavating sponge. Global Change Biol 20:1043–1054

Freeman CJ, Thacker RW (2011) Complex interactions between marine sponges and their symbiotic microbial communities. Limnol Oceanogr 56:1577–1586

Fromont J, Vanderklift MA, Kendrick GA (2006) Marine sponges of the Dampier Archipelago, Western Australia: patterns of species distributions, abundance and diversity. Biodiversity and Conservation 15:3731–3750

Gandin A, Debrenne F (2010) Distribution of the archaeocyath-calcimicrobial bioconstructions on the Early Cambrian shelves. Palaeoworld 19:222–241

Garson MJ, Clark RJ, Webb RI, Field KL, Charan RD, McCaffrey EJ (1999) Ecological role of cytotoxic alkaloids: *Haliclona* n. sp., an unusual sponge/dinoflagellate association. Mem Queensland Mus 44:205–213

Gilis M, Grauby O, Willenz Ph, Dubois P, Heresanu V, Baronnet A (2013) Biomineralization in living hypercalcified sponges: toward a shared mechanism? J Structural Bio 183:441–454

Glynn PW (1974) Rolling stones among the Scleractinia: mobile coralliths in the Gulf of Panamá. Proc 2nd Int Coral Reef Symp 2:183–198

González-Rivero M, Ferrari R, Schönberg CHL, Mumby PJ (2012) Impacts of macroalgal competition and parrotfish predation on growth of a common bioeroding sponge. Mar Ecol Prog Ser 444:133–142

Goreau TF, Hartman WD (1963) Boring sponges as controlling factors in the formation and maintenance of coral reefs. Mechanisms of Hard Tissue Destruction. Amer Assoc Adv Sci 75:25–54

Goreau TF, Hartman WD (1966) Sponge: effect on the form of reef corals. Science 151:343–344

Halfar J, Riegl R (2013) From coral framework to rodolith bed: sedimentary footprint of the 1982/1983 ENSO in the Galápagos. Coral Reefs 32:985

Hallock P (1988) The role of nutrient availability in bioerosion: consequences to carbonate buildups. Palaeogeogr Palaeoclimatol Palaeoecol 63:275–291

Hartman WD (1977) Sponges as reef builders and shapers. Studies in Geology 4:127–134

Hartman WD, Goreau TF (1970) Jamaican coralline sponges: their morphology, ecology, and fossil relatives. In: Fry WC (ed) The biology of the porifera: Zool Soc London Symp 25:205–243

Hartman WD, Goreau TF (1975) A pacific tabulate sponge, living representative of a new order of sclerosponges. Postilla 167:1–21

Hartman, WD, Wendt JW, Wiedenmayer F (1980) Living and fossil sponges. Sedimenta 8:1–274

Hill M, Allenby A, Ramsby B, Schönberg CHL, Hill A (2011) *Symbiodinium* diversity among host clionaid sponges from Caribbean and Pacific reefs: evidence of heteroplasmy and putative host-specific symbiont lineages. Mol Phylogen Evol 59:81–88

Holmes KE (1997) Eutrophication and its effect on bioeroding sponge communities. Proc. 8th Int Coral Reef Symp 2:1411–1416

Hooper JNA, Kennedy JA (2002a) Small-scale patterns of sponge biodiversity (Porifera) on Sunshine Coast reefs, eastern Australia. Invertebrate Systematics 16:637–653

Hooper JNA, Kennedy JA, Quinn RJ (2002b) Biodiversity "hotspots", patterns of richness and endemism, and taxonomic affinities of tropical Australian sponges. Biodiversity and Conservation 11: 851–885

Hourigan, TF, Stanton, FG, Motta, PJ, Kelley, CD, Carlson, B (1989) The feeding ecology of three species of Caribbean angelfishes (family Pomacanthidae). Envir Biol Fishes 24:105–116

Hubbard DK (1985) What do we mean by reef growth? Proc 5th Int Coral Reef Congr 6:433–438

Hubbard DK (1988) Controls of modern and fossil reef development: common ground for biological and geological research. Proc 6th Int Coral Reef Symp 1:243–252

Hubbard DK, Burke RB, Gill IP (1998) Where's the reef? The role of framework in the Holocene. Carbonates and Evaporites 13:3–9

Jackson JBC, Goreau TF, Hartman WD (1971) Recent brachiopod-coralline sponge communities and their paleoecological significance. Science 173:623–625

Kiessling W (2009) Geologic and biologic controls on the evolution of reefs. Ann Rev Ecol Evol Syst 40:173–192

Kiessling W, Aberhan M, Brenneis B, Wagner PJ (2007) Extinction trajectories of benthic organisms across the Triassic – Jurassic boundary. Palaeos 244:201–222

Kobluk DR (1981) Lower Cambrian cavity-dwelling endolithic (boring) sponges. Canadian Journal of Earth Science 18:972–980

Lang JC, Hartman WD, Goreau TF (1975) Sclerosponges: primary framework constructors on the Jamaican deep forereef. J Mar Res 33:223–231

Lawrence D (1969) The use of clionid sponges in palaeoenvironmental analyses. J Palaeontol 43:539–543

Lehnert H, Fischer H (1999) Distribution patterns of sponges and corals down to 107 m off North Jamaica. Mem Queensland Mus 44:307–316

León YM, Bjorndal KA (2002) Selective feeding in the hawksbill turtle, an important predator in coral reef ecosystems. Mar Ecol Prog Ser 245:249–258

Lesser MP (2006) Benthic-pelagic coupling on coral reefs: feeding and growth of Caribbean sponges. J Exp Mar Biol Ecol 328:277–288

Loh T-L, Pawlik JR (2014) Chemical defenses and resource trade-offs structure sponge communities on Caribbean coral reefs. Proc Natl Acad Sci 111:4151–4156

López-Victoria M, Zea S (2005) Current trends in space occupation by encrusting excavating sponges on Colombian coral reefs. Mar Ecol Prog Ser 26:33–41

López-Victoria M, Zea S, Weil E (2006) Competition for space between encrusting excavating Caribbean sponges and other coral reef organisms. Mar Ecol Prog Ser 312:113–121

Macdonald IA, Perry CT (2003) Biological degradation of coral framework in a turbid lagoon environment, Discovery Bay, north Jamaica. Coral Reefs 22:523–535

Maldonado M, Carmona MC, Uriz MJ, Cruzado A (1999) Decline in Mesozoic reef-building sponges explained by silicon limitation. Nature 401:785–788

Maldonado M, Ribes M, van Duyl FC (2012) Nutrient fluxes through sponges: biology, budgets, and ecological implications. Adv Mar Bio 62:113–182

Maliao RJ, Turingen RG, Lin J (2008) Phase-shift in coral reef communities in the Florida Keys National Marine Sanctuary (FKNMS), US. Mar Biol 154:841–853

Mallela J and Perry CT (2007) Calcium carbonate budgets for two coral reefs affected by different terrestrial runoff regimes, Rio Bueno, Jamaica. Coral Reefs 26:129–145

Marquez, J.C., Zea, S. 2012. Parrotfish mediation in coral mortality and bioerosion by the encrusting, excavating sponge *Cliona tenuis*. Mar Ecol 33:417–426

McLean EL, Yoshioka PM (2008) Substratum effects on the growth and survivorship of the sponge *Desmapsamma anchorata*. Carib J Sci 44:83–89

McMurray SE, Henkel TP, Pawlik JR (2010) Demographics of increasing populations of the giant barrel sponge *Xestospongia muta* in the Florida Keys. Ecol 91:560–570

Meylan A (1988) Spongivory in hawksbill turtles: A diet of glass. Science 239:393–395

Meylan A (1990) Nutritional characteristics of sponges in the diet of the hawksbill turtle. *Eretmochelys imbricata*. In: Rützler K (ed) New perspectives in sponge biology. Smithsonian Institution Press. Washington D.C. pp 472–477

Miller J, Muller E, Rogers C, Waara R, Atkinson A, Whelan KRT, Patterson M, Witcher B (2009) Coral disease following massive bleaching in 2005 causes 60 % decline in coral cover on reefs in the US Virgin Islands. Coral Reefs 28:925–937.

Mueller B, de Goeij JM, Vermej MJA, Mulders Y, van der Ent E, Ribes M, van Duyl FC (2014) Natural diet of coral-excavating sponges consists mainly of dissolved organic carbon (DOC). PloS One 9:e90152

Norström AV, Nyström M, Lokrantz J, Folke C (2009) Alternative states on coral reefs: beyond coral-macroalgal phase shifts. Mar Ecol Prog Ser 376:295–306

Padilla Verdín CJ, Carballo JL, Camacho ML (2010) A qualitative assessment of sponge-feeding organisms from the Mexican Pacific Coast. Open Mar Biol J 4:39–46

Pandolfi J, Kiessling W (2014) Gaining insights from past reefs to inform understanding of coral reef response to global climate change. Curr Opin Environ Sustainability 7:52–58

Pari N, Peyrot-Clausade M, Hutchings PA (2002) Bioerosion of experimental substrates on high islands and atoll lagoons (French Polynesia) during 5 years of exposure. J Exp Mar Bio Ecol 276:109–127

Pawlik JR (2011) The chemical ecology of sponges on Caribbean reefs: Natural products shape natural systems. BioScience 61:888–898

Perry CT (1998) Macroborers within coral framework at Discovery Bay, north Jamaica: species distribution and abundance, and effect on coral preservation. Coral Reefs 17:277–287

Perry CT (2000) Macroboring of Pleistocene coral communities, Falmouth Formation, Jamaica. Palaois 15:483–489

Perry CT, Edinger EN, Kench PS, Murphy GN, Smithers SG, Steneck RS, Mumby PJ (2012) Estimating rates of biologically driven coral reef framework production and erosion: a new census-based carbonate budget methodology and application to the reefs of Bonaire. Coral Reefs 31:853–868.

Peterson BJ, Chester CM, Jochem FJ, Fourqurean JW (2006) Potential role of sponge communities in controlling phytoplankton blooms in Florida Bay. Mar Ecol Prog Ser 328:93–103

Pleydell SM, Jones B (1988) Boring of various faunal elements in the Oligocene-Miocene Bluff Formation of Grand Cayman, British West Indies. J Paleont 62:348–367

Randall JE, Hartman WD (1968) Sponge-feeding fishes of the West Indies. Mar Biol 1:216–225

Reed JK, Pomponi SA (1997) Biodiversity and distribution of deep and shallow water sponges in the Bahamas. Proc 8th Int Coral Reef Symp 2:1687–1692

Reimer JD, Nozawa Y, Hirose E (2010) Domination and disappearance of the black sponge: a quarter century after the initial *Terpios* outbreak in southern Japan. Zoological Studies 50:394

Reiswig HM (1971) Particle feeding in natural populations of three marine demosponges. Biol Bull 141:568–591

Reiswig HM (1973) Population dynamics of three Jamaican Demospongiae. Bull Mar Sci 23:191–226

Reiswig HM (1974) Water transport, respiration, and energetics of three tropical marine sponges. J Exp Mar Biol Ecol 14:231–249

Reiswig HM (1981) Partial carbon and energy budgets of the bacteriosponge *Verongia fistularis* (Porifera: Demospongiae) in Barbados. Mar Ecol 2:273–293

Reitner J, Gautret P (1996) Skeletal formation in the modern but ultraconservative Chaetetid sponge *Spirastrella* (Acanthochaetetes) *wellsi* (Demospongiae, Porifera). Facies 34:193–208

Reitner J, Keupp H (1991) The fossil record of the haplosclerid excavating sponge *Aka* de Laubenfels. In: Reitner J, Keupp H (eds) Fossil and recent sponges. Springer Berlin: 102–120

Reitner J, Wörheide G, Lange R, Schumann-Kindel G (2001) Coralline demosponges – a geobiological portrait. Bull of the Tohoku University Museum 1:219–235

Richter C, Wunsch M, Rasheed M, Kötter I, Badran MI (2001) Endoscopic exploration of Red Sea coral reefs reveals dense populations of cavity-dwelling sponges. Nature 413:726–730

Rose CS, Risk MJ (1985) Increase in *Cliona delitrix* infestation of *Montastrea cavernosa* heads on organically polluted portions of the Grand Cayman fringing reef. Mar Ecol 6:345–363

Rowland SF (2001) Archaeocyaths – A history of phylogenetic interpretation. J Paleon 75:1065–1078

Rützler K (1978) Sponges in coral reefs. In: Stoddart DR, Johannes RE (eds) Coral reefs: research methods. Monographs on Oceanographic Methodology 5, UNESCO, Paris.

Rützler K (1990) Associations between Caribbean sponges and photosynthetic organisms. In: Rützler K (ed) New perspectives in sponge biology. Smithsonian Institution Press. Washington D.C. pp. 455–466

Rützler K (2002) Impact of crustose clionid sponges on Caribbean coral reefs. Acta Geologica Hispanica 37:61–72

Rützler K (2004) Sponges on coral reefs: A community shaped by competitive cooperation. Boll Mus Ist Biol Univ Genova 68:85–148

Rützler K (2012) The role of sponges in the Mesoamerican Barrier Reef ecosystem, Belize. Adv Mar Bio 61:211–271

Rützler K, Macintyre IG (1978) Siliceous sponge spicules in coral reef sediments. Mar Bio 49:147–159

Rützler K, Muzik K (1993) *Terpios hoshinota*, a new cyanobacteriosponge threatening Pacific reefs. Sciencias Marinas 57:395–403

Scalera-Liaci L, Sciscioli M, Lepore E, Gaino E (1999) Symbiotic zooxanthellae in *Cinachyra tarentina*, a non-boring demosponge. Endocytobiosis Cell Res 13:105–114

Scheibling RE (1979) The ecology of *Oreaster reticulatus* (L.) (Echinodermata: Asteroidea) in the Caribbean. PhD thesis, McGill University, Montreal

Schönberg CHL (2002) Substrate effects on the bioeroding demosponge *Cliona orientalis*. 1. Bioerosion rates. PSZNI Mar Ecol 23:313–326

Schönberg CHL (2003) Substrate effects on the bioeroding demosponge *Cliona orientalis*. 2. Substrate colonization and tissue growth. PSZNI Mar Ecol 24:59–74

Schönberg CHL (2008) A history of sponge bioerosion: from past myths and hypotheses to recent approaches. Current Developments in Bioerosion. pp. 165–202

Schönberg CHL, Fromont J (2012) Sponge gardens of Ningaloo Reef (Carnarvon Shelf, Western Australia) are biodiversity hotspots. Hydrobiologia 687:143–161

Schönberg CHL, Loh WKW (2005) Molecular identity of the unique symbiotic dinoflagellates found in the bioeroding demosponge *Cliona orientalis* Thiele, 1900. Mar Ecol Prog Ser 299:157–166

Schönberg CHL, Ortiz J-C (2009) Is sponge bioerosion increasing? Proc 11th Int Coral Reef Symp, Ft Lauderdale 1:527–530

Schönberg CHL, Suwa R (2007) Why bioeroding sponges may be better hosts for symbiotic dinoflagellates than many corals. In: Custódio MR, Lôbo-Hajdu G, Hajdu, E, Muricy G (eds) Porifera research: biodiversity, innovation and sustainability. Série Livros 28. Museu Nacional, Rio de Janiero pp 569–580

Schönberg CHL, Suwa R, Hidaka M (2008) Sponge and coral zooxanthellae in heat and light: preliminary results of photochemical efficiency monitored with pulse amplitude modulated fluorometry. Mar Ecol 29:247–258

Schönberg CHL, Tapanila L (2006) The bioeroding sponge *Aka paratypica*, a modern tracemaking analogue for the paleozoic ichnogenus *Entobia devonica*. Ichnos 13:147–157

Schönberg CHL, Wilkinson CR (2001) Induced colonization of corals by a clionid bioeroding sponge. Coral Reefs 20:69–76

Schönberg CHL, Wisshak M (2012) The perks of being endolithic. Aquat Biol 17:1–5

Schumacher H, Plewka M (1981) Mechanical resistance of reefbuilders through time. Oecologia 49:279–282

Schwartz SA, Budd AF, Carlon DB (2012) Molecules and fossils reveal punctuated diversification in Caribbean "faviid" corals. BMC Evolutionary Biology 12:123

Scoffin T, Stearn C, Boucher D, Frydl P, Hawkins CM, Hunter IG, MacGeachy JK (1980) Calcium carbonate budget of a fringing reef on the west coast of Barbados. I. Erosion, sediments and internal structure. Bull Mar Sci 30:475–508

Seguin, F, Le Brun, O, Hirst, R, Al-Thary, I, Dutrieux, E (2008) Large coral transplantation in Bal Haf (Yemen): an opportunity to save corals during the construction of a liquefied natural gas plant using innovative techniques. Proc 11th Int Coral Reef Symp, Ft Lauderdale 2:1272–1275

Simpson TR (1984) The cell biology of sponges. Springer Berlin, 677 pp

Soong K, Yang S-L, Chen CA (2009) A novel dispersal mechanism of a coral-threatening sponge, *Terios hoshinota* (Suberitidae, Porifera). Zoological Studies 48:596.

Stanley SM, Hardie LA (1998) Secular oscillations in the carbonate minerology of reef-building and sediment-producing organisms driven by tectonically forced shifts in seawater chemistry. Paleogeography, Paleoclimatology, Paleobiology 44:3–19.

Stevely JM, Sweat DE, Bert TM, Sim-Smith C, Kelly M (2011) Sponge mortality at Marathon and Long Key, Florida: Patterns of species response and population recovery. Proc. 53rd Gulf and Caribbean Fisheries Institute 63:384–400

Stock CW (2001) Stromatoporoidea, 1926–2000. J Paleo 75:1079–1089

Strimaitis AM (2012) Filter feeding ecology of erect branching sponges on Caribbean coral reefs. Master's Thesis, Department of Biological Science, Florida State University

Stubler AD, Furman BT, Peterson BJ (2014) Effects of *p*CO2 on the interaction between an excavating sponge, *Cliona varians*, and a hermatypic coral, *Porites furcata*. Mar Biol 161:1851-1859

Tapanila L, Copper P, Edinger E (2004) Environmental and substrate control on Paleozoic bioerosion in corals and stromatoporoids. Palaios 19:292–306

Thacker RW (2005) Impacts of shading on sponge-cyanobacteria symbioses: A comparison between host-specific and generalist associations. Integr Comp Biol 45:369–376

Trussell GC, Lesser MP, Patterson MR, Genovese SJ (2006) Depth-specific differences in growth of the reef sponge *Callyspongia vaginalis*: role of bottom-up effects. Mar Ecol Prog Ser 323:149–158

Vacelet J (1970) Les éponges Pharétronides actuelles. Zool Soc Lond Symp 25:189–204

Vacelet J, Willenz Ph, Hartman WD (2010) Living hypercalcified sponges. Treatise Online Number 1, Part E, Revised, Volume 4, Chapter 1

van Dam R, Diez CE (1997) Predation by hawksbill turtles at Mona Island, Puerto Rico. Proc 8th Int Coral Reef Symp 2:1421–1426

van Duyl FC, Moodley L, Nieuwland G, van Ijezerloo L, van Soest RWM, Houtekamer M, Meesters EH, Middelburg JJ (2011) Coral cavity sponges depend on reef-derived food resources: stable isotope and fatty acid constraints. Mar Biol 158:1653–1666

van Soest RWM (2009) New sciophilous sponges from the Caribbean (Porifera: Demospongiae). Zootaxa 2107:1–40

van Soest RWM, Boury-Esnault N, Vacelet J, Dohrmann M, Erpenbeck D, de Voogd NJ, da Santodomingo N, Vanhoorne B, Kelly M, Hooper JNA (2012) Global biodiversity of sponges (Porifera). PLoS One 7(4):e35105

Vicente VP (1990) Overgrowth activity by the encrusting sponge *Chondrilla nucula* on a coral reef in Puerto Rico. In: Rützler K (ed) New perspectives in sponge biology. Smithsonian Institution Press, Washington, D.C. pp 436–443

Wang J-T, Chen Y-Y, Meng P-J, Sune Y-H, Hsu C-H, Wei K-Y, Chen CA (2012) Diverse interactions between corals and the coral-killing sponge *Terpios hoshinota* (Suberitidae: Hadromerida). Zool Stud 51:150–159

Ward-Paige CA, Risk MJ, Sherwood OA, Jaap WC (2005) Clionid sponge surveys on the Florida reef tract suggest land-based nutrient inputs. Mar Poll Bull 51:570–579

Webster NS (2007) Sponge disease: a global threat? Envir Microbio 9:1363–1375

Webster NS, Taylor MW (2012) Marine sponges and their microbial symbionts: love and other relationships. Envir Microb 14:335–346

Weisz JB, Lindquist N, Martens CS (2008) Do associated microbial abundances impact marine demosponge pumping rates and tissue densities? Oecologia 155:367–376

West RR, Kershaw S (1991) Chaetetid habitats. In: Reitner J, Keupp H, eds. Fossil and Recent Sponges. Springer-Verlag Berlin Heidelberg Pp. 445–455

West RR, Vacelet J, Wood RA, Willenz Ph, Hartman WD (2010) Hypercalcified extant and fossil chaetetid-type and post-Devonian stromatoporoid-type Demospongiae: systematic descriptions. Treatise online number 58, part E, revised, volume 4, chapter 4A-B

Wilkinson CR (1983) Role of sponges in coral reef structural processes. In: Barnes DJ (ed) Perspectives on coral reefs. Brian Clouston, Publisher. pp 263–274

Wilkinson CR (1987) Interocean differences in size and nutrition of coral reef sponge populations. Science 236:1654–1657

Wilkinson CR, Cheshire AC (1990) Comparisons of sponge populations across the barrier reefs of Australia and Belize: evidence for higher productivity in the Caribbean. Mar Ecol Prog Ser 67:285–294

Willenz Ph, Hartman WD (1985) Calcification rate of *Ceratoporella nicholsoni* (Porifera: Sclerospongiae): An in situ study with calcein. Proc 5th Int Coral Reef Congr 5:113–118
Willenz Ph, Hartman WD (1989) Micromorphology and ultrastructure of Caribbean sclerosponges 1. *Ceratoporella nicholsoni* and *Stromatospongia norae* (Ceratoporellidae: Porifera). Mar Bio 103:387–401
Willenz Ph, Hartman WD (1999) Growth and regeneration rates of the calcareous skeleton of the Caribbean coralline sponge *Ceratoporella nicholsoni*: A long term survey. Memoirs of the Queensland Museum 44:675–685
Wisshak M, Schönberg CHL, Form A, Freiwald A (2012) Ocean acidification accelerates reef bioerosion. PLoS One 7:e45124
Wisshak M, Schönberg CHL, Form A, Freiwald A (2013) Effects of ocean acidification and global warming on bioerosion – lessons from a clionaid sponge. Aquatic Biol 19:111–127
Wisshak M, Schönberg CHL, Form A, Freiwald A (2014) Sponge bioerosion accelerated by ocean acidification across species and latitudes? Helgol Mar Res 68:253–262
Wood R (1987) Biology and revised systematics of some late Mesozoic stromatoporoids. Special papers in palaeontology 37:1–89
Wood R (1990) Reef-building sponges. American Scientist 78:224–235
Wood R (1993) Nutrients, predation and the history of reef-building. Palaios 8:526–543
Wood R (1995) The changing biology of reef-building. Palaios 10:517–527
Wood R (2011) Taphonomy of reefs through time. Topics in Geobiology 32:375–409
Wood R, Zhuravlev AY, Debrenne F (1992) Functional biology and ecology of Archaeocyatha. Palaios 7:131–156
Woodley JD, Chornesky EA, Clifford PA, Jackson JBC, Kaufman LS, Lang JC, Pearson MP, Porter JW, Rooney MC, Rylaarsdam KW, Tunnicliffe VJ, Wahle CW, Wulff JL, Curtis ASG, Dallmeyer MD, Jupp BP, Koehl MAR, Neigel J, Sides EM (1981) Hurricane Allen's impact on Jamaican coral reefs. Science 214:749–755
Wörheide G (1998) The reef cave dwelling ultraconservative coralline demosponge *Astrosclera willeyana* Lister 1900 from the Indo-Pacific: micromorphology, ultrastructure, biocalcification, isotope record, taxonomy, biogeography, phylogeny. Facies 38:1–88
Wulff JL (1984) Sponge-mediated coral reef growth and rejuvenation. Coral Reefs 3:157–163
Wulff JL (1994) Sponge-feeding by Caribbean angelfishes, trunkfishes, and filefishes. In: van Soest RWM, van Kempen TMG, Braekman J-C (eds) Sponges in time and space: biology, chemistry, paleontology. A.A. Balkema, Rotterdam, pp 265–271
Wulff JL (1995) Sponge-feeding by the Caribbean starfish *Oreaster reticulatus*. Mar Biol 123:313–325
Wulff JL (1997a) Mutually beneficial associations among species of coral reef sponges. Ecology 78:146–159
Wulff JL (1997b) Parrotfish predation on cryptic sponges of Caribbean coral reefs. Mar Biol 129:41–52
Wulff JL (1997c) Causes and consequences of differences in sponge diversity and abundance between the Caribbean and eastern Pacific at Panama. Proc 8th Int Coral Reef Symp 2:1377–1382
Wulff JL (2001) Assessing and monitoring coral reef sponges: Why and how? Bull Mar Sci 69:831–846
Wulff JL (2005) Trade-offs in resistance to competitors and predators, and their effects on the diversity of tropical marine sponges. J Anim Ecol 74:313–321
Wulff JL (2006a) Rapid diversity and abundance decline in a Caribbean coral reef sponge community. Biol Conserv 127:167–176
Wulff JL (2006b) Resistance vs. recovery: morphological strategies of coral reef sponges. Func Ecol 20:699–708
Wulff JL (2006c) A simple model of growth form-dependent recovery from disease in coral reef sponges, and implications for monitoring. Coral Reefs 25:419–426
Wulff JL (2006d) Sponge systematics by starfish: predators distinguish cryptic sympatric species of Caribbean fire sponges, *Tedania ignis* and *Tedania klausi* n. sp. (Demospongiae, Poecilosclerida). Biol Bull 211:83–94
Wulff JL (2006e) Ecological interactions of marine sponges. Canadian J Zool Spec Ser 84:146–166
Wulff JL (2008a) Collaboration among sponge species increases sponge diversity and abundance in a seagrass meadow. Mar Ecol 29:193–204
Wulff JL (2008b) Life history differences among coral reef sponges promote mutualism or exploitation of mutualism by influencing partner fidelity feedback. Amer Nat 171:597–609
Wulff JL (2010) Regeneration of sponges in ecological context: Is regeneration an integral part of life history and morphological strategies? Integr Comp Biol 50:494–505
Wulff J (2012) Ecological interactions and the distribution, abundance, and diversity of sponges. Adv Mar Biol 61:273–344
Wulff J (2013) Recovery of sponges after extreme mortality events: morphological and taxonomic patterns in regeneration versus recruitment. Integr Comp Biol 53:512–523
Wulff JL, Buss LW (1979) Do sponges help hold coral reefs together? Nature 281:474–475
Zea S (1994) Patterns of coral and sponge abundance in stressed coral reefs Santa Marta, Colombian Caribbean. In: van Soest RWM, van Kempen TMG, Braekman J-C (eds) Sponges in time and space: biology, chemistry, paleontology. A.A. Balkema, Rotterdam, p 257–264
Zea S (2001) Patterns of sponge (Porifera, Demospongiae) distribution in remote, oceanic reef complexes of the southwestern Caribbean. Revista de la Academia Colombiana de Ciencias Exactas, Físicas y Naturales 25:579–592

The Changing Face of Reef Building

6

Dennis K. Hubbard and Wolf-Christian Dullo

Abstract

Declining calcification and accelerating sea-level rise have brought us ever closer to the point where coral reefs may not be able to keep pace. Even if this is insufficient to change reef-community structure or totally overtop low reef islands in the twenty-first century, the impacts on reefs and the organisms that depend on them will still be profound. Patterns of sea-level rise have varied spatially in the past due to both local tectonics and regional crustal responses to deglaciation. The result has been regionally disparate sea-level histories that complicate our understanding of the links between past sea level and reef development.

At the same time, gaps remain in our understanding of how, and how fast, reefs build. Holocene reefs-accretion rates (generally <5 mm/year) are lower than previous estimates (10–15 mm/year), making coral reefs more vulnerable to rising sea level than has been assumed. Furthermore, the conflation of coral growth and reef accretion has provided an overly simplistic view of reef building that focuses on coral abundance and calcification. Protocols have been suggested to quantify the changing balance between carbonate production and bioerosion, but these still ignore the role of physical processes that redistribute and remove material from the reef, a scenario that will become even more important as the intensity of tropical storms increases. Holocene cores show that accretion does not mimic the depth dependence of calcification, suggesting that predictions based solely on biological assessments could be flawed.

Uniformitarianism, the idea that "the present is the key to the past", has been a fundamental tool for geologists trying to unravel the development of ancient reefs using their modern counterparts. As we try to separate anthropogenic change from natural variability that operates on cycles longer than human lifetimes, we might consider whether this concept could be reversed to help predict the fate of coral reefs – or to at least examine some of our critical assumptions about reef accretion and sea-level rise. This chapter considers some of our long-standing models of sea level and reef building, using recent data to provide a more complete picture of the factors involved in both the recent geologic past and the immediate future. The goal is to provide a better understanding of interactions between the two that might allow better models of ancient reefs while also providing more realistic answers to the question, "Will coral reefs keep up with rising sea level in the twenty-first century?"

D.K. Hubbard (✉)
Department of Geology, Oberlin College, 52 W. Lorain St., Oberlin, OH 44074, USA
e-mail: Dennis.Hubbard@oberlin.edu

W.-C. Dullo
Ocean Circulation und Climate Dynamics, Helmholtz-Zentrum für Ozeanforschung, Kiel, Germany
e-mail: cdullo@geomar.de

D.K. Hubbard et al. (eds.), *Coral Reefs at the Crossroads*,
Coral Reefs of the World 6, DOI 10.1007/978-94-017-7567-0_6

Keywords
Holocene • Reef accretion • Sea level • Reef drowning • Carbonate budget • Reef islands

6.1 Introduction

By most measures, the world's coral reefs are in decline and many have already been ecologically lost (Burke et al. 2011; De'ath et al. 2012; Jackson et al. 2014). Papers often point to changes in the late 1970s or early 1980s as the start of recent reef decline (e.g., Gladfelter 1982; Lessios et al. 1984; Aronson and Precht 2001; Gardner et al. 2003; Hughes et al. 2003; Carpenter et al. 2008). However, evidence suggests that measurable differences had already occurred over the preceding decades to centuries... or even longer (Jackson 1977; Pandolfi et al. 2005). Looking forward, impacts will include lower coral cover and diversity along individual reefs (Gardner et al. 2003; Pandolfi et al. 2011; Jackson et al. 2014; Chap. 7, this volume), large-scale bleaching and disease outbreaks (Glynn 1996; Harvell et al. 2004; Miller et al. 2009; Weil and Rogers 2011; Jackson et al. 2014; Descombes et al. 2015), depressed calcification (Kleypas et al. 1999; Aronson and Precht 2006; Hoegh-Guldberg et al. 2007; Jokiel 2011; Zeebe and Wolf-Gladrow 2011) and increasing physical damage by stronger storms (Emanuel 2005; Webster et al. 2005; Curry et al. 2006; Hetzinger et al. 2008). Myriad social impacts will include fisheries decline (McClanahan 2002), loss of wetlands (Mimura et al. 2007) and declining economic and ecological services (Moberg and Folke 1998; Cesar et al. 2003; Costanza et al. 2014).

6.1.1 Changing Perceptions... Changing Strategies

Before reef decline was obvious, we used broad surveys to document community structure and interactions among organisms on single reefs. Our questions tended to focus on "how reefs worked" and primary among the currencies were abundance, diversity and "stability". Geologists modified biological methods to better understand their ancient counterparts, but the comparisons were hampered by both evolutionary changes in reef biota and huge losses due to taphonomy and diagenesis. The resulting models were oversimplified and tended to focus on the role of calcification by in-place organisms in ancient reef building.

As decline became increasingly apparent, the focus on "reef function" was broadened to include monitoring that could quantify changes through time. Principal among the perceived stresses were sedimentation, pollution, nutrients, overfishing and other local factors operating "in all areas where human activities are concentrated" (Wilkinson 2000, p. 7). While bleaching had been reported nearly a century earlier (Mayer 1914), its significance was just emerging and reports of coral diseases were few. Conservation strategies were largely based on reefs being "discrete entities with water barriers between them "(Wilkinson 2000, p. 15) and a sense that the main drivers of decline were somewhere on this side of the horizon.

The negative effects of carbon dioxide and other Greenhouse Gasses were not yet fully obvious, but their levels were inexorably climbing. In 1958, Charles Keeling had measured CO_2 levels of 316 ppm atop Mona Loa, nearly 10 % higher than in 1900. It is perhaps telling that the National Science Foundation declined to fund his continued CO_2 measurements which they characterized as "routine" work,[1] a sentiment that continues today with respect to reef monitoring (Chap. 12).

By 1988, CO_2 concentrations had reached 350 ppm when James Hansen told congress, "Global warming has reached a level such that we can ascribe with a high degree of confidence a cause and effect relationship between the greenhouse effect and the observed warming" (Hansen 1988). As we surpass 400 ppm, temperature increases of 1.5–2.0 °C in this century have shifted from "possible" to "likely" to "unavoidable" based on scenarios like those described in the latest Synthesis Report of the Intergovernmental Panel on Climate Change (IPCC 2013). Bleaching and disease have triggered declines in coral cover on reefs from every ocean (Aronson and Precht 2006). In the US Virgin Islands, over 90 % of corals at five different sites showed evidence of bleaching or paling in response to elevated water temperatures in 2005. Including the disease that followed, this single event lowered coral cover from 21.4 % to only 8.3 % by late 2007 (Miller et al. 2009). Ocean pH has dropped by 30 % from pre-industrial levels (8.2) and is projected to fall to 7.8 by 2100 (Zeebe and Wolf-Gladrow 2011). Combined with lower coral cover and reduced carbonate saturation levels, this will result in a decline in the amount of calcium carbonate produced on reefs.

It is hard to imagine that all of this not impacting the ability of reefs to build at rates reported from the Holocene; and, at the same time, sea-level rise is accelerating. Since the first IPCC report in 1990, subsequent

[1] http://en.wikipedia.org/wiki/Charles_David_Keeling

measurements have tended to follow the more pessimistic projections, suggesting that the most recent proposal (0.45–0.82 m rise by 2100) may again be conservative. Some studies suggest that a rise of up to 2 m could be possible by century's end (e.g., Meehl et al. 2007; Rahmstorf 2007; Grinsted et al. 2009; Jevrejeva et al. 2009; Merrifield et al. 2009; Vermeer and Ramstorf 2009; Nichols and Casenave 2010; Jevrejeva et al. 2012).

As our appreciation of critical global factors grows (Chap. 11), we are asking questions like, "How fast will sea level rise?" and, "Will reefs and reef islands be able to keep up?" As sea-level rise accelerates, it is increasingly likely that some 2.2 million "climate refugees" may have to emigrate from low-lying islands by 2100 as they are flooded outright or simply left uninhabitable (Fletcher and Richmond 2010; Nicholls et al. 2011). All of this brings us back to the changing relationship between reef building and sea-level rise.

6.1.2 The Road Ahead

Whether we are trying to unravel reef history on a geologic scale or project the future of this important biophysical system in human terms, the best starting point is the obvious: reefs will lag behind rising sea level if either (a) accretion is slowed or (b) sea level accelerates. Unfortunately, human activities have increasingly impacted both. Global sea-level rise has continuously accelerated since the late 1800s (Church and White 2011) although this has varied significantly from place to place (Cabanes et al. 2001). At the same time, stresses like pollution, overfishing, elevated nutrient levels, changing acidity, bleaching, disease and growing human exploitation have led to lower coral cover and declining carbonate production (see summaries in Wilkinson 2008; Jackson et al. 2014).

One of the primary questions we consider below is what elements of the reef's carbonate budget are most critical to understanding the likely impacts of continuing environmental change. Glynn (1997) detailed the importance of the balance between carbonate production and bioerosion in reef building. Perry et al. (2012, 2013) recently suggested an expansion of existing monitoring protocols to include bioerosion and its changing relationship with carbonate production. This is an important step in better understanding the dynamics of reef building, but it continues to downplay the importance of non-biological processes such as increased physical disruption and the redistribution of detrital materials within and out of the reef. Rising sea-surface temperatures will increase the frequency (Hetzinger et al. 2008) and intensity (Webster et al. 2005) of tropical storms. This will disrupt reef structure and remove even more sediment (Hubbard et al. 1990; Hubbard 1992), shifting the mass balance within the reef toward export and removing a potentially important source of material for both reef and reef-island building.

As we struggle to understand recent changes in the world's reefs, the challenge is to place these events into a longer temporal context. The reefs avoided by Columbus were undoubtedly different from those first studied by Tom Goreau in Jamaica (Pauly 1995; Jackson 1977), and even our earliest monitoring records started well after the discipline of reef ecology emerged. Wendell Berry (1987) cautioned that, "we cannot know what we are doing until we know what nature would be doing if we were doing nothing." If we are to objectively confront conservation problems going forward, we need to distinguish between natural and anthropogenically driven change. Moreover, we need to carefully examine how well our models capture the complexities of reef building in both the geologic past and in the decades ahead. This involves a careful look at how reefs build, the vagaries of sea-level rise and how the two interact.

In this chapter, we start by briefly reviewing the basics of reef building and sea-level rise. We then use new data to examine some of the long-held assumptions about how reefs built in the past and consider how they might respond to projected changes in ocean processes in the twenty-first century. Finally, we discuss possible adjustments to our thinking if we are going to improve our ability to use science to effectively inform management.

6.2 What Do We (Think We) "Know"?

6.2.1 Reefs Without Us: The Late Quaternary

The natural balance between reef building and sea-level rise has controlled whether reefs have "kept up", "caught up" or "given up" (Davies and Montaggioni 1985; Neumann and Macintyre 1985). Each of these scenarios leaves behind a three-dimensional stratigraphic pattern and a vertical sequence of coral types that have allowed us to reconstruct the relationship between reef building and sea-level rise in the geologic record – at least over the last few million years when reef corals were similar to modern ones (Chaps. 8 and 9).

Throughout the early Holocene, sea level (blue line in Fig. 6.1) rose faster than most reefs could build vertically. As a result, they lagged behind and eventually "gave up". In the Caribbean, community structure shifted from branching *Acropora palmata* in shallow water to massive and then platy species as depth increased. As this scenario unfolded, reef response varied, depending on the sea-level scenario and the nature of the underlying substrate. If sea level was rising slowly enough and the substrate sloped gradually, the reef might have simply followed sea level upslope (**retrograde** scenario in Fig. 6.1). Where the slope was steep (e.g.,

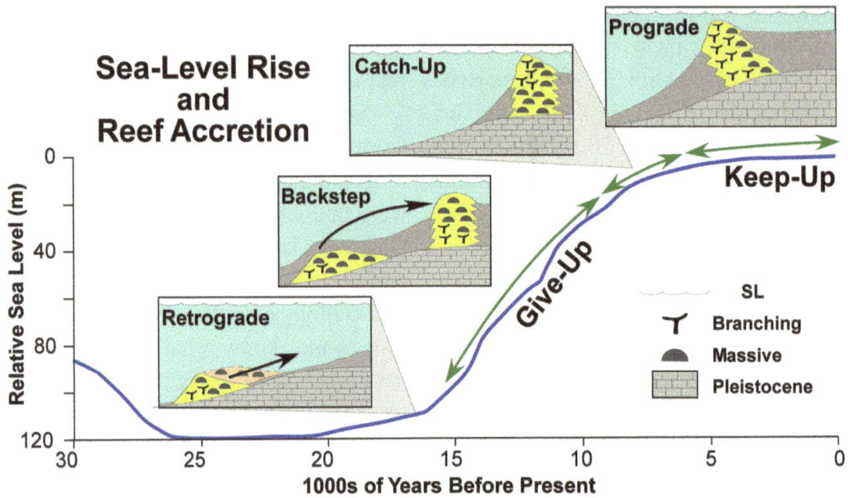

Fig. 6.1 Sea-level rise versus reef accretion. The *blue line* approximates the model-derived global sea-level curve of Peltier and Fairbanks (2006). From 16,000 to ca. 8000 ybp, sea level was rising faster than reefs could accrete and they were slowly left behind. Where slopes were gradual, reefs were able to retrograde upslope to offset rising sea level. Otherwise, they "backstepped" upslope or were abandoned altogether (drowned). The "give-up" reef signature associated with this scenario reflects deepening water and a shift from branching reef-crest species to progressively deeper-water corals. As sea-level rise slowed after ca. 8000 CalBP, reefs began to "catch-up". This left behind a vertical coral sequence reflecting progressively shallower water (i.e., massive corals overlain by branching species). Reefs that formed after sea level was rising more slowly than 3–4 mm/year were able to "keep up" and were dominated by shallow, reef-crest species. If carbonate production exceeded the creation of accommodation space, the reefs either prograded or build landward into the lagoon as reef flats widened (Modified from Hubbard (2014, 2015))

Barbados, Fairbanks 1989), the reef more likely **backstepped** to a shallower terrace where conditions were favorable to reef development. The new "backstepped" reef eventually suffered the same fate as its deeper predecessor unless rising sea level slowed.

Between 8000 and 6000 years ago, sea-level rise slowed dramatically (Fig. 6.1). Any reefs that were forming at water depths less than ca. 10 m were able to gradually "catch up" as the rate of sea-level rise was slower than reef-building capacity. While the specifics of the local history varied between oceans, community structure changed from deeper-water species toward those living at progressively shallower depths. Once reefs built to sea level, most were able to "keep up" and fast-growing shallow-water corals dominated the reef-crest community. Where sea level and accretion were well matched, the reef tended to aggrade vertically. As sea level slowed even more, lateral accretion became increasingly important as accommodation space was over-filled by reefs that expanded either into the lagoon or seaward along the forereef.

The term "give-up reef" is somewhat misleading as the reefs do not just suddenly cease to function; perhaps "giving up" might be a better term for reefs gradually losing the race. Even when water depth exceeds 30 m, mesophotic reefs can still support a functional community. In fact, because of the greater thermal stability at depth, these reefs have been proposed as refuge sites that might mitigate the effects of bleaching and disease in the twenty-first century (Lesser et al. 2009; Chollett and Mumby 2013). However, their morphology usually mimics underlying topography (Locker et al. 2010; Sherman et al. 2010), suggesting slow rates of carbonate production that are insufficient to create a significant geologic structure.

All of these patterns have allowed geologists to assign paleo-environments (and approximate water depths) to ancient reefs and to understand the relationship between reef accretion and sea level before the impact of *Homo sapiens* (Hubbard 2009; Chap. 7). Looking ahead, we can use geologically derived reef-accretion rates from the Holocene to make general projections about the fate of reefs in the face of accelerating sea-level rise. However, we must remember that anthropogenic influences will probably result in slower rates than what is reflected in reef cores.

6.2.2 The Variable Nature of Sea-Level Rise

For decades, reef scientists argued over the "correct sea-level curve" and the pros and cons of various proxies (e.g., corals, peats, microatolls). This is exemplified in even recent discussions of Caribbean sea-level history by Toscano and Macintyre (2003), Blanchon (2005) and Gischler (2006). We already understand that global sea-level rise is expressed differently from place to place due to local variations in tectonics, subsidence, and compaction or rebound close to melting glaciers. However, we hesitated to consider anything beyond these crude local relationships.

Even as the relative merits of various curves were being debated (e.g., Thom and Chappell 1975 for southern Australia versus Lighty et al. 1982 for the Caribbean), regional differences in crustal responses to deglaciation were already being discussed in the geophysical community (e.g., Clark et al. 1978; Lambeck and Chappel 2001). However, the latter have only recently been fully appreciated and integrated into our larger story of Quaternary reef building. On top of this, we are just beginning to understand the complexity of shorter-term sea-level patterns related to regional variability in climatic cycles and the processes that drive them.

6.2.3 Looking to the Future: How Good Are Our Reef Models?

We are becoming increasingly concerned about whether (or when) rising sea level will outpace coral reefs and how this could bear on the ecological viability of these important ecosystems and the organisms that rely on them, including *Homo sapiens*. Central to these questions are: (1) the processes that contribute to reef building, (2) factors that control sea level and (3) the changing relationship between the two.

On the reef, the main contributors of calcium carbonate are various species of coral and perhaps coralline algae. Because so much of a coral's energy is derived from photosynthesis (Pearse and Muscatine 1971; Muscatine 1990; see also Chap. 2), calcification generally declines exponentially from its maximum a few meters below sea level to a minimum at depths of 30–100 m depending primarily on water transparency (Chalker 1981; Tomascik et al. 1993; Yentsch et al. 2002). Bosscher and Schlager (1993) proposed a model based on "typical" reef-accretion rates of 10–14 mm/year (m/ky) and the assumption that reef building would decline exponentially with water depth. The persistence of these assumptions is evident in recent reef-building models (e.g., Nakamura and Nakamori 2011; Toomey et al. 2013; Woodroffe and Webster 2014). It has been suggested that, because Holocene reefs were able to build so fast, either a sudden and dramatic increase in the rate of sea-level rise (e.g., the meltwater pulses of Fairbanks 1989) or a severe decline in water quality were the only reasonable explanations for reefs drowning in the geologic past (Schlager 1981; Blanchon and Shaw 1995; Blanchon 2011). Looking ahead, this would argue that even doubling the rate of sea-level rise to ca. 7 mm/year might have little impact on the ability of reefs and reef islands to keep pace. The recent plight of Pacific islanders suggests that this is an overly optimistic view.

6.2.4 So... What Do We Still Need to Know?

Whether we are trying to understand reef building in the geologic past or predict the future of coral reefs in the face of increasing stresses both locally (sedimentation, nutrients, overfishing) and globally (climate change and ocean acidification), it is important to carefully examine our perceptions of both reef building and sea-level rise. Coral growth is only one component of reef accretion (Stearn et al. 1977; Land 1979; Hubbard et al. 1990). Nevertheless, it has been widely assumed that, because calcification provides the raw material, reef accretion will mimic the depth-related pattern of coral growth, albeit at slower rates (Bosscher and Schlager 1993; Toomey et al. 2013; Woodroffe and Webster 2014). This assumes that that the relationship between calcification, bioerosion and carbonate redistribution remains somewhat constant across the reef and that the latter play only secondary roles.

In the following pages, we start with a review of the "carbonate budget" as a way to quantitatively understand the complex pathways of calcium carbonate through the coral-reef system. The goal is to critically reexamine prevailing assumptions about reef building and the focus on biological processes that has emerged. We continue with an overview of how we use the geologic record to reconstruct both physical and biological processes over at least the past few million years when coral-community structure was similar to what exists today. We then examine commonly held ideas about sea-level rise, starting in deep geologic time and ending in the past few decades; the latter includes a discussion of spatial variability in recent sea-level rise over temporal scales of a few decades to several millennia. Finally, we end with a discussion of how all of this relates to our understanding of fossil-reef building and how modern ones might respond to climate change and accelerating sea-level rise. As part of this, we consider long-standing perceptions that may have skewed existing models and discuss ways to improve them, whether our goal is to more completely understand the past, the present or the future.

6.3 Carbonate Cycling and Reef Building

Reef building is a complex interplay of carbonate production, destruction and transport, as well as the reincorporation of sediment into the reef and myriad processes that cement and bind loose material back into the reef framework. Early discussions of carbonate cycling focused on the interaction of biological processes that create and destroy substrate. Stearn et al. (1977) and Scoffin et al. (1980) compared carbonate production, mostly by coral and coralline algae, to destruction by grazing and boring organisms on a fringing reef off Barbados. They concluded that, even in the mid-1970s, this system was producing less carbonate than was being destroyed by bioerosion.

Land (1979) built on these landmark studies, suggesting that quantifying the ultimate fate of bioeroded sediment was just as important as the rate at which it was produced. He described a conceptually simple equation to characterize the "reef budget":

$$P_G = P_N + S_R, \tag{6.1}$$

where:

P_G = Gross Carbonate Production: carbonate produced by all calcifying organisms on the reef

P_N = Net Carbonate Production: all carbonate (corals + sediment) ultimately incorporated in the reef

S_R = Sediment Removed: sediment produced by bioerosion and removed from the reef

As with earlier studies, Land (1979) approximated P_G using total reef cover and published production rates for corals and other calcifying organisms. P_N is most accurately determined from cores that quantify the vertical accretion rate and composition of the reef interior. His solution at Discovery Bay, Jamaica was two excavations in the side of a forereef channel and a short core (Land 1974). Radiometric dating of coral samples in the reef interior provided an estimate of the vertical accretion rate for the larger reef structure.

Land's "carbonate budget" equation (Eq. 6.1) recognized that some portion of the biologically produced sediment would be reincorporated into the reef and that some would be exported. He used painted segments of *Acropora cervicornis* placed in channels that crossed the deep reef as proxies for sediment transport downslope and into the adjacent basin. His approach highlighted, for the first time, the interplay between biological and physical parameters in reef building.

Using the concepts described by Land for the Jamaican forereef, Hubbard et al. (1990) developed a detailed carbonate budget for a section of reef on the northwest shelf of St. Croix in the US Virgin Islands. P_G was based on the abundance of coral and other calcifying species measured along 51 shore-parallel transects at depths between 2 and 40 m. Coral-growth rate was determined for specimens collected locally and X-rayed to reveal annual banding. This was converted to mass per year using bulk-density values determined from the sampled corals. Carbonate production rates for other calcifying species were taken from the literature. Bioerosion was derived from nearby studies of fish and urchins (Ogden 1977) plus estimates of infaunal bioerosion from the literature (MacGeachy and Stearn 1976; MacGeachy 1977; Moore and Shedd 1977). Seven vertical cores along two, shore-normal, transects revealed the accretionary history of the reef and the relative importance of solid reef framework, sediment and void space within the reef fabric. Sediment export was determined experimentally during periods of fair weather and tropical storms.

Because this study quantified each part of the budget, trying to "balance" it provided a way to objectively test our understanding of the factors involved in reef accretion and the relative importance of each. While calcification was the primary source of the raw materials for reef building, the ultimate fate of the sediment created by bioerosion was key to "balancing the budget". Much of the material broken down by physical and biological erosion was reincorporated into the reef (Hubbard et al. 1990, 1998). However, nearly 60 % was either redistributed to other parts of the reef or exported, especially during major storms (Hubbard 1992). Without direct measurements of the quantity of sediment reincorporated into the reef interior and the amount removed by waves and currents, the budget would have been incomplete.

In addition to providing details for the budget concept developed by Land (1979), the study on St. Croix also challenged the paradigm that reefs owe their rigidity to the predominance of in-place and interlocking organisms (Lowenstam 1950; Newell et al. 1953; Fagerstrom 1987; Stanley and Fagerstrom 1988). Less than half of the reef interior was coral, with some measure of this having been toppled and reincorporated into the framework. The remainder was either sand, rubble or void space, hardly a structure dominated by in-place coral as had been widely assumed. Nor was this a recipe for the rigid and wave-resistant structures that underpinned prevailing geological reef definitions. In addition, bioerosion was not manifested solely as substrate loss. Much of the sediment that was created was reincorporated into the reef and encrustation and cementation of detritus were critical parts of the reef-building process that could no longer be ignored. Just as carbonate production varies between and within sites over time, so do these other important reef-building processes, and an understanding of carbonate cycling is incomplete without reasonably accounting for all of them.

6.4 A Review of Sea-Level Basics

6.4.1 Phanerozoic Sea Level

Discussions of recent sea-level rise and fall have rightly focused on the roles of changing ice volume and the expansion and contraction of seawater as it warms or cools.

However, sea-level patterns and the processes that were responsible have varied dramatically over the past 550 million years in response to myriad factors. On the longest scale, cycles lasting hundreds of millions of years responded to the stability of large landmasses and the ocean basins that separated them. As new ocean crust is created along mid-ocean ridges, it displaces the seafloor upward, and water level with it. As the oceanic crust moves away from the spreading center, it cools, contracts and sinks. When spreading is rapid, large areas of the ocean floor are elevated and sea level rises. Conversely, when spreading is slow, the elevated seafloor is confined to the flanks of the ridges, and sea level is lower. There is also an atmospheric effect related to spreading. During active mid-ocean volcanism, large quantities of CO_2 are released and some of this reaches the atmosphere, causing warming, the melting of glaciers and sea-level rise.

In times of continental assembly (e.g., the formation of Pangea between 300 and 100 million years ago: P-TR: Fig. 6.2a), larger landmasses and quiescence along mid-ocean ridges resulted in lower sea level. In contrast, continental break-up and active sea-floor spreading in the Cretaceous, 145–66 my ago, reversed this pattern and the world ocean was at one of its highest elevations ever (K: Fig. 6.2a). The significance of such long-term shifts is expressed at evolutionary scales of millions of years. For example, closing of the Tethys Seaway in the late Mesozoic and the rise of the Isthmus of Panama in the Pliocene closed important connections that had previously encouraged more cosmopolitan reef faunas whose larvae moved freely between major ocean basins. Each of these events reduced connectivity and led to dramatic changes in circulation and coral-community structure.

At the same time, the cycling between "greenhouse" and "icehouse" climates associated with these longer cycles had an important bearing on carbonate chemistry (Stanley and Hardie 1998). In "greenhouse" times, temperature in shallow water was warmer, and the Mg/Ca ratio in seawater favored organisms that precipitated calcite. "Icehouse" times, like today, favor organisms with aragonite skeletons. Marine organisms are generally locked into the carbonate polymorph that was favored at the time they were evolving; in the case of scleractinean corals, this is aragonite. In the past, this has been an important factor in the waxing and waning of particular marine groups. Looking to the future, if warming increasingly favors calcite, it will become more difficult for corals to precipitate aragonite, regardless of pH.

At shorter temporal scales, changes in the shape of earth's orbit (periodicity of 100,000 year), tilt (41,000 year) and the precession of its rotation (26,000 year) have been responsible for the rhythmic changes in sea level explained by Milankovitch (1941). While these controls have existed throughout most of Earth history, the amplitude of glacial/interglacial sea-level cycles and the relative importance of these three factors have varied. Most recently, glacial cycles have had a 125,000-year periodicity (Fig. 6.2c). Before 1–1.5 mybp, however, the amplitude of sea-level rise and fall was smaller and the periodicity was more closely tied to the 41,000-year cycle (tilt). Causes of the change in periodicity are unclear, but the increased magnitude may be related to the shift toward icehouse conditions starting 35–30 million years ago (Pekar 2008). The increasing presence of huge

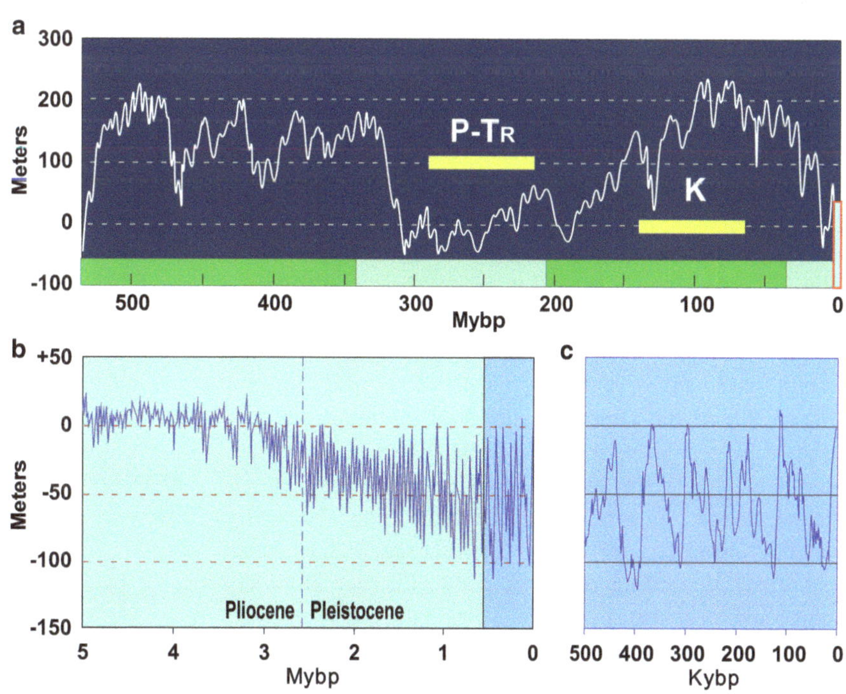

Fig. 6.2 (**a**) Glacio-eustatic sea level during the Phanerozoic (Adapted from Vail et al. 1977 and Prothero and Schwabb 1977). P-TR = Permo-Triassic; K = Cretaceous. Greenhouse (*green*) and icehouse intervals (*light blue*) are also shown. (**b**) Changing sea-level patterns for the last 5,000,000 years. Modified from deBoer et al. (2011); Hansen et al. (2013). (**c**) Rhythmic 125,000-year sea-level cycles typical of the late Quaternary (After Hansen et al. (2013) and Rohling et al. (2009))

ice fields would have provided an additional reservoir/repository for seawater not available in greenhouse times when changes in sea level were driven mostly by thermal expansion and contraction. This is consistent with both the general drop in sea level (lower CO_2 levels and cooler climate) and the increase in the magnitude of sea-level rise and fall seen in Fig. 6.2b. The latter may have encouraged a progressive shift in Caribbean reef-crest fauna from *Pocillopora* and *Stylophora* to more-rapidly calcifying acroporids (McNeil et al. 1997), dramatically increasing the accretionary potential of Caribbean reefs (Johnson et al. 2008; see also Chap. 9). Alternatively, the backreef/lagoonal preferences of *Pocillopora* and *Stylophora* may have made them more vulnerable to stranding in response to a rapid drop in sea-level (Chap. 7).

6.4.2 Sea Level in the Holocene

While our discussions of Holocene reef building often rely on a perceived "global" sea-level curve, this is an imperfect characterization of events near any one reef. Regional tectonics at colliding plate boundaries can lead to significant uplift. When combined with periodic sea-level cycles, this can create successively older reefs on elevated terraces, (e.g., Barbados: Mesolella et al. 1969; Papua, New Guinea: Chappell 1974). Conversely, subsidence related to islands moving away from hot spots plays an important role in the morphology of many Indo-Pacific reefs and their ability to keep up with rising sea level (Grigg 1982; DiCaprio et al. 2010). Lyell (1832, p. 290) suggested that atolls were "the crests of submarine volcanos, having the rims and bottoms of their craters overgrown by corals". While discounting their affinity to submerged craters, Darwin (1842) likewise invoked subsidence as an explanation for the evolution from Pacific fringing reefs to atolls. On shorter timescales, earthquakes can cause instantaneous vertical shifts measured in meters.

Less appreciated are the lithospheric responses to changing ice and water volumes during glaciation and deglaciation (Fig. 6.3). As glaciers expand in the northern polar region, the underlying crust subsides beneath an accumulating mass of ice. This creates a depression in front of the glaciated terrain and an offsetting crustal bulge just beyond (Fig. 6.3b). At the same time, uplift occurs at distant sites in response to a shrinking water mass. When glaciers melt, near-field effects are dominated by rebound closer to the glacial margin (Fig. 6.3c). At the same time, deepening water at far-field sites in the tropics triggers subsidence that varies depending on water depth, distance from the source and local shelf geometry. Toward the end of deglaciation (Fig. 6.3d), relaxation of the fore bulge creates space that is filled by water moving from distant equatorial areas, a process termed "equatorial siphoning" (Mitrovica and Peltier 1991; Mitrovica and Milne 2002).

These crustal responses to changing ice and water volumes were already recognized in the 1970s. Figure 6.4a, b illustrates the regional variability in sea-level rise predicted by Clark et al. (1978). In developing his Caribbean coral-reef model, Adey (1978) acknowledged such crustal responses but concluded that, "continental hydroisostasy (the sequence of events illustrated in Fig. 6.3) is not operating within the Holocene time frame". Since then, however, it has become increasingly apparent that differences in local sea-level history such as those seen in Fig. 6.4c are related to the phenomena just described. As will be discussed below, variability in the sea-level history of the Caribbean and other ocean basins has impacted not only the ability of individual reefs to keep up with rising sea level, but may have also affected regional differences in large-scale reef geometry as well as the history of human colonization. A full understanding of the relationship between sea-level rise and the accretionary history of a particular reef requires an adequate characterization of local patterns that may vary significantly from an average "global" curve.

6.4.3 Historic Sea-Level Change

Over the past three centuries, sea level has not only been rising but it has been accelerating at a rate of 0.013 ± 0.006 mm/year^2 (Church and White 2006). Figure 6.5 summarizes sea level since 1700. The early part of this record (blue) is based on a global network of tide gauges. Since 1993, satellites have provided synoptic measurements of sea-level rise[2] (red) that allow us to discern even smaller-scale spatial and temporal variations in the rate of sea-level rise (Nerem et al. 2010).

6.4.4 Regional Variations in Recent Sea-Level Rise

By 1993, the global average for sea-level rise was 3.2 ± 0.4 mm/year (Nerem et al. 2010). However, during that interval, sea level rose at an average rate of 9–15 mm/year in the Coral Triangle, 6–9 mm/year in the eastern Indian Ocean, 3–9 mm/year in the western Pacific and ~3 mm/year in the Caribbean and western Atlantic.[3] This regional variability (Fig. 6.6) has been attributed to ephemeral differences in regional wind and current patterns and the

[2] http://sealevel.colorado.edu/

[3] http://sealevel.colorado.edu/content/map-sea-level-trends

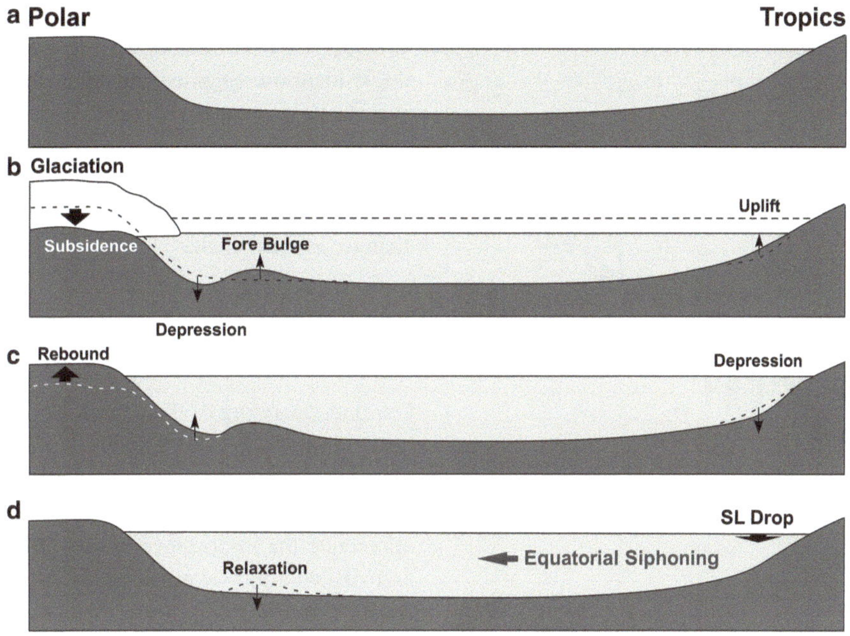

Fig. 6.3 Near- and far-field crustal responses to changing ice volume. (**a**). Simplified cross section from a site in the northern polar latitudes (*left*) to far-field tropical regions in the southern hemisphere (*right*). (**b**) Glaciation forces subsidence of the land beneath the ice field, creating a depression immediately in front of the glaciated area. This is partially offset by an upward crustal displacement along the nearby "fore bulge". At the same time, lower water levels in the tropics result in small uplift in far-field tropical areas. (**c**) As ice melts, the crust beneath and just in front of the old ice field rebounds. Continental areas far from the ice subside slightly as water depth increases. All of this results in significantly different rates of sea-level rise near the glacier, along distant continental margins and near mid-ocean reefs. (**d**) Finally, as the fore bulge relaxes, water levels drop, especially on southern-hemisphere reefs, due to "equatorial siphoning" into the northern hemisphere

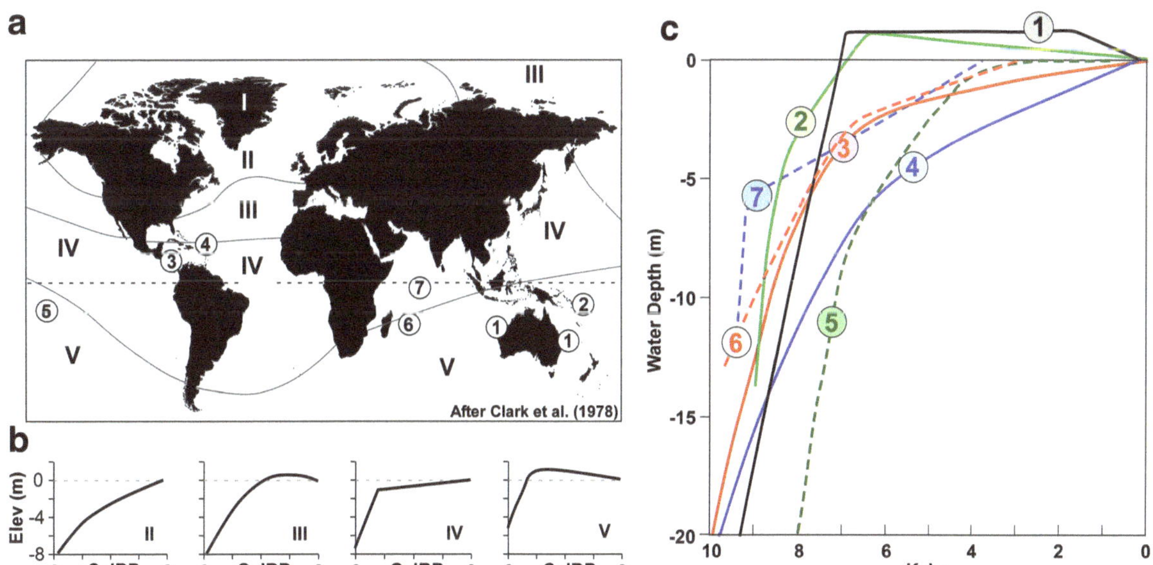

Fig. 6.4 Variability in local sea-level during deglaciation (redrawn and modified from Clark et al. 1978). (**a**). Map showing areas of different ocean-crustal response during deglaciation. (**b**). Representative sea-level curves predicted for each province in a. (**c**). Actual Holocene sea-level curves from six sites. *1* = Australia (Sloss et al. 2007); *2* = New Caledonia (After Cabioch et al. 1995); *3* = Belize (Gischler and Hudson 2004); *4* = Caribbean/Atlantic (corrected from Lighty et al. 1982); *5* = Tahiti (Thomas et al. 2009); *6* = Reunion (Peltier 1991); *7* = Seychelles (Camoin et al. 1997). The *circles* in (**a**) show the locations of the sites where these curves were determined

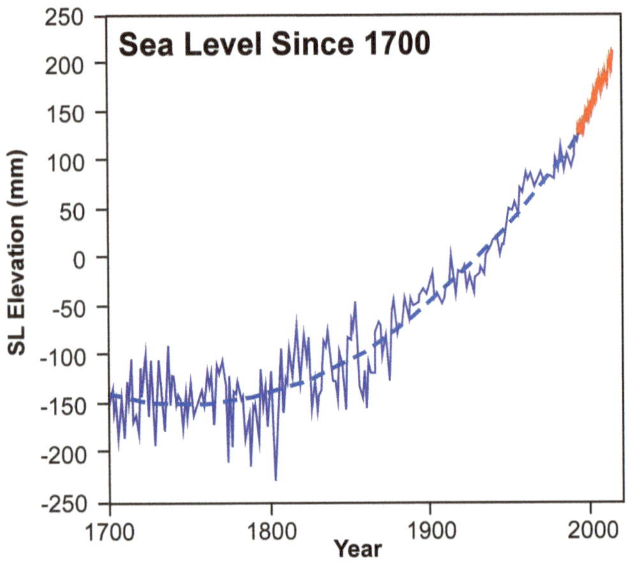

Fig. 6.5 Changes in sea level between 1970 and 2010. Earlier estimates (*blue*: redrawn from Jevrejeva et al. 2008 and Church and White 2011) are based primarily on tide-gauge data adjusted for local tectonic activity. Recent data from Topex-Poseidon satellites (*red*: http://sealevel.colorado.edu/) sit along the upper edge of the first IPCC projections. The dashed line is the best-fit polynomial curve of Jevrejeva et al. (2008)

relative intensity of large-scale meteorological cycles like ENSO (Church and White 2006, 2011; Becker et al. 2012).

Obviously, the faster rate of sea-level rise in the Coral Triangle cannot continue indefinitely. This suggests that the intensity and distribution of atmospheric pressure gradients must shift over time and have likely done so in the past. Chambers et al. (2012) noted a 60-year cyclicity in global sea level that is offset from basin to basin and may represent the instability that will drive future changes in circulation. Smith et al. (2015) suggest that the variability in climate metrics at both regional and global scales is presently changing faster than at any time in at least the last millennium. All of this makes it much more difficult for earth systems that adjust on evolutionary scales to keep up. Also, regional shifts in the rate of sea-level rise will make it increasingly difficult to characterize coastal risk at individual sites (Hinkel et al. 2015).

Looking to the past, when we try to correlate Holocene accretion by individual reefs to changes in sea level, these regional variations can be critically important. Even if they were not as marked in the Holocene as what we are witnessing today, we should still not lose sight of the fact that reefs responded to what was happening locally and that global sea-level curves will not reflect such potentially important site-to-site differences (Gischler 2006). Figure 6.4c illustrates the spatial variability in local sea-level history related to the factors described above.

Looking to the future, we must remember that global rates are likewise inadequate to capture the patterns at a single location. Engineering solutions in the Coral Triangle over the next decade will face different rates of sea-level rise than what is occurring in the Caribbean. At the same time, "winners" in the race with rising sea-level today may be "losers" tomorrow when the patterns summarized in Fig. 6.6 change, as they must.

6.4.5 The Lessons to Be Learned

Our understanding of the vagaries of sea-level rise has obviously improved. Nevertheless, we have been slow to embrace the spatial variability that was evident by at least the 1970s (e.g., Clark et al. 1978). While we have started to appreciate the importance of crustal responses to glaciation and deglaciation, even in areas far afield from major ice sheets (Peltier 1998; Becker et al. 2012), we still have a ways to go. Most recently, careful satellite measurements have revealed regional differences in sea level related to cyclical phenomena like ENSO (Cabanes et al. 2001). Looking back into the geologic record, we need to be mindful that complex local and regional processes confound the record of global change and that it is these local patterns to which each reef responded. Looking to the future, our discussion of short-term "winners" and "losers" in the twenty-first century must take the significant temporal and spatial variability of these patterns into account.

6.5 Corals Grow... Reefs Build

Bank accounts can certainly "grow", and there is little concern that anyone might envision this as an organic process. However, this is not the case with reefs. Many of our long-standing geological models of reef building stem from a presumption that it is solely, or at least dominantly, biological. Our earliest explanations of reef architecture required in-place and interlocking corals to create rigid and topographically elevated structures that stood above their surroundings (Lowenstam 1950; Newell et al. 1953; Fagerstrom 1987). At the same time, the term "reef growth" (e.g., Davies and Marshall 1980; Davies and Montaggioni 1985; Dullo 2005; Davies 2011; Perry et al. 2013) emphasized the biological process of calcification over the more complex mix of factors involved in "reef accretion" (Hubbard et al. 1990, 1998).

As an example, van Woesik et al. (2015) recently measured the rate of vertical coral growth in *Porites* microatolls on Palau and concluded that the ability of the "reef" to keep up with twenty-first century sea level will

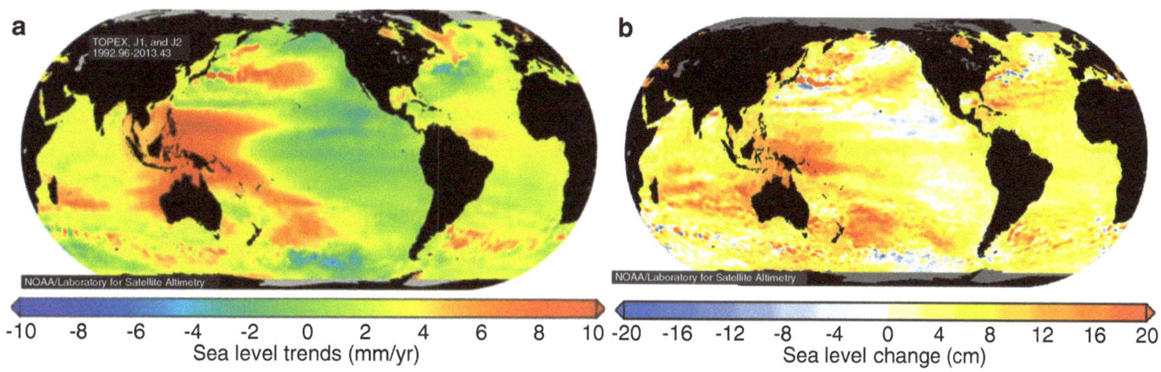

Fig. 6.6 Regional variations in the rate of sea-level rise based on satellite data. (**a**) Regional differences in the rate of sea-level rise (in mm/year) since 1992 (http://www.nesdis.noaa.gov/fourbox/07-15-13/). (**b**) Total change in sea level (in cm) since 1993 (http://www.star.nesdis.noaa.gov/sod/lsa/SeaLevelRise/)

depend on "(i) future rates of sea level rise, and (ii) future responses of microatoll growth in a warmer ocean." They further conclude that "to ensure the 'keep up' status of coral reefs with rising sea levels, reef management will need strategies to maximize living coral cover, which is commensurate with net reef accretion capacity." As with the earlier examples, these conclusions tie reef building in the twenty-first century directly to coral cover and calcification, ignoring bioerosion, sediment redistribution, cementation/encrustation and myriad other processes that contribute to reef accretion.

The conflation of coral growth and reef building is not limited to biological perspectives. The late Mike Lloyd once argued that the concept of uniformitarianism might not apply to reefs, joking that, "The present is the key to the late Pleistocene… perhaps" (email shared by Bill Precht). Because so few ancient reefs possessed the orderly and in-place structure assumed for their modern counterparts, it seemed logical that some major change in the reef-building process must have occurred in the Quaternary. However, all of this was based largely on supposition, as cores through modern reefs were still relatively rare and their inner structure had not actually been directly observed. In short, our modern-reef model was an intellectual construct based on a presumption that biologically dominated systems demanded an orderly framework.

As early cores increasingly showed the importance of detrital material and processes unrelated to coral growth, the idea that Holocene reefs might be more like their rubble-dominated forebears still remained controversial. Shinn (1995) had described the *three stages of discovery*: "1. you are wrong and I can prove it; 2. you are right but is it important? 3. you are right but didn't we know that all along?" Somewhere after countless cores showed that sediment and rubble were more dominant than coral in the majority of Caribbean reefs (Hubbard et al. 1988), the idea that calcification was only the start of reef building moved on to stage 3.

6.5.1 Changing Perspectives

Over the past decade, sea level has increasingly taken center stage and management concerns have broadened to consider the ability of coral reefs to keep up. The balance between carbonate production by corals and bioerosion seems to be tipping toward the latter (Alvarez-Filip et al. 2009) and it has been suggested that monitoring should be broadened to include direct measurements of substrate removal (Perry et al. 2008, 2012, 2013). While this will provide a much more complete understanding of reef dynamics, it still focuses on biological processes.

As discussed earlier, a more nuanced view started to develop in the 1970s that recognized a dynamic balance between the biological creation and destruction of substrate (Stearn et al. 1977; Scoffin et al. 1980), but also acknowledged the physical redistribution of the resulting detrital material within and from the reef (Land 1979; Hubbard et al. 1991; Hubbard 1992). What seems critical at this juncture is a quantitative comparison of all these processes. Reef models, whether biological or geological, presume that calcification and, therefore, reef building are depth related owing to their ultimate ties to light and calcification. If we are to effectively integrate physical and biological factors into a verifiable reef-building model, then these presumptions need to be tested. Further, as we try to triage limited resources to make good management decisions, we need to quantify the respective elements of the carbonate budget to identify gaps in our existing models and then determine whether they are important enough to create real problems as we make management decisions going forward.

In this section, we use recent data to re-examine traditional ideas about how reefs build. The focus is on the rate of reef accretion (limited here to the vertical dimension) and the relationship between coral growth, water depth and the upward aggradation of the reef surface. The principal goal is to provide a more complete picture of the relative roles of biological and physical processes in reef accretion, how they worked prior to recent anthropogenic insults and what they might tell us about the ability of the world's coral reefs to keep pace with accelerating sea-level rise in the decades ahead.

6.5.2 How Fast Do Reefs Build?

Based on early coring studies in the 1970s, it was suggested that Holocene reef-accretion rates of at least 10 mm/year were common (Adey et al. 1977; Adey 1978; Schlager 1981; Macintyre 1988; Bosscher and Schlager 1993) and that reefs at moderate depths could keep up with sea level rising as fast as 14 mm/year (Blanchon and Shaw 1995). Such rapid vertical accretion would have exceeded the rate of Holocene sea-level rise except during episodes of sudden glacial melting (e.g., the "meltwater pulses" of Fairbanks 1989) and it was proposed that extreme circumstances would have been required to trigger reef drowning in the geologic past (Schlager 1981; Blanchon 2011). Moreover, this suggests that reefs building this quickly should have no difficulty matching even the most pessimistic projections for twenty-first century sea-level rise (~4 mm/year).

Smith and Kinsey (1976) disagreed with these early rates based on their alkalinity measurements of carbonate production along the exposed forereef (3.6–4.0 kg/m^2-year) and in more protected environments (<1.0 kg/m^2-year). Using coral density (kg/m^3), they converted their measurements to accretion rates and concluded that, "coral reef communities as presently comprised would be unable to persist as three-dimensional structures if sea level were rising more rapidly than about 3–5 mm year^{-1}." Adey (1978) argued that this limitation represented a regional difference, again citing rates of Caribbean reef accretion between 9 and 15 mm/year.

Table 6.1 summarizes available data for reef accretion based on coring in all the major world oceans. The highest rate (10.13 mm/year) comes from three cores in the Philippines (Shen et al. 2010). Most rates are much lower and over two thirds of the reefs built more slowly than 5 mm/year. This clearly challenges earlier assumptions that reef-accretion rates of 10–14 mm/year are commonplace. Vertical aggradation rates for large carbonate platforms (Table 6.2; also see Bosscher 1993; Dullo 2005) are an order of magnitude less because carbonate will be spread over a much larger area relative to where it was created.

While the list of reefs provided here is far from exhaustive, we argue that they are representative. Figure 6.7 summarizes accretion rates derived from over 200 cores through modern reefs in the Caribbean (Hubbard 2009) and Indo-Pacific regions (Montaggioni 2005). In the Caribbean, 79 % of the reef cores reflect vertical reef building slower than 5 mm/year (Fig. 6.6a). In the Indo-Pacific (Fig. 6.6b), 65 % of the cores yielded accretion rates slower than 5 mm/year and only 18 % exceeded 7 mm/year.

Clearly, most Holocene reefs did not build as quickly as was assumed in the 1970s and 1980s when our existing reef models were developing. This has significant bearing on our discussion of reef accretion and sea-level rise both in the past and the immediate future. Based on the global melt-water curve of Peltier and Fairbanks (2006), any reefs building slower than 7.5 m/ky could not have kept pace with sea-level rise between 16,000 and 7000 years ago (Fig. 6.8). This includes the majority of Holocene reefs that have been cored, and reef back-stepping or drowning would have been the norm rather than a "paradoxical" event driven by extreme conditions as proposed by Adey et al. (1977), Schlager (1981), Bosscher (1992), Bosscher and Schlager (1993) and Blanchon (2005, 2010, 2011). Looking to the future, less than half of the cores summarized in Fig. 6.6 (33 % in the Caribbean; 48 % in the Indo-Pacific) record vertical reef-accretion rates in the Holocene that were faster than recent measurements of sea-level rise (3.3 mm/year: Church and White 2006). As depressing as this seems, it may be an optimistic estimate. The reef-building rates described here are from a time when anthropogenic stressors were largely absent. Therefore, it is likely that many reefs capable of building vertically at 3.3 mm/year throughout the latter Holocene are no longer capable of doing so.

6.5.3 Water Depth and Reef Building

Models of reef building have generally assumed a strong inverse relationship between reef accretion and water depth (Adey and Burke 1976; Adey 1978; James and Ginsburg 1979; Schlager 1981; Neumann and Macintyre 1985; Macintyre 1988). Both light intensity and wavelength change in progressively deeper water (Chalker 1981; Chalker et al. 1988), and Bosscher and Schlager (1993) used this to successfully model depth-related coral growth for what was then known as *Montastraea annularis* (now *Orbicella* spp.). They argued that, due to the link between light intensity, coral growth and reef accretion, the latter would similarly decrease with depth, albeit at slower rates. Chalker (1981) had shown that the depth-related decrease in light saturation followed a hyperbolic tangent function and Bosscher's models of light intensity, coral growth and reef accretion were based on this pattern. Both carbonate

Table 6.1 Average reef-accretion rates based on multiple cores from selected reefs

Location	Rate (mm/year)[a]	Source
Alcaran, Mexico	8.8	Macintyre et al. (1977)
Antigua	3.4	Macintyre et al. (1985)
Galeta, Panama	2.8	Macintyre and Glynn (1976)
Belize	6.0	Shinn et al. (1982)
Belize	2.8	Gischler and Hudson (2004)
Panama	5.0	Macintyre and Glynn (1976)
Barbados	7.3	Peltier and Fairbanks (2006)
St. Croix		
Lang Bank	1.1	Adey et al. (1977)
Lang Bank	5.6	Hubbard et al. (2013)
Buck Island	2.4	Hubbard et al. (2005)
Tague Bay	0.7	Burke et al. (1989)
Cane Bay	1.1	Hubbard et al. (1990)
Puerto Rico	4.1	Hubbard et al. (1997; unpubl. data)
Vieques	0.6	Shinn: in Lighty et al. (1982)
Florida, reefs	2.5	Shinn et al. (1977)
Florida, reefs	4.9	Shinn et al. (1981)
Florida, lagoon	1.3	Shinn et al. (1981)
Florida Bay	6.5	Lighty et al. (1978)
Indian Ocean	3.0	Kench et al. (2009)
Central GBR	8.0	Davies et al. (1985)
Southern GBR[b]		
Wreck Reef		
WK-1	9.06 (+18 %)	Davies and Hopley (1983)
WK-1	7.68	Dechnik et al. (2015)
WK-2	3.12 (+20 %)	Davies and Hopley (1983)
WK-2	2.59	Dechnik et al. (2015)
WK-3	7.30 (+8 %)	Davies and Hopley (1983)
WK-3	6.73	Dechnik et al. (2015)
WK-4	3.36 (+14 %)	Davies and Hopley (1983)
WK-4	2.77	Dechnik et al. (2015)
One Tree		
OT-1	1.70 (+33 %)	Davies and Hopley (1983)
OT-1	1.28	Dechnik et al. (2015)
OT-2	3.61 (+4 %)	Davies and Hopley (1983)
OT-2	3.47	Dechnik et al. (2015)
OT-3	0.72 (+67 %)	Davies and Hopley (1983)
OT-3	0.43	Dechnik et al. (2015)
OT-4	1.73 (+11 %)	Davies and Hopley (1983)
OT-4	1.56	Dechnik et al. (2015)
OT-5	3.85 (+11 %)	Davies and Hopley (1983)
OT-5	3.47	Dechnik et al. (2015)
OT-6	4.75 (+3 %)	Davies and Hopley (1983)
OT-6	4.6	Dechnik et al. (2015)
Fitzroy		
FIT-2	2.50 (+4 %)	Davies and Hopley (1983)
FIT-2	2.41	Dechnik et al. (2015)
FIT-3	4.71 (+4 %)	Davies and Hopley (1983)
FIT-3	4.55	Dechnik et al. (2015)
Fairfax		
FFX-1	4.54 (+12 %)	Davies and Hopley (1983)
FFX-1	4.04	Dechnik et al. (2015)
FFX-2	3.88 (+9 %)	Davies and Hopley (1983)
FFX-2	3.56	Dechnik et al. (2015)
FFX-3	8.91 (+2 %)	Davies and Hopley (1983)
FFX-3	8.74	Dechnik et al. (2015)
Ningaloo Reef	4.0	Twiggs and Collins (2010)
Cook Islands	2.2	Gray and Hein (2005)
Houtman Abrolhos	3.4	Eisenhauer et al. 1996
Maldives	0.7	Klostermann et al. (2014)
Maldives	3.3	Gischler et al. (2008)
Molokai Hawaii	1.9	Webster et al. (2004)
Aqaba	0.7–2.8	Dullo (2005) no avg. available
Sanganeb	1.6–9.6	Dullo (2005) no avg. available
Mayotte	2.8–8.6	Zinke et al. (2003)
Reunion	1.88	Camoin et al. (1997)
Tahiti	8.4	Camoin et al. (2012)
Philippines	10.13	Shen et al. (2010)

Except where indicted, rates are based on either U/Th or calibrated ^{14}C ages (CalBP)

[a]Average for multiple cores unless otherwise indicated

[b]Whole-core averages from individual cores in the southern GBR. Values from Davies and Hopley (1983) are based on conventional ^{14}C analyses; rates from Dechnik et al. (2015) are based on U/Th and corrected ^{14}C ages (CalBP). Values in parentheses show the inflation of accretion rate when using uncorrected ^{14}C ages

Table 6.2 Vertical aggradation rates for selected carbonate platforms

Location	Rate (mm/year)[a]	Source
Jurassic Apulia	0.02	Santantonio et al. (2012)
Tertiary Indonesia	0.2–0.3	Wilson et al. (2000)
Carboniferous Spain	0.1–0.5	Corrochano et al. (2012)
J-Tr West Australia	0.1	Erskine and Vail (1987)

[a]For a more extensive list, see Bosscher (1992)

production (P) and "reef growth" (G) started at some maximum value near the surface (P_m and G_m, respectively in Fig. 6.9a) and decreased with depth.

This relationship is still at the heart of many reef-building models. For example, Toomey et al. (2013) addressed reef geometry and accretion patterns in seven Pacific reef systems using a model (Fig. 6.9b) essentially identical to that of Bosscher (1992). They used a maximum accretion rate (G_{max} in Fig. 6.9b) of 10 mm/year, again based on the papers discussed in Sect. 6.5.2. The usefulness of this approach and the conclusions derived from such models will depend on the validity of the depth-related pattern of reef building that has been widely accepted. Similarly, any reef budget that assumes reef building will mimic the depth-related pattern for coral growth relies on strong and direct ties between coral growth and reef accretion.

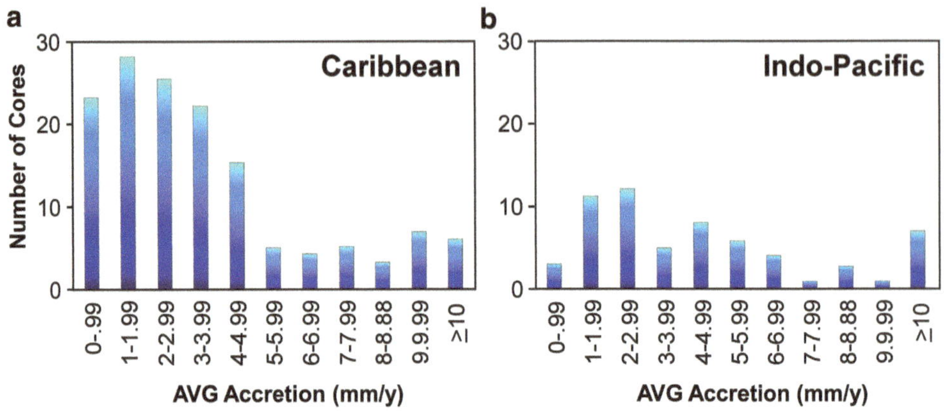

Fig. 6.7 Regional patterns of Holocene reef accretion. (**a**) Caribbean reef-accretion rates plotted from rates in Appendix 1 of Hubbard (2009); n = 143. (**b**) Indo-Pacific reef-accretion plotted from rates in Montaggioni (2005); n = 60

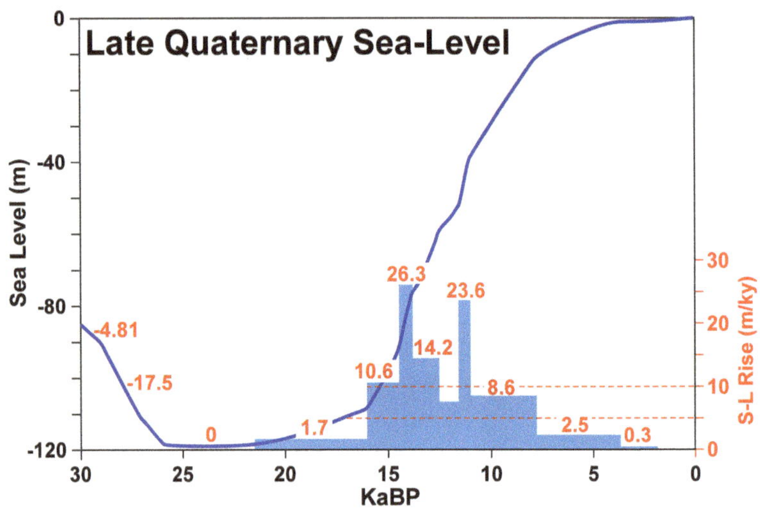

Fig. 6.8 Sea-level rise for the past 30,000 years. The dark blue line approximates global sea level, based on the ICE-5G (VM2) model in Peltier and Fairbanks (2006). The light blue histograms and red numbers show the calculated rates of sea-level rise for intervals since ~22,000 ybp. Note that sea level rose faster than 5 m/ky continuously between 16,000 and 8000 years before present. Thus, reef drowning was probably a common phenomenon during this interval even without the meltwater pulses that have been either proposed or documented

A closer look at available data suggests that vertical aggradation may not follow the depth-related pattern that has been widely assumed. Figure 6.10a summarizes Holocene reef-accretion rates for the Great Barrier Reef (a) and the Caribbean (b). Paleo-water depth was determined by subtracting the difference between sea level today and at the time of deposition (ΔSL) from the depth (d) of the sampled core interval below present sea level. Based on traditional ^{14}C ages, Hopley et al. (2007) reported maximum reef-accretion rates of 4–8 mm/year (m/ky) on the Great Barrier Reef (Fig. 6.10a). They suggested that the lower accretion rates at depths shallower than 3 m were a response to increased water turbidity near the surface. However, a similar pattern was also present along the outer reef, some 50–80 km from shore. Because reef building in shallower paleo-depths in the GBR occurred after sea level had stabilized between 8000 and 6000 years ago, it seems equally plausible that reef building was limited by more intense wave action in shallower water, causing greater damage and higher export of detrital materials (Davies and Hopley 1983). Regardless of the cause, accretion rates do not drop off consistently until water depths exceed 15 m. Also, accretion rates for reefs deeper than 15 m were not significantly lower than those in 1–4 m of water.

It is worth noting that these rates were originally published in the late 1980s and were based on uncorrected ^{14}C ages. While the magnitude varies depending on the age being converted, accretion rates based on calibrated ^{14}C ages are generally lower than those based on uncorrected ages from the same samples. Dechnik et al. (2015) provided a more recent assessment of reef accretion in the southern Great Barrier Reef based on corrected radiocarbon ages and additional U/Th analyses. They used many of the same cores included in Fig. 6.10a and this provides an opportunity to better constrain the patterns summarized by Hopley et al. (2007) and Davies (2011).

Accretion rates based on their calibrated ^{14}C and U/Th ages (Table 6.1: Southern GBR) were generally between 2 and 20 % slower than those based on uncorrected ^{14}C dates from the same samples (Davies and Hopley 1983). If

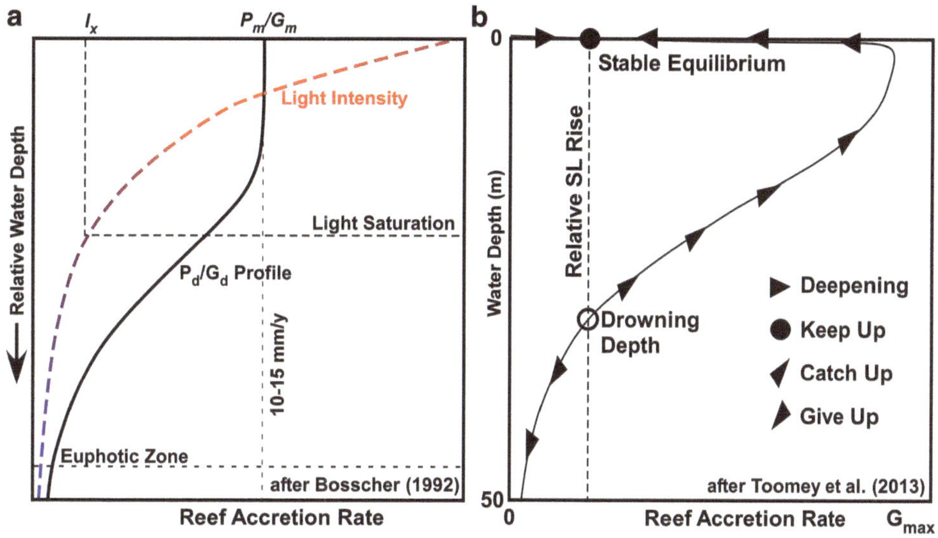

Fig. 6.9 Models of reef accretion versus water depth. (**a**) Model of Bosscher (1992) used to describe depth-related patterns of carbonate production (P_d) and "reef growth" (G_d). Carbonate production declines from a near-surface maximum (P_m) according to the hyperbolic tangent function. "Reef growth" at any depth (G_d) follows a similar path, starting with optimal accretion rates of 10–15 mm/year (G_m). The model, therefore, relies on both the high accretion rates widely assumed in the early 1980s and a depth-related decline in reef accretion (Abstracted from Bosscher 1992). (**b**) Recent model of depth-related reef accretion based on the same assumptions (Toomey et al. 2013). At the "drowning depth", reef building drops below the rate of relative sea-level rise (actual SL rise plus local subsidence) and the reef gradually falls behind. Shallower than that point (set here at ca. 30 m), reef building accelerates and reefs in shallower water will "catch up" with sea level faster than their deeper-water counterparts. Near sea level, "stable equilibrium" will be reached because the reef cannot build above the sea surface. This model was used to quantitatively explain the distribution of reef types (e.g., atolls, barrier and fringing reefs) and, like earlier models, assumes rapid reef building (G_m = 10 mm/year) and a strong depth-related decline in the rate of reef accretion (Abstracted from Toomey et al. 2013 for consistency with (**a**))

the two anomalously high values (33 and 67 %: Table 6.1) are ignored, the average decline in accretion rate falls between 9 and 11 %. The shaded yellow area in Fig. 6.10a shows the range in GBR accretion rates based on a 10 % reduction in accretion rates due to the conversion from uncorrected ^{14}C to CalBP. Rates are reduced and ties to water depth are even less pronounced.

In the western Atlantic, data from 151 intervals in 70 cores show little relationship between reef accretion and either water depth or coral type (Fig. 6.10b). Furthermore, *A. palmata* occurred at significantly greater depths than has been assumed (Lighty et al. 1982), probably due to high wave and current conditions at the site where these cores were taken (Hubbard et al. 2013). Gischler (2008) found a similar pattern for Belize and suggested that reefs at depths greater than 10–15 m and dominated by massive corals may actually build faster than their shallower counterparts.

While photo-inhibition can lead to reduced coral-growth and calcification rates very close to the surface, the overall pattern is strongly depth dependent. However, recent analyses of our growing database from reef cores suggest that reef building does not follow suit. When combined with the previous discussion of how fast reefs actually built vertically throughout the Holocene, this realization argues for a careful reconsideration of geologic models that depend on rapid and depth-dependent reef accretion. Looking to the future, this suggests that our focus on coral abundance and rates of calcification may ignore important processes that will determine whether or not reefs will be able to keep up with rising sea level. Possible explanations for these seeming paradoxical patterns and their bearing on our monitoring methods are considered below.

6.5.4 A Possible Role for Bioerosion

Calcification by corals and other skeletal organisms provides the raw materials for reef building. However, if the depth-related pattern associated with carbonate production is not reflected in reef accretion, then we need to identify other possible factors and determine their importance, at least qualitatively. Ties between bioerosion and marine primary productivity (i.e., plankton and algae eaten by many bioeroders) have been noted (Highsmith 1980; Edinger 2003; Glynn and Manzello 2015), presumably driven by light intensity. Also, depth-related patterns of macro- and micro-bioerosion have been documented using both experimental substrates and

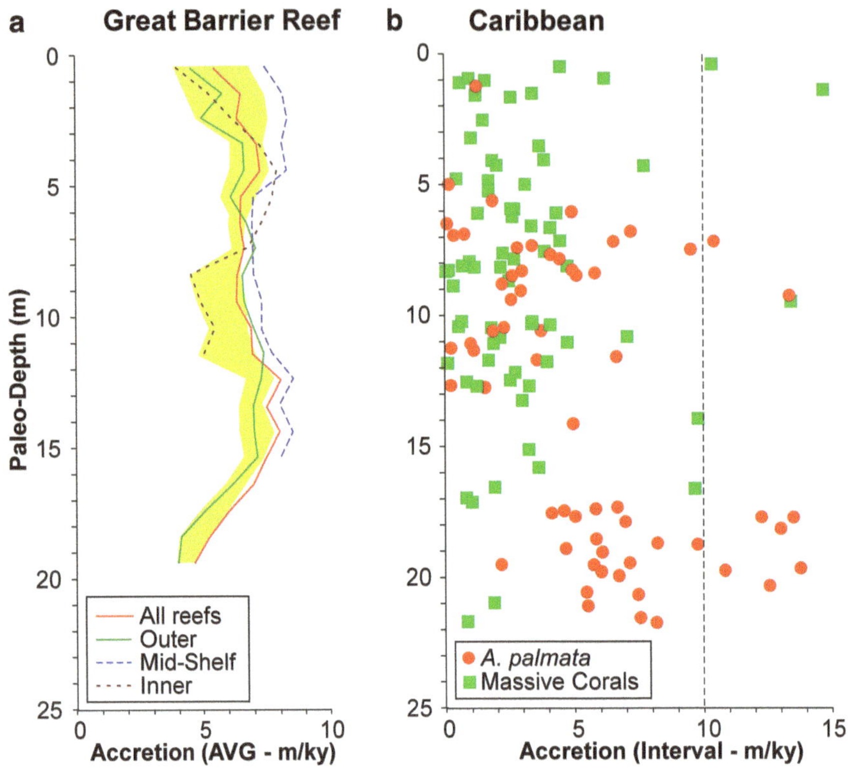

Fig. 6.10 Vertical reef accretion versus paleo-water depth. (**a**) Reef accretion rates averaged for entire cores from inner, middle and outer reefs of the Great Barrier Reef (After Hopley et al. 2007 and Davies 2011). Rates for each reef tract are based on uncorrected ^{14}C ages reported in Hopley (1989) and Partain and Hopley (1989). The *yellow shaded* area represents the range of rates based on a 10 % reduction in accretion when sample ages are corrected (CalBP) from the raw ^{14}C values on which the original curves were based. Reef-accretion rates consistently fall in the 4–7 m/ky (mm/year) range and any relationship with water depth is weak. (**b**) Vertical accretion within 151 core intervals from Caribbean and western Atlantic reefs (plotted from data in Appendix 1 of Hubbard 2009). Rates are based on calibrated ^{14}C ages (CalBP). Note the lack of a strong accretionary dependence on either paleo-water depth or coral type (*A. palmata* = *red circles*; massive corals = *green squares*)

natural samples (Schumacher et al. 1995; Pandolfiand Greenstein 1997; Perry 1998; Vogel et al. 2000; Farber et al. 2015). The biomass of grazing fishes and urchins is greater in shallow water (Weinstein et al. 2014) and substrate removal is strongly dependent on the size of the fishes and urchins involved (Perry et al. 2012). All of this argues for a depth-related pattern of bioerosion.

Based on this, Hubbard (2009) suggested that parallel declines in both carbonate production and bioerosion with depth might offset each other and explain the lack of a strong depth-related signal in available reef-accretion data. However, a study of bioerosion off St. John in the U.S. Virgin Islands showed no depth-related decrease in the rate of bioerosion to a water depth of 20 m (Whitcher 2011). Weinstein et al. (2014) likewise found no significant difference in substrate loss down to a depth of 30 m off nearby St. Thomas, and macro-boring densities were identical at depths of 9.0 and 21.0 m. This suggests that, while differences in the types of bioeroders correlate strongly with reef environment and water depth, the actual rate of bioerosion might not follow suit – at least within the upper 30 m of the water column where carbonate production by reef corals is generally at its greatest.

6.5.5 Sediment Redistribution and Export

While carbonate production and bioerosion are obvious factors in reef building, they seem insufficient to explain the lack of a strong depth-related signal in reef accretion. Land (1979) proposed that the fate of bioeroded sediment is as important as how much is created in the first place. While some of the detritus is retained in the reef, much of it is deposited elsewhere (Hubbard et al. 1974; Davies and Hopley 1983; Hubbard et al. 1991; Hubbard 1992; Hubbard et al. 2001).

Perry et al. (2013) recognized sediment production and redistribution in their protocol to assess the changing carbonate budget on modern reefs. However, their primary focus was the balance between carbonate productions and destruction; they suggested that sediment export be universally set at 50 % of total bioerosion. This assumes that the oceanographic

Table 6.3 Sediment export from Salt River submarine canyon

Conditions	Export rate (kg/day)	(kg/m-day)[a]	Reference
Fair weather	33	0.3	Hubbard (1986)
Tropical storm	440	3.5	
25-year storm	360,000	2880.0	
Hurricane Hugo	2,000,000	16000.0	Hubbard (1992)

[a]Assumes that transport was uniform across the 125-m wide canyon floor

conditions responsible for sediment export on the north shore of St. Croix, where this ratio was determined (Hubbard et al. 1990), are reasonable proxies for all reefs. How problematic this is will depend on (a) the variability in oceanographic processes between reefs, and (b) how sensitive sediment-export rates are to these differences.

Table 6.3 summarizes off-shelf sediment transport from Salt River submarine canyon on the north coast of St. Croix (Fig. 6.11a). Transport was measured during fair weather and tropical-storm conditions using dyed sand deployed in multiple experiments throughout the canyon (Hubbard 1986). In 1989, Hurricane Hugo passed over St. Croix and removed tremendous quantities of sediment (Hubbard 1992). The canyon walls exposed immediately after the storm were stark white and stood out from the algal and coral encrusted substrate that sat above the sediment surface prior to the storm (dashed line in Fig. 6.11b). In addition, a network of tightly stretched polypropylene navigation lines provided a three-dimensional reference for the pre-storm sediment surface across the entire canyon. Together with the monitored transport experiments, these facilitated an accurate measurement of the total volume of sediment removed by strong down-canyon currents as the hurricane passed overhead.

The data summarized in Table 6.3 clearly show that rate of sediment removal associated with fair-weather conditions (33 kg/day) increased significantly during even small storms that might occur 2–5 times per year (440 kg/day), and by four orders of magnitude during storms with a 25-year return frequency. The volume of sediment removed during Hurricane Hugo (2,000,000 kg/day) was five orders of magnitude greater than the fair-weather rate, and export during the passage of this one storm matched the total sediment removal for at least the previous 2–300 years (Hubbard 1992).

Salt River submarine canyon represents a unique geometry that probably intensified off-shelf transport. Nevertheless, the data demonstrate how sensitive sediment export can be to even relatively small changes in wave energy. Table 6.4 summarizes measurements made along two straight platform margins where factors triggering sediment transport should not be so pronounced. Dyed sediment experiments were again used to measure export under a variety of conditions (Fig. 6.11c). In addition, sediment traps were deployed just below the terminus of several reef channels where they intersected the vertical drop-off to the deep basin beyond.

Sand channels off the southern margin of Grand Bahama Bank occurred at a greater depth (60 m) than at Cane Bay (30 m: Fig. 6.11) and the magnitude of export was understandably lower where wave-generated currents would have been attenuated. Nevertheless, similar patterns emerged at both sites: even small tropical storms increased off-shelf transport by more than an order of magnitude, and both responded similarly to Salt River canyon.

As was the case at Salt River, Hurricane Hugo left a clear "bathtub ring" marking the position of the pre-storm sediment surface in shelf-edge channels off Cane Bay (Fig. 6.11d). The volume of sediment removed by the slow-moving Category 4 hurricane was four orders of magnitude greater than during moderate tropical storms, and 250,000 times greater than the rates associated with fair weather. Anchors exposed in Cane Bay by this major storm (Fig. 6.11d) date to the period of sugar cane production on the island in the 1700s, consistent with estimates that Hurricane Hugo represented a 200- to 300-year storm.

6.5.6 Relevance to the Carbonate Budget

Based on the study conducted in Cane Bay, 58 % of the carbonate produced annually on the reef was reduced to sediment by bioerosion (Hubbard et al. 1990). Over 40 % of this was transported into the deep basin in front of the island through channels crossing the shelf. Sediment loss represented a quarter of the entire carbonate budget and the volumes of sediment production and transport were clearly significant in the overall picture of reef accretion. As just discussed, even small changes in wave climate produced order-of-magnitude increases in sediment export and a single major hurricane might removed as much sediment as day-to-day processes and current associated with more modest storms over several centuries.

The proposed addition of bioerosion to existing monitoring protocols by Perry et al. (2013) is an important step toward a more complete assessment of the changing carbonate budget as reefs continue to decline. However, the data just discussed demonstrate the sensitivity of sediment export to even small changes in wave regime, suggesting its importance to a complete understanding of reef building in both the past and near future. In addition, wave climate has been shown to exert significant control over the reef-crest community and forereef zonation across the Caribbean (Adey and Burke 1977; Geister 1977) and the Great Barrier Reef (Done 1982). Hubbard (1997, 2011, 2015) added storms to this picture and demonstrated that both intensity and

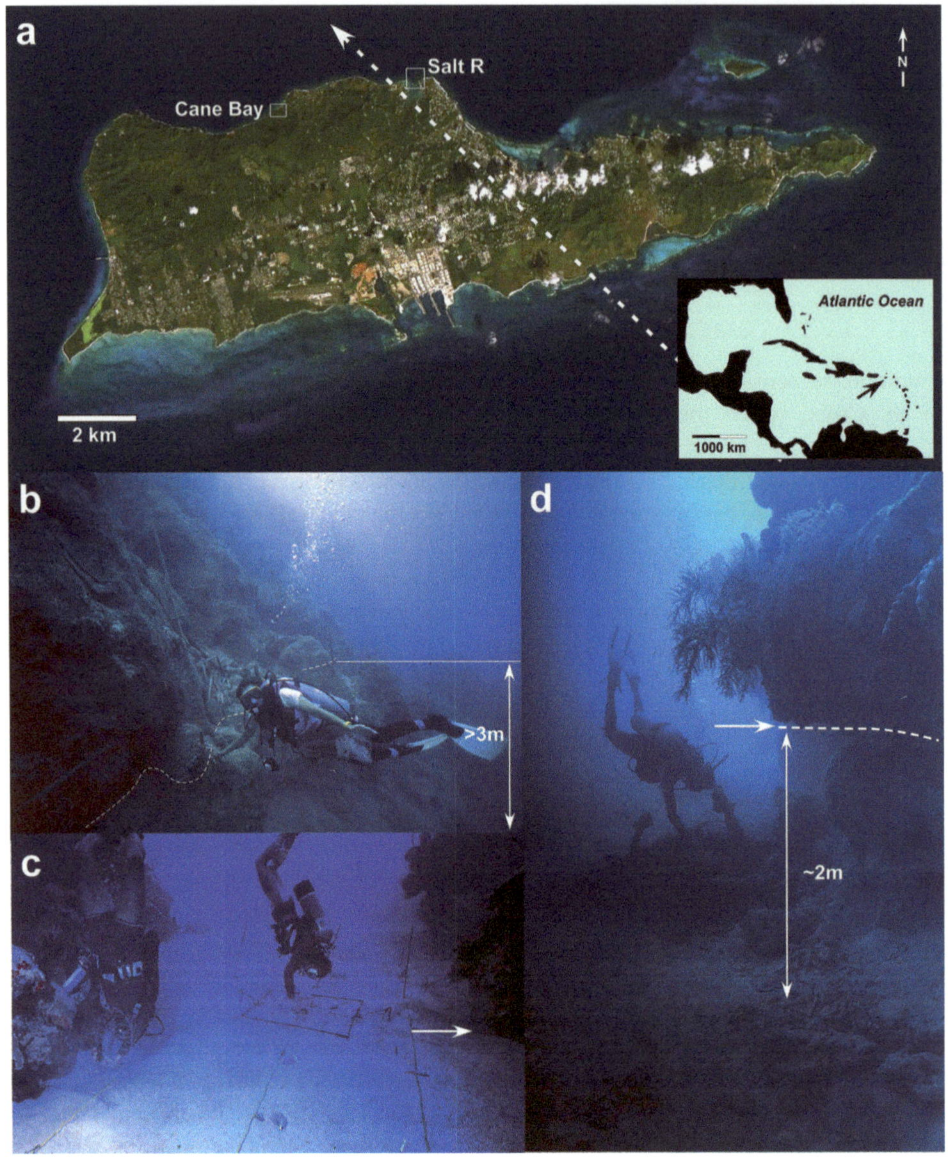

Fig. 6.11 Storm-related sediment transport on St. Croix. (**a**) Landsat image showing the locations of Salt River submarine canyon and Cane Bay on St Croix (landsat.usgs.gov). Note the narrow shelf around much of the island. The inset shows the island location (arrow) in the Caribbean. The dashed line approximates the path of Hurricane Hugo across the island in 1989. (**b**) Underwater photograph along the west wall of Salt River canyon after Hurricane Hugo (d ~ 30 m: © D Hubbard). The dashed line shows the pre-storm sediment level before >3 m of erosion at this site. (**c**) Underwater photograph of a shelf-edge reef channel in Cane Bay (d ~ 25 m: © D Hubbard). Divers are collecting dyed sediment along a grid to quantify sediment transport. (**d**) Post-hurricane photograph of the west side of the same channel in 1989 after Hurricane Hugo (d ~ 25 m: © D Hubbard). The arrows in **c** and **d** show the same point before and after the storm. Note the anchors left by eighteenth century vessels and exposed by storm-related sand removal nearly 300 years later

Table 6.4 Sediment export from Caribbean shelf margins

Location	Conditions	Export rate (kg/m-day)	Reference
(b) Bahamas	Fair weather	0.04	Hubbard et al. (1974)
d = 60 m	Tropical storm	0.27	Hubbard et al. (1974)
	Tropical storm	0.33	Hubbard et al. (1974)
(c) Cane Bay (USVI)	Fair weather	0.18[a]	Sadd (1984) and Hubbard et al. (1990)
d ~ 27 m)	Tropical storm	3.14[a]	Sadd (1984) and Hubbard et al. (1990)
	Hurricane Hugo	48,000[a]	Hubbard (1992)

[a]Rates in Hubbard et al. (1990) were totals for a 7-m wide channel (= rate/7 m)

frequency vary across the Caribbean at least as much as fair-weather wind conditions. The most extensive algal-ridge systems are found in the Windward Islands (Adey and Burke 1976, 1977) where major hurricanes (>Category 3) are most common. At the same time, high day-to-day waves discourage grazers, allowing thick algal ridges to accumulate. In contrast, reef crests dominated by massive corals are more common in Florida and the NW Bahamas where lower wave energy is less efficient at clearing sediment from intolerant *A. palmata* (Adey and Burke 1977). Prior to the

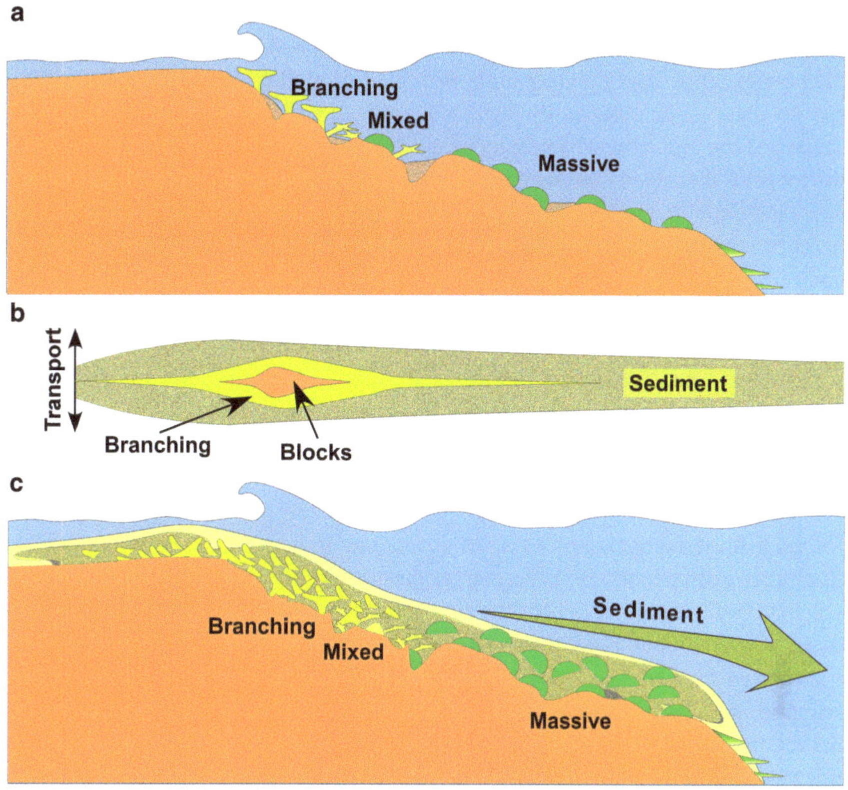

Fig. 6.12 Simplified model of Caribbean reef building. (**a**) Caribbean reef zonation prior to recent reef decline. (**b**) Relative transportability of major reef components (large reef blocks; branching corals; sediment). Large blocks will generally remain in-situ. Branching corals (*yellow*) will shift laterally as larger fragments. However, the expanded facies will generally remain in tact. Massive species will not move far owing to their shape and the lower wave energy at depth. Large quantities of sediment will move landward, downslope or into the deep basin. (**c**) Accretion patterns over time. Facies generally mimic zonation but shallower branching and mixed coral zones are expanded downslope. Significant volumes of sediment are redistributed or lost from the reef altogether. The end result is significant downslope transport despite the general preservation of gross zonation patterns

most recent decline, reefs in the NE Caribbean were well zoned and highly variable from site to site, depending on the exposure to open-ocean swell and storm waves. Similar variability has been described in the Great Barrier Reef (Done 1982) and typhoon frequency and exposure to Indo-Pacific swell has a profound effect on reef type, zonation and sediment transport.

Day-to-day winds and wave conditions as well as major storms and oceanic swell are responsible for significant changes in reef type and, therefore, the dominant carbonate producers on different reefs. At the same time, this variability in processes will significantly change the transport regime and the fate of any sediment from reef to reef. Given that even small increases in wave energy can increase sediment export by orders of magnitude, the variability in wind strength and storm importance across the western Atlantic and elsewhere argues that a single equation for sediment export (50 % of total bioerosion: Perry et al. 2013) ignores a quantitatively important part of the carbonate budget both within and between reefs. Looking ahead, as the importance of major storms increases (Emanuel 2005; Webster et al. 2005; Curry et al. 2006; Hetzinger et al. 2008) and water depth over lagging reef crests increasingly allows more energy to reach protected areas behind (Storlazzi et al. 2004; Ogston and Field 2010; Storlazzi et al. 2011; Ferrario et al. 2013), higher sediment export rates will probably be as important as declining carbonate production in compromising the ability of the world's coral reefs to keep up with rising sea level. At the same time, greater sediment losses from carbonate platforms and atolls could significantly reduce the volume of material available for island building.

6.5.7 The Zonation Conundrum

One factor that has obscured the importance of physical processes in reef building is the well-defined zonation pattern preserved in many ancient reefs, at least over the past few million years. Mesolella et al. (1969) noted the similarity between the zonation patterns in uplifted Pleistocene terraces on Barbados and what was seen on many modern Caribbean reefs prior to the recent decline. Similar patterns can be found in the Pleistocene reefs of Curacao (Pandolfi et al. 2001), the north-central Caribbean (Klaus and Budd 2003), the Indo-Pacific (Pandolfi 1996, 1999) and the Great Barrier Reef (Webster and Davies 2003) as well as in well-preserved Holocene reefs exposed in the western Dominican Republic (Mann et al. 1984; Taylor et al. 1985; Hubbard et al. 2008b) and in cores. If the lack of depth-related accretion described above is the result of wholesale transport, then how can reef zonation possibly be preserved?

This paradox can be resolved if we consider the durability and transportability of different reef components. Figure 6.12 illustrates this principle using a simplified Caribbean reef, but the general patterns should be the same in other regions.

Massive corals and larger sections of the reef edifice will be dislodged only during major storms unless bioerosion reduces their competence. All of this contrasts with shallow branching species that will be more easily disrupted owing to both their shape and the greater wave intensity near the reef crest. Also, the longer a dislodged clast remains in place, whether a coral or a sand grain, the greater the likelihood it will be stabilized by encrustation or cementation. Thus, sediment and smaller coral fragments will move more often and farther than larger blocks and whole massive colonies. Because the processes responsible for downslope transport generally decrease with water depth, the ratio between imported and exported material will progressively increase down the forereef, leading to greater deposition and muting of the relationship between calcification and accretion.

Combined with gravity-driven transport along the forereef, these factors result in a net downslope movement of all reef components. However, larger clasts (and in-place colonies) will remain in or near the environment where they lived, while smaller coral fragments and sediment will be transported further downslope or onto the reef flat. In this scenario, downslope transport will be reflected in a broader branching-coral facies (perhaps combining with the upper mixed-coral zone) and an increasing importance of sediment in downslope facies. As a result, zonation can be preserved even though the depth-related pattern of accretion will be significantly muted, as reflected in Fig. 6.10.

6.6 The Path Forward

6.6.1 What Should We Be Measuring?

Calcification creates most of the material found in reefs, and bioerosion is the primary agent of substrate destruction. Addressing the changing balance between the two provides an important opportunity to better understand how reefs build and how they might respond to increasing anthropogenic stresses and rising sea level. However, biological processes (calcification and bioerosion) provide only part of the picture. Reef type across the Caribbean responds significantly to wave energy (Adey and Burke 1977; Geister 1977) and storm frequency (Hubbard et al. (2008a). Likewise, wave climate and local sea-level history were important drivers of past accretion. Finally, looking to the future, the intensity of storms will increase. This will significantly impact the likelihood of sediment export and redistribution. Thus, accurately characterizing both the magnitude of sediment retention in the reef and the role that it plays in reef building (i.e., does it simply fill cryptic spaces or do greater sediment retention, cementation and encrustation contribute to faster accretion?) is critical to understanding the relationship between storm-related transport and reef success either in the distant past or in the twenty-first century.

6.6.2 Can Reefs Keep Up?

While the data provided in Table 6.1 and Fig. 6.7 are far from exhaustive, they leave little room for optimism. Half of the cores reflect reef building slower than the present-day rate of sea-level rise. However, understanding which reefs will and won't keep pace is difficult. Without critical information on physical processes in monitoring protocols, our understanding of the changing balance between construction and destruction will be incomplete. In addition, North Carolina's unique approach notwithstanding,[4] sea level is not only rising, but it is accelerating (Church and White 2006). Longer-term processes like subsidence or uplift are reasonably well understood. However, regional differences like those seen in Fig. 6.6, and the likelihood that they will not be temporally stable, makes picking winners and losers all the more difficult.

What we do know is that the patterns discussed above are probably optimistic. Even reefs that built throughout the Holocene at rates faster than sea level will rise by 2100 will probably be less capable of doing so as anthropogenic stresses intensify. A more acidic ocean will make calcification more difficult even though corals have developed unique ways to isolate sites of calcification from ambient seawater. The proton flux mechanisms described in Chap. 2 come at a cost, and the currency available to corals is limited. On top of this, storm intensity will disrupt reef structure more often and material export is likely to accelerate. Because even small increases in ambient wave energy will have disproportionate impacts on the balance of the carbonate budget (Tables 6.3 and 6.4), this is all the more distressing.

If we are going to use reef monitoring as part of a larger strategy for making management decisions related to rising sea level, we clearly need to consider protocols that take into account factors beyond biological processes that create or destroy substrate. Sediment export rivals the balance between calcification and bioerosion in the carbonate budget. The physical oceanographic regime varies significantly across and between regions and even modest changes in wave energy and storm frequency can reap seemingly disproportionate increases in export. On top of this, increasing storm intensity and the lost capacity of reefs to act as natural breakwaters as they lag behind rising sea level (Storlazzi et al. 2004; Ogston and Field 2010; Storlazzi et al. 2011) will have consequences that rival the measured losses of coral

[4] "The Coastal Resources Commission and the Division of Coastal Management of the Department of Environment and Natural Resources shall not define rates of sea-level change for regulatory purposes...." - House Bill 819, General Assembly of North Carolina, Sect. 113A-107.1 Sea-level policy.

cover and the anticipated declines in calcification as temperature and acidity rise.

Quantitatively measuring sediment export as part of every monitoring project will be difficult, costly and probably impractical. However, both existing and, hopefully, new measurements can be combined with thoughtful modeling efforts to improve on a one-size-fits-all model for the removal and redistribution of bioeroded sediment. Reef building is more than just coral growth. If we are going to project the capacity of reefs to keep pace with accelerating sea-level rise and increasing physical disruption in the coming decades, then we clearly need to move beyond counting corals and modeling physiological responses to environmental change.

6.6.3 How Will This Impact Those Who Depend on Reefs?

In the western Atlantic, shallow-water reefs were declining even before the 1970s when Black Band Disease and White Plague were noted in the western Atlantic (Antonius 1973; Garrett and Ducklow 1975; Dustan 1977). Since Gladfelter (1982) reported White Band Disease on the island of St. Croix, the abundance of *Acropora* spp. has declined from well over 50 % to less than 5 % on nearby Buck Island (Bythell et al. 1989; Aronson and Precht 2001), and similar scenarios have been reported from virtually every ocean (Gardner et al. 2003; Bruckner and Bruckner 2007; Bruno and Selig 2007; Miller et al. 2009; Weil and Rogers 2011; Anderson et al. 2013; Jackson et al. 2014).

The IPCC projects that 38 % of beaches will be lost on small Caribbean islands by the end of the century due to sea-level rise alone. An average of 12 % of the mangroves on Pacific islands may disappear and up to a third of wetlands could be converted to open water (Mimura et al. 2007). Ten percent of the world's population resides within 10 m (vertically) of sea level (McGranahan et al. 2007) and 500 million people rely to some extent on reefs for their livelihood. The fate of 30 million residents of atolls and low islands hinges on sediment contributed by reef and lagoon organisms (Wilkinson 2008; McKoy et al. 2010). An estimated 78,000 climate refugees will relocate from low-lying areas in the short run (Ware 2005) and rising sea level may displace as many as 2.2 million inhabitants of small Caribbean and Indo-Pacific islands by 2100 (Nicholls et al. 2011).

The effects of sea-level rise will be disproportionately felt by low-lying island nations. By 6000 years ago, sea level was 1–2 m higher than today throughout much of Pacific Oceania and human colonization was triggered by the emergence and expansion of island landscapes as sea level fell (Dickinson 2003). At rates projected by the latest IPCC report, these islands could be drowned or rendered uninhabitable in well under 200 years, ravaging populations that took six millennia to develop. In the Caribbean, many of the low, nearshore cays are dominated by *A. palmata* ramparts derived from live colonies along the forereef. Off La Parguera on the southwest corner of Puerto Rico, the origins of Isla Turromote date to 8200 CalBP when the dominant reef corals were massive *Orbicella* spp. Cores through the island show that the dominant species shifted to *A. palmata* as the reef caught up with sea level starting 3600 years ago (Hubbard et al. 1997). The recent demise of the branching-coral community will make it increasingly difficult for this and other islands in the area to keep pace as sea level continues to rise (Williams et al. 1999). As these critical offshore buffers are lost, erosion along the mainland is likely to increase.

These and similar patterns will be exacerbated by the likelihood of higher waves (Webster et al. 2005; Hemer et al. 2013) and the disproportionate loss of protection as water depth over the reef increases. In Hawaii, Ogston and Field (2010) showed that the energy passing over a reef crest that is presently 0.5 m below sea level could triple if depth increases by as little as 20 cm. All of this will be made worse by declining rugosity due to the loss of branching reef-crest corals (Alvarez-Filip et al. 2009) which will further reduce wave attenuation. Effective management cannot focus on one of these factors at the expense of others, and strategies that do so will be doomed to eventual failure. The problems and their answers are too complex to be viewed from the miopia of a single discipline.

6.6.4 Now More Than Ever

We stand at a crossroads where we need, more than ever, to find ways to look at problems from multiple perspectives and ask questions that are more profound than anything we could imagine within our own disciplinary boundaries. Bill Precht recently reminded me of the conclusions by Gladfelter et al. (1977) following their initial description of *A. palmata* decline in the US Virgin Islands the year I arrived at the West Indies Lab (no correlation, I hope): "at least two agents potentially harmful to the living corals of Buck Island reef do exist: 1) an unknown agent causing the "white-band disease" of *A. palmata*, and 2) parrotfishes, which because of protection might reach abnormally high densities in the Buck Island Lagoon" (p. XI-5). While the first was prescient, the latter contrasts dramatically with the International Coral Reef Initiative's recent call to "adopt conservation and fisheries management strategies that lead to the restoration of parrotfish populations and so restore the balance between algae and coral that characterises healthy coral reefs" (Jackson et al. 2014, p. 107).

These seemingly disparate views highlight the need to think beyond "fish problems" or "coral problems" and consider questions that embrace the larger biophysical system as a whole. In the Florida Keys, coral losses have continued at similar rates in areas with and without protected status (Toth et al. 2014). Does this suggest that the decline in coral cover might have continued even if there were significant numbers of parrotfish present? If so, would the Keys be facing an even more negative carbonate budget consistent with Bill Gladfelter's concerns four decades ago? Since 1977, parrotfish abundances declined and then partially rebounded at Buck Island. By 2007, coral cover remained low even in the face of reduced macroalgal abundance (Burpee 2008). A similar situation existed on the unprotected Tague Bay forereef only a kilometer away, except that urchins had rebounded rather than grazing fish along the unprotected forereef (Arienzo 2008). Does this mean that grazer stability is not the holy grail that we seek, or have their numbers simply not risen enough yet – or is it something different altogether?

As we consider the fate of coral reefs, the organisms that build them and those that rely on them, the solutions for both social and scientific problems will depend on how we frame the questions. Fewer grazing fish may translate to both fewer corals and less bioerosion. Is this "good" or "bad"? Is the latest "coral reef problem" one of coral growth or reef building? Is bioerosion "good" for reef islands because it supplies sand or "bad" because it lowers the platforms on which they build? All of this is a matter of perspective. For example, Webb and Kench (2010) have argued that the 27 Pacific islands they have monitored are, "geomorphically resilient landforms that thus far have predominantly remained stable or grown in area over the last 20–60 year." However, this downplays island retreat and accelerating erosion due to lost natural breakwaters, contamination of water supplies by increased flooding and myriad other indirect impacts of rising sea level that will probably render low-lying reef islands uninhabitable long before the last sand dune has been overtopped by the sea (Fletcher and Richmond 2010; Hubbard et al. 2014). At what point do we declare that a reef has "given up" or an island has "drowned"? Whatever the answers are, we will probably still be arguing over terms like "coral growth" versus "calcification"... or "linear extension".... and whether reefs can "grow" and even "what is a reef" by the time we figure them out.

References

Adey WH (1978) Coral reef morphogenesis: a multidimensional model. Science 202: 831–837

Adey WH, Burke RB (1976) Holocene bioherms (algal ridges and bank-barrier refs) of the Caribbean. GSA Bull. 87:95–109

Adey WH, Burke RB (1977) Holocene Bioherms of Lesser Antilles - geologic control of development. in: Frost S, Weiss M, Saunders J (eds.). Reefs and related carbonates – ecology and sedimentology. AAPG Studies in Geology 4:67–81

Adey WH, Macintyre IG, Stuckenrath R, Dill RF (1977) Relict barrier reef system off St. Croix: its implications with respect to late Cenozoic coral reef development in the western Atlantic. Third International Coral Reef Symposium 2:15–21

Alvarez-Filip L, Dulvy NK, Gill JA, Cote IM, Watkinson AR (2009) Flattening of Caribbean coral reefs: region-wide declines in architectural complexity. Proceedings of the Royal Society 276:3019–3025

Anderson DA, Armstrong RA, Weil E (2013) Hyperspectral sensing of disease stress in the Caribbean reef-building coral Orbicella faveolata - perspectives for the field of coral disease modeling. PLoS One 8: doi: 10.1371/journal.pone.0081478

Antonius A (1973) New observations on coral destruction in reefs. Tenth Meeting of the Association of Island Marine Laboratories of the Caribbean (abstract), University of Puerto Rico, Mayaguez, PR. 10: 3

Arienzo, M (2008) Temporal and spatial distribution of corals at Tague Bay reef and Buck Island National Monument reef, St. Croix USVI. Proc 21st Keck SYmposium, p 157–161, Amherst MA

Aronson RB, Precht WF (2001) White-band disease and the changing face of Caribbean coral reefs. Hydrobiologia 460:25–38

Aronson RB, Precht WF (2006) Conservation, precaution, and Caribbean reefs. Coral Reefs 25:441–450

Becker M, Meyssignac B, Letetrel C, Llovel W, Cazenave A, Delcroix T (2012) Sea level variations at tropical Pacific islands since 1950. Global and Planetary Change, 80-81:85–98

Berry W (1987) Home Economics. North Point Press, New York

Blanchon P (2005) Comments on 'Corrected western Atlantic sea-level curve for the last 11,000 years based on calibrated ^{14}C dates from Acropora palmata framework and intertidal mangrove peat'. Coral Reefs 24:183–186.

Blanchon P (2010) Reef demise and backstepping during the last interglacial, northeast Yucatan. Coral Reefs 29:481–498

Blanchon P (2011) Back-stepping In: Hopley D (ed) Encyclopedia of Modern Coral Reefs: Dordrecht, The Netherlands, Springer, p 77–84

Blanchon P, Shaw J (1995) Reef drowning during the last deglaciation: evidence for catastrophic sea-level rise and ice-sheet collapse. Geology 23:4–8

Bosscher H (1992) Growth potential of coral reefs and carbonate platforms. Proefschrift Vrie Universiteit, Amsterdam, 2005 157p

Bosscher H, Schlager W (1993) Computer simulation of reef growth. Sedimentology 39:503–512

Bruckner AW, Bruckner RJ (2007) The recent decline of *Montastraea annularis* (complex) coral populations in western Curaçao: a cause for concern? Int J Trop Biol 54:45–58

Bruno JF, Selig ER (2007) Regional decline of coral cover in the Indo-Pacific: timing, extent and subregional comparisons. PLoS One 2 (8): e711. doi: 10.1371/journal.pone.0000711

Burke L, Reytar K, Spalding M, Perry A (2011) Reefs at Risk Revisited. World Resources Institute, 1–112

Burke R, Adey W, Macintyre I (1989) Overview of the Holocene history, architecture and structural components of Tague reef and lagoon, in: Hubbard, DK (ed.) Terrestrial and marine geology of St. Croix, U.S. Virgin Islands. Special Pub 8, West Indies Laboratory: p 105–10.

Burpee A (2008) Tropical cascades and sediment production in marine reserves. Proc. 21st Keck Symposium, http://www.keckgeology.org/symposium-21, p. 167–171, Amherst MA

Bythell JC, Gladfelter EH, Gladfelter WH, French KE, Hillis Z (1989) Buck Island Reef National Monument - changes in modern ref community structure since 1976. In: Hubbard DK (ed.) Terrestrial and marine geology of St. Croix, U.S. Virgin Islands. West Indies Lab. Spec Pub. 8, Fairleigh Dickinson Univ. p145–153

Cabanes C, Cazenave A, Le Provost C (2001) Sea level rise during past 40 years determined from satellite and in situ observations. Science 294:840–842

Cabioch G, Montaggioni F, Faure G (1995). Holocene initiation and development of New Caledonian fringing reefs. South-West Pacific. Coral Reefs 14:131–140.

Camoin GF, Colonna M, Montaggioni LF, Casanova J, Faure G, Thomassin BA (1997) Holocene sea level changes and reef development in the southwestern Indian Ocean. Coral Reefs 16:247–259

Camoin GF, Seard C, Deschamps P, Webster JM, Abbey E, Braga JC, Iryu Y, Durand N, E. Bard E, Hamelin B, Yokoyama Y, Thomas AL, Henderson GM, Dussouillez P (2012) Reef response to sea-level and environmental changes during the last deglaciation: Integrated Ocean Drilling Program Expedition 310, Tahiti Sea Level. Geology 40:643–646

Carpenter KE, Abrar M, Aeby G, Aronson RB, Banks S, Bruckner A, Chiriboga A, Cortes J, Delbeek JC, DeVantier L, Edgar GJ, Edwards AJ, Fenner D, Guzman HM, Hoeksema BW, Hodgson G, Johan O, Licuanan WY, Livingstone SR, Lovell ER, Moore JA, Obura DO, Ochavillo D, Polidoro BA, Precht WF, Quibilan MC, Reboton C, Richards ZT, Rogers AD, Sanciangco J, Sheppard A, Sheppard C, Smith J, Stuart S, Turak E, Veron JEN, Wallace C, Weil E, Wood E (2008). One-third of reef-building corals face elevated extinction risk from climate change and local impacts. Science 321:560–563

Cesar H, Burke L, Pet-Soede L (2003) The economics of worldwide coral reef degradation. Cesar Environmental Consulting (CEEC), Arnham, Netherlands, 23p

Chalker BE (1981) Simulating light-saturation curves for photosynthesis and calcification by reef-building corals. Mar Biol 63:135–141

Chalker BE, Barnes DJ, Dunlop WC, Jokiel PI (1988) Light and reef building corals. Interdisc. Sci. Rev. 13:222–237

Chambers D, Merrifield M, Nerem RS (2012) Sea level oscillation of 60 years. Geophys Res Let 39: 10.1029/2012GL052885

Chappell J (1974) Geology of coral terraces, Huon Peninsula, New Guinea: A study of Quaternary tectonic movements and sea-level changes Bull Geol Soc Am 85:553–570

Chollett I, Mumby P (2013) Reefs at last resort: locating and assessing thermal refugia in the wider Caribbean. Biol Conserv 167:179-186.

Church JA, White NJ (2006) A 20th century acceleration in global sea-level rise. Geophysical Research Letters 33: L01602 4 p. doi:10.1029/2005GL024826

Church JA, White NJ (2011) Sea-level rise from the late 19th to the early 21st century. Surveys in Geophysics 32:585–602

Clark JA, Farrell WE, Peltier WR (1978) Global changes in postglacial sea level: a numerical calculation. Quaternary Research 9:265–287

Corrochano D, Barba P, Colmenero JR (2012) Glacioeustatic cyclicity of a Pennsylvanian carbonate platform in a foreland basin setting: An example from the Bachende Formation of the Cantabrian Zone (NW Spain). Sedimentary Geology 245:76–93

Costanza R, De Groot R, Sutton P, van der Ploeg S, Anderson SJ, Kubiszewski I, Farber S, Turner RK (2014) Changes in the global value of ecosystem services. Global Environmental Change 26:152–158

Curry JA, Webster PJ, Holland GJ (2006) Mixing politics and science in testing the hypothesis that greenhouse warming is causing a global increase in hurricane intensity. Bull Amer Meteor Soc 87:1025–1037

Darwin C (1842) The structure and distribution of coral reefs. London, Smith Elder and Co. 215 p

Davies P, Montaggioni L (1985) Reef growth and sea level change: the environmental signature. Proc 5th Intl Coral Reef Symp 3:477–511

Davies PJ (2011) Great Barrier Reef: origin, evolution, and modern development. In: Hopley D (ed.) Encyclopedia of modern coral reefs. Springer, the Netherlands. p 504–534

Davies PJ, Hopley D (1983) Growth fabrics and growth rates of Holocene reefs in the Great Barrier Reef. BMR J of Australian Geol and Geophys 8:237–251

Davies PJ, Marshall JF (1980) A model of epicontinental reef growth. Nature 287:37–38

Davies PJ, Marshall JF, Hopley D (1985) Relationship between reef growth and sea level in the Great Barrier Reef. Proc. Fifth Coral Reef Symp. 3:95–103

De'ath G, Fabricius KE, Sweatman H, Puotinen M (2012) The 27–year decline of coral cover on the Great Barrier Reef and its causes. Procceedings of the National Academy of Science 109:17995–17999

deBoer B, van deWal RSW, Bintanja R, Lourens L, Tuenter E (2010), Cenozoic global ice-volume and temperature simulations with 1-D ice-sheet models forced by benthic $\delta^{18}O$ records, Ann. Glaciol., 51 23–33

Dechnik B, Webster JM, Davies PJ, Braga J-C, Reimer PJ (2015) Holocene "turn-on" and evolution of the southern Great Barrier Reef: revisiting reef cores from the Capricorn Bunker Group. Marine Geology 363:174–190

Descombes P, Wisz MA, Leprieur F, Heine C, Olsen S, Swingedouw D, Kulbicki M, Mouillot D, Pellissier L (2015) Forecasted coral reef decline in marine biodiversity hotspots under climate change. Global Climate Change doi: 10.1111/gcb.12868

DiCaprio L, Müller RD, Gurnis M (2010) A dynamic process for drowning carbonate reefs on the northeastern Australian margin. Geology 38:11–14

Dickinson WR (2003) Impact of Mid-Holocene hydro-isostatic highstand in regional sea level on habitability of islands in Pacific Oceania. Journal of Coastal Research 19:489–502

Done TJ (1982) Patterns in the distribution of coral communities across the central Great Barrier Reef. Coral Reefs 1:95–107

Dullo, W-C (2005) Coral growth and reef growth: a brief review. Facies 51:33–48

Dustan P (1977) Vitality of reef coral populations off Key Largo, Florida: recruitment and mortality. Env Geol 2:1:51–58

Edinger EN (2003) Fossilization Processes: Bioerosion. In: Briggs DEG, Crowther PR, editors. Palaeobiology II, Blackwell Publishing company, Berlin, Germany. 273–276.

Eisenhauer A, Zhu ZR, Collins LB, Wyrwoll KH, Eichstatte R (1996) The last interglacial sea level change: New evidence from the Abrolhos islands, West Australia, Geol. Rundschau 85:606–614

Emanuel KA (2005) Increasing destructiveness of tropical cyclones over the past 30 years. Nature 436:686–688

Erskine RD, Vail PR (1987) Seismic stratigraphy of the Exmouth Plateau. In: Bally, AW (ed.) Atlas of Seismic Stratigraphy. AAPG Studies in Geology 27:163–173

Fagerstrom A (1987) The Evolution of Reef Communities. John Wiley & Sons, New York, 600 p

Fairbanks RG (1989) A 17,000-year glacio-eustatic sea level record: influence of glacial melting rates on the Younger Dryas event and deep-ocean circulation. Nature 342:637–642

Farber C, Wisshak M, Pyko I, Bellou N, Freiwald Z (2015) Effects of water depth, seasonal exposure and substrate orientation on microbial bioerosion in the Ionian Sea (eastern Mediterranean). Plos One, doi:10.1371/journal.pone.0126495

Ferrario F, Beck MW, Storlazzi CD, Fiorenza M, Shepard CC, Airoldi L (2013) The effectiveness of coral reefs for coastal hazard risk reduction and adaptation. Nature Communications, 5:3794 | doi: 10.1038/ncomms4794

Fletcher CH, Richmond BM (2010) Climate change in the Federated States of Micronesia: food and water security, climate risk management and adaptive strategies. Report of Findings, Univ. of Hawaii Sea Grant College Program, 28 p

Gardner TA, Cote IM, Gill JA, Grant A, Watkinson AR (2003) Long-Term Region-Wide Declines in Caribbean Corals. Science 301:958–960

Garrett P, Ducklow H (1975) Coral Diseases in Bermuda. Nature 253:349–350
Geister, J., 1977. The influence of wave exposure on ecological zonation of Caribbean coral reefs. Proc 3rd Intl Coral Reef Symp 1:23–29.
Gischler E (2006) Comment on 'Corrected western Atlantic sea-level curve for the last 11,000 years based on calibrated 14C dates from Acropora palmate framework and intertidal mangrove peat' Coral Reefs 2:273–279
Gischler E (2008) Acretion patterns in Holocene tropical coral reefs: do massive coral reefs in deeper water with slowly growing corals accrete faster than shallower branched corals with rapidly growing corals? Intl J Earth Sci 97:851–859
Gischler E, Hudson JH (2004) Holocene development of the Belize barrier reef. Sedimentary Geology 164:223–236
Gischler E, Hudson JH, Pisera A (2008) Late Quaternary reef growth and sea level in the Maldives (Indian Ocean). Mar Geol 250:104–113
Gladfelter WB (1982) White-Band Disease in *Acropora palmata*: implications for the structure and growth of shallow reefs. Bull Mar Sci 32:639–643
Gladfelter WB, Gladfelter EH, Monahan RK, Ogden JC, Dill RF (1977) Environmental studies of Buck Island Reef National Monument. US Dept. of Interior, National Park Service Report, Washington DC, 135 p
Glynn P (1996) Coral reef bleaching: facts, hypotheses and implications, . Global Change Biology 2:495–509
Glynn PW (1997) Bioerosion and coral-reef growth: a dynamic balance. In: Birkeland (ed.) Life and death of coral reefs. Chapman & Hall, New York, p 68–95
Glynn PW, Manzello DP (2015) Bioerosion and coral reef growth: a dynamic balance. In: Birkeland C. (ed.) Coral reefs in the anthropocene. Springer, Dordrecht, p 67–97
Gray SC, Hein JR (2005) Lagoonal reef accretion and sea-level history from three atolls in the Cook Islands, central South Pacific. J Coast Res 42:253–264
Grigg RW (1982) Darwin Point: a threshold for atoll formation. Coral Reefs 1:29–34
Grinsted A, Moore JC, Jevrejeva S (2009) Reconstructing sea level from paleo and projected temperatures 200 to 2100 AD. Clim Dyn 34:461–472
Hansen (1988) The greenhouse effect: impacts on current global temperature and regional heat waves. Statement to the United States Senate Committee on Energy and Natural Resources. June 23, 1988, 6p
Hansen J, Sato M, Russell G, Kharecha P (2013) Climate sensitivity, sea level and atmospheric carbon dioxide. Phil Trans Royal Soc A 371:
Harvell CD, Aronson R, Baron N, Connell J, and others (2004) The rising tide of ocean diseases: unsolved problems and research priorities. Front Ecol Environ 2:375–382
Hemer MA, Fan Y, Mori N, Semedo A, Wang XL (2013) Projected changes in wave climate from a multi-model ensemble, Nature Climate Change. Nature Climate Change 3:471–476
Hetzinger S, Pfeiffer M, Dullo W-C, Keenlyside N, Latif M, Zinke J (2008) Caribbean coral tracks Atlantic Multidecadal Oscillation and past hurricane activity. Geology 36:11–14
Highsmith RC (1980) Geographic patterns of coral bioerosion: a productivity hypothesis. J Exp Mar Biol Ecol 46:177–196
Hinkel J, Jaeger C, Nichols RJ, Lowe J, Renn O, Peijun S (2015) Sea-level rise scenarios and coastal risk management. Nature Climate Change 5:188–190
Hoegh-Guldberg O, Mumby P, Hooten AJ, Steneck RS, Greenfield P, Gomez E, Harvell CD, Sale PF, Edwards AJ, Caldeira K, Knowlton N, Eakin CM, Iglesias-Prieto R, Muthiga N, Bradbury RH, Dubi A, Hatziolos ME (2007) Coral reefs under rapid climate change and ocean acidification. Science 318:1737–1742
Hopley D (1989) Coral reefs: zonation, zonality and gradients. Essener Geog Arbeit 18: 79–123
Hopley D, Smithers SG, Parnell KE (2007) The geomorphology of the Great Barrier Reef: development, diversity and change. Cambridge University Press, 532 p
Hubbard D, Gischler E, Davies P, Montaggioni L, Camoin L, Dullo W-C, Storlazzi C, Field M, Fletcher C, Grossman E, SHeppard C, Lescinsky H, Fenner D, McManus J, Scheffers S (2014) Island Outlook: warm and swampy. Science 345:1461
Hubbard DK. (1986) Sedimentation as a control of reef development: St. Croix. U.S.V.I. Coral Reefs 5:117–125
Hubbard DK (1992) Hurricane-induced sediment transport in open-shelf tropical systems - an example from St. Croix, U.S. Virgin Islands. J Sedim Petrol 62:946–960
Hubbard DK (1997) Dynamic processes of coral-reef development, In: Birkeland C (Ed.) Life and death of Coral Reefs, Chapman and Hall Publishers, New York, p 43–67
Hubbard DK (2009) Depth-related and species-related patterns of Holocene reef accretion in the Caribbean and western Atlantic: a critical assessment of existing models. In: Swart PK, Eberli G, McKenzie J (eds) Perspectives in Carbonate Geology. Wiley-Blackwell, p 1–18
Hubbard DK (2011) Eastern Caribbean reefs. In: Hopley D, Encyclopedia of modern coral reefs, Springer, the Netherlands, p 338–348
Hubbard DK (2014) Holocene accretion rates and styles for Caribbean coral reefs: lessons for the past and future. In: Verwer Km Playton TE, Harris PM (eds.) Deposits, architecture and controls of carbonate margin, slope and basin settings. SEPM Spec. Publ. 105:264–281, Tulsa OK
Hubbard DK (2015) Reef biology and geology - not just a matter of scale. In: Birkeland C (ed.) Coral reefs in the anthropocene. p 43–66, Springer
Hubbard DK, Burke RB, Gill IP, Ramirez W, Sherman C (2008a) Coral-reef geology: Puerto Rico and the US Virgin Islands, in: Riegl BM, Dodge RE (eds.) Coral Reefs of the USA. Springer, p 263–302
Hubbard DK, Burke RB, Gill IP (1988) Where's the reef: the role of framework in the Holocene. Carbonates and Evaporites 13:1–9
Hubbard DK, Burke RB, Gill IP (1998) Where's the reef: the role of framework in the Holocene. Carbonates and Evaporites 13:3–9
Hubbard DK, Gill IP, Burke RB, Morelock J (1997) Holocene reef backstepping - southwestern Puerto Rico shelf. Proc. 8th Intl Coral Reef Symp 2:1779–1784
Hubbard DK, Gill IP, Burke RB (2001) The role of framework in modern reefs and its application to ancient systems. In: Stanley G. (Ed), The History and Sedimentology of Ancient Reef Systems, Kluwer Academic/Plenum Publishers, New York, p 351–386
Hubbard DK, Gill IP, Burke RB (2013) Holocene reef building on eastern St. Croix, US Virgin Islands: Lang Bank revisited. Coral Reefs, doi: 10.1007/s00338-013-1041-1
Hubbard DK, Miller AI, Scaturo D (1990) Production and cycling of calcium carbonate in a shelf-edge reef system (St. Croic, U.S. Virgin Islands): applications to the nature of ref systems in the fossil record. Journal of Sedimentary Petrology 60:335–360
Hubbard DK, Parsons KM, Bythell JC Walker ND (1991) The effects of Hurricane Hugo on the reefs and associated environments of St. Croix, U.S. Virgin Islands - a preliminary assessment. J. Coast Res 8: 3–48.
Hubbard DK, Ramirez W, Cuevas D, Erickson T, Estep A (2008b) Holocene reef accretion along the north side of Bahia Enriquillo (western Dominican Republic): unique insights into patterns of reef development in response to sea-level rise. Proc. 11th Intl Coral Reef Symp 1:43–47

Hubbard DK, Ward LG, FitzGerald DM, Hine AC (1974) Bank Margin Morphology and sedimentation, Lucaya, Grand Bahama Island. Tech Rpt No 7-CRD, Dept. of Geology, Univ. of S. Carolina, 36p

Hubbard DK, Zankl H, Van Heerden I, Gill IP (2005) Holocene Reef Development Along the Northeastern St. Croix Shelf, Buck Island, U.S. Virgin Islands. J Sed Res 75:97–113

Hughes TP, Baird AH, Bellwood DR, Card M, Connolly SR, Folke C, Grosberg R, Hoegh-Guldberg O, Jackson JBC, Kleymas J, Lough JM, Marshall P, Nystrom M, Palumbi SR, Pandolfi JM, Rosen B, Roughgarden J (2003) Climate change human impacts, and the resilience of coral reefs. Science 301:929–933

IPCC (2013) Climate change 2013, fifth assessment report of the Intergovernmental Panel on Climate Change. Cambridge University Press. New York NY, 1535 p

Jackson JBC (1977) Reefs since Columbus. Coral Reefs 16. Coral Reefs 16:S23–S32

Jackson J. Donovan M, Cramer K, Lam V (2014) Status and trends of Caribbean coral reefs: 1970–2012. Global Coral Reef Monitoring Network, IUCN, Gland Switzerland, 304 p

James NP, Ginsburg RN (1979) The seaward margin of Belize barrier and atoll reefs. IAS Spec. Pub. 3. Blackwell Scientific Publications, Oxford, London, Edinburgh, Melbourne

Jevrejeva S, Grinsted A, Moore JC (2009) Anthropogenic forcing dominates sea level rise since 1850. Geophys Res Let 36, doi: 10. 1029/2009GL040216

Jevrejeva S, Moore JC, Grinsted A, Woodworth PL (2008), Recent global sea level acceleration started over 200 years ago. Geophys Res Lett 35, doi: 10.1029/2008GL033611

Jevrejeva S, Moore JC, Grinsted A (2012) Sea level projections to AD2500 with a new generation of climate change scenarios. Global and Planetary Change 80–81:14–20

Johnson KG, Jackson JBC, Budd AF (2008) Caribbean reef development was independent of coral diversity over 28 million years. Science 319:1521–1523

Jokiel PL (2011) Ocean acidification and control of reef calcification by boundary layer lamination of proton flux. Bull Mar Sci 87:639–657

Kench PS. Smithers SG, McLean RF, Nichol SL (2009) Holocene reef growth in the Maldives: evidence of a mid-Holocene sea-level highstand in the central Indian Ocean. GSA Bull 37:455–458

Klaus, JS, Budd AF (2003) Comparison of Caribbean coral reef communities before and after Plio-Pleistocene faunal turnover: analyses of two Dominican Republic reef sequences. Palaios 18:3–21

Kleypas JA, Buddemeier RW, Archer D, Gattuso J-P, Langdon C, Opdyke BM (1999) Geochemical consequences of increased atmospheric carbon dioxide on coral reefs. Science 284:118–120

Klostermann L, Gischler E, Storz D, Hudson JH (2014) Sedimentary record of late Holocene event beds in a mid-ocean atoll lagoon, Maldives, Indian Ocean: Potential for deposition by tsunamis. Mar Geol 348:37–43

Lambeck K, Chappell J (2001) Sea level change through the last glacial cycle. Science 292:679–686

Land L (1979) The fate of reef-derived sediment on the North Jamaican island slope. Marine Geology 29:55–71

Land LS (1974) Growth rate of a West Indian (Jamaican) reef. Proc. 2nd Intl Coral Reef Symp. 2:409–412

Lesser PM, Slattery M, Leichter JJ (2009) Ecology of mesophotic reefs. J Exp. Mar Biol and Ecol 375:1–8

Lessios HA, Robertson DR, Cubit JD (1984) Spread of *Diadema* mass mortality through the Caribbean. Science 226:335–337

Lighty RG, Macintyre IG, Stuckenrath P (1978) Submerged early Holocene barrier reef, southeastern Florida shelf. Nature 276:59–60

Lighty RG, Macintyre IG, Stuckenrath R (1982) *Acropora palmata* reef framework: a reliable indicator of the sea level in the Western Atlantic. Coral Reefs 1:125–130

Locker SD, Armstrong RA, Battista TA, Rooney JJ, Sherman C, Zawada DG (2010) Geomorphology od mesophotic coral ecosystems: current perspectives on morphology, distribution and mapping strategies. Coral Reefs 29:329–345

Lowenstam HA (1950) Niagaran Reefs of the Great Lakes Area. GSA Memoir 67:215–248

Lyell C (1832) Principles of geology, being an attempt to explain the former changes of the Earth's surface, by reference to causes now in operation. London: John Murray 2: 286

MacGeachy JK, (1977) Factors controlling sponge boring in Barbados reef corals. Proc 3rd Intl Coral Reef Symp 2:477–483

MacGeachy JK, Stearn C W (1976). Boring by macro- organisms in the coral *Montastrea annularis* on Barbados reefs. Int Revue ges Hydrobiol 61:715–745

Macintyre I, Multer HG, Zankl H, Hubbard D, Weiss M, Stuckenrath R (1985) Growth and depositional facies of a windward reef, Nonsuch reef complex, Nonsuch Bay, Antigua, W.I. Proc 5th Intl Coral Reef Symp 6:605–610

Macintyre IG (1988) Modern coral reefs of western Atlantic: new geological perspective. American Association of Petroleum Geologists Bulletin 72:1360–1369

Macintyre IG, Burke RB, Stuckenrath R (1977) Thickest recorded Holocene reef section, Isla Perez core hole, Alacran Reef, Mexico. Geology 5:749–754

Macintyre IG, Glynn PW (1976) Evolution of modern Caribbean fringing reef, Galeta Point. Bulletin of the American Association of Petroleum Geologists 60:1054–1072

Mann P, Taylor F, Burke K, Kulstad R (1984) Subaerially exposed Holocene coral reef, Enriquillo Valley, Dominican Republic. GSA Bull 95:1084–92

Mayer AG (1914) The effects of temperature upon marine animals. Carnegie Institute of Washington Papers in Marine Biology 6:3–24

McClanahan T (2002) The near future of coral reefs. Environmental Conservation 29:460–483

McGranahan G, Balk D, Anderson B (2007) The rising tide: assessing the risks of climate change and human settlements in low elevation coastal zones. Environment and Urbanization 19:17–37

McKoy H, Kennedy DM, Kench PS (2010) Sand cay evolution on reef platforms, Mamanuca Islands, Fiji. Marine Geology 269:61–73

McNeil DF, Budd AF, Borne PF (1997) Earlier (late Pliocene) first appearance of the Caribbean reef-building coral Acropora palmata: stratigraphic and evolutionary implications. Geology 25:891–894

Meehl GA, Collins WD, Friedlingstein P, Gaye AT, Gregory JM, Kitoh A, Knutti R, Murphy JM, Noda A, Raper SCB, Watterson IG, Weaver AJ, Zhao Z-C (2007) Global Climate Projections. In: Solomon S, Qin D, Manning M, Chen Z, Marquis M, Averyt KB, Tignor M, Miller HL (eds) The Physical Science Basis Contribution of Working Group I to the Fourth Assessment Report of the Intergovernmental Panel on Climate Change. Cambridge University Press, Cambridge, United Kingdom and New York, NY, USA,

Merrifield MA, Merrifield ST, Mitchum GT (2009), An anomalous recent acceleration of global sea level rise, J. Clim. 22:5772–5781

Mesolella KJ, Matthews RK, Broecker SW, Thurber DL (1969). The astronomical theory of climatic change: Barbados data. J Geol 77:250–274

Milankovitch M (1941) Canon of Insolation and the Ice Age Problem. Royal Serbian Academy Special Publication

Miller J, Muller E, Rogers C, Waara R, Atkinson A, Whelan KRT, Patterson M, Witcher B (2009) Coral disease following massive bleaching in 2005 causes 60 % decline in coral cover on reefs in the US Virgin Islands. Coral Reefs 28:925–937

Mimura N, Nurse L, McLean RF, Agard J, Briguglio L, Lefale P, Payet R, Sem G (2007) Small islands. Climate Change 2007: Impacts, Adaptation and Vulnerability. Contribution of Working Group II to the Fourth Assessment Report of the Intergovernmental

Panel on Climate Change. In: Parry ML, Canziani OF, Palutikof JP, Linden PJvd, Hanson CE (eds), Cambridge, UK 687–716
Mitrovica, JX and Milne, G.A. (2002) On the origin of late Holocene sea-level highstands within equatorial ocean basins', Quat Sci Rev 2:2179–90
Mitrovica, J.X. and Peltier, W.R. (1991) On postglacial geoid subsidence over the equatorial oceans. J Geophys Res 96:20,053-71
Moberg F, Folke C (1998) Ecological goods an services of coral reef ecosystems. Ecol Econ 29:215–233
Montaggioni LF (2005) History of Indo-Pacific coral reef systems since the last glaciation: development patterns and controlling factors. Earth-Sci Rev. 71:1–75
Moore CH, Shedd WW (1977) Effective rates of sponge nioerosion as a function of carbonate production. Proc. 3rd Intl Coral Reef Symp 2:499–506
Muscatine L (1990) The role of symbiotic algae in carbon and energy flux in reef corals. In: Dubinsky Z (ed.) Coral reefs of the world, Ecosystems of the world, Elsevier, Amsterdam, 25:85–87
Nakamura T, Nakamori T (2011) A simulation model for coral reef formation: reef topographies and growth patterns responding to relative sea-level histories. In: Wright LL (ed.) Sea level rise, coastal engineering, shorelines and tides. Nova Science Publishers, London, p 251–261
Nerem RS, Chambers D, Choe C, Mitchum GT (2010) Estimating mean sea level change from TOPEX and Jason altimeter mission. Marine Geodesy 33: 435
Neumann AC, Macintyre IG (1985) Reef responses to sea level rise: keep-up, catch-up or give-up. Proc. 5th Intl Coral Reef Symp. 3:105–110
Newell ND, Rigby JK, Fischer AG, Whiteman AJ, Hicks LE, Bradley J. (1953) The Permian Reef Complex of the Guadalupe Mountains Region, Texas and New Mexico: A Study in Paleoecology. WH Freeman, San Francisco. 236 p
Nicholls RJ, Marinova N, Lowe JA, Brown S, Vellinga P, Gusmão Dd, Hinkel J, Tol RSJ (2011) Sea-level rise and its possible impacts given a 'beyond 4 °C world' in the twenty-first century. Philosophical Transactions of the Royal Society 369:161–181
Nichols RJ, Casenave A (2010) Sea-level rise and its impact on coastal zones. Science 328: 1517–1520
Ogden JC (1977) Carbonate-sediment production by parrotfish and sea urchins on Caribbean reefs. In: Frost SH, Weiss MP, Saunders JB (eds.) Reefs and related carbonates. Ecology and Sedimentology, AAPG Studies in Geology 4:281–288
Ogston AS, Field ME (2010) Predictions of turbidity due to enhanced sediment resuspension resulting from sea-level rise on a fringing coral reef: evidence from Molokai, Hawaii. Journal of Coastal Research 26:1027–1037
Pandolfi JM (1996) Limited membership in Pleistocene reef coral assemblages from the Huon Peninsula, Papua New Guinea: Constancy during global change. Paleobiology 22:152–176
Pandolfi JM (1999) Response of Pleistocene coral reefs to environmental change over long temporal scales. Amer Zool 39:113–130
Pandolfi JM, Connolly SR, Marshall DJ, Cohen AL (2011) Projecting coral reef futures under global warming and ocean acidification. Science 333: 418–42
Pandolfi JM, Greenstein BJ (1997) Taphonomic Alteration of Reef Corals: Effects of Reef Environment and Coral Growth Form. I. The Great Barrier Reef. Palaios 12:27–42
Pandolfi JM, Jackson JBC, Baron N, Bradbury RH, Guzman HM, Hughes TP, Kappe CV, Micheli F, Ogden JC, Possingham HP, Sala E (2005) Are U.S. coral reefs on the slippery slope to slime? Science 307: 1725–1726
Pandolfi JM, Jackson JBC, Geister J (2001) Geologically sudden natural extinction of two widespread Late Pleistocene Caribbean coral reefs, In: Chettham AH, Jackson JBC, Lidgard S, McKinney (eds.) Processes from pattern in the fossil record, Univ. of Chicago Press, Chicago IL, p 120–158
Partain BY, Hopley D (1989) Morphology and development of the Cape Tribulation fringing reefs, GretBarrier Reef, Australia. Tech Mem 21, Great Barrier Reef Marine Park Authority, Townsville Australia
Pauly D (1995) Anecdotes and the shifting baseline syndrome of fisheries. Trends in Ecology and Evolution 10:430
Pearse VB, Muscatine L (1971) Role of symbiotic algae (zooxanthellae) in coral calcification. Biol. Bull. 141:350–363
Pekar SF (2008) Climate change: when did the icehouse cometh? Nature 455:602–603
Peltier WR (1991) The ICE-3G model of late Pleistocene deglaciation: construction, verification and applications. In: Sabadini R, Lambeck K, Boschi E (eds) Glacial isostacy, sea-level and Mmantle reology. NATO ASI Series C, 334, Kluwer, Dordrecht, p 95–119
Peltier WR (1998) Postglacial variations in the level of the sea : implications for climate dynamics and solid-Earth geophysics. Rev of Geophys 36:603–689
Peltier WR, Fairbanks RW (2006) Global glacial ice volume and last glacial maximum duration from an extended Barbados sea level record. Quat Sci Rev 25:3322–3337
Perry CT (1998) Macroborers within coral framework at Discovery Bay, north Jamaica: species distribution and abundance, and effects on coral preservation. Coral Reefs 17:277–287.
Perry CT, Edinger EN, Kench PS, Mumby PJ, Murphy G, Steneck RS, Smithers SG. (2012) Estimating rates of biologically driven coral reef framework production and erosion: a new census-based carbonate budget methodology and applications to the reefs of Bonaire. Coral Reefs 31:853–868
Perry CT, Murphy GN, Kench PS, Smithers SG, Edinger EN, Steneck RS, Mumby PJ (2013) Caribbean-wide decline in carbonate production threatens coral reef growth. Nature Communications. doi: http://dx.doi.org/10.1038/ncomms2409 (Open Access)
Perry CT, Spencer T, Kench PS (2008) Carbonate budgets and reef production states: a geomorphic perspective on the ecological phase-shift concept. Coral Reefs 27:853–866
Prothero DR, Schwabb F (1977) Sedimentary geology. WH Freeman & Co. New York, 527 p
Rahmstorf S (2007) A Semi-Empirical Approach to Projecting Future Sea-Level Rise. Science 315:368–370
Rohling EJ, Grant K, Bolshaw M, Riberts AP, Sidall M, Hambelen C, Kuera M (2009) Antarctic temperature and global sea level closely coupled over the past five glacial cycles. Nat Geosci 2:500–504
Sadd JL (1984) Sediment transport and $CaCO_3$ budget on a fringing reef, Cane Bay, St. Croix, U.S. Virgin Islands: Bull Mar Sci 35:221–238
Santantonio M, Scrocca D, Lipparini L (2012) The Ombrina-Rospo Playeau (Apulian platform): evolution of a carbonate platform and its margins during the Jurassic and Cretaceous. Mar Pet Geol 42:4–29
Schlager W (1981) The paradox of drowned reefs and carbonate platforms. Geological Society of America Bulletin 92:197–211
Schumacher H, Kiene w, Dullo W-C (1995) Factors controlling Holocene reef growth: an interdisciplinary approach. Facies 32:145–188
Scoffin TP, Stearn CW, Boucher D, Frydl P, Hawkins CM, Hunter IG, MacGeachy JK (1980) Calcium carbonate budget of a fringing reef on the west coast of Barbados: part II, erosion, sediments and internal structure. Bulletin of Marine Science 30:47–508
Shen C-C, Siringan FP, Lin K, Dai C-F, Gong S-Y, (2010) Sea-level rise and coral-reef development of northwestern Luzon since 9,9 ka. Paleogeography, Paleoclimatology, Paleoecology. 292:465–473
Sherman C, Nemeth M, Ruíz H, Bejarano I, Appeldoorn R, Pagán F, Schärer M, Weil E (2010) Geomorphology and benthic cover of

mesophotic coral ecosystems of the upper insular slope of southwest Puerto Rico. Coral Reefs 29:347–360
Shinn, E. A., 1995, Marine versus meteoric cementation: the three stages of discovery, Proc. 1st SEPM Congress on Sedimentary Geology, St. Petersburg Beach, FL 1: 113
Shinn EA, Hudson JH, Halley RB, Lidz B, Robbin DM, Macintyre IG (1982) Geology and sediment accumulation rates at Carrie Bow Cay, Belize. In: Rutzler K, Macintyre IG (Eds) The Atlantic Barrier Reef Ecosystem at Carrie Bow Cay, Belize. Smithsonian Contributions to the Marine Sciences 12, Smithsonian Institution Press, Washington, DC, 539 p
Shinn EA, Hudson JH, Halley RB, Lidz B (1977) Topographic control and accumulation rate of some Holocene coral reefs, South Florida and Dry Tortugas, Proc. 3rd Intl Coral Reef Symp. 2: 1–7
Shinn EA, Hudson JH, Robbin DM, Lidz B (1981) Spurs and grooves revisited - construction versus erosion, Looe Key Reef, Florida. Proc. 4th Intl Coral Reef Symp. 1:475–483
Sloss CR, Murray-Wallace CV, Jones BG (2007) Holocene sea-level change on the southeast coast of Australia: a review. The Holocene 17:999–1014
Smith SV, Edmonds J, Hartin CA, Mundra A, Calvin K (2015) Near-term acceleration in the rate of temperature change. Nature Climate Change, doi:10.1038/nclimate2552
Smith SV, Kinsey DW (1976) Calcium Carbonate Production, Coral Reef Growth, and Sea Level Change. Science 194:937–939
Stanley GD, Fagerstrom JA (1988) Ancient reef ecosystems: and introduction to the volume. Palaios 3:110
Stanley SM, Hardie LA (1998) Secular oscillations in the carbonate mineralogy of reef-building and sediment-producing organisms driven by tectonically forced shifts in seawater chemistry. Paleogr Paleoclim Peleoecol 144:3–19
Stearn CW, Scoffin TP, Martindale W (1977) Calcium carbonate budget of a fringing reef on the west coast of Barbados - zonation and productivity. Bulletin of Marine Science 27:479–510
Storlazzi C D, Elias E, Field M E, Presto M K (2011) Numerical modelling of the impact of sea-level rise on fringing coral reef hydrodynamics and sediment transport. Coral Reefs 30:83–96
Storlazzi CD, Ogston AS, Bothner MH, Field ME, Presto MK (2004) Wave- and tidally-driven flow and sediment flux across a fringing coral reef: south-central Molokai, Hawaii. Cont Shelf Res 24:1397–1419
Taylor FW, Mann P, Valastro S, Burke K (1985) Stratigraphy and radiocarbon chronology of a subaerially exposed Holocene coral reef, Dominican Republic. J of Geology 93: 311–332
Thom, B.G. and Chappell, J. 1975: Holocene sea-levels relative to Australia. *Search* 6:90–9
Thomas AL, Henderson GM, Deschamps P, Yokoyama Y, Mason AJ, Bard E, Hamelin B, Durand N, Camoin G (2009) Penultemate deglacial sea-level timing from uranium/Thorium dating of Tahitian corals. Science 324:1186–1189
Tomascik T, Surarsono T, Mah A (1993) Case histories: a historical perspective of the natural and anthropogenic impacts in the Indonesian archipelago with a focus on the Kepulauan Seribu, Java Sea. In: Ginsburg RN (ed.) Global aspects of coral reefs. University of Miami Rosenstiel School of Marine and Atmospheric Sciences, Miami FL, J26–J32
Toomey M, Ashton AD, Perron JT (2013) Profiles of ocean island coral reefs controlled by sea-level history and carbonate accumulation rates. Geology 41:731–734
Toscano MA, Macintyre IG (2003) Corrected western Atlantic sea-level curve for the last 11,000 years based on calibrated C14 dates from Acropora palmata framework and intertidal mangrove peat. Coral Reefs 22:257–270
Toth LT, van Woesik R, Murdoch TJT, Smith SR, Ogden JC, Precht WF, Aronson RB ((2014) Do no-take reserves benefit Florida's corals? 14 years of change and stasis in the Florida Keys National Marine Sanctuary. Coral Reefs 33:565–577
Twiggs EJ, Collins LB (2010) Development and demise of a fringing coral reef during Holocene environmental change, eastern Ningaloo Reef, Western Australia. Marine Geology. 275:20–36
Vail PR, Mitchum RM Jr, Thompson S III (1977) Seismic stratigraphy and global changes of sea level, part four: global cycles of relative changes of sea level. Amer Assn Petr Geol Mem 26: 83–98
van Woesik R, Golbuu Y, Roff G (2015) Keep up or drown: adjustment of western Pacific coral reefs to sea-level rise in the 21st century. Royal Society Open Science. doi: 10.1098/rsos.150181
Vermeer M, Ramstorf S (2009) Global sea level linked to global temperature. PNAS 106:21527-21532
Vogel K, Gektidis M, Golubic S, Kiene W, Radtke G (2000) Experimental studies on microbial bioerosion at Lee Stocking Iskand, Bahamas and One Tree Island, Great Barrier Reef, Australia: implications for paleoecological reconstructions. Lethaia 33:190–204
Ware H (2005) Demography, migration and conflict in the Pacific. Journal of Peace Research 42:435–454
Webb AP, Kench PS (2010) The dynamic response of reef islands to sea-level rise: evidence from multi-decadal analysis of island change in the central Pacific. Global and Planetary Change 72:2354–246
Webster JM, Clague DA, Riker-Coleman K, Gallup C, Braga JC, Potts D, Moore JG, Winterer EL, Paull CK (2004), Drowning of the 150 m reef off Hawaii: A casualty of global meltwater pulse 1A?, Geology 32:249–252
Webster JM, Davies PJ (2003) Coral variation in two deep drill cores: significance for the Pleistocene development of the Great Barrier Reef. Sed Geol 159:61–80
Webster PJ, Holland GJ, Curry JA, Chang HR (2005) Changes in Tropical Cyclone Number, Duration, and Intensity in a Warming Environment. Science 309:1844–1846
Weil E, Rogers CR (2011) Coral reef diseases in the Atlantic-Caribbean. In: Dubinsky Z, Strambler N (ed.) Coral reefs: an ecosystem in transition. Springer, p 465–492
Weinstein DK, Klaus JS, Smith TB (2014) Mesophotic bioerosion: structural impact and changes with depth on U.S. Virgin Island deep reefs. Geomorphology 222:14–24
Whitcher E (2011) Macrobioerosion rates of in-situ coral colonies: St. John, U.S. Virgin Islands. 24th Keck Geology Symposium. p 219–225
Wilkinson C (2000) Status of the Coral Reefs of the World: 2000. Australian Institute of Marine Science, Townsville Australia, 363 p
Wilkinson C (2008) Status of the Coral Reefs of the World: 2008. Global Coral Reef Monitoring Network and Reef and Rainforest Research Centre, Townsville, Australia 296p
Williams EH, Bartels PJ, Bunkley-Williams L (1999) Predicted disappearance of coral-reef ramparts: a direct result of major ecological disturbances. Glob Change Biol. 5:839–845
Wilson MEJ, Bosence DWJ, Libong A (2000) Tertiary syntectonic carbonate platform development in Indonesia. Sedimentology 47:395–419
Woodroffe C, Webster JM (2014) Coral reefs and sea-level change. Marine Geologydoi.org/ 10.1016/j.margeo.2013.12.006
Yentsch CS, Yentsch CM, Cullen JJ, Lapointe P, Phinney DA, Yentsch SW (2002) Sunlight and water transparency: cornerstones in coral research. J. Exp Mar Biol Ecol 268: 171–183
Zeebe EE, Wolf-Gladrow DA (2011) CO_2 in seawater equilibrium, kinetics and isotopes. Elsevier Oceanogr Ser No 65, Elsevier B. V., 346 p
Zinke J, Reijmer J, Thomassin B, Dullo W-C, Grootes, PM, Erlenkeuser H (2003) Postglacial flooding history of Mayotte lagoon (Comoro archipelago, southwest Indian Ocean). Marine Geology 194: 181–196

Stability of Reef-Coral Assemblages in the Quaternary

7

William F. Precht and Richard B. Aronson

Abstract

At small spatial and temporal scales reefs are non-equilibrial, dynamic, disturbance-dominated ecosystems. At larger scales, however, the community structure of coral reefs appears stable; coral assemblages from the same environments exhibit striking similarities in species composition and dominance on time scales of decades to hundreds of thousands years. Thus, community membership over time is commonly more stable and persistent than that expected by chance alone. In both the Caribbean and Indo-Pacific, patterns of zonation observed on modern reefs are faithfully recorded in fossil reef sequences preserved through successive Pleistocene high-stands of sea level. Many paleoecologists, however, view the changes in both sea level and sea-surface temperature (SST) recorded during Pleistocene glaciations as major disturbance events requiring community reassembly *de novo* after each event. Why did reef communities respond in a repetitive fashion to the frequent and supposedly large environmental fluctuations of the Ice Ages?

Two major hypotheses have been developed to explain the observed stability of reef assemblages through the Quaternary. The first invokes interspecific interaction or interdependence as an emergent property, which stabilizes community composition for long periods even in the face of environmental change. The second recognizes that the persistence of communities includes or implies persistently stable environments and faunal tracking of environments even when conditions vary. The null model for persistence-stability is that similar community types should recur whenever and wherever similar environmental conditions exist, so long as the same general species pool is available for recruitment.

Analyzing reef facies preserved within a sequence-stratigraphic framework allows us to test the null model on Quaternary reefs. We propose here that reassembly was unnecessary, because reef communities were able to track even the most rapid changes in sea level, producing recurrent biofacies largely through asexual and sexual recruitment from local populations. Analysis of climate change and accompanying tropical SSTs associated with glacial cycles shows they were not sufficient to cause coral populations or the coral reefs they build to disappear and then to reorganize anew. We are, therefore, unable to reject the null model. Incremental faunal tracking of suitable habitats through time and in regional space is the likely mechanism conferring persistence-stability in these coral assemblages.

W.F. Precht (✉)
Division of Marine and Coastal Programs, Dial Cordy and Associates, Inc., Miami, FL 33014, USA
e-mail: bprecht@dialcordy.com

R.B. Aronson
Department of Biological Sciences, Florida Institute of Technology, 150 West University Boulevard, Melbourne, FL 32951, USA
e-mail: raronson@fit.edu

D.K. Hubbard et al. (eds.), *Coral Reefs at the Crossroads*,
Coral Reefs of the World 6, DOI 10.1007/978-94-017-7567-0_7

Keywords
Quaternary • Coral reefs • Stability • Glacial cycles • Sequence stratigraphy

7.1 Introduction

The various deposits of the same facies areas and similarly the sum of the rocks of different facies areas are formed beside each other in space, though in cross-section we see them lying on top of each other. Translated from Johannes Walther (1894)

Beginning with the classic works of James Hutton and Charles Lyell in the nineteenth century, the Principle of Uniformitarianism has formed the cornerstone of modern geological and biological thought (Albritton 1975). The Principle states that the physical processes that now operate on earth have acted in a similar manner throughout geologic time. The resulting doctrine that the present is a key to the past provides the basis for most of our understanding and interpretations in sedimentary geology, including the seminal coral-reef work of Darwin (1842). The field of paleoecology has developed primarily through the scaling up of paleobiological results to the point at which patterns and processes putatively become scale-dependent (Gould 1981, 1985; Aronson 1994). Like neoecology, paleoecology focuses on communities and environments; the paleobiology of fossil assemblages and the geology of the rocks in which they occur yield clues to their environments of formation (Simpson 1970). One of the most important outgrowths of uniformitarian thinking was the comparative approach to understanding sedimentary processes and products, pioneered by Johannes Walther (reviewed in Middleton 1973; see also Ginsburg et al. 1994). Vaughan (1940) was the first reef paleoecologist to highlight the importance of Walther's comparative methodology, stating, "There should be continuous shuttling from the studies of the modern to studies of the ancient and back again from the ancient to the modern." Thus, our ability as geologists to reconstruct past environments requires us to think, at least conceptually, like biologists doing field work in the fossil record (Vermeij and Herbert 2004).

An extension of Walther's comparative approach of comparing the living two-dimensional surface of coral reefs with the third dimension of time was first reported empirically by Mesolella (1967; Mesolella et al. 1970) for the exposed Pleistocene reef terraces in Barbados. Soon thereafter, Land (1974) used explosives underwater to expose the internal facies mosaics of the Holocene reef deposits lying directly beneath their living counterparts on the fore reef at Discovery Bay, Jamaica. Earlier, Shinn (1963) had used explosives to expose the internal structure of Holocene reef spurs at Key Largo Dry Rocks and Molasses Reef in the Florida Keys. These studies were the first to glimpse the geologic architecture and building blocks of Quaternary reefs. Ginsburg (1974) noted at that time the paucity of information on the anatomy and development of Quaternary reefs and platforms. He challenged geologists to use the comparative approach as a tool to answer many of the pressing questions in coral reef geology and biology. Also around that time, Ian Macintyre developed a hydraulically powered, portable, submersible drill for collecting cores from coral-reef habitats ranging from the exposed reef flat to the deep fore reef. This new tool enabled scientists to obtain detailed information on the growth-history of Holocene reef structures (Macintyre 1975; Hubbard 2011). Since then, an extensive literature has highlighted comparative examples of Quaternary reefs in space and time (to mention but a few: Geister 1980, 1983; James and Macintyre 1985; Boss and Liddell 1987; Hunter and Jones 1996; Aronson and Precht 2001; Precht and Aronson 2006). These studies and others have shown that, almost without exception, Pleistocene and Holocene fossil-reef sections exhibit patterns of species composition and zonation similar to living reefs at the same locations (Fig. 7.1), at least prior to the 1980s (but see Precht and Miller 2007 for a counter-example). Pandolfi (2001, 2002) suggested that these Quaternary data point to a high degree of order and predictability in coral assemblages over broad spatial and temporal scales. In light of this pattern, we will use a uniformitarian approach to answer an important question of causality: Why do coral reefs appear so persistent through time and space?

7.2 Stability and Persistence of Coral Reefs

In 1992, Jeremy Jackson published the first paper in a series of closely related manuscripts based on the premise that the community structure of coral reefs was more-or-less stable during the last half-million years of Earth history (Jackson 1992; Jackson 1994a, 1994b; Jackson and Budd 1996; Jackson et al. 1996; Budd and Johnson 1997; Pandolfi and Jackson 1997; Pandolfi 1999; Pandolfi et al. 1999; Jackson and Johnson 2000; Knowlton and Jackson 2001; Pandolfi and Jackson 2001; Pandolfi 2002, Jackson and Erwin 2006; Pandolfi and Jackson 2006; Pandolfi and Greenstein 2007; Pandolfi and Jackson 2007; Johnson et al. 2008; Pandolfi 2011). Using examples from both the Caribbean and Indo-Pacific, these papers concluded that reef-zonation patterns preserved through the Pleistocene high-stands of sea level

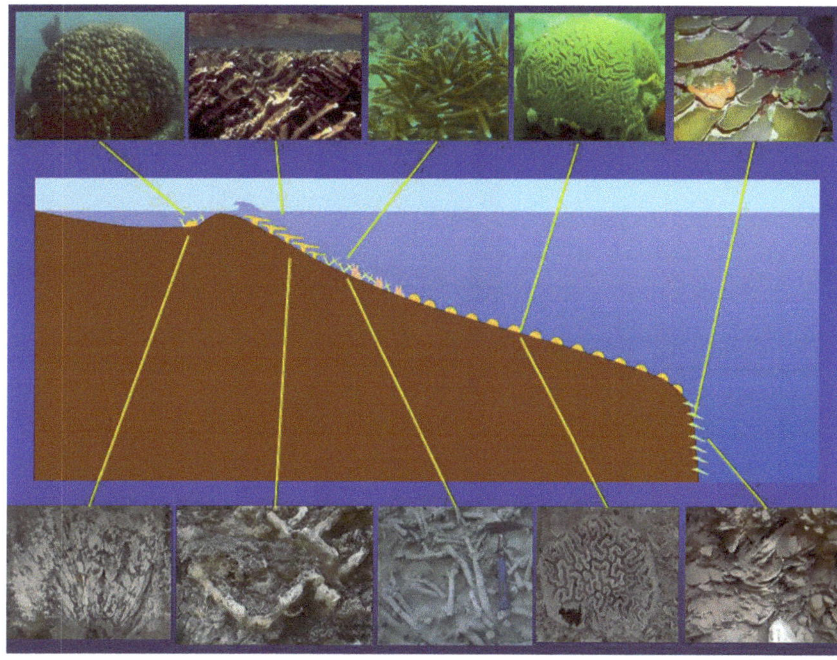

Fig. 7.1 Comparative facies diagram of a generalized Caribbean reef with corresponding living and fossil examples. Zonation scheme follows that of Goreau (1959) (Graphic cartoon courtesy of Dennis K. Hubbard. Individual photographs taken by WF Precht)

were very similar to the patterns observed on modern reefs. Why did reef communities respond in a repetitive fashion to the frequent and large environmental fluctuations of the Ice Ages? The corals must have survived somewhere during the Quaternary glacial episodes in order for similar high-stand reef assemblages to be recorded time and time again during interglacial periods (Pandolfi 2002; Tager et al. 2010), but where? These questions revived the longstanding debate over whether ecological communities in general are stable through time and space. Jackson and Erwin (2006) concluded that coral reefs are remarkably persistent, even through the "exceptionally large environmental fluctuations associated with the waxing and waning of ice sheets" during the Pleistocene Epoch.

There are two explanations for the observed stability of reef assemblages through time (Ivany 1996). The first invokes interspecific interaction or interdependence as an emergent property, which stabilizes community composition for long periods even in the face of environmental change (Miller 1997). The second recognizes that the persistence of communities includes or implies persistently stable environments and faunal tracking of environments, even when conditions vary (Miller 1996; Brett 1998; Brett et al. 2007). According to Jackson and colleagues, the latter explanation is not tenable because Quaternary variations in temperature and sea level were too hostile—too large and too rapid—for faunal tracking to confer community persistence. The main elements of their model are as follows.

1. Multiple, exposed reef terraces, which were deposited during the sea-level maxima of sequential interglacial cycles, imply that coral-reef communities reassembled in a similar fashion after each change in global sea level.
2. Each exposed reef terrace represents a new reef community and, therefore, is an independent, natural experiment in larval recruitment and biotic assembly; autogenic drivers are most important.
3. The persistence of coral assemblages through multiple episodes of global climatic fluctuation in the Pleistocene argues against environmental variability as generating the primary or exclusive forces that structure reef communities through time and regional space; these allogenic drivers are far less important.

In this chapter, we critically examine explanations for why the community membership of Pleistocene coral reefs has persisted despite the extreme environmental changes that have been assumed during glacial–interglacial cycles.

7.3 The Pattern

A population can respond to environmental change by adjusting its location or habitat within a region. It is hardly surprising that the composition of terrestrial plant assemblages changed radically when the massive North American ice sheets retreated by thousands of kilometers in a few thousand years after the Last Glacial Maximum (LGM). In some cases, the result of biotic response to changing environmental conditions was a plant assemblage unlike anything observed in the modern (Overpeck et al. 1992; Williams and Jackson 2007). In stark contrast,

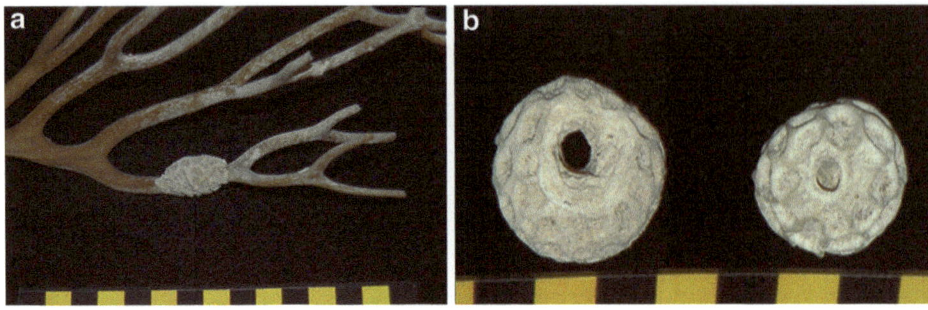

Fig. 7.2 The coral *Favia fragum* living on an octocoral branch (**a**). Resultant doughnut-hole feature in fossil and subfossil *F. fragum* colonies once the octocorals are taphonomically removed (**b**) (Collected and photographed by WFP)

taxonomic composition and diversity on tropical coral reefs have remained remarkably constant through the repeated glacial–interglacial cycles of the Pleistocene. The most striking feature has been the prominence of the genus *Acropora* in time and space (Jackson 1992; Aronson and Precht 2001). The Holocene dominance of branching *Acropora* in the Caribbean, in combination with its predominance in fossil-reef outcrops, has been taken to imply that shallow-water coral assemblages have remained largely constant in the Caribbean/Atlantic province since about the mid-Pleistocene (but see Klaus et al. 2012 for evidence that domination by acroporids pre-dates the mid-Pleistocene). Geister (1980), however, has shown that this dominance by *Acropora* is far from universal; there is a spectrum of shallow-water reef types in the Pleistocene, which follows a predictable zonation scheme related to variability in local environmental conditions (Geister 1977).

In the Pacific, Pandolfi (1996) examined paleocommunity composition based on the presence or absence of coral species on nine sequentially uplifted, Pleistocene reef terraces on the Huon Peninsula, Papua New Guinea. The interval examined spanned 95 ky (thousand years) and included nine glacial cycles. Pandolfi noted that similarity in species composition was greater across space than it was through time. Thus, Pleistocene reef-coral assemblages were more distinct between reefs of the same age from different places than between reefs formed at different times at the same location. According to Pandolfi (1996), local environmental parameters had a greater influence on reef-coral composition than changes in global climate and sea level.

In Jamaica, limestone outcrops spanning 0–6 m above present sea level represent coral reefs deposited ~125 ky (thousand years) ago during the last major interglacial sea-level highstand of the Pleistocene (Marine Isotope Stage 5e; Boss and Liddell 1987). The 125-ky bank-barrier reef exposed along the eastern margin of Rio Bueno Harbor, on the north coast of the island, is dominated by coral assemblages of *Acropora cervicornis* and the *Obicella annularis* species complex in the fore-reef facies, and *A. palmata* and *Porites porites* in the reef-crest facies (Liddell et al. 1984), following the classic coral-zonation scheme of Goreau (1959) for modern reefs. The same suite of species characterized the living reef community in the waters just below at Rio Bueno and at nearby Discovery Bay prior to the 1980s (Goreau 1959; Precht and Hoyt 1991).

The living reef assemblage at Negril, on Jamaica's west coast, is quite different. Here, the coral assemblage is typical of hardground habitats found throughout the Caribbean. The Negril shelf is populated primarily by gorgonians with lesser amounts of massive hard corals, including the *O. annularis* complex, *Siderastrea siderea*, *Diploria* spp., the pillar coral *Dendrogyra cylindrus*, and small, scattered thickets of branching *A. cervicornis*. Examination of the emergent, 125-ky Pleistocene deposits exposed in the cliffs just above reveals the same suite of coral species. Gorgonians are not preserved due to their taphonomic vulnerability; however, small colonies of the coral *Favia fragum* exhibiting a unique, "doughnut-hole" feature are common fossils in the Negril outcrops. The hole is characteristic of *Favia* colonies that, when alive, grew on gorgonian branches (Fig. 7.2; see also Rodríguez-Martínez and Jordán-Dahlgren 1999). Their abundance in the Pleistocene of Negril implies that gorgonians were common at the time. As on the Huon Peninsula (Pandolfi 1996), the difference between locations in Jamaica is far greater than the difference between 125 ky and the present (see Aronson and Precht 2001 for additional examples).

7.4 The Problem

Coral reefs are among the most frequently cited examples of community-level destruction stemming from changes in sea level (Valentine and Jablonski 1991). Many paleoecologists view the rapid changes in sea level and sea-surface temperatures (SSTs) during Pleistocene glaciations as major disturbance events (Taviani 1998), requiring community assembly *de novo* after each event. As Webster et al. (2004) noted, "Despite major environmental perturbations (i.e. relative sea-level and temperature changes) the platforms and the shallow water coral reefs exposed at the top have been able to re-establish themselves time and time again over the last 450 ka."

We argue that there is no evidence to support the contention that Quaternary coral assemblages in the tropics significantly waxed and waned as sea level rose and fell. Although some of these glacial–interglacial events were dramatic in geologic time (Lambeck et al. 2002; Cutler et al. 2003), they were still slow compared to the turnover rates of the corals and their abilities to recruit to adjacent environments made habitable by sea-level change. We propose here that reassembly was unnecessary. As the summations of individual taxa tracking sea level, the reef communities they comprise were able to track even the fastest changes in sea level throughout the Pleistocene. The null model for persistence-stability is that similar community types should recur whenever and wherever similar environmental conditions exist so long as the same general species pool is available for recruitment (Bennington and Bambach 1996). It is incumbent on those advocating an alternate hypothesis for assemblage stability throughout the Quaternary to be able to reject that null model.

7.5 Climatic Variability in the Tropics During Glacial—Interglacial Cycles

During the Quaternary Period, the Earth's climate system has been driven by long-term and periodic variations in solar radiation, which in turn have been forced by orbital variations: the Milankovitch Cycles. Earth's climate has oscillated between a glacial mode that favored the expansion of glaciers and ice sheets at middle and high latitudes, and a briefer, warmer interglacial mode characterized by the rapid disintegration of those glaciers and ice sheets. These glacial–interglacial cycles have driven fluctuations in eustatic sea level in the world's oceans (Fig. 7.3). During the interglacial high-stands of sea level, corals developed thick, well-zoned, reef sequences on insular shelves in the tropics. The data from coral reefs (e.g. Bender et al. 1979) are corroborated by global temperature estimates based on a compilation of oxygen isotope ratios of planktonic foraminifera taken from deep-sea cores (Lambeck et al. 2002).

The Milankovitch theory is based on two premises: (1) that global climate is controlled by the seasonal and latitudinal distribution of solar insolation; and (2) that the expansion and recession of northern-continental ice sheets during the Quaternary has been regulated by slow variations in the Earth's orbital parameters (Milankovitch 1941; Berger 1978). The primary Milankovitch cycles have periodicities of 100 ky (eccentricity), 41 ky (obliquity), and 21 ky (precession; Hays et al. 1976). Global climatic changes of the Quaternary have been the direct result of predictable, quantifiable periodicities of the Earth's position as it wobbles on its rotational axis and orbits in a variable path around the Sun.

Although it has been clear for some time that exposed reef terraces on tectonically uplifted islands can be correlated with Milankovich cycles (Broecker 1966; Broecker et al. 1968; Mesolella et al. 1969; Veeh and Chappell 1970; CLIMAP 1976, 1981; Berger et al. 1984; Imbrie et al. 1984; Imbrie and Imbrie 1986; Pillans et al. 1998), there are minor, suborbital cycles superimposed on the main climate cycles (Paillard 2001; Lambeck et al. 2002; Potter et al. 2004; Thompson and Goldstein 2005), creating subtle variations in sea level and climate that also influence the resultant, preserved coral biofaces in both space and time (Precht and Harris 2008). These subtleties include a double high sea-stand and resultant shallowing-upwards parasequences preserved during the last interglacial, which can be correlated across the tropics (Precht 1993a; Sherman et al. 1993; White et al. 1998; Wilson et al. 1998; Blanchon and Eisenhauer 2000; Hearty et al. 2007; Blanchon et al. 2009).

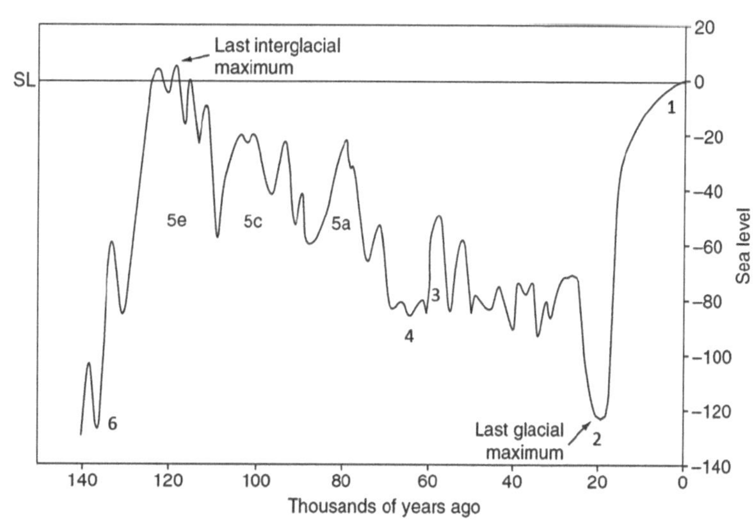

Fig. 7.3 Generalized eustatic sea-level curve for the last 140 ky (modified from Precht and Miller 2007). Marine Isotope Stages are labeled after Emiliani (1972). Episodes of reef growth exposed on tectonically uplifted islands occurred during times of sea-level maxima. Curve is based on data from numerous sources, including Chapell (1983), Chappell and Shackelton (1986), Buddemeier and Kinzie (1998), and Lisiecki and Raymo (2005). SL denotes modern sea level

7.6 Temperature

Temperature has long been considered the main control on reef distribution (Dana 1843). Reef-corals and other zooxanthellate organisms live close to their upper thermal-tolerance limits and are largely confined to the shallow waters of the photic zone, with the optimal range for coral growth around 25–29 °C (Vaughan and Wells 1943). Cold-temperature tolerances are not well defined for corals but early experiments documented 16 °C as stressful to most corals and exposure to temperatures below 15 °C as often lethal (Mayer 1914, 1915; Lirman et al. 2011). The present-day latitudinal limits of coral-reef distribution generally coincide with the 18 °C monthly-minimum seawater isotherm (Johannes et al. 1983; Kleypas et al. 1999; but see Coles and Fadlallah 1991; Yamano et al. 2001).

Initial studies indicated that SSTs calculated for the tropical oceans were remarkably stable through the Pleistocene glacial–interglacial cycles, with no more than 4–5 °C variability from the peak warmth of the last major interglacial (~125 ky) to the peak cooling of the last glacial maximum (LGM; Emiliani 1966; CLIMAP 1976, 1981; Crowley 2000). In recent years, however, debate has centered on the nature of the thermal response of tropical seas to Pleistocene glacial and interglacial episodes (McCulloch et al. 1999). The records of tropical SSTs during the glacial maxima remain somewhat controversial (Montaggioni and Braithwaite 2009).

The LGM occurred sometime around 20–26 kybp. At that time, global sea level was ~120 m lower than today (Fairbanks 1989; Peltier 2002). CLIMAP hindcast models indicated that minimum sea-surface temperatures during the LGM were less than 2 °C cooler than those associated with modern tropical seas (but see Emiliani and Ericson 1991). Since then, mostly through the use of coral proxy records (Guilderson et al. 1994, 2001; Beck et al. 1997) and terrestrial paleoclimate data (Rind and Peteet 1985; Farrera et al. 1999), controversy has arisen as to whether tropical SSTs could have been as much as 4–5 °C colder during this period (Colinvaux et al. 1996; Beck et al. 1997; Guilderson et al. 2001; Gagan et al. 2000, 2004; Powers et al. 2005; Hippler et al. 2006; Felis et al. 2014). Solow and Huppert (2004) showed, however, that error in the calibration of oxygen-isotope data from corals has led to a substantial bias in the LGM temperature reconstruction. Although some studies have shown the tropical ocean to be slightly cooler at the LGM than previously estimated by CLIMAP, the broad consensus of multi-proxy data, faunal records, and global climate models suggests an overall drop of no more than ~2.5 °C in the tropics (Anderson et al. 1989; Broccoli 2000; Crowley 2000; Lea et al. 2000; Pflaumann et al. 2003; Ballantyne et al. 2005; Annan and Hargreaves 2012), with the least cooling occurring in the western tropical and subtropical gyres (the warm pools) of all the ocean basins (Thunell et al. 1994; Mix et al. 1999; Trend-Staid and Prell 2002; Niebler et al. 2003; Schmidt et al. 2006).

On longer time scales, data indicate that in the western Coral Sea SSTs have changed by 1.5 °C or less over the past 800 ky (Lawrence and Herbert 2005). Reconstructions of SST for the past 1.25 my (million years) based on marine sediments from the eastern equatorial Pacific partially support these findings, with glacial–interglacial temperature variability ranging from about 1.0 to 4.5 °C (Liu and Herbert 2004). These studies, coupled with the analysis of the temperature-dependent production of alkenone molecules ($U_{37}^{K'}$) by marine organisms (Sonzogni et al. 1998; Herbert and Schuffert 2000; Rosell-Melé et al. 2004) and Mg/Ca-based paleo-temperature reconstructions from all ocean basins (Barker et al. 2005), lead to the conclusion that SSTs in the western tropics and subtropics of all ocean basins were essentially invariant from modern values, departing by no more than 1.0–3.0 °C (see also de Rosenthal and Lohmann 2002; de Garidel-Thoron et al. 2005; Schmidt et al. 2006; Tachikawa et al. 2009; Herbert et al. 2010). Even the worst-case scenario of SSTs dropping by as much as 5–6 °C during glacial maxima (Felis et al. 2014) would not have pushed reef-building corals outside of their thermal-tolerance limits within the tropics (Kleypas 1997). Sea temperatures simply were not low enough to terminate coral growth or reef development through the major glacial–interglacial cycles of the Pleistocene.

There was no need for the surviving reef-coral species of the tropical Caribbean to be pre-adapted to cool temperatures, to contract their ranges to lower latitudes, or to pass through a thermal extinction filter of glacial cooling in the early Pleistocene, as proposed by Budd et al. (1994), Jackson and Johnson (2000) and Getty et al. (2001). Van Woesik et al. (2012) likewise argued in favor of an extinction peak coinciding with a period of global cooling (glaciation) and eustatic sea-level drop that changed oceanic circulation patterns. However, O'Dea et al. (2007) specifically noted the lack of coincidence of coral extinction events with the onset of Northern Hemisphere glaciation. It is more likely that environmental changes related to the final closure of the Isthmus of Panama were responsible for the culling of coral diversity in the Caribbean prior to and during the early Pleistocene. The fact that Indo-Pacific corals did not experience similar extinctions under essentially the same sea-level and temperature trajectories (Paulay 1990; Budd et al. 1994) argues against those factors as the primary causal agents. Benzie (1999) argued from genetic analysis that the resurgence of coral populations in response to isolation during the Quaternary low-stands must have occurred repeatedly from the same refugia during repeated transgressive events.

During the glacial maxima, polar faunas expanded their ranges toward the equator, but there is little evidence for contraction of tropical faunas (Vermeij 1992; Crowley 2000; Pflaumann et al. 2003; Valentine et al. 2008). Kleypas (1997) calculated that lower temperatures during the LGM reduced the total area of coral reefs only at their latitudinal extremes in the subtropics, with no contraction in the tropics. There is compelling evidence of substantial reef development during several Pleistocene glacial intervals, including the LGM (Harris and Davies 1989; Macintyre et al. 1991; Colonna et al. 1996; Kleypas 1997; Cutler et al. 2003; Peltier and Fairbanks 2006; Montaggioni 2005; Cabioch et al. 2008a, 2008b). Thus, climate change associated with glacial cycles apparently has not been sufficient to cause coral reefs to disappear from the tropics and the coral assemblages then to reorganize.

7.7 Reefs on the Edge

Logically, we would expect the abundance of a taxon in the fossil record to have been greatest near the center of its geographic range and progressively decrease to the margins as conditions became less favorable (Brown 1984; Enquist et al. 1995). Dynesius and Jansson (2000) noted that most species have experienced and survived many Milankovitch-scale climatic oscillations by tracking their habitats through space (see also Bartlein and Prentice 1989). Because the coral reefs of south Florida are at the latitudinal extreme of reef development in the western Atlantic, it is not surprising that cold-sensitive species such as the acroporid corals have contracted and expanded in response to changing environmental conditions. In southeastern Florida, a series of submerged, shore parallel, fossil-reef terraces reveals a nearly continuous barrier-reef system that extended northward from what today is Miami to Palm Beach County during the latest Pleistocene and the early to middle Holocene (Banks et al. 2007). During the Holocene thermal maximum (HTM) from 10.5 to 5.4 ky (COHMAP 1988; Ruddiman and Mix 1993; Lin et al. 1997; Kerwin et al. 1999; Haug et al. 2001), SSTs were warmer than today in the western Atlantic, allowing *Acropora*-dominated reefs to expand northward (Lighty 1977; Lighty et al. 1978; Precht and Aronson 2004). In apparent response to climatic cooling in the late Holocene (deMenocal et al. 2000; Jessen et al. 2005), the northern limits of the *Acropora* species contracted 150 km south to Fowey Rocks (Precht and Aronson 2004). In historical times, Fowey Rocks was the northernmost emergent reef of the Florida reef tract, as well as the northernmost extent of *A. palmata* (Vaughan 1914; Jaap 1984; Porter 1987). A similar expansion and contraction of *A. palmata* during the HTM has been observed on reefs of the Flower Garden Banks in the northern Gulf of Mexico (Precht et al. 2014). The HTM also correlates with the latitudinal expansion of coral-reef and mangrove ecosystems in the Pacific (Veron 1992; Mildenhall 2001; Twiggs and Collins 2010; Woodroffe et al. 2010; Hongo 2012). Evidence from both terrestrial and coastal regions shows that warming during this interval allowed many species to migrate poleward (Clarke et al. 1967; COHMAP 1988; Salvigsen et al. 1992; Hjort et al. 1995; Dyke et al. 1996; Dahlgren et al. 2000; Carbotte et al. 2004; Jansen et al. 2009).

Similar range expansions and contractions have been identified in reef deposits formed during the last major interglacial. One example can be found in the Pleistocene coral assemblage at Rottnest Island off Western Australia (32 °S). The living reef at this locality has some 25 species of zooxanthellate corals. Most are at the southern limit of their range, with *Acropora* spp. being exceedingly rare or absent (Marsh 1993); however, ~125 ky ago, when sea temperatures were a few degrees higher, both staghorn and tabular *Acropora* spp. were much more common (Szabo 1979). These paleoecological examples of species replacements and range expansions, especially those involving acroporids, emphasize the varied, individual responses of coral species and the fluidity with which they can reconstitute reef communities in the face of environmental change (see also Greenstein and Pandolfi 2007).

Although reefs at their latitudinal extremes have responded rapidly to climatic shifts, results from coring and outcrop studies in the tropical Caribbean show the persistence of coral assemblages through time (Aronson and Precht 2001). The evidence from these *Acropora*-dominated reefs supports the notion that tropical oceanic climates have been buffered from extreme climatic variability throughout the Holocene (Macintyre et al. 1977; Fairbanks 1989; Aronson and Precht 1997; Gill et al. 1999; Wapnick et al. 2004; Greer et al. 2009). Similar results have been recorded from the Indo-Pacific (Chappell and Polach 1991; Camoin et al. 1997; Cabioch et al. 1999a, 1999b, 2003; Montaggioni et al. 1997). Not surprisingly, reefs in thermally reactive, subtropical areas are more likely than tropical reefs to change in species composition as the climate warms or cools.

7.8 Sea Level

Coral reefs, their resultant ecological communities, and the preserved sedimentary facies record complex interactions among myriad factors including tectonics, topography, eustatic changes in sea level, $CaCO_3$ production, sediment supply, ocean chemistry, temperature, physical conditions (agitation), taphonomy, paleoecology, and diagenesis. Changes in relative sea level associated with glacial–

interglacial cycles produce spatial shifts in reef facies that can generally be correlated within and between reef sequences. The term 'sequence' here refers to the smallest set of genetically related strata bounded by unconformities or their correlative conformities deposited during a single sea-level cycle (low-stand to high-stand and back to low-stand; Mitchum 1977). The carbonate sequence is comprised of four distinct depositional units or systems tracts (Catuneanu et al. 2009, 2010). First, the transgressive systems tract (TST) is deposited during the rising part of a relative sea-level cycle, the base of which is a marine flooding surface. Second, the highstand systems tract (HST) is deposited during and shortly after the sea-level maximum. Third, the falling systems tract (FST) is usually deposited as a veneer while sea level is falling. During the FST, subaerial exposure results in an unconformity across the upper portions of the platform. Fourth, the lowstand systems tract (LST) is composed of autochthonous reef sediment that was formed and deposited during sea-level lowering (forced regression), when the shoreline was displaced below the contemporary shelf-edge. Within sequences are smaller parasequence-scale depositional units. Parasequences are upward-shallowing successions of facies bounded by marine flooding surfaces, which are commonly nested within sequences and systems tracts (van Wagoner et al. 1988) and are fractally scaled (Schlager 2004).

In the Quaternary, these reef sequences record the advance and retreat of the ice sheets. Within each sequence, the vertical stacking of reef facies mirrors the horizontal zonation of reef facies coexisting on a synchronous depositional surface. This pattern is referred to as Walther's Law (Middleton 1973). It is similar to the space-for-time substitution used by ecologists to explain decadal- to centennial-scale successional dynamics over temporal scales at which even long-term monitoring data are insufficient to explain the mechanisms of community change (Pickett 1989). Some geologists (e.g., Walker and Alberstadt 1975; James 1983; Copper 1988; Pandolfi 2011) have mistakenly interpreted the substitution of one facies for another due to spatial changes in habitat over time as autogenic ecological succession (but see Hoffman and Narkiewicz 1977; Rollins et al. 1979; Gould 1980; Viau 1983; Miller 1986; Karlson 1999; Mewis and Kiessling 2013). Understanding the dynamics of Walther's Law both spatially and temporally with respect to rising or falling sea level (Kerans and Tinker 1997) is, therefore, crucial to testing the null model of the persistence-stability of Pleistocene reef communities.

During interglacial periods, as sea level has risen, reefs have responded by moving or expanding upslope, generally tracking sea level. Goreau (1969) coined the expression "post-Pleistocene urban renewal of coral reefs" to describe the emergence of Holocene reefs associated with the flooding of platforms during the last deglaciation. Because accommodation space increases as sea level rises (Masse and Montaggioni 2001), reefs build vertically to fill this space, resulting in characteristic geometries and internal facies mosaics (Neumann and Macintyre 1985). The accretion rates of these reefal sequences generally follow a logistic model (Schlager 2005). The base (TST) of an individual sequence is characterized by a transgressive surface: a lithological discontinuity across which there is a shift of facies that indicates an abrupt increase in water depth. Some authors have suggested that this discontinuity represents as much as a 2-ky lag between platform flooding and reef initiation (Adey 1978; Kendall and Schlager 1981; Hopley 1994; Eberli 2013). This 'start-up' phase is followed by a progressive and often rapid shallowing of reef facies characteristic of a 'catch-up' style of reef development. As the reef continues to shallow upward, the accommodation space across the platform becomes more restricted and the community structure changes accordingly to a 'keep-up' mode of reef development (Neumann and Macintyre 1985; Macintyre 2007).

A minimum sea-level curve since the LGM was developed for the Caribbean using *A. palmata* sampled from well-documented shallow reef-crest framework (Lighty et al. 1982; Fairbanks 1989; Toscano and Macintyre 2003). Toscano and Macintyre (2005) pointed out that sea-level reconstructions require sampling of *A. palmata* specifically from reef-crest facies, thus permitting the assumption that the samples had grown within the restricted depth range (<1–5 m) typical for this facies. Sampling a restricted range of paleodepths is important, because *A. palmata* is able to recruit and survive to depths in excess of 20 m (Zimmer et al. 2006). In the Indo-Pacific there are a number of *Acropora* species used as paleodepth indicators for sea-level reconstructions (Montaggioni and Faure 1997; Cabioch et al. 1999b; Hongo and Kayanne 2010; Woodroffe and Webster 2014). It is clear from the curve in Fig. 7.4 that the maximum rate of rise in sea level during the last 18 ky never exceeded the maximum growth rate of branching *Acropora* colonies, which can increase linearly between 10 and 20 cm y^{-1}. Coral growth, however, does not equal vertical reef accretion (Dullo 2005; Hubbard 2009; Chap. 6). No matter how rapidly individual corals can grow, at times the rate of sea-level rise outpaces the ability of carbonate deposition to keep up (Schlager 1981). This is especially true on steep slopes. The result is a shift or 'backstepping' of the reef facies to more shallow (more shoreward) positions on the shelf (Lighty et al. 1978; Hubbard et al. 1997; Macintyre 1988, 2007; Cabioch et al. 2008b).

A key point is that rapid sea-level rise *per se* does not kill corals or coral reefs; it just alters the resulting preservation of the reef facies in space and time (Dullo 2005). As the shallowest reef flourishes, the former position of the

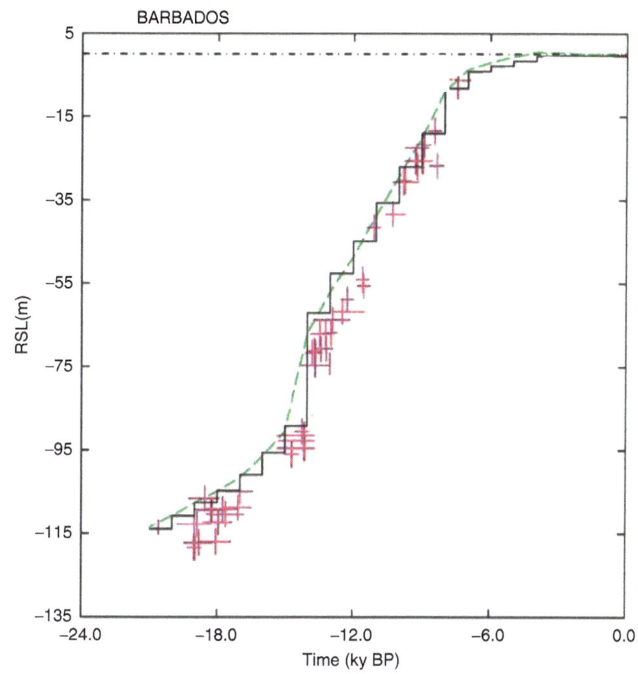

Fig. 7.4 Smoothed sea-level curve from last glacial maximum to the present using the coral-derived, relative sea-level record from Barbados of Fairbanks (1989) and corrected for tectonic uplift. The eustatic function is the black, discontinuous, step-like curve, whereas the green, smooth curve is the theoretical prediction (From Peltier 2002). Purple crosses are dates obtained from *Acropora palmata*. RSL, relative sea level (Figure courtesy of W.R. Peltier)

previous reef-margin facies becomes incipiently drowned and is sequentially capped by deeper-water species (the "give-up reefs" of Neumann and Macintyre 1985). In many locations throughout the Caribbean and western Atlantic, this upslope migration of reef facies has led to a step-like pattern of submerged reefs and reef terraces (Macintyre 1972; Lighty 1977; Fairbanks 1989), in some cases leading to apparent gaps in the reef record (Hubbard et al. 2004; Hubbard 2008, 2014).

The ultimate cause, timing, and duration of individual submerged reef terraces preserved within particular TSTs have been debated (i.e., Blanchon 2005; Toscano and Macintyre 2005). Regardless, it is important to note that, although some of these spatial jumps from terrace to terrace may appear geologically rapid, they are still slow in an ecological sense. The youngest dates from the top of one terrace often overlap or are coeval with the oldest dates at the bottom of the next (Shinn et al. 1981; Fairbanks 1989; Lidz et al. 1997; Toscano and Lundberg 1998; Stathakopoulos and Riegl 2015; see Montaggioni et al. 1997 for a Pacific example), producing a relatively smooth sea-level curve for that locality (Fairbanks 1989; Bard et al. 1990; Peltier 2002; Toscano and Macintyre 2003, 2005; but see Blanchon and Shaw 1995).

For example, the sea-level curve derived from dates of *A. palmata* from the Barbados cores allows for the calculation of the rate of glacial-meltwater discharge into the Atlantic (Fairbanks 1990). Rates of sea-level rise were rapid during this period, punctuated by two periods of extremely rapid rise (meltwater pulses) in the latest Pleistocene. During these meltwater pulses, sea-level rise apparently outstripped the accumulation rates of *A. palmata* reef facies, resulting in incipiently drowned, or "give-up" reefs followed by continued backstepping of the shallow *Acropora*-dominated reef communities to more shoreward positions higher on the shelves. In fact, Fairbanks's (1989) coring data from Barbados revealed a nearly continuous record of *A. palmata*, dating from 17.1 ky at −117 m to 7.8 ky at −20 m. This record also indicates that there is no apparent time lag from when sea level flooded the exposed platform to the initiation of reef growth. These examples show that the rate of sea-level rise is inconsequential to coral growth and recruitment but is directly related to the development of reefal lithofacies preserved within each of these parasequences.

In the Caribbean, the exposed Pleistocene reef-terrace deposits on emergent islands represent zoned, shallow and often-prograding communities characterized by *Acropora*- and *Orbicella*-dominated coral assemblages, with keep-up motifs preserved during sea-level maxima (HST; Humphrey and Kimbell 1990). The tops of the *A. palmata* facies are used as biostratigraphic markers for the upper limits of sea level in these sequences (Mesolella 1967; Mesolella et al. 1970). Where *A. palmata* is absent, the absolute water depths for reef facies can be calculated using other shallow-water indicators that accurately express the positions of the low-tide datum (Fig. 7.5; see Precht and Miller 2007 for additional examples). These HST-reefs are continuous with the backstepping reefal parasequences, which formed earlier during deposition of reef facies along and within the TST. The maximum flooding surfaces form the boundaries between the TSTs and the HSTs (Schlager 1992, 2005). The facies contained within each of the Pleistocene parasequences, from TST to HST, are gradational and Waltherian. Therefore, isolated reef-terrace deposits at the apexes of the sea-level cycle (HSTs) do not represent new reefs, nor do they constitute independent natural experiments in larval recruitment and assembly, as Jackson and Erwin (2006) asserted.

Conversely, when sea level was falling during glacial intervals, the reef community and its constituent species relocated downslope in a stepwise fashion, resulting in shifts in the biotic facies (FSTs: Kendall and Schlager 1981; Paulay 1990; Montaggioni and Braithwaite 2009; Schlager and Warrlich 2009). Reefs predominated in all areas during the FSTs, with shallow-water facies down-stepping over the deeper-water facies of the previous TSTs and HSTs

Fig. 7.5 Micro-atoll of living *Orbicella annularis* colony from South Carysfort Reef in the Florida Keys National Marine Sanctuary. This growth-form is an excellent marker for spring low-tide levels. Note dead, bioeroded top and slightly raised living rim of colony (**a**). Outcrop photograph of micro-atoll structure in *O. annularis* as a response to growth to sea level, Pleistocene Key Largo Limestone, Florida Keys. Coral is in growth position. Note the highly bored/bioeroded top surface and raised margin of the colony in cross-sectional view (**b**). These paired photographs show the utility of the comparative approach in paleoecologic investigations (Photographs taken by WFP)

(Schlager and Warrlich 2012; see also Hinestrosa et al. 2014). Lagoonal facies are generally absent in the FSTs and LSTs because they are stranded at the top of the HSTs (Pomar and Ward 1994). The lagoonal faunas were extirpated while the coral-dominated fore-reef communities persisted, albeit displaced downslope (see Paulay 1990). Because accommodation space was constantly being lost during FSTs, fossil FST reefs and their associated facies are often poorly preserved or eroded (Kendall and Schlager 1981; Eberli 2013). Not all systems tracts need be present in each sequence (Catuneanu et al. 2009), and poor preservation of FSTs due to the loss of accommodation space should not be construed as the absence of coral reefs, requiring subsequent biotic reorganization.

Well-established LST reefs have been described for the LGM (Kleypas 1997; Webster et al. 2004; Peltier and Fairbanks 2006; Tager et al. 2010). Pomar (1991) similarly noted in Miocene reef deposits in Mallorca, Spain that down-stepping of facies was observed during sequential lowstands of sea level. In these cases, the downward shift of the coral zones measured the amount of sea-level fall. Tracking of reef communities and carbonate facies in response to spatially shifting environments appears to be the rule, not the exception, throughout the tropics during the Pleistocene (Chappell 2002).

One additional model of possible reef decline associated with glacial low-stands is the smothering of reef communities by sediment eroding from the exposed platforms (Veron 1995). This mechanism may be plausible in some siliciclastic-dominated settings; however, in mixed carbonate-siliciclastic environments the siliciclastic deposits are usually funneled through discrete channels and bypass the shelf-margin reefs (Ferro et al. 1999; Fielding et al. 2003). Even in volcanic-island settings, extraordinary thicknesses of (shallow-marine) platform and peri-platform carbonates, as well as rapid facies changes between volcanic and carbonate rocks, have been observed in the deep past (Soja 1996) and the Quaternary (Camoin et al. 2007) negating this as a likely explanation of reef decline. Most importantly, Dravis (1996) noted that, on carbonate-dominated platforms, subaerial exposure and resulting meteoric diagenesis lead to geologically instantaneous lithification of the carbonate sediments, preventing their erosion by wind or water. Once cemented, these deposits are further stabilized by vegetational processes, inhibiting their transport off the platform and into adjacent deeper-water basins. Thus, low-stand sediment-shedding from exposed platforms is an unlikely mechanism for reef collapse.

7.9 A Coral's-Eye View

Johnson et al. (2008) suggested that exceptionally rapid growth and high rates of fragmentation by *Acropora* spp. allowed these species to keep up with the sudden rises in sea level that have occurred repeatedly during the past million years. But what about all the other coral species that coexisted with the *Acropora* species? How did they keep pace? At the generational level of the coral holobiont, even the most rapid rate of sea-level rise or fall is inconsequential to its survival, growth, and recruitment. For example, imagine that a colony of *Orbicella annularis*, a long-lived, broadcast-spawning species found over a wide range of depths, recruits to a reef at 10 m water depth. Throughout its 300-year lifespan this coral has an average linear-extension rate of ~12 mm/year (Bosscher and Meesters 1992; van Veghel and Bosscher 1995). While it is alive, assume that sea level rises at its maximal rate of ~21 mm/year (Meltwater Pulse 1A; Fairbanks 1989). In this worst-case scenario, when the coral colony dies it is 3.6 m tall. The top of the colony is sitting in 12.7 m of water, which is only 2.7 m deeper than the depth at which it started.

As a second example, imagine that a colony of *Porites astreoides*, a short-lived, brooding species that is one of the commonest coral species on Caribbean reefs (Green et al. 2008) and which also has a broad depth distribution, recruits to a reef habitat at 3 m water depth. Throughout its 20-year lifespan, this coral has an average linear-extension rate of ~5 mm/year (Kissling 1977; Huston 1985). While it is alive, sea level rises at its maximal rate of ~21 mm/year. When the coral colony dies it is 10 cm tall and is sitting in 3.42 m of water, a deepening of only 32 cm for the top of the colony and only 42 cm for its base. Neither the *Orbicella* nor the *Porites* colony is removed from its preferred depth range or habitat over the course of its lifetime, and reproduction and recruitment are providing more than enough offspring to take over in a more shoreward position when this single colony dies.

Now reconsider the *O. annularis* colony, but use the average rate of sea-level rise for the last 18 ky: 6.6 mm/year. The average rate of coral growth exceeds the average rate of sea-level rise. After 300 years, the top of the colony sits in 8.4 m of water, which is 1.6 m shallower than its starting depth. In fact, it is precisely the ability of *O. annularis* to produce massive colonies that form the building-blocks of reef framework that led Bosscher and Schlager (1993) to recommend using the growth rates of the *O. annularis* species complex as a general proxy for reef growth in the Caribbean. They noted that although the growth rates of branching *Acropora* spp. may exceed the growth rates of massive corals by an order of magnitude, much of this skeletal growth is turned into sediment and rubble that fills the interstices of the framework.

These three examples may help explain why reef-corals in general, not just the acroporids, were able to keep pace with sea-level rise since the LGM. Even under the most rapid pulses of sea level, ecological space continued to exist for Caribbean corals. Even though reefs might not be able to build at rates sufficient to fill all the available accommodation space, it is unlikely that changing sea level *per se* resulted in conditions unsuitable for local larval recruitment.

Because individual coral colonies in moderate depths were likely not affected by even the most rapid fluctuations of sea level in the Pleistocene, coral populations would have been able to alter their spatial distributions incrementally over long periods. The summations of these incremental shifts, displayed in the fossil record, have been prematurely interpreted as collapse and reassembly. Similar shifts have been observed for co-occurring bivalve species (Gardiner 2001). Faunal tracking is the likely mechanism conferring persistence-stability in these coral assemblages. That is, the reef communities simply moved with their preferred environments, producing recurrent biofacies in time and space. Habitat tracking applied at the level of biofacies, however, is not a group phenomenon, as envisioned by Brett et al. (2007). A simpler and more accurate explanation is that most species independently tracked their environmental preferences, likely by local asexual and sexual recruitment. The recurrence of reef biofacies represents the summation of coral species' independent responses to environmental fluctuations (see Hoffman 1979), not Clementsian dynamics, community integration, or ecological locking, as suggested by Jackson (1994a).

In Papua New Guinea, Tager et al. (2010) showed that low-stand communities during the LGM were distinct in coral-species composition from their high-stand counterparts. Pandolfi (2011), in referencing Tager et al. (2010), argued that habitat tracking does not provide a reasonable explanation for the pattern of similar assemblages recurring in successive high-stand reefs because of differences in coral composition between low-stand and high-stand assemblages. Unfortunately, Tager et al. (2010) did not compare reefs from similar physiographic or oceanographic settings. They noted:

> ... it is likely that the difference in the species composition and temporal dynamics of coral community structure between the highstand and lowstand reefs is simply a reflection of environmental differences between the Huon Gulf and Huon Peninsula, rather than ecological differences manifested in the different meta-population structure between the reef types.

Because environmental conditions were so different between the locations, Tager's conclusions are exactly what one would have predicted *a priori* using Pandolfi's (1996) model for constancy in Pleistocene reef-coral assemblages, discussed earlier.

What about back-reef and lagoonal habitats that are stranded, exposed, and isolated during sea-level fall? Corals recruit to bathymetric highs within lagoonal habitats following sea-level rise; reef growth is aggradational (Macintyre et al. 1977; Aronson and Precht 1997), often exaggerating structural underpinnings in the subsurface (Halley et al. 1977). Lagoonal-reef complexes generally have low species richness and diversity compared with their fore-reef counterparts and are composed of species suited to their local conditions. For instance, in Belize lagoonal-reef communities differ markedly in coral composition and zonation depending on their positions on the shelf, their locations relative to wave and current energy, their depths, and their ages (Precht 1993b; Burke 1994). Even though the same general species pool is available for recruitment, the zonation patterns expressed on these reefs clearly represent environmental controls on their distribution (Aronson et al. 1998).

7.10 The Shattering of Ecological and Evolutionary Stability?

Bennett (1990) postulated that Milankovitch-scale events drive the reorganization of communities on time scales of 20–100 ky. High rates of coral extinction resulting from glacial–interglacial cycles would constitute evidence that the stability of Quaternary reefs was compromised during rapid and extreme events. Pandolfi (1999) speculated that an overall reduction in habitable reef area over large spatial scales at the LGM were responsible for the rapid extinction of two widespread coral species in the Caribbean. Holland (2012) has shown, however, that the habitable shallow-water area calculated as a function of sea level reveals a complicated relationship, in which sea-level rise does not consistently generate an increase in shelf area, nor does sea-level fall consistently reduce shelf area. It is more likely that habitat preferences of the species in question, *Pocillopora* cf. *palmata* and the organ-pipe morph of *Orbicella annularis*, were responsible for their sequential losses subsequent to their abundance during the last interglacial maximum. Where observed in outcrop, both of these species were abundant in rear-zone, back-reef, and lagoonal settings, making them particularly vulnerable to sea-level fall and consequent stranding (*sensu* Paulay 1990).

7.11 Summary

The metaphor "to move at a glacial pace" means to move extremely slowly (Free Dictionary 2014). In contrast, some reef paleoecologists interpret the rate and scale of environmental changes associated with actual Pleistocene glacial cycles as punctuated, rapid, and extreme. Specifically, they view the changes in sea level and SSTs as major disturbance events causing community dissolution and requiring subsequent reorganization and reassembly. In the tropics, the putative results are fossil reefs preserved during successive interglacial high-stands. The coral assemblages reveal striking, recurrent similarity in both species composition and dominance. To address questions of dynamics and process, we took the alternative approach of looking to the record of reefs preserved through entire glacial cycles, not just at the apex of the intervening interglacial periods.

The Quaternary fossil record, preserved as biofacies within reef parasequences, suggests that the habitat tolerances and preferences of most coral taxa have remained relatively constant. Consequently, gradients of species distribution have persisted through repeated glacial–interglacial (regressive–transgressive) cycles. Community reassembly was unnecessary, because even the most rapid changes in sea level through the Pleistocene were inconsequential to growth and recruitment on the time scale of the individual coral colony. We are unable to reject the null model for persistence-stability on Quaternary reefs. At scales from decades to millennia, the persistence of coral species and the assemblages they comprise has been met through the capacity of those corals and their resulting sedimentary facies incrementally to track favorable environments that have shifted spatially over time. Thus, from the standpoint of corals and the reefs they build, the metaphor of moving at a glacial pace is both appropriate and scientifically accurate.

Acknowledgements We thank Hemmo Bosscher, Paul (Mitch) Harris, Dennis K. Hubbard, Joanie Kleypas, Ian Macintyre, Steven Miller, David Thomas, Marguerite Toscano, and Lauren Toth for advice and discussion. Many of the ideas developed in this chapter stemmed from conversations between the senior author and the late Cesare Emiliani. Special thanks to Dennis K. Hubbard for all his help and for providing a suitable bio-geological venue for this manuscript. We also thank two anonymous reviewers for their comments and suggestions. This paper is dedicated to our friend, colleague, and mentor Ian Macintyre of the Smithsonian Institution. Ian has led a generation of reef scientists with his extraordinary field skills, exemplary scholarship, unflagging dedication, uncompromising integrity, and kind disposition. Our research on coral reefs has been supported over the years by the Smithsonian Institution, the National Geographic Society, and Northeastern University's Three Seas Program, and the U.S. National Science Foundation (most recently grant OCE-1535007 to R.B.A.). This is contribution number 84 from the Institute for Research on Global Climate Change at the Florida Institute of Technology.

References

Adey WH (1978) Coral reef morphogenesis: a multidimensional model. Science 202:831–837

Albritton CC (1975) Philosophy of Geohistory: 1785–1970. Stroudsburgh, PA: Dowden, Hutchinson & Ross 386 pp.

Anderson DM, Prell WL, Barratt NJ (1989). Estimates of sea surface temperature in the Coral Sea at the last glacial maximum. Paleoceanography 4:615–627

Annan JD, Hargreaves JC (2012) A new global reconstruction of temperature changes at the Last Glacial Maximum. Climate of the Past Discussions 8:5029–5051

Aronson RB (1994) Scale-independent biological processes in the marine environment. Oceanogr Mar Biol Ann Rev 32:435–460

Aronson RB, Precht WF (1997) Stasis, biological disturbance, and community structure of a Holocene coral reef. Paleobiology 23:326–346

Aronson RB, Precht WF (2001) Evolutionary paleoecology of Caribbean coral reefs. In: Allmon WD, Bottjer DJ (eds) Evolutionary Paleoecology: The Ecological Context of Macroevolutionary Change. New York: Columbia University Press, pp 171–233

Aronson RB, Precht WF, Macintyre IG (1998) Extrinsic control of species replacement on a Holocene reef in Belize: the role of coral disease. Coral Reefs 17:223–230

Ballantyne AP, Lavine M, Crowley TJ, Liu J, Baker PB (2005) Meta-analysis of tropical surface temperatures during the Last Glacial Maximum, Geophys Res Lett 32:L05712

Banks KW, Riegl BM, Shinn EA, Piller WE, Dodge RE (2007) Geomorphology of the southeast Florida continental reef tract (Miami-Dade, Broward, and Palm Beach counties, USA). Coral Reefs 26:617–633

Bard E, Hamelin B, Fairbanks RG, Zindler A (1990) Calibration of the C^{14} timescale over the past 30,000 years using mass spectrometric U-Th ages from Barbados corals. Nature 345:405–410

Barker S, Cacho I, Benway H, Tachikawa K (2005) Planktonic foraminiferal Mg/Ca as a proxy for past oceanic temperatures: a methodological overview and data compilation for the Last Glacial Maximum. Quat Sci Rev 24:821–834

Bartlein PJ, Prentice IC (1989) Orbital variations, climate and paleoecology. Trends Ecol Evol 4:195–199

Beck JW, Récy J, Taylor F, Edwards RL, Cabioch G (1997) Abrupt changes in early Holocene tropical sea surface temperature derived from coral records. Nature 385:705–707

Bender ML, Fairbanks RG, Taylor FW, Matthews RK, Goddard JG, Broecker WS (1979) Uranium-series dating of the Pleistocene reef tracts of Barbados, West Indies. Geol Soc Am Bull 90:577–594

Bennett KD (1990) Milankovitch cycles and their effects on species in ecological and evolutionary time. Paleobiology 16:11–21

Bennington JB, Bambach RK (1996) Statistical testing for paleocommunity recurrence: Are similar fossil assemblages ever the same? Palaeogeog Palaeoclim Palaeoecol 127:107–133

Benzie JA (1999) Genetic structure of coral reef organisms: ghosts of dispersal past. Am Zool 39:131–145

Berger A (1978) Long-term variations of daily insolation and Quaternary climatic changes. J Atmos Sci 35:2362–2367

Berger A, Imbrie J, Hays J, Kukla G, Saltzman B (1984) Milankovitch and climate: understanding the response to astronomical forcing. In: Milankovitch and Climate: Understanding the Response to Astronomical Forcing (Vol. 1). Proceedings of the NATO Advanced Research Workshop held 30 November–4 December, 1982 in Palisades, NY. Dordrecht: D. Reidel Publishing, pp 191–222

Blanchon, P (2005) Comments on "Corrected western Atlantic sea-level curve for the last 11,000 years based on calibrated 14 C dates from *Acropora palmata* framework and intertidal mangrove peat" by Toscano and Macintyre [Coral Reefs (2003) 22:257–270]. Coral Reefs 24:183–186

Blanchon P, Eisenhauer A (2000) Multi-stage reef development on Barbados during the Last Interglaciation. Quat Sci Rev 20:1093–1112

Blanchon P, Eisenhauer A, Fietzke J, Liebetrau V (2009) Rapid sea-level rise and reef back-stepping at the close of the last interglacial highstand. Nature 458:881–884

Blanchon P, Shaw J (1995) Reef drowning during the last deglaciation: Evidence for catastrophic sea-level rise and ice-sheet collapse. Geology 23:4–8

Boss SK, Liddell WD (1987) Back-reef and fore-reef analogs in the Pleistocene of North Jamaica: implications for facies recognition and sediment flux in fossil reefs. Palaios 2:219–228

Bosscher H, Meesters EH (1992) Depth related changes in the growth rate of *Montastrea annularis*. Proc 7th Intl Coral Reef Symp 1:507–512

Bosscher H, Schlager W (1993) Accumulation rates of carbonate platforms. J Geol 101:345–355

Brett CE (1998) Sequence stratigraphy, paleoecology, and evolution; biotic clues and responses to sea-level fluctuations. Palaios 13:241–262

Brett CE, Hendy AJ, Bartholomew AJ, Bonelli JR, McLaughlin PI (2007) Response of shallow marine biotas to sea-level fluctuations: A review of faunal replacement and the process of habitat tracking. Palaios 22:228–244

Broccoli AJ (2000) Tropical cooling at the Last Glacial Maximum: An atmosphere-mixed layer ocean model simulation. J Clim 13:951–976

Broecker W (1966) Absolute dating and the astronomical theory of glaciation. Science 151:299–304

Broecker WS, Thurber DL, Goddard J, Ku TL, Matthews RK, Mesolella KJ (1968) Milankovitch hypothesis supported by precise dating of coral reefs and deep-sea sediments. Science 159:297–300

Brown JH (1984) On the relationship between abundance and distribution of species. Am Nat 124:255–279

Budd AF, Johnson KG (1997) Coral reef community dynamics over 8 million years of evolutionary time: Stasis and turnover. Proc 8th Intl Coral Reef Symp 1:423–428

Budd AF, Johnson KG, Stemann TA (1994) Plio-Pleistocene extinctions and the origin of the modern Caribbean reef-coral fauna. In: Ginsburg RN (ed) Proc Colloq on Global Aspects of Coral Reefs: Health, Hazards and History, 1993. Miami: Rosenstiel School of Marine and Atmospheric Science, University of Miami, pp 7–13

Buddemeier R, Kinzie III RA (1998) Reef science: asking all the wrong questions in all the wrong places? Reef Encounter 23:29–34

Burke RB (1994) How have Holocene sea level rise and antecedent topography influenced Belize barrier reef development? In: Ginsburg RN (ed) Proc Colloq on Global Aspects of Coral Reefs: Health, Hazards and History, 1993. Rosenstiel School of Marine and Atmospheric Science, University of Miami, pp 14–20

Cabioch G, Banks-Cutler KA, Beck WJ, Burr GS, Corrège T, Edwards, R, Taylor FW (2003) Continuous reef growth during the last 23 calkyr BP in a tectonically active zone (Vanuatu, Southwest Pacific). Quat Sci Rev 22:1771–1786

Cabioch G, Camoin GF, Montaggioni LF (1999a) Postglacial growth history of a French Polynesian barrier reef tract, Tahiti, central Pacific. Sedimentology 46:985–1000

Cabioch LF, Montaggioni LF, Faure G, Ribaud-Laurenti A (1999b) Reef coralgal assemblages as recorders of paleobathymetry and sea level changes in the Indo-Pacific province. Quat Sci Rev 18:1681–1695

Cabioch, G, Montaggioni L, Frank N, Seard C, Sallé E, Payri C, Paterne M (2008b) Successive reef depositional events along the Marquesas foreslopes (French Polynesia) since 26 ka. Mar Geol 254:18–34

Cabioch G, Montaggioni L, Thouveny N, Frank N, Sato T, Chazottes V, Sémah AM (2008a) The chronology and structure of the western New Caledonian barrier reef tracts. Palaeogeog Palaeoclim Palaeoecol 268:91–105

Camoin GF, Colonna M, Montaggioni LF, Casanova J, Faure G, Thomassin BA (1997) Holocene sea level changes and reef development in the southwestern Indian Ocean. Coral Reefs 16:247–259

Camoin GF, Iryu Y, McInroy DB (2007) IODP Expedition 310 reconstructs sea level, climatic, and environmental changes in the South Pacific during the last deglaciation. Scientific Drilling 5:4–12

Carbotte SM, Bell RE, Ryan WBF, McHugh C, Slagle A, Nitsche F, Rubenstone J (2004) Environmental change and oyster colonization within the Hudson River estuary linked to Holocene climate. Geo-Mar Lett 24:212–224

Catuneanu O, Abreu V, Bhattacharya JP, Blum MD, Dalrymple RW, Eriksson PG, Fielding CR, Fisher WL, Galloway WE, Gibling MR, Giles KA, Holbrook JM, Jordan R, Kendall CGStC, Macurda B, Martinsen OJ, Miall AD, Neal JE, Nummedal D, Pomar L, Posamentier HW, Pratt BR, Sarg JF, Shanley KW, Steel RJ, Strasser A, Tucker ME, Winker C (2009) Towards the standardization of sequence stratigraphy. Earth-Sci Rev 92:1–33

Catuneanu O, Bhattacharya JP, Blum MD, Dalrymple RW, Eriksson PG, Fielding CR, Fisher WL, Galloway P, Gianolla WE, Gibling MR, Giles, Holbrook JM, Jordan R, Kendall CGStC, Macurda B, Martinsen OJ, Miall AD, Neal JE, Nummedal D, Posamentier HW, Pratt BR, Shanley KW, Steel RJ, Strasser A, Tucker ME (2010) Sequence stratigraphy: common ground after three decades of development. First Break – Stratigraphy 28:21–38

Chapell J (1983) A revised sea-level record for the last 300,000 years from Papua New Guinea. Search 14:99–101

Chappell J (2002) Sea level changes forced ice breakouts in the Last Glacial cycle: new results from coral terraces. Quat Sci Rev 21:1229–1240

Chappell J, Polach H (1991) Post-glacial sea-level rise from a coral record at Huon Peninsula, Papua New Guinea. Nature 349:147–149

Chappell J, Shackelton N (1986). Oxygen isotopes and sea level. Nature 324:137-140

Clarke AH, Stanley DJ, Medcof JC, Drinnan RE (1967) Ancient oyster and bay scallop shells from Sable Island. Nature 215:1146–1148

CLIMAP (1976) The surface of the Ice-Age Earth. Science 191:1131–1137

CLIMAP (1981) Seasonal reconstructions of the Earth's surface at the last glacial maximum. Geol Soc Am Tech Rep MC-36

COHMAP (1988) Climatic changes of the last 18,000 years: observations and model simulations. Science 241:1043–1052

Coles SL, Fadlallah YH (1991) Reef coral survival and mortality at low temperatures in the Arabian Gulf: new species-specific lower temperature limits. Coral Reefs 9:231–237

Colinvaux PA, De Oliveira PE, Moreno JE, Miller MC, Bush MB (1996) A long pollen record from lowland Amazonia forest and cooling in glacial times. Science 274:85–88

Colonna M, Casanova J, Dullo WC, Camoin G (1996) Sea-Level Changes and $\delta\ ^{18}O$ Record for the Past 34,000 yr from Mayotte Reef, Indian Ocean. Quat Res 46:335–339

Copper P (1988) Ecological succession in Phanerozoic reef ecosystems: Is it real? Palaios 3:136–151

Crowley TJ (2000) CLIMAP SSTs re-revisited. Clim Dynam 16:241–255

Cutler KB, Edwards RL, Taylor FW, Cheng H, Adkins J, Gallup CD, Bloom AL (2003) Rapid sea-level fall and deep-ocean temperature change since the last interglacial period. Earth Plan Sci Lett 206:253–271

Dahlgren TG, Weinberg JR, Halanych KM (2000) Phylogeography of the ocean quahog (*Arctica islandica*): influences of paleoclimate on genetic diversity and species range. Mar Biol 137:487–495

Dana JD (1843) On the temperature limiting the distribution of corals. Am J Sci 45:130–131

Darwin C (1842) The structure and distribution of coral reefs: being the first part of the geology of the voyage of the Beagle, under the command of Capt. Fitzroy, RN during the years 1832 to 1836. London: Smith, Elder and Company 213 pp.

de Garidel-Thoron T, Rosenthal Y, Bassinot F, Beaufort L (2005) Stable sea surface temperatures in the western Pacific warm pool over the past 1.75 million years. Nature 433:294–298

deMenocal P, Ortiz J, Guilderson T, Adkins J, Sarnthein M, Baker L, Yarusinsky M (2000) Abrupt onset and termination of the African Humid Period: rapid climate responses to gradual insolation forcing. Quat Sci Rev 19:347–361

Dravis JJ (1996) Rapidity of freshwater calcite cementation—implications for carbonate diagenesis and sequence stratigraphy. Sed Geol 107:1–10

Dullo C (2005) Coral growth and reef growth: a brief review. Coral Reefs 51: 33–48

Dyke AS, Dale JE, McNelly RN (1996) Marine molluscs as indicators of environmental change in glaciated North America and Greenland during the last 18,000 years. Géographie Physique et Quaternaire 50:125–184

Dynesius M, Jansson R (2000) Evolutionary consequences of changes in species' geographical distributions driven by Milankovitch climate oscillations. Proc Natl Acad Sci USA 97:9115–9120

Eberli GP (2013) The uncertainties involved in extracting amplitude and frequency of orbitally driven sea-level fluctuations from shallow-water carbonate cycles. Sedimentology 60:64–84

Emiliani C (1966) Paleotemperature analysis of Caribbean cores P6304-8 and P6304-9 and a generalized temperature curve for the past 425,000 years. J Geol 74:109–124

Emiliani C (1972) Quaternary paleotemperatures and the duration of the high-temperature intervals. Science 178:398–401

Emiliani C, Ericson DB (1991) The glacial/interglacial temperature range of the surface water of the oceans at low latitudes. In: Taylor HP, O'Neil JR, Kaplan IR (eds) Stable Isotope Geochemistry: A Tribute to Samuel Epstein. Geochem Soc Spec Publ 3:223–228

Enquist BJ, Jordan MA, Brown JH (1995) Connections between ecology, biogeography, and paleobiology: relationship between local abundance and geographic distribution in fossil and recent molluscs. Evol Ecol 9:586–604

Fairbanks RG (1989) A 17, 000-year glacio-eustatic sea level record: influence of glacial melting rates on the Younger Dryas event and deep-ocean circulation. Nature 342:637–642

Fairbanks RG (1990) The age and origin of the "Younger Dryas climate event" in Greenland ice cores. Paleoceanography 5:937–948

Farrera I, Harrison SP, Prentice IC, Ramstein G, Guiot J, Bartlein PJ, Yu G (1999) Tropical climates at the Last Glacial Maximum: a new synthesis of terrestrial palaeoclimate data. I. Vegetation, lake-levels and geochemistry. Clim Dynam 15:823–856

Felis T, Braddock HVM, Linsley K, Tudhope AW, Gagan MK, Suzuki A, Inoue M, Thomas AL, Esat TM, Thompson WG, Tiwari M, Potts DC, Mudelsee M, Yokoyama Y, Webster JM (2014) Steeper meridional temperature gradient in the Great Barrier Reef during the last deglaciation. Nature Comm 5:4102 DOI:10.1038/ncomms5102.

Ferro CE, Droxler AW, Anderson JB, Mucciarone D (1999) Late Quaternary shift of mixed siliciclastic-carbonate environments induced by glacial eustatic sea-level fluctuations in Belize. Advances in Carbonate Sequence Stratigraphy Application to Reservoirs Outcrops and Models Tulsa, OK: SEPM Spec Pub No 63, pp 385–411

Fielding CR, Trueman JD, Dickens GR, Page M (2003) Anatomy of the buried Burdekin River channel across the Great Barrier Reef shelf: how does a major river operate on a tropical mixed siliciclastic/carbonate margin during sea level lowstand? Sed Geol 157:291–301

Free Dictionary (2014) Definition: Glacial. http://www.thefreedictionary.com/glacial

Gagan MK, Ayliffe LK, Beck JW, Cole JE, Druffel ERM, Dunbar RB, Schrag DP (2000) New views of tropical paleoclimates from corals. Quat Sci Rev 19:45–64

Gagan MK, Hendy EJ, Haberle SG, Hantoro WS (2004) Post-glacial evolution of the Indo-Pacific warm pool and El Nino-Southern Oscillation. Quat Intl 118:127–143

Gardiner L (2001) Stability of Late Pleistocene reef mollusks from San Salvador Island, Bahamas. Palaios 16:372–386

Geister J (1977) The influence of wave exposure on the ecological zonation of Caribbean coral reefs. Proc 3rd Intl Coral Reef Symp 1:23–29

Geister J (1980) Calm-water reefs and rough-water reefs of the Caribbean Pleistocene. Acta Palaeontologia Polonica 25:541–556

Geister J (1983) Holozane westindische korallenriffe: Geomorphologie, okologie und fazies. Facies 9:173–284.

Getty SR, Asmerom Y, Quinn TM, Budd AF (2001) Accelerated Pleistocene coral extinctions in the Caribbean Basin shown by uranium-lead (U-Pb) dating. Geology 29:639–642

Gill I, Hubbard D, Dickson JAD (1999) Corals, reefs, and six millennia of Holocene climate. 11th Bathurst Meeting, Cambridge, J Conf Abs 4:921

Ginsburg RN (1974) Introduction to comparative sedimentology of carbonates. AAPG Bull 58:781–786

Ginsburg RN, Gischler E, Schlager W (1994) Johannes Walther on reefs. Pioneering concepts of biogeology 1885-1910. Geological Milestones Series 2. Miami: Rosenstiel School of Marine and Atmospheric Science, University of Miami.

Goreau TF (1959) The ecology of Jamaican coral reefs I. Species composition and zonation. Ecology 40:67–90

Goreau TF (1969) Post Pleistocene urban renewal in coral reefs. Micronesica 5:323–326

Gould SJ (1980) The promise of paleobiology as a nomothetic, evolutionary discipline. Paleobiology 6:96–118

Gould SJ (1981) Palaeontology plus ecology as palaeobiology. In: May RM (ed) Theoretical ecology: principles and applications, 2nd ed. Sunderland, MA: Sinauer Associates, pp 295–317

Gould SJ (1985) The paradox of the first tier: an agenda for paleobiology. Paleobiology 11:2–12

Green DH, Edmunds PJ, Carpenter RC (2008) Increasing relative abundance of *Porites astreoides* on Caribbean reefs mediated by an overall decline in coral cover. Mar Ecol Prog Ser 359:1–10

Greenstein BJ, Pandolfi JM (2007) Escaping the heat: range shifts of reef coral taxa in coastal Western Australia. Global Change Biol 14:513–528

Greer L, Jackson JE, Curran HA, Guilderson T, Teneva L (2009) How vulnerable is *Acropora cervicornis* to environmental change? Lessons from the early to middle Holocene. Geology 37:263–266

Guilderson TP, Fairbanks RG, Rubenstone JL (1994) Tropical temperature variations since 20,000 years ago: modulating interhemispheric climate change. Science 263:663–665

Guilderson TP, Fairbanks RG, Rubenstone JL (2001) Tropical Atlantic coral oxygen isotopes: glacial–interglacial sea surface temperatures and climate change. Mar Geol 172:75–89

Halley RB, Shinn EA, Hudson JH, Lidz B (1977) Recent and relict topography of Boo Bee patch reef, Belize. Proc 3rd Intl Coral Reef Symp, Miami 2:29–35

Harris PT, Davies PJ (1989) Submerged reefs and terraces on the shelf edge of the Great Barrier Reef, Australia: Morphology, occurrence and implications for reef evolution. Coral Reefs 8:97–98

Haug GH, Hughen KA, Sigman DM, Peterson LC, Röhl U (2001) Southward migration of the Intertropical Convergence Zone through the Holocene. Science 293:1304–1308

Hays JD, Imbrie J, Shackleton NJ (1976) Variations in the Earth's orbit: pacemaker of the ice ages. Science 194:1121–1132

Hearty PJ, Hollin JT, Neumann AC, O'Leary MJ, McCulloch M (2007) Global sea-level fluctuations during the Last Interglaciation (MIS 5e). Quat Sci Rev 26:2090–2112

Herbert TD, Peterson LC, Lawrence KT, Liu Z (2010) Tropical ocean temperatures over the past 3.5 million years. Science 328:1530–1534

Herbert TD, Schuffert JD (2000) Alkenone unsaturation estimates of sea-surface temperatures at site 1002 over a full glacial cycle. Proc Ocean Drill Prog, Sci Results 165:239–247

Hinestrosa G, Webster JM, Beaman RJ, Anderson LM (2014) Seismic stratigraphy and development of the shelf-edge reefs of the Great Barrier Reef, Australia. Mar Geol 353:1–20

Hippler D, Eisenhauer A, Nägler TF (2006) Tropical Atlantic SST history inferred from Ca isotope thermometry over the last 140 ka. Geochimica et Cosmochimica Acta 70:90–100

Hjort C, Mangerud J, Adrielsson L, Bondevik S, Landvik JY, Salvigsen O (1995) Radiocarbon dated common mussels *Mytilus edulis* from eastern Svalbard and the Holocene marine climatic optimum. Polar Res 14:239–243

Hoffman A (1979) Community paleoecology as an epiphenomenal science. Paleobiology 5:357–379

Hoffman A, Narkiewicz M (1977) Developmental pattern of Lower to Middle Paleozoic banks and reefs. Neues Jahrb Geol Palaont Monatsh 1977:272–283

Holland SM (2012) Sea level change and the area of shallow-marine habitat: implications for marine biodiversity. Paleobiology 38:205–217

Hongo C (2012) Holocene key coral species in the Northwest Pacific: indicators of reef formation and reef ecosystem responses to global climate change and anthropogenic stresses in the near future. Quat Sci Rev 35:82–99

Hongo C, Kayanne H (2010) Holocene sea-level record from corals: Reliability of paleodepth indicators at Ishigaki Island, Ryukyu Islands, Japan. Palaeogeog Palaeoclim Palaeoecol 287:143–151

Hopley D (1994) Continental shelf reef systems. In: Carter RWG, Woodroffe CD (eds) Coastal Evolution, Late Quaternary Shoreline Morphodynamics. New York: Cambridge Univ. Press, pp 303–340

Hubbard DK (2008) A new Caribbean reef model: a view from ye shoulders of giants. Proc 11th Intl Coral Reef Symp, Ft. Lauderdale 1:2–6

Hubbard DK (2009) Depth Related and species related patterns of Holocene reef accretion in the Caribbean and western Atlantic: a critical assessment of existing models. In: Swart PK, Eberli GP, McKenzie JA (eds) Perspectives in carbonate geology: a tribute to the career of Robert Nathan Ginsburg. IAS Spec Pub 41, pp 1–9

Hubbard DK (2011) Eastern Caribbean coral reefs. In: Hopley D. (ed.) Encyclopedia of modern coral reefs: structure, form and process. Springer, the Netherlands, 338–348

Hubbard DK (2014) Holocene accretion rates and styles for Caribbean coral reefs: lessons for the past and future. In: Verwer K, Playton TE, Harris PM (eds) Deposits, architecture, and controls of carbonate margin, slope, and basinal settings. Tulsa, OK: SEPM Spec Pub 105, pp 264–281

Hubbard DK, Gill IP, Burke RB, Morelock J (1997) Holocene reef backstepping—southwestern Puerto Rico shelf. Proc 8th Intl Coral Reef Symp 2:1779–1784

Hubbard DK, Gill, IP, Toscano MA (2004) The Holocene *Acropora palmata* record and recent coral-reef decline. In: Lewis RD, Panuska BC (eds) Proceedings of the 11th Symposium on the Geology of the Bahamas and Other Carbonate Regions, San Salvador, Bahamas: Gerace Research Center, pp 1–11

Humphrey JD, Kimbell TN (1990) Sedimentology and sequence stratigraphy of upper Pleistocene carbonates of southeastern Barbados, West Indies. AAPG Bull 74:1671–1684

Hunter IG, Jones B (1996) Coral associations of the Pleistocene Ironshore Formation, Grand Cayman. Coral Reefs 15:249–267

Huston MA (1985) Patterns of species diversity on coral reefs. Ann Rev Ecol Sys16:149–177

Imbrie J, Hays JD, Martinson DG, McIntyre A, Mix AC, Morley JJ, Pisias NG, Prell WL, Shackleton NJ (1984) The orbital theory of Pleistocene climate: Support from a revised chronology of the marine ^{18}O record, In: Berger AL (ed) Milankovitch and climate, Part 1, Norwell, MA: D. Riedel, pp 269–305

Imbrie J, Imbrie KP (1986) Ice ages: solving the mystery. Cambridge MA: Harvard University Press.

Ivany LC (1996) Coordinated stasis or coordinated turnover? Exploring intrinsic vs. extrinsic controls on pattern. Palaeogeog Palaeoclim Palaeoecol 127:239–256

Jaap WC (1984) The ecology of the south Florida coral reefs: a community profile. USFWS OBS-82/08 and MMS 84-0038

Jackson JBC (1992) Pleistocene perspectives on coral reef community structure. Am Zool 32:719–731

Jackson JBC (1994a) Community unity? Science 264:1412–1413

Jackson JBC (1994b) Constancy and change of life in the sea [and discussion]. Proc R Soc B 344:55–60

Jackson JBC, Budd AF (1996) Evolution and environment: Introduction and overview. In: Jackson JBC, Budd AF, Coates AG (eds) Evolution and Environment in Tropical America. Chicago: University of Chicago Press, pp. 1–20

Jackson JBC, Budd AF, Pandolfi JM (1996) The shifting balance of natural communities. Evolutionary Paleobiology. Chicago: University of Chicago Press, pp 89–122

Jackson JBC, Erwin DH (2006) What can we learn about ecology and evolution from the fossil record? Trends Ecol Evol 21:322–328

Jackson JBC, Johnson KG (2000) Life in the last few million years. Paleobiology 26:221–234

James NP (1983) Reefs. In: Scholle PA, Bebout DG, Moore CH (eds) Carbonate depositional environments. Tulsa, OK: AAPG Memoir 33, pp 345–440

James NP, Macintyre IG (1985) Carbonate depositional environments: modern and ancient. Part I: Reefs, zonation, depositional facies, and diagenesis. Colo Sch Mines Quart 80:1–70

Jansen E, Andersson C, Moros M, Nisancioglu KH, Nyland BF, Telford RJ (2009) The early to mid-Holocene Thermal Optimum in the north Atlantic In: Battarbeem RW, Binney HA (eds) Natural climate variability and global warming: a Holocene perspective. Oxford, UK: Wiley Blackwell, pp 123–137

Jessen CA, Rundgren M, Björck S, Hammarlund D (2005) Abrupt climatic changes and an unstable transition into a late Holocene Thermal Decline: a multiproxy lacustrine record from southern Sweden. J Quat Sci 20:349–362

Johannes RE, Wiebe WJ, Crossland CJ, Rimmer DW, Smith SV (1983) Latitudinal limits of coral reef growth. Mar Ecol Prog Ser 11:105–111

Johnson KG, Jackson J B, Budd AF (2008) Caribbean reef development was independent of coral diversity over 28 million years. Science 319:1521–1523

Karlson RH (1999) Dynamics of coral communities. Population and community biology series 23, Dordrecht, The Netherlands: Kluwer Academic Publishers, 260 p

Kendall CGStC, Schlager W (1981) Carbonates and relative changes in sea level. Mar Geol 44:181–212

Kerans C, Tinker SW (1997) Sequence stratigraphy and characterization of carbonate reservoirs. Tulsa, OK: SEPM Short Course Notes No. 40, 130 p

Kerwin MW, Overpeck JT, Webb RS, DeVernal A, Rind DH, Healy RJ (1999) The role of oceanic forcing in mid-Holocene Northern Hemisphere climatic change. Paleoceanography 14:200–210

Kissling DL (1977) Population structure characteristics for some Paleozoic and modern colonial corals. 2nd Intl Symp Corals Fossil Coral Reefs, Paris, September 1975, Paris: Memoires Du B.R.G.M. No 89, pp 497–506

Klaus JS, McNeill DF, Budd AF, Coates AG (2012) Neogene reef coral assemblages of the Bocas del Toro region, Panama: the rise of *Acropora palmata*. Coral Reefs 31:91–103

Kleypas JA (1997) Modeled estimates of global reef habitat and carbonate production since the last glacial maximum. Paleoceanography 12:533–545

Kleypas JA, McManus J, Menez LB (1999) Environmental limits to coral reef development: Where do we draw the line? Am Zool 39:146–159

Knowlton N, Jackson JBC (2001) The ecology of coral reefs. In: Bertness MD, Gaines S, Hay ME (eds) Marine community ecology. Sunderland, MA: Sinauer, pp 395–422

Lambeck K, Esat TM, Potter EK (2002) Links between climate and sea levels for the past three million years. Nature 419:199–206

Land LS (1974) Growth rate of a West Indian (Jamaican) reef. Proc 2nd Intl Coral Reef Symp 2:409–412

Lawrence KT, Herbert TD (2005) Late Quaternary sea-surface temperatures in the western Coral Sea: Implications for the growth of the Australian Great Barrier Reef. Geology 33:677–680

Lea DW, Pak DK, Spero HJ (2000) Climate impact of late Quaternary equatorial Pacific sea surface temperature variations. Science 289:1719–1724

Liddell WD, Ohlhorst SL, Coates AG (1984) Modern and Ancient Carbonate Environments of Jamaica: Sedimenta X. Miami: Rosenstiel School of Marine and Atmospheric Science, University of Miami, 101 pp.

Lidz BH, Shinn EA, Hansen ME, Halley RB, Harris MW, Locker SD, Hine AC (1997) Sedimentary and biological environments, depth to Pleistocene bedrock, and Holocene sediment and reef thickness, Key Largo, south Florida. USGS Misc Invest Ser Map I-2505

Lighty RG (1977) Relict shelf-edge Holocene coral reef: southeast coast of Florida. Proc 3rd Intl Coral Reef Symp 2:215–221

Lighty RG, Macintyre IG, Stuckenrath R (1978) Submerged early Holocene barrier reef south-east Florida shelf. Nature 276:59–60

Lighty RG, Macintyre IG, Stuckenrath R (1982) *Acropora palmata* reef framework: a reliable indicator of sea level in the western Atlantic for the past 10,000 years. Coral Reefs 1:125–130

Lin H.-L, Peterson LC, Overpeck JT, Trumbore SE, Murray DW (1997) Late Quaternary climate change from δ18O records of multiple species of planktonic foraminifera: high-resolution records from the anoxic Cariaco Basin, Venezuela. Paleoceanography12:415–427

Lirman D, Schopmeyer S, Manzello D, Gramer LJ, Precht WF, Muller-Karger F, Banks K, Barnes B, Bartels E, Bourque A, Byrne J, Duquesnel J, Fisher L, Gilliam D, Hendee J, Johnson M, Maxwell K, McDevitt E, Monty J, Rueda D, Ruzicka R, Thanner SL (2011) Severe 2010 cold-water event caused unprecedented mortality to corals of the Florida reef tract and reversed previous survivorship patterns. PLoS ONE 6:e23047

Lisiecki LE, Raymo ME (2005) A Pliocene Pleistocene stack of 57 globally distributed benthic $\delta^{18}O$ records. Paleoceanography 20:PA1003

Liu Z, Herbert TD (2004) High-latitude influence on the eastern equatorial Pacific climate in the early Pleistocene epoch. Nature 427:720–723

Macintyre IG (1972) Submerged reefs of the eastern Caribbean. AAPG Bull 56:720–738

Macintyre IG (1975) A diver-operated hydraulic drill for coring submerged substrates. Atoll Res Bull 185:21–26

Macintyre IG (1988) Modern coral reefs of western Atlantic: new geological perspective. AAPG Bull 72:1360–1369

Macintyre IG (2007) Demise, regeneration, and survival of some western Atlantic reefs during the Holocene transgression. In: Aronson RB (ed) Geological approaches to coral reef ecology. New York: Springer-Verlag, pp 181–200

Macintyre IG, Burke RB, Stuckenrath R (1977) Thickest recorded Holocene reef section, Isla Perez core hole, Alacran Reef, Mexico. Geology 5:749–754

Macintyre IG, Rutzler K, Norris JN, Smith KP, Cairns SD, Bucher KE, Stenick RS (1991) An early Holocene reef in the western Atlantic: Submersible investigations of a deep relict reef off the west coast of Barbados, W. I. Coral Reefs 10:167–174

Marsh LM (1993) The occurrence and growth of *Acropora* in extra-tropical waters off Perth, Western Australia. Proc 7th Intl Coral Reef Symp, Guam 2:1233–1238

Masse J, Montaggioni L (2001) Growth history of shallow-water carbonates: control of accommodation on ecological and depositional processes. Intl J Earth Sci (Geol Rdsch) 90:452–469

Mayer AG (1914) The effects of temperature upon tropical marine animals. Papers Tortugas Laboratory, Carnegie Inst Washington 6:1–14

Mayer AG (1915) The lower temperature at which reef corals lose their ability to capture food. Yearb Carnegie Inst Washington Publ 183:1–24

McCulloch MT, Tudhope AW, Esat TM, Mortimer GE, Chappell J, Pillans B, Omura A (1999) Coral record of equatorial sea-surface temperatures during the penultimate deglaciation at Huon Peninsula. Science 283:202–204

Mesolella KJ (1967) Zonation of uplifted Pleistocene coral reefs on Barbados, West Indies. Science 156:638–640

Mesolella KJ, Matthews RK, Broecker WS, Thurber DL (1969) The astronomical theory of climatic change: Barbados data. J Geol 77:250–274

Mesolella KJ, Sealy HA, Matthews RK (1970) Facies geometries within Pleistocene reefs of Barbados, West Indies. AAPG Bull 54:1899–1917

Mewis H, Kiessling W (2013) Environmentally controlled succession in a late Pleistocene coral reef (Sinai, Egypt). Coral Reefs 32:49–58

Middleton GV (1973) Johannes Walther's law of the correlation of facies. Geol Soc Amer Bull 84:979–988

Milankovitch M (1941) Kanon der Erdbestrahlung und seine Andwendung auf das Eiszeiten-problem (Canon of insolation and the ice-age problem). R. Serbian Acad. 132, Belgrade. Belgrade Serbia

Mildenhall DC (2001) Middle Holocene mangroves in Hawke's Bay, New Zealand. New Zealand J Botany 39:517–521

Miller AI (1997) Coordinated stasis or coincident relative stability? Paleobiology 23:155–164

Miller III W (1986) Paleoecology of benthic community replacement. Lethaia 19:225–231

Miller III W (1996) Ecology of coordinated stasis. Palaeogeog Palaeoclim Palaeoecol 127:177–190

Mitchum RM Jr (1977) Seismic stratigraphy and global changes of sea level, part II: glossary of terms used in seismic stratigraphy. In: Payton CE (ed) Seismic stratigraphy — applications to hydrocarbon exploration. Tulsa, OK: AAPG Mem 26, pp 205–212

Mix AC, Lund DC, Pisias NG, Boden P, Bornmalm L, Lyle M, Pike J (1999) Rapid climate oscillations in the northeast Pacific during the last deglaciation reflect Northern and Southern Hemisphere sources. Geophys Mon, Am Geophys Union 112:127–148

Montaggioni LF (2005) History of Indo-Pacific coral reef systems since the last glaciation: development patterns and controlling factors. Earth-Sci Rev 71:1–75

Montaggioni LF, Braithwaite CJR (2009) Quaternary coral reef systems: history, development processes and controlling factors. Amsterdam: Elsevier, 532 pp

Montaggioni LF, Cabioch G, Camoin GF, Bard E, Ribaud-Laurenti A, Faure G, D'ejardin P, R'ecy J (1997) Continuous record of reef growth over the past 14 ka on the mid-Pacific island of Tahiti. Geology 25:555–558

Montaggioni LF, Faure G (1997) Response of reef coral communities to sea-level rise: a Holocene model from Mauritius (Western Indian Ocean). Sedimentology 44:1053–1070

Neumann AC, Macintyre IG (1985) Reef response to sea level rise: keep-up, catch-up or give-up. Proc 5th Intl Coral Reef Congr 3:105–110

Niebler HS, Arz HW, Donner B, Mulitza S, Pätzold J, Wefer G (2003) Sea surface temperatures in the equatorial and South Atlantic Ocean during the Last Glacial Maximum (23–19 ka). Paleoceanography 18:1069

O'Dea A, Jackson JB, Fortunato H, Smith JT, D'Croz L, Johnson KG, Todd JA (2007) Environmental change preceded Caribbean extinction by 2 million years. Proc Natl Acad Sci USA 104:5501–5506

Overpeck JT, Webb RS, Webb T (1992) Mapping eastern North American vegetation change of the past 18 ka: No-analogs and the future. Geology 20:1071–1074

Paillard D (2001) Glacial cycles: Toward a new paradigm. Rev Geophys 39:325–346

Pandolfi JM (1996) Limited membership in Pleistocene reef coral assemblages from the Huon Peninsula, Papua New Guinea: constancy during global change. Paleobiology 22:152–176

Pandolfi JM (1999) Response of Pleistocene coral reefs to environmental change over long temporal scales. Am Zool 39:113–130

Pandolfi, JM (2001). Numerical and taxonomic scale of analysis in paleoecological data sets: examples from neo-tropical Pleistocene reef coral communities. J Paleont 75: 546–563

Pandolfi JM (2002) Coral community dynamics at multiple scales. Coral Reefs 21:13–23

Pandolfi JM (2011) The paleoecology of coral reefs. In: Dubinsky Z, Stambler N (eds) Coral reefs: an ecosystem in transition. New York: Springer-Verlag, pp 13–24

Pandolfi JM, Greenstein BJ (2007) Using the past to understand the future: Paleoecology of coral reefs. In: Marshal P, Johnson J, Wachenfeld D (eds) Assessing climate change vulnerability of the Great Barrier Reef. Great Barrier Reef Marine Park Authority and the Australian Greenhouse Office, Townsville, Australia. pp 717–744

Pandolfi JM, Jackson JBC (1997) The maintenance of diversity on coral reefs: examples from the fossil record. Proc 8th Intl Coral Reef Symp 1:397–404

Pandolfi JM, Jackson JBC (2001) Community structure of Pleistocene coral reefs of Curaçao, Netherlands Antilles. Ecol Mon 71:49–67

Pandolfi JM, Jackson JBC (2006) Ecological persistence interrupted in Caribbean coral reefs. Ecol Lett 9:818–826

Pandolfi J, Jackson JBC (2007) Broad-scale patterns in Pleistocene coral reef communities from the Caribbean: Implications for ecology and management. In: Aronson RB (ed) Geological approaches to coral reef ecology. New York: Springer-Verlag, pp 201–236

Pandolfi JM, Llewellyn G, Jackson JBC (1999) Pleistocene reef environments, constituent grains, and coral community structure: Curacao, Netherlands Antilles. Coral Reefs 18:107–122
Paulay G (1990) Effects of late Cenozoic sea-level fluctuations on the bivalve faunas of tropical oceanic islands. Paleobiology 16:415–434
Peltier WR (2002) On eustatic sea level history: Last Glacial Maximum to Holocene. Quat Sci Rev 21:377–396
Peltier WR, Fairbanks RG (2006) Global glacial ice volume and Last Glacial Maximum duration from an extended Barbados sea level record. Quat Sci Rev 25:3322–3337
Pflaumann U, Sarnthein M, Chapman M, d'Abreu L, Funnell B, Huels M, Weinelt M (2003) Glacial North Atlantic: Sea-surface conditions reconstructed by GLAMAP 2000. Paleoceanography 18:1065
Pickett ST (1989). Space-for-time substitution as an alternative to long-term studies. In: Likens GE (ed) Long-term studies in ecology: approaches and alternatives. New York: Springer-Verlag, pp 110–135
Pillans B, Chappell J, Naish TR (1998) A review of the Milankovitch climatic beat: template for Plio–Pleistocene sea-level changes and sequence stratigraphy. Sed Geol 122:5–21
Pomar L (1991) Reef geometries, erosion surfaces and high-frequency sea-level changes, upper Miocene Reef Complex, Mallorca, Spain. Sedimentology 38:243–269
Pomar L, Ward WC (1994) Response of a late Miocene Mediterranean reef platform to high-frequency eustasy. Geology 22:131–134
Porter JW (1987) Species profiles: life histories and environmental requirements of coastal fishes and invertebrates (south Florida) - reef-building corals. USFWS Biol Rep 82, pp 1–23
Potter EK, Esat TM, Schellmann G, Radtke U, Lambeck K, McCulloch MT (2004) Suborbital-period sea-level oscillations during marine isotope substages 5a and 5c. Earth Planet Sci Lett 225:191–204
Powers LA, Johnson TC, Werne JP, Castañeda IS, Hopmans EC, Damsté JSS, Schouten S (2005) Large temperature variability in the southern African tropics since the Last Glacial Maximum. Geophys Res Lett 32:L08706
Precht WF (1993a) Stratigraphic evidence from reef studies for a double-high sea stand during the last interglacial maximum. AAPG Bull 77:1473
Precht WF (1993b) Holocene Coral Patch Reef Ecology and Sedimentary Architecture, Northern Belize, Central America: Comment. Palaios 8:499–503
Precht WF, Aronson RB (2004) Climate flickers and range shifts of reef corals. Front Ecol Env 6: 307–313
Precht WF, Aronson RB (2006) Death and resurrection of Caribbean coral reefs: a palaeoecological approach. In: Côté IM, Reynolds JD (eds) Coral reef conservation. Cambridge, UK: Cambridge University Press, pp 40–77
Precht WF, Deslarzes KJ, Hickerson EL, Schmahl GP, Nuttall MF, Aronson RB (2014) Back to the future: The history of acroporid corals at the Flower Garden Banks, Gulf of Mexico, USA. Mar Geol 349:152–161
Precht WF, Harris PM (2008) Using Quaternary reefs and platforms as a comparative tool. Search and Discovery Article #50082 http://www.searchanddiscovery.com/documents/2008/08028harris22/images/harris22.pdf
Precht WF, Hoyt WH (1991) Reef facies distribution patterns, Pleistocene (125 ka) Falmouth Formation, Rio Bueno, Jamaica, W. I. AAPG Bull 75:656–657
Precht WF, Miller SL (2007) Ecological shifts along the Florida Reef Tract: the past as the key to the future. In: Aronson RB (ed) Geological approaches to coral reef ecology. New York: Springer-Verlag, pp 237–312
Rind D, Peteet D (1985) Terrestrial conditions at the Last Glacial Maximum and CLIMAP sea-surface temperature estimates: Are they consistent? Quat Res 24:1–22
Rodríguez-Martínez RE, Jordán-Dahlgren E (1999) Epibiotic and free-living *Porites astreoides*. Coral Reefs 18:159–161
Rollins HB, Carothers M, Donahue J (1979) Transgression, regression and fossil community succession. Lethaia 12:89–104
Rosell-Melé A, Bard E, Emeis KC, Grieger B, Hewitt C, Müller PJ, Schneider RR (2004) Sea surface temperature anomalies in the oceans at the LGM estimated from the alkenone-U 37 K′ index: comparison with GCMs. Geophys Res Lett 31:L03208
Rosenthal Y, Lohmann GP (2002) Accurate estimation of sea surface temperatures using dissolution-corrected calibrations for Mg/Ca paleothermometry. Paleoceanography 17:16-116-6
Ruddiman WF, Mix AC (1993) The north and equatorial Atlantic at 9000 and 6000 yr BP. In: Wright HE (ed) Global climates since the Last Glacial Maximum. Minneapolis, MN: University of Minnesota Press, pp 94–124
Salvigsen O, Forman SL, Miller GH (1992) Thermophilous molluscs on Svalbard during the Holocene and their paleoclimatic implications. Polar Res 11:1–10
Schlager W (1981) The paradox of drowned reefs and carbonate platforms. Geol Soc Am Bull 92:197–211
Schlager W (1992) Sedimentology and sequence stratigraphy of reefs and carbonate platforms, Tulsa, OK: AAPG Continuing Education Course Note Series #34, 71 pp. .
Schlager W (2004) Fractal nature of stratigraphic sequences. Geology 32:185–188
Schlager W (2005) Carbonate sedimentology and sequence stratigraphy. Tulsa, OK: SEPM Concepts in Sedimentology and Paleontology No 8
Schlager W, Warrlich G (2009) Record of sea-level fall in tropical carbonates. Basin Res 21:209–224
Schlager W, Warrlich GM (2012) Falling-stage systems tract in tropical carbonate rocks. In: Swart PK, Eberli GP, McKenzie JA (eds) Perspectives in carbonate geology: a tribute to the career of Robert Nathan Ginsburg. IAS Spec Pub 41, pp 305–325
Schmidt MW, Vautravers MJ, Spero HJ (2006) Western Caribbean sea surface temperatures during the late Quaternary. Geochem Geophys Geosyst 17(2):1–17
Sherman CE, Glenn CR, Jones AT, Burnett WC, Schwarcz HP (1993) New evidence for two highstands of the sea during the last interglacial, oxygen isotope substage 5e. Geology 21:1079–1082
Shinn E (1963) Spur and groove formation on the Florida reef tract. J Sed Pet 33:291–303
Shinn EA. Hudson JH, Robbin DM, Lidz BH (1981) Spurs and grooves revisited: Construction versus erosion, Looe Key Reef, Florida. Proc 4th Intl Coral Reef Symp, Manila 1:475–483
Simpson GG (1970) Uniformitarianism. An inquiry into principle, theory and method in geohistory and biohistory. In: Hecht MK, Steere WC (eds) Essays in evolution and genetics in honor of Theodosius Dobzhansky. New York: Appleton, pp 43–96
Soja CM (1996) Island-arc carbonates: Characterization and recognition in the ancient geologic record. Earth-Sci Rev 41:31–65
Solow AR, Huppert A (2004) A potential bias in coral reconstruction of sea surface temperature. Geophys Res Lett 31:L06308
Sonzogni C, Bard E, Rostek F (1998) Tropical sea-surface temperatures during the last glacial period: a view based on alkenones in Indian Ocean sediments. Quat Sci Rev 17:1185–1201
Stathakopoulos A, Riegl BM (2015) Accretion history of mid-Holocene coral reefs from the Southeast Florida continental reef tract, USA. Coral Reefs 34:173–187
Szabo BJ (1979) Uranium-series age of coral reef growth on Rottnest Island, Western Australia. Mar Geol 29:M11–M15
Tachikawa K, Vidal L, Sonzogni C, Bard E (2009) Glacial/interglacial sea surface temperature changes in the Southwest Pacific ocean over the past 360 ka. Quat Sci Rev 28:1160–1170
Tager D, Webster JM, Potts DC, Renema W, Braga JC, Pandolfi JM (2010) Community dynamics of Pleistocene coral reefs during alternative climatic regimes. Ecology 91:191–200

Taviani M (1998) Stable tropics not so stable: climatically driven extinctions of reef-associated molluscan assemblages (Red Sea and western Indian Ocean; last interglaciation to present). In: Camoin GF, Davies PJ (eds) Reefs and Carbonate Platforms in the Pacific and Indian Oceans, IAS Special Publication 25, Blackwell Science, pp 69–76

Thompson WG, Goldstein SL (2005) Open-system coral ages reveal persistent suborbital sea level cycles. Science 308:401–404

Thunell R, Anderson D, Gellar D, Miao Q (1994) Sea-surface temperature estimates for the tropical western Pacific during the last glaciation and their implications for the Pacific warm pool. Quat Res 41:255–264

Toscano MA, Lundberg J (1998) Early Holocene sea-level record from submerged fossil reefs on the southeast Florida margin. Geology 26:255–258

Toscano MA, Macintyre IG (2003) Corrected western Atlantic sea-level curve for the last 11,000 years based on calibrated ^{14}C dates from *Acropora palmata* framework and intertidal mangrove peat. Coral Reefs 22:257–270

Toscano MA, Macintyre IG (2005) Blanchon P, Comment on Toscano MA and Macintyre IG (2003): Corrected western Atlantic sea-level curve for the last 11,000 years based on calibrated ^{14}C dates from *Acropora palmata* framework and intertidal mangrove peat [Coral Reefs 22: 257–270]. Coral Reefs 24:187–190

Trend-Staid M, Prell WL (2002) Sea surface temperature at the Last Glacial Maximum: a reconstruction using the modern analog technique. Paleoceanography 17:1065

Twiggs EJ, Collins LB (2010) Development and demise of a fringing coral reef during Holocene environmental change, eastern Ningaloo Reef, Western Australia. Mar Geol 275:20–36

Valentine JW, Jablonski D (1991) Biotic effects of sea level change: the Pleistocene test. J Geophys Res 96(B4):6873–6878

Valentine JW, Jablonski D, Krug AZ, Roy K (2008) Incumbency, diversity, and latitudinal gradients. Paleobiology 34:169–178

van Veghel MLJ, Bosscher H (1995) Variation in linear growth and skeletal density within the polymorphic reef building coral *Montastrea annularis*. Bull Mar Sci 56:902–908

van Wagoner JC, Posamentier HW, Mitchum RM, Vail PR, Sarg JF, Loutit TS, Hardenbol J (1988) An overview of sequence stratigraphy and key definitions. In: Wilgus CK, Hastings BS, Kendall CGStC, Posamentier HW, Ross CA, Van Wagoner JC (eds) Sea level changes – an integrated approach. Tulsa, OK: SEPM Spec Pub 42, pp 39–45

van Woesik R, Franklin EC, O'Leary J, McClanahan TR, Klaus JS, Budd AF (2012) Hosts of the Plio-Pleistocene past reflect modern-day coral vulnerability. Proc Royal Soc B 279:2448–2456

Vaughan TW (1914) Investigations of the geology and geologic processes of the reef tracts and adjacent areas of the Bahamas and Florida. Carnegie Institution of Washington Yearbook 12:1–183

Vaughan TW (1940) Ecology of modern marine organisms with reference to paleogeography. Geol Soc Am Bull 51:433–468

Vaughan TW, Wells JW (1943) Revision of the suborders, families, and genera of the Scleractinia. Geol Soc Am Sp Paper 44:1–363

Veeh HH, Chappell J (1970) Astronomical theory of climatic change: support from New Guinea. Science 167:862–865

Vermeij GJ (1992) Trans-equatorial connections between biotas in the temperate eastern Atlantic. Mar Biol 112:343–348

Vermeij GJ, Herbert GS (2004) Measuring relative abundance in fossil and living assemblages. Paleobiology 30:1–4

Veron JEN (1992) Environmental control of Holocene changes to the world's most northern hermatypic coral outcrop. Pac Sci 46:405–425

Veron JEN (1995) Corals in space and time: the biogeography and evolution of the Scleractinia. Ithaca: Cornell University Press, 321 p

Viau C (1983) Depositional sequences, facies and evolution of the upper Devonian Swan Hills Reef buildup, Central Alberta, Canada. In: Harris PM (ed.) Carbonate Buildups – a Core Workshop. Tulsa, OK: SEPM Core Workshop 4, pp 113–143

Walker K, Alberstadt LP (1975) Ecological succession as an aspect of structure in fossil communities. Paleobiology 1:238–257

Walther J (1894) Einleitung in die geologie als historische wissenschaft: Beobachtungen über die bildung der gesteine und ihrer organischen einschlüsse. Jena

Wapnick CM, Precht WF, Aronson RB (2004) Millennial-scale dynamics of staghorn coral in Discovery Bay, Jamaica. Ecol Lett 7:354–361

Webster JM, Wallace L, Silver E, Potts D, Braga JC, Renema W, Riker-Coleman K, Gallup C (2004) Coralgal composition of drowned carbonate platforms in the Huon Gulf, Papua New Guinea; implications for lowstand reef development and drowning. Mar Geol 204:59–89

White B, Curran HA, Wilson MA (1998) Bahamian coral reefs yield evidence of a brief sea-level lowstand during the last interglacial. Carb Evap 13:10–22

Williams JW, Jackson ST (2007) Novel climates, no-analog communities, and ecological surprises. Front Ecol Env 5:475–482

Wilson MA, Curran HA, White B (1998) Paleontological evidence of a brief global sea level event during the last interglacial. Lethaia 31:241–250

Woodroffe CD, Brooke BP, Linklater M, Kennedy DM, Jones BG, Buchanan C, Mleczko R, Hua Q, Zhao J-X (2010) Response of coral reefs to climate change: expansion and demise of the southernmost Pacific coral reef. Geophys Res Lett 37 (L15602):1–6

Woodroffe CD, Webster JM (2014) Coral Reefs and Sea-Level Change. Mar Geol 352:248–267

Yamano H, Hori K, Yamauchi M, Yamagawa O, Ohmura A (2001) Highest-latitude coral reef at Iki Island, Japan. Coral Reefs 20:9–12

Zimmer B, Precht W, Hickerson E, Sinclair J (2006) Discovery of *Acropora palmata* at the Flower Garden Banks National Marine Sanctuary, northwestern Gulf of Mexico. Coral Reefs 25:192

Reefs Through Time: An Evolutionary View

8

Jere H. Lipps and George D. Stanley, Jr.

Abstract

Although reef-like structures formed in the Neoproterozoic, reefs built by metazoans did not appear until the early Paleozoic. From then until the Recent, reefs diversified, underwent extinctions many times and then diversified again. Reef-inhabiting organisms included many different groups from algae to vertebrates as well as enigmatic, extinct suprageneric taxa. Evolution of these groups continued unabated and sometimes resulted in significant changes in the communities making up reefs. These reef groups varied over geologic time with extinction events commonly marking dramatic changes in the biotas. Paleozoic reefs consisted of sponges, corals, foraminifera, algae, bryozoans, and brachiopods, among others. The major extinction event at the end of the Paleozoic eliminated these forms as reef constituents and new groups (e.g., the first scleractinian corals) appeared in the Triassic. The Mesozoic was dominated by sponges, corals, rudist bivalves, and algae, most of which were eliminated in the end-Cretaceous extinction event. The Cenozoic reef biotas included red algae, foraminifera, sponges, corals, various invertebrates, and fish.

Throughout the Phanerozoic, these biotas were eliminated by extinction events of differing magnitude. Each event corresponded to warming due to rising greenhouse gases (CO_2 and CH_4), and ocean acidification caused by lowered pH and anoxia of shallow waters that took severe tolls on reef organisms. These extinction events caused the decline of reef organisms and the reefs they built, resulting in decreased diversity and slower carbonate deposition. Photosymbiotic reef ecosystems failed during extinctions and these failures may have been driven, at least in part, by either the demise of the symbiosis or the extinction of symbionts.

Reefs were generally absent in post-extinction times due to different ecologies. The ancestors of the next radiation of reef organisms must have been present somewhere—perhaps in deeper water, remote seamounts or isolated shallow seas. These ancestral faunas gave rise to radiations of reef organisms following several million years of depauperate and unusual biotas. Once underway, these radiations were relatively rapid and were responses to an amelioration of the extinction conditions and an increase in ecological opportunities. They did not restore the same taxa; rather new organisms at the familial or generic levels commonly diversified in most groups.

J.H. Lipps (✉)
John D. Cooper Archaeological and Paleontological Center, Santa Ana, CA 92701, USA

Museum of Paleontology, University of California, Berkeley, CA 94720, USA
e-mail: jlipps@berkeley.edu

G.D. Stanley, Jr.
Department of Geosciences and Paleontology Center, The University of Montana, Missoula, MT 59812, USA
e-mail: george.stanley@umontana.edu

D.K. Hubbard et al. (eds.), *Coral Reefs at the Crossroads*,
Coral Reefs of the World 6, DOI 10.1007/978-94-017-7567-0_8

The extinction events of the past do not bode well for the survival of modern reefs because of impending anthropogenic changes. Although humans have caused reef destruction through pollution, sedimentation, nutrient influx, and physical damage, increasing global warming and ocean acidification caused by CO_2 and CH_4 emissions are the principal threats to modern reef ecosystems with severe degradation or even extinction possibilities. Scientific and political will to change these inputs soon are essential to the survival of reefs, as well as other aspects of modern civilization.

Keywords
Reefs • Geologic history • Radiations • Extinctions • Photosymbiosis • CO_2 • Algae • Animals • Protists

Who has turned the wonderworld of the seas into underwater cemeteries bereft of color and life?[1]

8.1 Introduction

Reefs and carbonate platforms are common in the geologic record. Reef-like structures are present as early as the Archaean (4.0–2.5 billion years or Ga), were most likely built by photosynthetic cyanobacteria, and they became more common in the Proterozoic (2.5 Ga–542 million years ago or Ma). The earliest structures can be called biologic reefs, but they lacked the eukaryotes and metazoans that characterized later communities from the Cambrian (543 Ma) onward. These formed complex and biologically diverse reefs and carbonate banks that developed in the Paleozoic (542–251 Ma) and continued to the Recent with interspersed extinction events followed by diversification of phylogenetically different biotas. Each period of reef radiation, expansion and extinction has been related to sudden environmental (Fig. 8.1) or biological change (Kiessling 2009, 2011).

Photosymbiosis was likely a major evolutionary driver since the early Paleozoic (Stanley 2006; see also Chap. 3) and certainly by the Triassic (251–201 Ma). It is the primary energy source for modern shallow-water reefs and platforms today. Indeed, reefs are photosynthetically driven ecosystems where the chief constructional elements are larger organisms that rely on photoendosymbionts to support the high rates of calcification required to build them (Chap. 2). Clear patterns in modern systems can be traced back for hundreds of millions of years (Cowen 1983, 1988; Stanley 2006). Over this long time, such ecosystems range from the simplest microbe-dominated mounds in the Precambrian to complex and dynamic reefs of the Phanerozoic. Although the species, genera and families may have differed through time, they all built large reef structures and produced prodigious amounts of biogenic carbonate.

This chapter details the major evolutionary and extinction events in the history of reefs, focusing on the changes in reef-community structure and biodiversity. We address the major controls of both reef and carbonate-platform development over that time, interpret the causes of extinctions, discuss the radiation and development of reef organisms after extinctions, and apply those to a view of the future of reefs using geologic patterns to provide natural context. The scale is large but the goal is to chronicle the evolution of reefs through time in response to large-scale episodic changes in environmental conditions that favored photosymbiosis and competition among major reef builders and other marine taxa. A more detailed consideration of oceanographic process interacting with these evolutionary trends is presented in Chap. 9. Together, these two chapters relate the evolution of major reef dwellers to the changing world ocean and provide a view of natural change before the rise of modern *Homo sapiens*.

8.1.1 What Are Carbonate Reefs and Platforms?

Carbonate reefs and platforms at both global and local scales are assemblages of abundant and highly diverse organisms from all domains of life (Paulay 1997; Reaka-Kudla 1997; Stanley 2001; Pandolfi 2002; Karlson et al. 2004; Knowlton et al. 2010; Knowlton and Jackson 2011), having complex biological interactions (Meyer et al. 2005; Knowlton and Jackson 2008; Dixon and Hay 2012; Dixon et al. 2014). Together, these organisms built carbonate structures of all shapes and sizes rising above the sea floor, from small knobs and pinnacles, through atolls to huge barrier reefs bordering continents; all of these are considered reefs, as long ago described by Darwin (1842) and Dana (1872). The

[1] *What Is Happening To Our Beautiful Land?* A Pastoral Letter on Ecology. The Catholic Bishops' Conference of the Philippines 1988. http://cbcponline.net/documents/1980s/1988-ecology.html

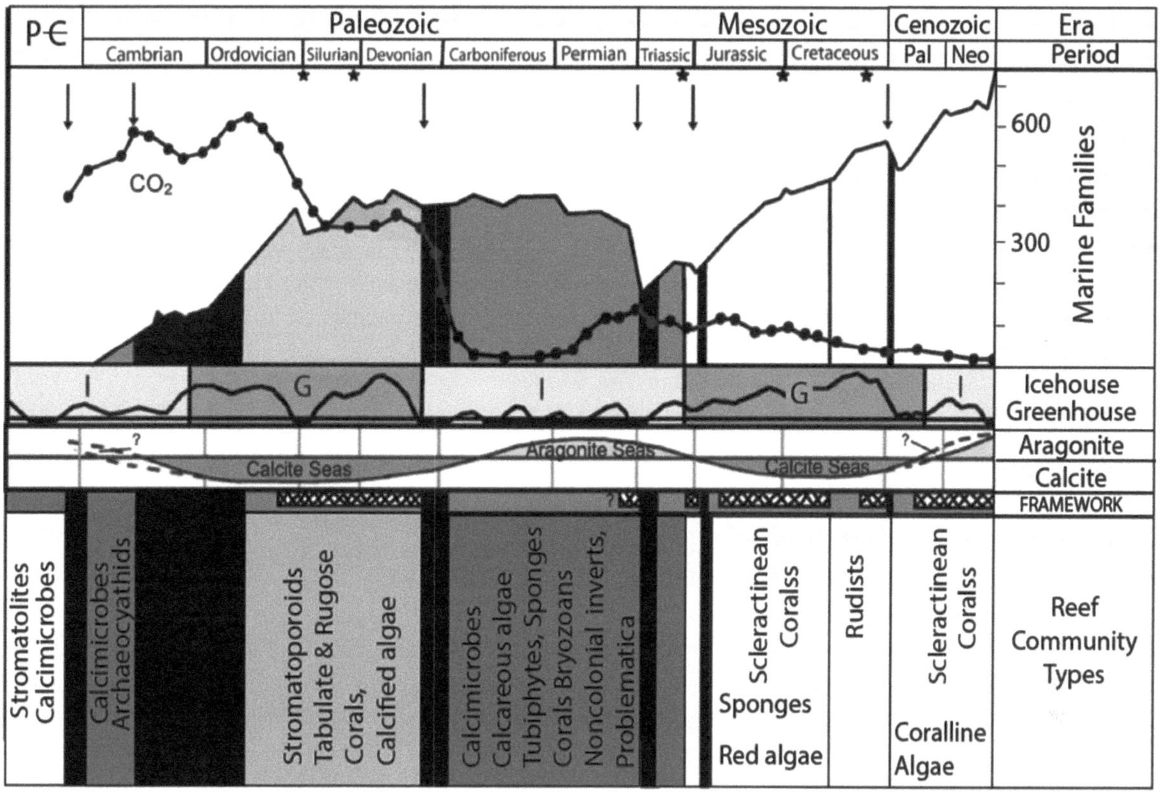

Fig. 8.1 Reef systems through time and various associated parameters or metrics. Below Sepkoski's familial diversity curve for marine life, are various stable evolutionary reef community types. *Black vertical lines* depict times of either reef demise or poor development of framework reefs. Along top, below geologic periods, *arrows* indicate major mass extinctions with stars for second order extinctions. The *curve* shows generalized relative levels of atmospheric CO_2 (GEOCARB III; Berner and Kothavala 2001). Shown below the curve are icehouse—greenhouse cycles with generalized, relative temperature curve (polar ice indicated), and aragonite-calcite transitions. Hypercalcification and extensive framework-building is indicated by XXXXs in the row labeled "Framework" (*right*), which roughly coincides with episodes of photosymbiosis (Modified from Stanley 1991)

organisms themselves are dependent on the maintenance of a physical platform in the face of the dynamic, commonly intense and always variable conditions near the air-water interface and of geological processes operating to enhance or degrade them (Chap. 6).

Carbonate platforms are submarine or intertidal shelves whose elevation is maintained by active carbonate deposition. The calcareous sediment is produced predominantly by organisms growing on the platforms, on smaller patch reefs, and on or in the sediment. Many of the organisms that live or have lived in these environments possess photosymbionts that enhance the deposition of skeletal $CaCO_3$ (Muscatine 1990; see also Chap. 3); hence they contribute to the production of both the sediments and rock of reefs and platforms.

Photosymbiosis between photosynthetic microorganisms and non-photosynthetic larger animal and protist hosts has been a powerful ecologic factor for hundreds of millions of years (Stanley 2006; see also Chaps. 3 and 9). The ability of these complex systems to keep up with rising sea level reflects a balance between carbonate production and sea-level rise that is gradually tipping in favor of the latter (Chap. 6). At growth rates of millimeters per year (meters per thousand years), some carbonate platforms and reefs have the potential to keep up or even exceed any rise of sea level (Schlager 1981), but others do not (Davies and Hopley 1983; Dullo 2005; Montaggioni 2005; Hubbard 2009; Hubbard et al. 2013; Hubbard 2014; see also Chap. 6). On Indo-Pacific reefs over the last 23 ka, rates of vertical accretion ranged between 0.2 to 40 mm/year and lateral accretion ranged from 55 to 90 mm/year (Montaggioni 2005). The future success of reefs will depend in part on the changing balance between processes that create carbonate substrate and those that degrade it and export it from the system.

8.1.2 Are Reefs Fragile Ecosystems?

In the vast literature on reefs, statements about the fragility of living reefs are common, and so most ecologists and environmentalists view them as vulnerable to changes in physical, chemical and biological parameters (Graham et al. 2006). Most blogs and news releases reinforce this idea. However, many of the ancient reefs studied by paleontologists and geologists over long time frames were amazingly persistent and, in fact, appear to have been both tough and resilient. However, these views are based on different time frames. Many observations of modern reef damage, whether by humans or natural occurrences, are immediately apparent, but the reefs may recover over decades and not be observed over the long term. These reefs have adaptability that allows them to endure and recover from various insults (Lipps 2011; Spalding and Brown 2015). While some damage may be significant and cause serious harm to reefs (e.g., physical destruction by massive hurricanes or bleaching and dying due to high seawater temperatures), given decades, the organisms may recover if conditions return to normal before widespread mortality can occur (Salvat 1987; Salvat et al. 2008; Colin 2009; Spalding and Brown 2015). However, under persistent, long term and repeated damage, recovery is most likely not possible, especially when recruitment and growth potential have been minimized.

From a geological perspective, lengthy reef intervals of millions of years indicate relative stability and hardly any overall change at all. Paleozoic and Mesozoic reef ecosystems persisted through long intervals as did the reefs of the Neogene to Holocene, and have resolutely survived warming, major tectonic upheavals and sea-level changes amounting to hundreds of meters. While short-term events have damaged reefs during those long intervals, recovery, time averaging and other taphonomic processes have masked them. Thus, the short-term variability, obvious at human timescales, has been obscured in the geologic record. The result is long periods when reefs were still mostly intact and biodiversity was not severely reduced at geologic timescales, and diversification continued slowly.

Mass extinction events nevertheless did occur, and these caused whole reef ecosystems to collapse, followed by lengthy reef eclipses. As first articulated by N.D. Newell (1971), these have significant impact on both reef organisms and structures in times when environmental changes exceeded the abilities of large numbers of species to survive. Eventually, usually within a few million years, reefs were again plentiful following a radiation of diverse but different taxa. In the following pages, we address the waxing and waning of reef builders and utilize the evidence now accumulating to explain the natural environmental and biological conditions that led to long-term absences of reef organisms and the structures they built. We end with a discussion of what might lie in the near future if anthropogenic stresses resembling natural ones occurring in the past last for centuries or millennia.

8.2 Ancient Reefs

Reefs or reef-like build-ups on the sea floor first appeared in the Archaean (>3.5 Ga), nearly synchronous with the earliest records of life. Since then, reef builders have evolved in response to changing environmental conditions as well as an increasingly competitive world as life in the sea diversified. The following sections chronicle this history and focus on changes in reef builders, the structures they created and the environmental conditions to which they responded.

8.2.1 Precambrian Reefs: Earth's Most Ancient Photosynthetic Reefs (3.4–0.541 Ga)

Until the end of the Neoproterozoic, when the first multicellular organisms appeared in Earth's oceans, Precambrian seas had biologic reefs produced by photosynthetic cyanobacteria. The earliest of these structures occur around 3.4 Ga (Allwood et al. 2007), and these older Archean mounds were a few meters high and tens of meters wide. They had reef-like characteristics in a broad sense (Walter 1983; Schopf 1992) exhibiting many characteristics of later Phanerozoic reefs. Later ones were globally widespread, especially during middle and late Proterozoic time. The fabric of these Archaean and Proterozoic structures included sea-floor precipitates whereas mid-Proterozoic carbonate rocks contain complex clastic textures including ooids, grapestone (clusters of ooids) and molar-tooth structures (Bishop and Sumner 2006), perhaps indicating changes in sea-floor chemistry. These led to the expansion of calcimicrobial reefs that included rigid stromatolites (Grotzinger 1990; Grotzinger and James 2012) and later, a variety of *Girvanella* and *Renalcis*-like coccoid algae that calcified in shallow and well-lit tropical settings (Copper 2001). Such microbial reef-like mounds were oases of life on the ancient seafloor and photosynthetic cyanobacteria took in CO_2 and deposited $CaCO_3$. Near the end of the Archean and during the transition into the Proterozoic, these microbes produced large carbonate platforms (Grotzinger 1994).

By 1.63 Ga, the first evidence for eukaryotes is clear but they may have appeared as early as the Paleoproterozoic, around 2 Ga (Butterfield 2015). Single-celled eukaryotes, possibly including dinoflagellates, were diverse in the Neoproterozoic between 1.5 and 0.541 Ga (Butterfield and Rainbird 1998; Lipps 2006). Near the end of that period, tube-dwelling metazoans built reef structures (Grotzinger et al. 2000; Wood and Curtis 2014), but most likely these reef assemblages did not include photosymbiotic associations between algae and larger heterotrophic protists or metazoans. Animals and heterotrophic crown protists did not appear until after the oceans reached a critical threshold of oxygen during the mid-Proterozoic (Planavsky et al. 2014).

8.2.2 Early Paleozoic Reefs: The Beginnings of Reefs (541–443 Ma)

Various calcifying organisms dominated the long-lived reef systems of the Phanerozoic (541 Ma to present) only to succumb to severe decreases in reef biodiversity during mass extinction events (Fig. 8.1). These reef systems responded to long-term climatic changes associated with icehouse-greenhouse cycles and sudden emissions of CO_2, chiefly caused by large volcanic eruptions into the atmosphere and oceans (Courtillot and Renne 2003). They were also likely affected by geochemical changes in ocean chemistry (Stanley and Hardie 1998; see also Chap. 9) which at certain times made the secretion of skeletons composed of aragonite over calcite more favorable energetically (Fig. 8.1).

During the Early Cambrian, most continents were arrayed along the equator (Landing et al. 2013a, 2013b) where abundant archeocyathid-cyanobacterial reef communities inhabited the shallow waters. The reefs expanded throughout the tropical regions following sea level rise and existed for more than 10 million years. In some places, these reefs reached huge sizes up to the 200–300 km wide and 1500 km long reef structure known as the *"Great Siberian Reef Complex"* (Rowland and Hicks 2004). Whether these Cambrian reefs were photosymbiotic or not is unclear. Some archeocyathids, interpreted now as sponges (Rowland 2001), had small sizes, occupied cryptic habitats, and lack geochemical evidence of algal fractionation (Wood 1999) while others built reefs in well-lit waters. Like modern corals, some archaeocyathids with expansive discoidal shapes may have harbored algal or cyanobacterial symbionts (Rowland and Shapiro 2002). The variability of growth form, the association with extensive reef structures in tropical environments, an affinity to shallow-water habitats, and seemingly rapid carbonate skeletal construction all support a photosynthetic, and presumably symbiotic, life style (Cowen 1983, 1988). Fluctuating nutrient levels may have deterred photosymbiosis (Wood 1999; Zhuravlev 2001) even though some archeocyathids may have preferred oligotrophic settings and ecological conditions conducive to photosymbiosis (Rowland and Shapiro 2002). In these instances, free-living cyanobacteria growing with the archaeocyathids were photosynthetic, although no evidence indicates they were actual symbionts.

In the later Early Cambrian, as sea level was falling, calcifying metazoans and archeocyathids underwent a major extinction, followed by a global decline in reef building that lasted about 40 million years (Zhuravlev 2001; Rowland and Hicks 2004). During this interval, thrombolites (like stromatolites except with clotted rather than layered textures and clear associations with organic cements), calcareous algae, metazoans, and some other calcified organisms which cannot be classified continued to build reef-like associations that persisted for the remainder of the Cambrian. Enigmatic corals, some classified as tabulates, were present but not important as reef builders. At least seven hypotheses have been proposed to explain this decline, including high levels of atmospheric CO_2 and global warming (Karhu and Epstein 1986; Rowland and Hicks 2004).

The *"Great Ordovician Biodiversification Event"* (Webby 2002) was characterized by the diversification of reef biotas that replaced the waning Cambrian community (Fig. 8.1). The earliest known reefs in south China and the Early Ordovician shift from microbial to metazoan reef building were facilitated by skeletal-dominated organisms that were part of that diversification (Adachi et al. 2011). Some mounds, patches and stromatolites in the Early Ordovician (Fig. 8.2) were built by calcifying microbes as well as algae, like *Renalcis* and *Girvanella,* and the stromatoporoid *Pulchrilamina* (Adachi et al. 2012). Major reef expansion occurred in the Middle Ordovician and included complex communities of bryozoans, lithistid sponges and stromatoporoids. This diversification was likely driven by biological interactions among various species living in association rather than by physical or chemical environmental events. By the Late Ordovician, the "Paleozoic Fauna" had diversified and formed reef complexes dominated by corals, stromatoporoids and a variety of filter-feeding organisms such as bryozoans and brachiopods that rimmed shallow platforms in different provinces (Elias et al. 2013). Latest Ordovician (Hirnantian) reefs were characterized by stromatoporoids and impressive framework produced by tabulate and rugose corals. Many of these organisms may have harbored photosymbionts based on their morphologies and distributions (Chap. 3). Reef and other non-reef organisms endured complex environmental excursions from cooling to warming with extinctions occurring under both conditions (Melchin et al. 2013). Possible glacial cooling during and after the late Hirnantian may have led to extensive reef failure and extinction of many corals and stromatoporoids due to the lowering of sea level and resulting habitat loss (Brenchley et al. 1995; Finnegan et al. 2012). Warming seas and related phenomena may have been responsible as well (see Sect. 8.3.4).

Fig. 8.2 Ordovician reef mound near Beatty, Nevada (Photo by J. H. Lipps 2003)

8.2.3 Mid-Paleozoic Reefs. The First Coral-Reef Ecosystems (443–359 Ma)

After the end-Ordovician extinction, a long period of species diversification in reef ecosystems was dominated by the coral-stromatoporoid-calcified algae association (Copper 1994; Brunton et al. 1997). This continued from the Silurian to the Devonian, lasting for some 80 million years and coinciding with a super-greenhouse interval as well as high sea levels and elevated atmospheric CO_2 (Copper 2002b). These reef builders lived under some of the highest temperatures of the Phanerozoic and inhabited latitudes as high as 45°–55° (Copper 2011). The Middle Devonian so called "mega-reefs" developed into globally distributed and large-scale systems (Copper and Scotese 2003). These ecosystems were pan-tropical and displayed characteristics of modern reefs but they were dominated by stromatoporoids, red algae and possibly photosymbiotic tabulate and rugose corals (Fig. 8.3). Among the 200 genera of corals in this interval, many display high corallite integration levels, large colony size and rapid growth—indicators that both corals and stromatoporoids possessed photosymbionts (Copper 2002a; see also Chap. 3).

Very large reef systems filled the expanded tropics of the Silurian and some barrier systems were thousands of kilometers long, exceeding even their present-day counterparts. Eight or more global and climatically-controlled episodes of reef building in tropical to subtropical, shallow-water settings occurred in the Silurian (Brunton et al. 1997). Not until the late Devonian mass extinction did this long-lived reef ecosystem disappear (Copper 1994). During this event, reefs declined and terminated globally at the Frasnian-Famennian boundary (Fig. 8.1). During the succeeding Famennian time, the once wide latitudinal reef belt shrank and reef builders were limited to scattered sponge-microbial biotic associations with some rare lithistid and stromatoporoid patches. This reduction in species diversity has been attributed to decreased rates of speciation rather than a major environmental change (Stigall 2012). Global cooling has been associated with this event as has global warming (Van Geldern et al. 2006). Major framework builders were absent for over 20 million years of Famennian time during the aftermath of one of the largest known mass extinctions (Copper 2002b).

8.2.4 Late Paleozoic Reefs After Extinction (359–252 my)

After the Devonian, a lengthy icehouse interval ("aragonite seas" in Fig. 8.1) was generally characterized by a lack of large-scale, coral-dominated reefs and, after the mid-Paleozoic, framework reefs were rare (Webby 2002). Instead, reefs were dominated by calcified algae, sponges, bryozoans, non-colonial invertebrates and other organisms whose relationships are unknown ("problematic"). In the Early Carboniferous (Waulsortian) banks (smaller reef mounds and mud mounds), presumably built by deeper-water communities, were common (Fagerstrom 1987; Lees 1988). These mounds were composed of cyanobacteria, occasional corals, echinoderms and bryozoans. Chaetetid sponge mounds are also known from both the Carboniferous and Permian. An unusual framework of corals has been described in the Early Carboniferous of Japan (Fagerstrom 1987), but generally such reefs were rare. Subsequent mounds and reef-like structures of the Late Carboniferous

Fig. 8.3 A Middle Silurian reef reconstruction, illustrating a coral-dominated ecosystem, is characterized by large colonies of tabulate and rugose corals along with crinoids, bryozoans, brachiopods, and other invertebrate taxa. This was the closest approximation to scleractinian-dominated reefs at that time (Courtesy of Terry Chase)

and Early Permian had some coral as well as abundant calcified algae, large foraminifera, brachiopods, chaetetid sponges, bryozoans and many "problematic" organisms (West 1988). Calcimicrobes and thrombolites also were present but most mounds and buildups lacked significant organic framework.

Following this sparse interval, the Middle to Late Permian was a time of highly differentiated reefs represented by many dominant groups (Flügel and Stanley 1984). Late Permian reefs were present in Tethys, south China and Japan. Lower integrated corals (cerioid and uniserial-dendroid colonies) were common in the mid-Permian (Guadalupian) while reefs of the latest Permian (Lopingian) were characterized by simple thickets of branching, waagenophyllid rugosan corals. Late Permian structures were mostly microbial and without framework builders (Grotzinger and Knoll 1995).

A large and diverse assortment of lightly calcifying organisms existed on reefs of the Late Permian which, unlike today's reefs, were characterized by many different community types. These included calcimicrobes, foraminifera, chambered sponges, bryozoans, *Tubiphytes*, algae, and many non-colonial calcified organisms such as reef-adapted richthofenid brachiopods and crinoids. Small patches of rugose corals were present but large, extensive framework builders were missing. Nevertheless, the general lack of larger frame-building species did not preclude the accumulation of impressive reef-related facies such as the Capitan Reef in western Texas and southwestern New Mexico (Newell et al. 1953).

The structure of most Permian reef-like ecosystems differed from those of later Mesozoic to Cenozoic time in paleoecologic structure, composition and the importance of synsedimentary reef cements that helped bind the reef rock. Those organisms living on reefs before the end-Permian extinction included forms that possibly utilized a photosymbiotic lifestyle (Chap. 3) such as the large fusulinid foraminifera in the Verbeekinidae and Schwagerinidae (Groves and Yue 2009; Zhang and Payne 2012), richthofenid brachiopods (Cowen 1983, 1988), large sponges, a diversity of corals, and giant bivalves (Isozaki and Aljinović 2009).

At the end of the Permian, the largest mass extinction that ever affected marine ecosystems eliminated all corals (including Rugosa and Tabulata), fusulinids, stromatoporoids and calcified algae and the reefs they built, as well as much of the terrestrial biota. A global decrease in carbonate sedimentation and the demise of many calcifying marine groups occurred (Benton 2003; Erwin 2006). The end-Permian mass extinction coincided with enormous volcanism contributing CO_2 to the atmosphere and oceans, resulting in dramatic warming (Svensen et al. 2009; Joachimski et al. 2012), ocean acidification (Clarkson et al. 2015) and anoxia (Payne and Clapham 2012; He et al. 2014) moving from deep to shallow water (He et al. 2014). Although seemingly well studied, a number of critical observations are still required to confirm any hypotheses for the cause of the extinction (Shen and Bowring 2014).

8.2.5 Mesozoic Reefs: The First Modern Coral Reefs? (252–66 Ma)

The Mesozoic was a time of major fluctuations in the CO_2 content of the atmosphere (Fletcher et al. 2008) with changes ranging from ~420 parts/million (ppm) near the Triassic-Jurassic boundary (200 Ma) to ~1130 ppm in the mid-Cretaceous (100 Ma) falling to ~680 ppm across the Tertiary boundary at 60 Ma in the mid-Paleocene. Reefs thrived through most of these changes and periods of high CO_2 as new organisms making new reef types evolved. Extinctions occurred several times during the Mesozoic and at its end.

Following the end-Permian collapse, reefs and most skeletonized ecosystems did not return for 7–8 million years (Flügel and Stanley 1984; Stanley 1988; Weidlich 2002) and carbonate deposition notably decreased (Woods 2014). The lengthy interval without corals may reflect the many millions of years required to evolve new reef organisms or may have involved marine conditions that were inimical to either their radiation or preservation. Most likely reef-building organisms did not diversify during the Early Triassic due to an extended period of unusual oceanic chemistry such as ocean acidification and/or anoxia (Payne and Clapham 2012; Wignall and Twitchett 1996; Woods 2014; see also Chap. 9). Reefs and reef-like structures did not return until Middle to Late Triassic time but some limited reefal accumulations occurred (Brayard et al. 2011). These were formed by cyanobacterial microbialites, calcified algae, sponges, *Tubiphytes*-like forms and some other small "problematic" organisms in the western US.

The first scleractinian corals appeared in the Middle Triassic (Anisian) but they did not build reefs until the Late Triassic after reef ecosystems had already diversified and possessed a variety of complex corallum morphologies (Stanley 1981, 1988; Pruss and Bottjer 2005). This is not expected of a newly evolved group and suggests either a prior, but unpreserved, evolutionary history or a very rapid radiation of new taxa. The earliest Middle Triassic corals were not reef builders and likely did not harbor photosymbionts. These forms were able to live in deeper water while corals with symbionts were restricted to the surface of the ocean, a pattern that persists among scleractinian corals through the Mesozoic and Cenozoic (Kiessling and Kocsis 2015). However, later in the Middle Triassic, small reefs formed and evidence from various criteria (Chap. 3; Stanley 2005) indicates photosymbiosis. Ladinian corals (~242–235 Ma) had large size, complex shapes and annual growth bands nearly identical to those of modern reef-building corals (Stanley and Helmle 2010). The sudden and apparently world-wide appearance of this new group of corals (like other kinds of animals as well) in the Middle Triassic was previously explained by either survival of seed stock from rugosan ancestors or by origination from zoantharian forms lacking preservable skeletons (Oliver 1996). A third possibility is that they reappeared and diversified from unidentified refugia elsewhere in the world's oceans, a phenomenon that is of growing interest to modern reef ecologists.

Scleractinia as a group are genetically deeply rooted in the Paleozoic. While the geologic record shows them first in the Middle Triassic, molecular-clock models project scleractinian corals into the Carboniferous (Romano and Palumbi 1996) and the Permian (Simpson et al. 2011). The obvious closeness of scleractinian corals to some living anemone-like forms (Stanley and Fautin 2001) suggests that, like some other invertebrate groups, scleractinian corals have variable calcification capabilities and could have existed in both soft-bodied anemone-like and calcified coral forms, a scenario known as the "naked coral hypothesis" (Stanley 2003). If so, coral calcification may also have been ephemeral in deeper geologic time, with uncalcified forms living alongside calcified ones during the Paleozoic. The rare occurrences of Ordovician and Permian scleractiniamorphs may represent short-term adaptations to calcification by soft-bodied anemone-like forms (Stanley 2003). With the possible development of scleractinian corals in the Carboniferous or Permian, this scenario may explain the anomalously rapid speciation of scleractineans in the Late Triassic. Diversity may have proceeded at a slower pace, but its orderly progression was not preserved until more moderate ocean chemistry facilitated calcification. According to this scenario, the Middle Triassic appearances of calcified scleractinians would have been a response to moderating seawater chemistry following Early Triassic post-extinction perturbations.

The "naked coral hypothesis" is supported by calcification experiments (Fine and Tchernov 2007) in which corals lost and then regained their skeletons, as well as findings based on molecular results (Medina et al. 2006). Other groups, from foraminifera to brachiopods, also experienced extinctions only to radiate in the Triassic from a few ancestral taxa surviving the extinction events (Stanley 2011). As is true of other extinction episodes, the timing of both the extinctions and radiations of different groups of organisms is variable, probably due to different timing of first and last occurrences (Signor and Lipps 1982).

Triassic reef ecosystems developed in three steps (Stanley 1988). After the Early Triassic metazoan reef gap following the Permo-Triassic extinction, the first Middle Triassic patch reefs, mounds and shelf-edge buildups were both taxonomically and paleoecologically diverse with microbes, encrusting foraminifera, *Tubiphytes*, calcareous algae, sponges, corals, bryozoans, bivalves, serpulids, crinoids and diverse encrusting "microproblematica". Internal reef cements, so important in the framework of Permian reefs, were not as prevalent in these Middle Triassic reefs but they increased later in the Middle Triassic and early Late Triassic (Flügel and Senowbari-Daryan 2001). At this time reef mounds, patch reefs and shelf-edge buildups increased in size while calcified sponges contributed to the framework structure. This was followed by a smaller global extinction or biotic turnover among reef faunas between the Late Triassic Carnian to Norian stages (Flügel 2002) that was characterized by changes in corals and corresponding to shifts in terrestrial biotas (Roniewicz 2010).

During the succeeding latest Triassic (mid-Norian to Rhaetian time), corals, chambered sponges and calcified demosponges became more abundant on reefs (Stanley 1988). Together with diverse calcareous algae and encrusting "microproblematica", corals produced extensive ramp and shelf-edge reef complexes in the Tethys Seaway. The Late Triassic (Norian to Rhaetian) global reef expansion (Flügel 2002) was also characterized by an increasing volume of corals over sponges in reefs, with corals creating framework in patch reefs as well as in shelf-edge and backreef settings of the Tethys. Referred to as the "modernization" of reef ecosystems, this was the first time corals had dominated since the Devonian; these corals were ancestors of modern taxa (Stanley 1988). Smaller-scale reefs with corals and chambered sponges also developed on isolated volcanic islands of the Panthalassan Ocean (Stanley 1988).

The rise in diversity and abundance of scleractinians in reefs during the Late Triassic may have correlated with the spread of coral photosymbiosis (Stanley 1981, 1988). Flugel (2002) proposed that the Late Triassic increase in reefs was driven by reef-building corals and was associated with a long period of global warmth. Alternatively, the rise of photosymbiotic corals during the Late Triassic reef expansion may be related to a draw-down of CO_2 brought on by global cooling (Kiessling 2010).

Photosymbiosis within Late Triassic reefs of the Alps has been questioned based upon sedimentological and paleoecological characteristics as well as nutrient models (Stanton and Flügel 1987; Stanton 2006). However, complex ecological interactions, colony size, growth rates and shapes of the corals are consistent with these reef organisms harboring photosymbionts (Cowen 1983, 1988; see also Chap. 3). This is further supported by analyses of stable isotopes in corals (Stanley 1988; Stanley and Swart 1995; Muscatine et al. 2005; Kiessling 2010).

While zonation within coral-dominated reefs of the Late Triassic (Rhaetian) was not as well-developed as in modern ones, morphological adaptations to light requirements of symbionts and reef structure resemble patterns in present-day, light-controlled environments. The crests of many modern reefs are dominated by large monotypic stands of multiserial branching corals including pocilloporids and acroporids, with a diversity of solitary and colonial corals inhabiting other biotopes. Late Triassic reefs were similarly dominated by uniserial phaceloid corals and a diversity of other organisms inhabiting forereef and backreef environments (Fig. 8.4).

Red coralline algae, so intricately associated with coral reefs of today, did not live on early Mesozoic reefs; instead red solenoporacean calcified algae were present (Bernecker 2005). Upper Triassic reefs, constructed by scleractinian corals and calcified chambered sponges, contained a plethora of organisms with no modern counterparts, including tabulozoans, chaetetids, spongiomorphs, chambered sponges, and many different "microproblematica" (Hodges and Stanley 2015). Triassic reef ecosystems of the Tethys began to decline near the end of the period and then collapsed suddenly during the end-Triassic mass extinction (Lathulière and Marchal 2009).

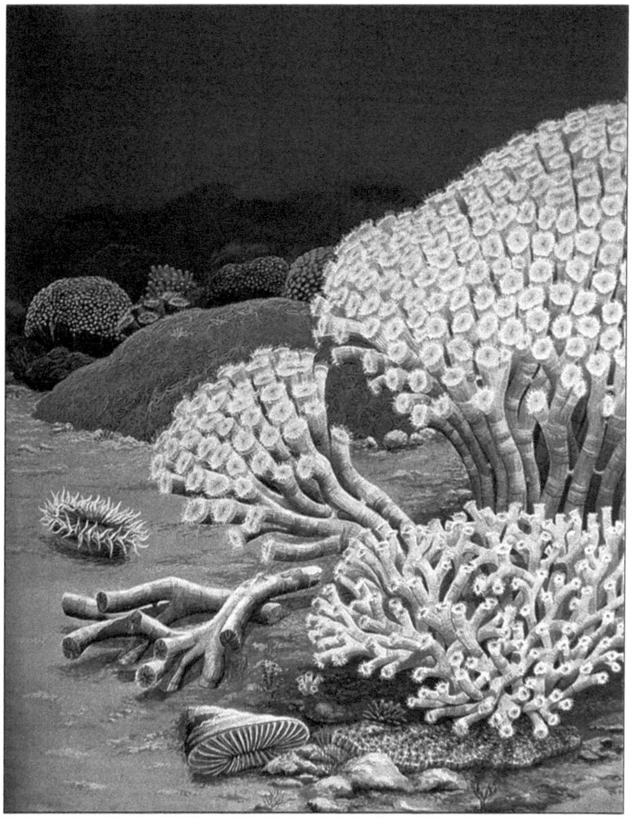

Fig. 8.4 Reconstruction of an Upper Triassic Tethyan reef from the Alps illustrating some typical scleractinian corals. Picture depicts the stands of uniserial branching coral *Retiophyllia*, one of the dominant reef builders. Also shown are other taxa of more highly integrated corals of this time, along with calcareous sponges that lived with corals (Drawing, courtesy of Geoff Kelley)

Coral recovery began soon after this extinction but proceeded slowly (Fig. 8.1). In the Early Jurassic, a few small coral reefs and carbonate deposits composed of large lithiotid bivalves are known (Fraser et al. 2004). This was a time of intermittent global warming which correlates with other marine extinction events (Svensen et al. 2007; Jourdan et al. 2008; Sella et al. 2014; Burgess et al. 2015). Six warming events and associated marine anoxia likely inhibited reef development after the Triassic collapse, at least locally in the Tethys Sea (Caruthers et al. 2014).

Fig. 8.5 The luxuriant photosymbiotic ecosystem of a Cretaceous reef in Austria growing on the shallow seafloor in the presence of upright rudistid bivalves living with abundant flat, platy and pinnacle-growing corals whose size and growth morphologies reflect adaptations to light (Painting, courtesy of Jose Garcia)

Reefs diversified in Middle to Late Jurassic time, again centered in the Tethys realm. Middle Jurassic reefs were paleoecologically and paleogeographically complex and varied in composition, being controlled by sea-level fluctuation, climate, sedimentation, nutrients and water depth (Leinfelder 2001). In contrast to modern coral-dominated reefs, those in the Jurassic were constructed in different water-depths by a variety of different organisms including sponges, microbes and corals. Coral reefs inhabited shallow water while hexactinellid sponges lived in deeper water and "lithistid" demosponges formed mud-mounds and biostromes in both deep and shallow water.

Middle-Late Jurassic reefs underwent major biotic diversifications at least twice, each time accompanied by geographic reef expansion. Reef development reached a maximum in the Late Jurassic when the diversity of coral genera exceeded that of the Triassic. The coral reefs were constructed by phaceloid taxa (e.g. uniserial *Aplosmilia, Thecosmilia* and stylinids) and branching multiserial corals (e.g. *Thamnasteria*). Some framework-building colonies of *Aplosmilia* reached over 4 m in height while others were composed of dense, platy microsolenid corals.

Like modern reefs, those of the Late Jurassic (Oxfordian) were strongly controlled by light, water energy and sediment. Eight highly varied reef types are known (Insalaco et al. 2008): 1) microsolenid biostromes, (2) marly-facies biostromes of platy microsolenids, (3) reef thickets of high-growing and dense phaceloid colonies, (4) microbial-coral reefs of massive, branching colonies; (5) large reefs of high diversity in bioclastic facies, (6) small, low-diversity reefs in mixed carbonate/siliciclastic facies, (7) microbial-coral reefs with massive colonies, and (8) reefal thickets in sand shoals and coral debris channels. Where corals dominated these reefs, evidence of photosymbiosis included complex growth morphology and annual growth bands similar to those seen in living photosymbiotic corals. Where sedimentation and nutrient influx could be shown (Leinfelder 2001), coral growth was depressed, similar to patterns observed in modern reefs.

During the Jurassic to Cretaceous transition, rudistid bivalves appeared and reefs became dominated by coral-stromatoporoid-rudistid buildups. However, by the end of the Early Cretaceous, rudistids gradually replaced both corals and stromatoporoids (Scott 1988). Corals coexisted for tens of millions of years alongside rudistids in reef communities (Fig. 8.5) but later, in the Early Cretaceous, rudistids exceeded scleractinian corals as framework builders and they evolved morphologies convergent on those of corals. During the Early Cretaceous (Aptian and Albian), rudistid constructions accounted for more than 60 % of patch reefs and buildups along shelf edges. The abundance of corals in shallow tropical reefs decreased until the Late Cretaceous, but they were common in deeper water and in smaller numbers within some rudistid buildups (Scott et al. 1990; Baron-Szabo 2003, 2008).

During the Maastrichtian greenhouse climate near the end of the Cretaceous, rudistid buildups were common worldwide, but they were particularly large in the Caribbean (Johnson and Kauffman 2001; Johnson 2002). Unlike some changes in other reef ecosystems over geologic time, the transition from corals to rudists was not associated with a mass extinction (Fig. 8.1). Debate continues over the extent of rigid framework in rudistid reefs and the cause of the gradual replacement of corals by rudistids is similarly unclear. Some argue for competition while others favor different adaptive preferences of the two groups (Kauffman and Johnson 1988; Gili et al. 1994; Johnson 2002).

Since hypercalcifying rudistids were composed of both calcite and aragonite, changes in seawater chemistry (e.g., Mg/Ca ratios) may explain their success over aragonitic corals (Stanley and Hardie 1998). In this scenario, the Cretaceous was a time when calcite was the favored mineralogy (see "Aragonite Threshold" in Fig. 8.1). Biological interactions, such as competition for space, predation, nutrient acquisition, were at least partly responsible for those changes (Kiessling 2009).

As at many previous times, the Cretaceous-Paleogene transition involved an extinction event followed by an interval devoid of high-diversity biotas and reefs. The extinction has been attributed to the impact of an asteroid into the carbonate banks of the Chicxulub area of SE Mexico. At the same time, the erupting Deccan Traps in India had been expelling huge amounts of volcanic CO_2 for much of the latest Cretaceous. In fact, the later eruptions of the Deccan Traps may have been triggered in part by the asteroid impact (Richards et al. 2015). The impact itself would have also generated massive amounts of CO_2 as the carbonates were vaporized. The instantaneous input of carbon into the atmosphere would have been about 4600 gigatons, resulting in an average warming of the earth by ~7.5 °C (Beerling et al. 2002). The sudden atmospheric and oceanic warming would have been sufficient to change ecologic conditions so profoundly that reefs and the other diverse ecosystems would have been extinguished (Schoene et al. 2015). The dramatic end-Cretaceous shift in reef-community structure would have had a disproportionate effect on corals possessing symbionts as opposed to asymbiotic species (Rosen and Turnšek 1989; Kiessling and Baron-Szabo 2004). Reef construction by rudistids ceased near the end of the Cretaceous (Steuber et al. 2002). Whatever the driver of that extinction, notable reduction in carbonate production, accompanied high atmospheric CO_2 and greenhouse conditions continued for many millions of years from the Cretaceous into the Cenozoic (Hallock 1997).

8.2.6 Cenozoic Reefs (66 Ma–Present)

Following the end-Cretaceous mass extinction, only eleven zoothanellate-like coral genera survived into the early Cenozoic—Paleogene, but a complete severance of coral photosymbiosis did not take place (Rosen 2000). Diversity was low in the Paleocene following the mass extinction although some potentially reef-building corals were present. As in the end of the Cretaceous, a decoupling between diversification of corals and their ability to build reefs still occurred in the early Paleocene. During that greenhouse time, reefs declined during short pulses of warming eventually leading to the Paleocene-Eocene Thermal Maximum (PETM), an event that has recently been proposed as an analog for future global warming.

In some regions of Europe, middle Paleocene coral-algal patch reefs and barrier reefs adapted to the conditions of that time and persisted (Zamagni et al. 2012), although many did not. Factors retarding reef building during this time may have included high pCO_2, ocean warming or changing chemistry, acidification and high nutrient supply (Perrin and Kiessling 2010).

By Eocene time, corals started to diversify, and possible symbiont-bearing corals began forming reef-like structures. Different tropical reef ecosystems emerging during the Cenozoic included many symbiont-bearing coral species. In addition to their co-evolution with red calcareous algae, Cenozoic corals evolved fast-growing, porous skeletons that were capable of rapid linear extension and quick recovery following disturbances. The rise of coralline algae also helped bind the reef framework, increasing the preservation of photosynthetically produced carbonate and strengthening the reef structure.

The Cenozoic was marked by large and sudden climate changes connected with warming-cooling cycles, plate tectonic movements, atmospheric CO_2 increases (Tripati et al. 2009) and changes in ocean circulation. In the early Eocene, following the climatically-induced high temperatures of the Paleocene, coral reefs again diversified but then underwent a decline in diversity during the Eocene-Oligocene and in the Oligocene-Miocene transition. Like marine benthic and planktic biota in other habitats, the ancestors of modern reef organisms developed in the Oligocene to Miocene. Corals and the reefs they built expanded geographically and became ecologically more similar to modern counterparts. This was especially true in the Caribbean region, where five major episodes of diversification are recognized (Budd 2000): (1) middle to late Eocene, (2) Oligocene to early Miocene, (3) early to middle Miocene, (4) late Miocene to early Pliocene, and (5) late Pliocene to Recent. These intervals were punctuated by diversity declines during climate warming, periods of rising or falling sea level and changing ocean circulation.

A major adaptive radiation of photosymbiotic corals and their reefs took place in the early-middle Miocene when the Mediterranean, Caribbean and Indo-Pacific provinces were forming (Perrin 2002). Probable photosymbiotic corals and reefs experienced a local decline in the Mediterranean at the end of the Miocene, but in the Caribbean region, the Miocene was a time of development of corals and reefs (Budd 2000). A notable reduction in coral species diversity did not correspond to a decline in reef development in the Caribbean (Johnson et al. 2008). In fact the largest reef development occurred after coral diversity was reduced by 50 % and fast-growing acroporids increasingly dominated shallow environs.

Photosymbiotic corals and reefs changed in concert with the climatic variations underway during the Cenozoic (Perin 2002). Photosymbionts apparently became more diverse and efficient in their association with reef organisms. Five of the many *Symbiodinium* (dinoflagellate) clades arose during this time and their presumed rate of molecular evolution places their origination near the time of Eocene cooling (Pochon et al. 2006). These molecular data suggest that major diversification of the present-day *Symbiodinium* clades also coincided with the Miocene acme of corals and reefs. Whether the symbionts diversified in response to increased availability of habitats within the increasing number of hosts or whether their presence initiated the symbiosis is unknown.

Toward the end of the Cenozoic, reef ecosystems changed significantly in association with climate shifts (e.g. drying that triggered the Messinian "crises") and large-scale tectonic events, like the rise of the Isthmus of Panama that led to the separation of Pacific and Caribbean marine provinces. The rise of the Panamanian land bridge, dated by both molecular and geologic evidence at about 2.8 Ma (Lessios 2008), changed coral and reef ecology, cutting off the supply of larvae to eastern Pacific reefs (Leigh et al. 2014). Even during such events and the glacial and interglacial climate and sea-level changes of the Pleistocene, photosymbiotic reef organisms survived, demonstrating the resiliency of those reef communities. More recently, however, the intensification of El Niño events have reduced or eliminated many eastern Pacific reef faunas (Hueerkamp et al. 2001).

Overall, reefs of the Pleistocene along with their constituent organisms appeared resilient. They were able to endure global changes and could reestablish themselves quickly or keep up with sea-level change. Because CO_2 levels never rose higher than about 280 ppm (Köhler et al. 2010), maximum temperatures, even during warmer interglacial periods, were within the tolerances of reef organisms.

What promoted this ability to adapt to such conditions is presently a subject of great interest and study. While we traditionally think of the corals as masters of the reef, this role may well be held by their photosymbionts. Part of the ability of reef organisms to resist such perturbations as CO_2 rise and global warming may be in the thermal adaptation within the clades of *Symbiodinium* (Rowan 2004), particularly *Symbiodinium* Clade D (Ladner et al. 2012). Although the symbionts increase calcification rates in corals, without them the corals die which may happen during extreme warming events. Other factors may have contributed to this resistance as well as symbiosis or in concert with symbiosis. Major buildups of Caribbean reefs, for example, may have been facilitated by fast-growing, high calcification, low-diversity assemblages dominated by *Acropora palmata* which could keep up with sea level rise during deglaciations of the Pleistocene while reefs with slow growing species could not keep up (Johnson et al. 2008; see also Chap. 6).

8.3 Extinction of Reef Organisms and the Reefs They Built in Geologic Time

Mass extinctions interrupted long periods of high diversity, continuous low-level speciation and apparent stability of reef-building organisms (Kiessling 2009; Stanley and Lipps 2011). The five great mass extinctions as well as at least 13 lesser ones (Bambach 2006) in the Phanerozoic are each different in terms of their intensity and the extent of their repercussions—they were polyphyletic, widely distributed, crossed ecologic boundaries from land to the deep sea and commonly affected marine and terrestrial biotas. Both the major and minor events appear to be related to large volcanic eruptions (Courtillot and Renne 2003), differing in the extent of volcanism and associated consequences on the biosphere. The biotic and environmental patterns for most major extinction events indicate world-wide phenomena while the minor extinctions had fewer consequences. Extinction dynamics resulted in a gradation of impacts depending on the particular circumstances of volcanism and geographic settings. In all cases, extinctions occurred suddenly (perhaps over 100–10,000 s of years) in the geologic record and lasted for a relatively short time (a few million years). Certain other time periods may have also had low diversity (Kiessling 2002), like the Carboniferous, but these were not sudden mass extinctions and were likely caused by different biological processes.

Reef-building taxa did not go extinct alone. Other groups of marine organisms from protists to vertebrates as well as terrestrial biotas disappeared nearly simultaneously. The five major mass extinctions eliminated entire ecosystems through world-wide environmental perturbations. Starting even before the Cambrian and the radiation of metazoans and protists and clearly in the Phanerozoic, 13 (more likely 25 or so) lesser extinctions occurred during which fewer taxa in selected functional groups and in more limited ecosystems were extinguished. While the extinction events were complex, the primary cause may have been similar in most cases, varying mostly in magnitude or length of time. Other factors may also have played a role in the extinctions, hence they were multidimensional in their effects particularly on reef organisms that were differentially affected by those processes.

Hypotheses for reef extinctions must be compatible with nearly simultaneous extinctions in other non-reef groups either directly or indirectly. At each of the great five mass extinctions, reef organisms and reefs disappeared. Many extinction hypotheses relating to reefs, some better developed than others, have been proposed (see Bambach 2006; Veron 2008; Kiessling 2009; Stanley and Lipps 2011 for reviews). They include impacts by extraterrestrial bolides, impact-induced global fires or dust, sea-level changes, climate cooling, massive volcanic eruptions, CO_2 variations in the atmosphere and oceans, climate warming, acidification of the oceans and oceanic anoxia (Moffitt et al. 2015), among others. Some of these, of course, are consequences of others (e.g. high CO_2 and ocean acidification) which might complicate or support interpretations. Such massive extinction events across so many ecologies may not be simple in detail but are likely driven by similar global phenomena with multiple consequences.

8.3.1 Extraterrestrial Impacts

These do not cause extinction directly by the impacting bolide, except for those species whose ranges are immediately affected. Other factors induced by impacts besides "nuclear winter" scenarios (difficult to document), global fires (Robertson et al. 2013) or dust (Toon et al. 1997) may have triggered extinctions. The fires, although thought to be terrible for terrestrial organisms, would have little or no direct effect on marine organisms far at sea or deeper than a few meters (Robertson et al. 2013) and evidence for dust in the atmosphere shutting down photosynthesis around the world (Pope 2002) does not support these as extinction mechanisms. A possible consequence of the end-Cretaceous impact event could have been a sudden increase in CO_2 created by the intense heating of the limestone at the impact site (Beerling et al. 2002).

8.3.2 Sea-Level Changes

Sea-level changes may have great influence on biotas inhabiting epicontinental seaways or continental margin environments where habitats may be significantly reduced or even eliminated, but for marine organisms in general, sea-level change is not a significant driver of extinction or diversity increases. Of course, the elimination of shallow-water habitats that are unique would decrease diversity, just as the development of those habitats as sea level rose would raise diversity. But along coasts or deeper seaways, as long as sea-level rise is not too fast (Chap. 6), the faunas can simply track sea level rises and falls without causing significant loss or growth of biodiversity because habitats will shift but still remain similar (Chap. 7). Even as epicontinental seaways emptied, the biotas could track habitat redistributions. While world-wide diversity would remain the same, diversity would decline in the seaways. Since these seaways and continental margins are well represented in the geologic record, changes in diversity would be well preserved, although perhaps not representative of the world's oceans. Pleistocene coral reefs world-wide, for a recent example, did not experience extinctions as they dealt with a number of sea level declines and rises of up to 120 m over the past 500,000 years (Pandolfi 1996, 2002). In particular, the record of sea level rise of 122 m in the last 13,000 years produced significant shifting of environments but no major extinctions in marine faunas or floras (see Chap. 6 for a discussion of the factors involved in changing sea level).

8.3.3 CO_2 Decline and Climate Cooling

As CO_2 declines through weathering, burial of organisms (particularly the Paleozoic coal forests and more recent plankton) and other processes, the climate cools over the long term. Cooling is commonly associated with several reef and biotic extinctions in the geologic record from the Neoproterozoic to the Cenozoic. Cold negatively impacts reef faunas by restricting their ranges or eliminating their habitats. However, cold temperatures on the earth disproportionately affect high latitudes leaving plenty of marine and terrestrial environments without significant change in lower latitudes. Today, except for the recent warming over the past century, the world is cooler than it has ever been in the last 65 million years (Zachos et al. 2001), yet no massive extinctions in the seas have occurred. Nor did coral reefs become less abundant in the Pleistocene cold episodes (Pandolfi 2002). Cold climates, if they involve emplacement of ice on the continents, would also involve sea-level drops which may well change the distribution of geographic diversity and lead to extinction if the drops in temperature and the resulting latitudinal shifts occurred too fast for the organisms to keep up (see Chap. 9 for a discussion of the rate of change as a control on adaptation, migration and acclimatization), but that did not happen in the Pleistocene.

8.3.4 Volcanism, CO_2 Increases, Climate Warming, Ocean Acidification and Anoxia

On earth, as CO_2 rises in the atmosphere due to volcanic releases, the climate warms. The oceans also become oxygen deficient and more acidic as CO_2 joins with water to form carbonic acid in the sea, which in turn, lowers the pH.

Warming climates and oceans present greater difficulties for animals and plants than cooling climates. Warming that caused extinctions was likely sudden and impacted many other ecologic factors in the sea—deep ocean warming, more homogeneous oceans with depth and latitude, fewer biogeographic zones with depth and latitude, anoxia, sea level increases, and upwelling declines (Lipps 1970). In most environments on land or in the sea, organisms cannot migrate to cooler areas in high altitudes, higher latitudes or deeper water because all of these places eventually warm as well. Indeed, because the tropics heat less than other places, the temperature increase at higher latitudes is much greater than Earth's average. These result in a more homogeneous world in which cooler and denser-water habitats would be eliminated (Lipps 1970). More homogeneous oceans also mean sluggish currents, fewer or degraded water masses, less upwelling and decreased productivity as nutrients are not replenished to the photic zone. All of these would enhance extinction probability.

What then could have influenced all of these factors that have been correlated with extinction events in the fossil record? A logical hypothesis and one that has been gaining support is a sudden increase in CO_2 and CH_4 in the atmosphere and ocean leading to a suddenly warmer world. This then leads to acidification of the oceans and development of anoxia or expansion of the anoxic zones that commonly occur along continental shelves and in shelf basins. Support for this hypothesis comes from the Permo-Triassic extinction events (Joachimski et al. 2012; Payne and Clapham 2012; Sun et al. 2012), the end-Triassic extinction (Blackburn et al. 2013), the end-Cretaceous (Beerling et al. 2002), the Paleocene-Eocene (Zachos et al. 2006), the changed distribution of reef corals during the last Pleistocene interglacial (Kiessling et al. 2012) and the increasing warming and ocean acidification that is underway now due to rising anthropogenic CO_2 and CH_4 emissions (Barnosky et al. 2011; Payne and Clapham 2012; Barnosky 2014; McCauley et al. 2015).

The chief source of CO_2 in the past has been massive volcanic release of CO_2. For most of the major extinction events, such evidence is prominent. Gradual increases in CO_2 may not lead to massive extinction if the biotas can evolve through selection, but sudden events over 100 s to 10 s of thousands of years would cause extinction because the biotas could not evolutionarily adapt to such rapid changes. For the K/T "impact" event, which occurred in the midst of high CO_2 and warm climates, the chief factor in this extinction may have been a sudden increase in CO_2 through the impact heating of and the chemical disassociation of limestone ($CaCO_3$). But even in this case, an impact source for CO_2 might not be needed (but would be contributory) if it occurred during periods of sudden widespread volcanism such as that of the Deccan traps of India (Chenet et al. 2007; Schoene et al. 2015).

In modern seas, reefs and their photosymbiotic biota are particularly sensitive to CO_2 increases and the resulting warming of the shallow marine waters that trigger a loss of their symbionts. Today, ocean acidification due to rising CO_2 (Kleypas and Yates 2009) has been implicated as a cause in many of the past mass extinctions affecting the marine environment (Hönisch et al. 2012; Clarkson et al. 2015). Since the Industrial Revolution, the burning of fossil fuels has increased CO_2 levels in the atmosphere at a rising rate, leading to ocean warming, acidification (Kylpas and Yates 2009) and anoxia. Ocean pH has decreased by 0.11 of a pH unit or approximately 30 % (Zeebe and Wolf-Gladrow 2001). Increases in the partial pressure of CO_2 in seawater affects carbonate solubility and saturation, decreasing $CaCO_2$ precipitation by calcifying animals. Living photosymbiotic corals show a high degree of sensitivity to declining pH in seawater and corals exposed to higher CO_2 levels in the future , will likely develop thinner, more fragile skeletons and declining growth rates (Hoegh-Guldberg et al. 2007; Hoegh-Guldberg 2014; Spalding and Brown 2015). This, in turn, would hamper their ability to build reefs. Levels of atmospheric CO_2 today are above 400 ppm, the highest level in the past 800,000 years or perhaps even the past 20 million years (Tripati et al. 2009). This change from 280 ppm before the Industrial Revolution is occurring at rates that may be among the fastest ever experienced during prior extinction events. The geologic history of reef change and global collapse during mass extinctions reveals a high correspondence with elevated CO_2 (Veron 2008, 2011; Stanley 2015); ocean acidification is likewise a likely contributor to the mass extinctions of the last 300 million years (Hönisch et al. 2012; see also Chap. 3).

CO_2 releases by large-volume volcanic activity, such as those associated with flood basalts, have been identified with past extinctions (Courtillot and Renne 2003). Some of them persisted for 100,000 s to millions of years, but the change could still have been rapid enough that ecosystems could not adapt or recover. Thus, a sudden CO_2 release, whether from volcanoes or human activities, that results in global warming, anoxia and ocean acidification is a viable hypothesis to account for extinctions across all environments and taxonomic groups. If human contributions of greenhouse gases continue at the rates now underway, extinctions on the magnitude of at least some smaller past events may occur.

8.3.5 The Role of Photoendosymbiosis in Extinctions and Diversifications

Does the success or failure of photosymbiosis provide an adequate explanation for speciation and extinctions of reef-building organisms throughout geologic time? It certainly provided opportunities for improved growth, metabolism and ecology for marine, and particularly reef, organisms. Among many invertebrates, this also opened up ecological opportunities and selective advantages that did not previously exist (Kiessling et al. 2010).

The evolution and extinction of photosymbionts correspond to the history of reefs with its long-lived ecosystems, extinctions and periods lacking reef organisms, reefs and carbonate sedimentation. These have been well summarized globally (Cowen 1988; Wood 1999; Stanley 2001; Stanley 2003; Kiessling 2009; Stanley and Lipps 2011; Spalding and Brown 2015; see also Chaps. 3 and 9) and with specific reference to the Great Barrier Reef of Australia (Hopley et al. 2007; Veron 2008) and the western Atlantic (Johnson et al. 2008). Photosymbionts may have first occurred in late Proterozoic and continued to operate throughout geologic time in many, if not most, ancient and modern tropical to subtropical reef organisms. These symbionts could have been any of a number of single-celled algae and cyanobacteria since all are known to function as photosymbionts within marine hosts today and most have long evolutionary histories. Particular taxonomic groups of organisms may be inferred to have hosted photosymbionts (Cowen 1983, 1988; Rosen 2000; see also Chap. 3). The criteria include host characteristics of relatively large size, abundant massive skeletal $CaCO_3$, modular construction of skeletons, and tropical paleogeographic distributions among others (Chap. 3).

Some radiations resulted in communities dominated by organisms quite similar in morphology and function to the previous ones, while others were characterized by a restructured and quite different reef ecosystem. The collapse of many ancient reefs has been attributed to the breakdown of photosymbioses (Talent 1988), followed by decline and extinction in response to rapid increases of CO_2, increased ocean warming, ocean acidification, and anoxia of surface waters—all factors inimical to living reefs today. While nutrients, sedimentation, predation, and other factors may have controlled the local development of reefs (Wood 1993), sunlight, climate and temperature rank high in controlling photosymbiosis in today's tropical marine environments.

As sea-surface temperatures peak today, symbionts decline in number in the tissues or cells of the hosts allowing the white carbonate skeletons or shells to show through—a process known as "bleaching" or "paling" (Kleppel et al. 1989; Glynn 1996). Simultaneous bleaching events occur in corals, bivalves and foraminifera on the same reef when temperatures are high (or, in some instances, low: Baker et al. 2008). Without reacquiring new symbionts to provide the required nutrition, hosts starve to death (Van Oppen and Lough 2009; see also Chap. 3). In single-celled foraminifera, bleaching is associated with a decline in number of symbionts and cytoplasmic cell degradation (Talge and Hallock 2003). After bleaching episodes, some corals reacquire symbionts from other clades that may be better adapted to the new regime of thermal stress. According to this "adaptive bleaching hypothesis" (Fautin and Buddemeier 2004), the expulsion and reacquisition of new zooxanthellae may be a process by which corals deal with increased thermal stress. Whether or not shifting symbionts are a recent phenomenon or one that developed in the geologic past is unknown, although the flexibility of the symbionts could be a driving force controlling growth in reef corals (Little et al. 2004).

Nutrient availability commonly sets limits on reefs. Today's photosymbiont-bearing, reef-building organisms are uncommon in regions of high nutrient concentrations because of increased numbers of plankton that decrease the amount of light received at different depths and increased growth of large, fleshy algae that excludes corals and other organisms (Hallock and Schlager 1986). Because of these, photosymbiosis is also excluded or reduced, and this would lead to extinction of some modern and ancient reefs. Nutrients and biodiversity in oceans have varied locally and globally in the geologic past.

8.3.6 Extinctions

Several alternative but not necessarily exclusive hypotheses have been proposed for the extinction of reef organisms as well as other biotas. An acceptable hypothesis must account for the global extent, the impact on both terrestrial and marine ecosystems, the geochemical and sedimentological characteristics, and the aftermath of the extinction events. The hypothesis of sudden but enduring CO_2 releases by volcanism and resulting ocean acidification and anoxia supports a scenario of negative worldwide impacts on all ecosystems causing species declines as well as characteristic sedimentary records. High CO_2 by itself does not necessarily cause extinctions; rather it is the geologically sudden rise, creating environmental changes that were too fast for organisms to adapt that mattered. Such sudden releases resulted from massive volcanism occurring over short time intervals, or, in modern times, the emission of CO_2 by the burning of wood, gas, oil and coal.

8.4 The Future of Reefs

The future of reefs looks grim, chiefly because of climate and ocean changes caused by anthropogenic increases in atmospheric CO_2 and CH_4 (IPCC 2014), as well as increasing pollution, sedimentation, physical damage and exploitation of reefs (Knowlton and Jackson 2008; Salvat et al. 2008; Lipps 2011; Hoegh-Guldberg 2014; Spalding and Brown 2015) in the relatively short time interval of the Anthropocene.

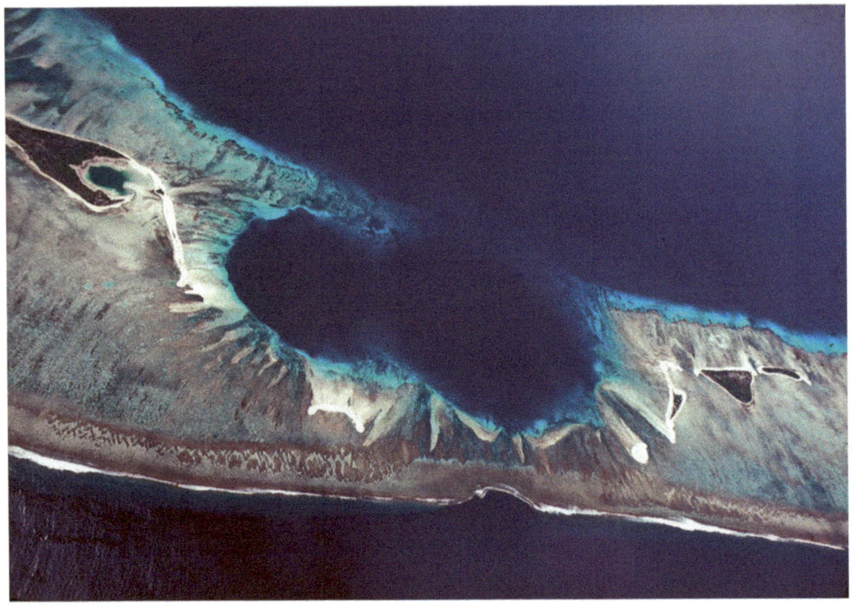

Fig. 8.6 The reef flat on the northern part of Eniwetak Atoll Marshall Islands photographed from an altitude of about 10,000 ft and looking slightly to the south. This reef and several islands were the sites of nuclear bomb tests in the 1950s (Simon and Robinson 1997), which created massive craters and ejecta blankets of broken reef rock and sediment. Redwing Seminole atomic bomb (June 6, 1956) lies in Bogon Island to the east (left) of two hydrogen bomb craters (IVY MIKE November 1, 1952, the largest crater (1.8 km in diameter) on the west (*right*) and the KOA May 12, 1958 crater (1.5 km in diameter) to the east (*left*) of MIKE). MIKE and KOA, as ground-based explosions, each eliminated a small sand island. A large slab of the reef (the crescent indentation north of Mike) broke off the reef sometime between 1952 and 1958 and included about 300 m of reef crest and 60 m of reef flat and the forereef for as deep as can be seen. Even this kind of human destruction of reefs pales in comparison to the damage or extinction that can be inflicted by sudden global warming because of the longer times and much larger geographic extent (Photo courtesy of Patrick L. Colin (see Colin 1987 for further information))

Both the fossil and modern records show that reefs can recover from small-scale events in a relatively short time of a few years or decades (Salvat et al. 2008) as long as those stresses soon abate and structural integrity is preserved (Graham and Nash 2013)—and even to some major events like the many hurricanes or the nuclear test explosions on reefs (Simon and Robinson 1997). Reefs in the geologic record are resistant to these levels of damage because long-lasting damage was local (Fig. 8.6) and not over wide areas (Lipps 2011). Short-term (i.e., 1–3 weeks) warming events that cause bleaching may be reversed if conditions return to normal quickly enough (Spalding and Brown 2015). Even where longer exposures lead to coral death over wide areas, reefs can recover within a dozen or so years if conditions return to normal and there is an adequate larval pool to fuel repopulation (Salvat 1987; Salvat et al. 2008). However, increasing El Niño intensities and extended periods of bleaching may result in either immediate death or subsequent attack by various diseases (Miller et al. 2009), thus turning short-term decline into permanent loss (Hubbard et al. 2013). Attempts to "restore" reefs degraded by humans commonly increase the level of damage by emplacing artificial objects that then become projectiles in storms which further disrupt already marginalized reefs (Lipps 2011).

While these phenomena are not good or desirable, they do not appear to cause extinction of species or declines in the reef structure. Instead, the geologic record indicates that sudden but long and enduring changes are required to cause declines, disruptions and extinctions of reef organisms, hence the reef structure as well. The only ones of worldwide impact involve warming of the climate (Kiessling 2011) and subsequent warming of the oceans, changes in surface and subsurface circulation of the oceans due to that warming (Lipps 1970), acidification due to the additional CO_2 in the oceans (Kleypas and Yates 2009; Barbeitos et al. 2010) and anoxia (Moffit et al. 2015). As the world's oceans continue to absorb CO_2, they will become increasingly acidic. While the recent decrease in pH of 0.11 may not seem like much, the pH scale is logarithmic and this corresponds to a 30 % increase in acidity, more than enough to dissolve minute calcified life that are an essential part of the oceanic food chain (Beaufort et al. 2011). Currently, scleractinian corals and other, more-heavily calcified organisms are already affected by ocean acidification. The decline in $CaCO_3$ production, coupled with an increase in carbonate dissolution will boost future reef decline (Pandolfi et al. 2005). Increasing warming will also contribute to sea-level rise through the thermal expansion of seawater and collapse of ice masses in Greenland and West Antarctica into the sea (Alley et al. 2005). As sea-level rise continues to accelerate, reefs may fall behind, threatening communities that depend on them (Chap. 6). Coral reefs are also under attack by bleaching, nutrification, overfishing and diseases and could collapse globally as early as the end of the twenty-first century (Bellwood et al. 2004; Hoegh-Guldberg et al. 2007; Carpenter et al. 2008).

All of these events can decrease biotic diversity, cause loss of symbionts from reef organisms and, if the conditions persist, reestablishment of the symbioses and reef building is likely impossible. The "naked coral hypothesis" (Stanley 2003) indicates how changing seawater acidity and carbonate saturation level fall might select for "naked" corals in the future. Reefs today will not survive in the hotter and more deleterious oceans expected even within the next 50 years without becoming quite different in their biotic and structural characteristics (Hughes et al. 2003). Reefs of the future could have lower biodiversity, less structural complexity, lower calcification rates, and even "naked" corals.

Humans can choose now to preserve reefs as we know them and avoid millions of years without reefs or with considerably changed reefs with new consortia of organisms we might not recognize once extinction ensues. The level of CO_2 in the present atmosphere hit 400 ppm in March 2013, and the rate of increase continues to rise, enhanced by the release of CH_4 caused by warming marine sediments, permafrost, and marshes that contain this gas. The logical choice to preserve reefs is to control, or better, eliminate the increase in CO_2 and CH_4 emissions, the primary drivers of global warming, ocean acidification and ocean anoxia. However, CO_2 levels are likely to increase well beyond even 500 or more ppm, as will CH_4 levels in the next century or two, making reefs very different, if they are not extinguished altogether. Given our apparent inability to significantly reduce emissions, we must plan alternative strategies to preserve reefs in some form. Managing this scenario will require great effort and it will not be an easy task, either scientifically or politically. It will require a commitment of will, ingenuity, money, and effort. Nevertheless, because much more than reefs are at stake, the results will likely prove beneficial to all in the future.

A major hindrance to reducing the CO_2 content of the atmosphere and oceans is the continued cacophony of certain people that promulgate myths, disinformation, and false ideas about the problem (Oreskes and Conway 2010; Pilkey and Pilkey 2011). The problem of disruption by increasing CO_2 is not just with reefs, but also with our own societies and civilizations (Ehrlich and Ehrlich 2012). Everyone has much at stake in the future, even if they do not recognize it. The geologic record shows us how dramatic the changes can be if we fail to work together to solve these problems.

Acknowledgements We are indebted to numerous people with whom we have interacted about reefs. Lipps particularly thanks Drs. Loraine Casazza, Michele Weber, Scott Fay, Ken Severin, Bernard Ormsby, Yan Song, Ted Delaca, Malcolm Erskin, Pat Colin, Martin Langer, John Pandolfi, Steve Rowland, and A. Rozanov for working with him in the field on modern and fossil reefs as well as the many students who followed him into the field and asked penetrating questions. He also thanks Professors Brent Mishler and Carole Hickman, the late Professor David Stoddart and Mr. Frank Murphy for long time co-teaching with him on the reefs of French Polynesia. We thank Jose Garcia for the artistic reconstructions and Kallie Moore for photographs and management of the collections. Stanley and Lipps acknowledge support by grants from the National Science Foundation (DBI-0749683 and EAR 84-08001). Co-editors Dennis K. Hubbard and Caroline Roberts provided very useful suggestions that have improved this chapter, for which we thank them.

References

Adachi N, Ezaki Y, Liu J (2011) Early Ordovician shift in reef construction from microbial to metazoan reefs. Palaios 26:106–114

Adachi N, Liu J, Ezaki Y (2012) Early Ordovician stromatoporoid *Pulchrilamina spinosa* from South China: geobiological significance and implications for the early development of skeletal-dominated reefs. Paleontol Res 16:59–69

Alley RB, Clark PU, Huybrechts P, Joughin I (2005) Ice-sheet and sea-level changes. Science 310:456–460

Allwood AC, Walter MR, Burch IW, Kamber BS (2007) 3.43 billion-year-old stromatolite reef from the Pilbara Craton of Western Australia: ecosystem-scale insights to early life on Earth. Precamb Res 158:198–227

Baker A, Glynn P, Riegl B (2008) Climate change and coral reef bleaching: an ecological assessment of long-term impacts, recovery trends and future outlook. Estuar Coast Shelf Sci 80:435–471

Bambach RK (2006) Phanerozoic biodiversity mass extinctions. Annu Rev Earth Planet Sci 34:127–55

Barbeitos MS, Romano SL, Lasker H R (2010) Repeated loss of coloniality and symbiosis in scleractinian corals. Proc Nat Acad Sci 107:11877–11882

Barnosky AD (2014) Dodging Extinction. Univ California Press, Berkeley, 256 p

Barnosky AD, Matzke N, Tomiya S, Wogan GOU, Swartz B, Quental T, Marshall C, McGuire JL, Lindsey EL, Maguire KC, Mersey B, Ferrer EA (2011) Has the Earth's sixth mass extinction already arrived? Nature 471:51–57

Baron-Szabo RC (2003) Corals of the Theresienstein reef (Upper Turonian-Coniacian, Salzburg, Austria). Bull Biol Soc Wash 10:257–268

Baron-Szabo RC (2008) Corals of the K/T-boundary: scleractinian corals of the suborders Dendrophylliina, Caryophylliina, Fungiina, Microsolenina, and Stylinina. Zootaxa 1952:1–244

Beaufort L, Probert I, de Garidel-Thoron T, Bendif EM, Ruiz-Pino D, Metzl M, Goyet C, Buchet, N, Coupel P, Grelaud M, Rost B, Rickaby REM, de Vargas C. (2011) Sensitivity of coccolithophores to carbonate chemistry and ocean acidification. Nature 476:80–83

Beerling DJ, Lomax BH, Royer DL, Upchurch, GR Jr., Kump LR (2002) An atmospheric pCO_2 reconstruction across the Cretaceous-Tertiary boundary from leaf megafossils. Proc Nat Acad Sci 99:7836–7840

Bellwood DR, Hughes TP, Folke C, Nystro M (2004) Confronting the coral reef crisis. Nature 429:833

Benton MJ (2003) When life nearly died: the greatest mass extinction of all time. Thames & Hudson, London, 336 p

Bernecker M (2005) Late Triassic reefs from the northwest and south Tethys: distribution, setting, and biotic composition. Facies 51:457–468

Berner RA, Kothavala Z (2001) GEOCARB III: a revised model of atmospheric CO_2 over Phanerozoic time. Amer J Sci 301:182–204

Bishop JW, Sumner DY (2006) Molar tooth structures of the Neoarchean Monteville Formation, Transvaal Supergroup, South Africa. I: constraints on microcrystalline $CaCO_3$ precipitation. Sedimentology 53:1049–1068

Blackburn TJ, Olsen PE, Bowring SA, McLean NM, Kent DV, Puffer J, McHone G, Rasbury ET, Et-Touhami M (2013) Zircon U-Pb Geochronology links the end-Triassic extinction with the Central Atlantic Magmatic Province. Science 340:941–945

Brayard A, Vennin E, Olivier N, Bylund KG, Jenks J, Stephen DA, Bucher H, Hofmann R, Goudemand N, Escarguel G (2011) Transient metazoan reefs in the aftermath of the end-Permian mass extinction. Nature Geosci 4:693–697

Brenchley PJ, Carden GAF, Marshall JD (1995) Environmental changes associated with the "first strike" of the late Ordovician mass extinction. Modern Geol 20:69–82

Brunton FR, Copper P, Dixon OA (1997) Silurian reef-building episodes. Proc 8th Int Coral Reef Symp 2:1643–1650

Budd AF (2000) Diversity and extinction in the Cenozoic history of Caribbean reefs. Coral Reefs 19:25–35

Burgess SD, Bowring SA, Fleming TH, Elliot, DH (2015) High-precision geochronology links the Ferrar large igneous province with early-Jurassic ocean anoxia and biotic crisis. Earth Planet Sci Let 415:90–99

Butterfield NJ (2015) Early evolution of the Eukaryota. Palaeontol 58:5–17

Butterfield NJ, Rainbird RH (1998) Diverse organic-walled fossils including possible dinoflagellates from the early Neoproterozoic of Arctic Canada. Geol 26:963–966

Carpenter KE, Abrar M, Aeby G, Aronson RB, Banks S, Bruckner A, Chiriboga A, Cortes J, Delbeek JC, DeVantier L, Edgar GJ, Edwards AJ, Fenner D, Guzman HM, Hoeksema BW, Hodgson G, Johan O, Licuanan WY, Livingstone SR, Lovell ER, Moore JA, Obura DO, Ochavillo D, Polidoro BA, Precht WF, Quibilan MC, Reboton C, Richards ZT, Rogers AD, Sanciangco J, Sheppard A, Sheppard C, Smith J, Stuart S, Turak E, Veron JEN, Wallace C, Weil E, Wood E (2008) One-third of reef-building corals face elevated extinction risk from climate change and local impacts. Science 321:560–563

Caruthers AH, Smith PL, Gröcke DR (2014) The Pliensbachian-Toarcian (Early Jurassic) extinction: a North American perspective. Geol Soc Amer Spec Papers 505:225–243

Chenet A-L, Quidelleur X, Fluteau F, Courtillot V (2007) 40K/40Ar dating of the main Deccan large igneous province: further evidence of KTB age and short duration. Earth Planet Sci Lett 263:1–15

Clarkson MO, Kasemann SA, Wood RA, Lenton TM, Daines SJ, Richoz S, Ohnemueller F, Meixner A, Poulton SW, Tipper ET (2015) Ocean acidification and the Permo-Triassic mass extinction. Science 348:229–232

Colin PL (1987) Physiography of Enewetak Atoll. In Devaney DM (ed) The natural history of Enewetak Atoll. The natural history of Enewetak Atoll. Oak Ridge, Tenn: U.S. Dept. of Energy, Off Energy Research, Off Health Environ Res, Ecol Res Div, pp. 27–38

Colin PL (2009) Marine environments of Palau. Indo-Pacific Press, San Diego 414 p

Copper P (1994) Ancient reef ecosystem expansion and collapse. Coral Reefs 13:3–11

Copper P (2001) Evolution, radiations, and extinctions in Proterozoic to Mid-Paleozoic reefs. In: Stanley GD Jr. (ed) The history and sedimentology of ancient reef systems. Kluwer Academic/Plenum Pub, New York, pp. 89–119

Copper P (2002a) Reef development at the Frasnian/Famennian mass extinction boundary. Palaeogeogr Palaeoclim Palaeoecol 181:27–65

Copper P (2002b) Silurian and Devonian reefs: 80 million years of global greenhouse between two ice ages. SEPM Spec Publ 72:181–238

Copper P (2011) 100 million years of reef prosperity and collapse: Ordovician to Devonian interval. Paleontol Soc Papers 17:15–32

Copper P, Scotese CR (2003) Megareefs in Middle Devonian supergreenhouse climates. Geol Soc Amer Spec Pap 370:209–230

Courtillot VE, Renne PR (2003) On the ages of flood basalt events. C. R. Geoscience 335:113–140

Cowen R (1983) Algal symbiosis and its recognition in the fossil record. In: Tevesz MJS, McCall PL (eds) Biotic interactions in Recent and fossil benthic communities. Plenum Press, New York, pp. 432–478

Cowen R (1988) The role of algal symbiosis in reefs through time. Palaios 3:221–227

Dana JD (1872) Corals and coral islands. Dodd Mead Pub, New York, 349 p

Darwin, C (1842) The geology of the voyage of the Beagle. Part 1. The structure and distribution of coral reefs. Smith Elder Co, London 214 p

Davies PJ, Hopley D (1983) Growth fabrics and growth rates of Holocene reefs in the Great Barrier Reef. BMR J of Austral Geol Geophys 8:237–251

Dixon DL, Abrego D, Hay ME (2014) chemically mediated behavior of recruiting corals and fishes: a tipping point that may limit reef recovery. Science 345:892–897

Dixon DL, Hay ME (2012) Corals chemically cue mutualistic fishes to remove competing seaweeds. Science 338:804–807

Dullo W-C (2005) Coral growth and reef growth: a brief review. Coral Reefs 51:953–958

Ehrlich P, Ehrlich AH (2012) Can a collapse of global civilization be avoided? Proc. Royal Soc B 280: doi: 10.1098/rspb.2012.2845

Elias RJ, Young GA, Lee D-J, Bae B-Y (2013) Coral biogeography in the Late Ordovician (Cincinnatian) of Laurentia. In: Harper DAT, Servais T. (eds) Early Palaeozoic biogeography and palaeogeography. Geol Soc London Mem 38:97–115.

Erwin DH (2006) Extinction: how life on earth nearly ended 250 million years ago. Princeton Univ Press, Princeton, 320 p

Fagerstrom JA (1987) The evolution of reef communities. Wiley, New York, 600 p

Fautin DG, Buddemeier RW (2004) Adaptive bleaching: a general phenomenon. Hydrobiol 530/531:459–467

Fine M, Tchernov D (2007) Scleractinian coral species survive and recover from decalcification. Science 315:1811

Finnegan S, Heimc NA, Petersc SE, Fischera WW (2012) Climate change and the selective signature of the Late Ordovician mass extinction. Proc Nat Acad Sci 109:6829–6834

Fletcher BJ, Brentnall SJ, Anderson CW, Berner RA, Beerling DJ (2008) Atmospheric carbon dioxide linked with Mesozoic and early Cenozoic climate change. Nature Geosci 1:43–48

Flügel, E. (2002) Triassic reef patterns. In Kiessling, W, Flügel, E and Golonka, J (eds) Phanerozoic reef patterns. Tulsa, SEPM (Society for Sedimentary Geology) Special Publication 72:391–463

Flügel E, Senowbari-Daryan B (2001) Triassic reefs of the Tethys. In: Stanley GD Jr. (ed) The history and sedimentology of ancient reef systems. Kluwer Academic/Plenum Pub, New York, pp. 217–249

Flügel E, Stanley GD Jr (1984) Reorganization, development and evolution of post-Permian reefs and reef organisms. Palaeontogr America 54:177–186

Fraser NM, Bottjer DG, Fischer AG (2004) Dissecting "*Lithiotis*" bivalves: implications for the Early Jurassic reef eclipse. Palaios 19:51–67

Gili E, Masse J-P, Skelton PW (1994) Rudists as gregarious sediment-dwellers, not reef-builders, on Cretaceous carbonate platforms. Palaeogeogr Palaeoclimatol Palaeoecol 118:245–267

Glynn P (1996) Coral reef bleaching: facts, hypotheses and implications. Glob Change Biol 2:495–509

Graham NAJ, Nash KL (2013) The importance of structural complexity in coral reef ecosystems. Coral Reefs 32:315–326

Graham NAJ, Wilson KS, Jennings SK, Polunin S, Bijoux JP, Robinson J (2006) Dynamic fragility of oceanic coral reef ecosystems. Proc Nat Acad Sci 103:8425–8429

Grotzinger JP (1990) Geochemical model for Proterozoic stromatolite decline. Amer J Sci 290A:80–103

Grotzinger JP (1994) Trends in Precambrian carbonate sediments and their implication for the understanding of the evolution of life. In: Bengtson S (ed) Early life on Earth. Columbia Univ Press, New York, pp. 245–258

Grotzinger JP, James NP (2012) Precambrian carbonates: evolution of understanding. In: Grotzinger JP, James NP (eds) Carbonate sedimentation and diagenesis in the evolving Precambrian world. SEPM Spec Publ 67:3–22

Grotzinger JP, Knoll AH (1995) Anomalous carbonate precipitates: is the Precambrian the key to the Permian? Palaios 10:578–596

Grotzinger JP, Watters WA, Knoll AH (2000) Calcified metazoans in thrombolite-stromatolite reefs of the terminal Proterozoic Nama Group, Namibia. Paleobiol 26: 334–359

Groves JR, Yue W (2009) Foraminiferal diversification during the late Paleozoic ice age. Paleobiol 3:367–392

Hallock P (1997) Reefs and reef limestones in Earth history. In: Birkeland C (ed) Life and death of coral reefs. Chapman Hall, New York, pp. 13–42

Hallock P, Schlager W (1986) Nutrient excess and the demise of coral reefs and carbonate platforms. Palaios 1:389–398

He WH, Shi GR, Twitchett RJ, Zhang Y, Zhang K-X, Song H- J, Yue M-L, Wu S-B, Wu H-T, Yang T-L, Xiao Y-F. (2014) Late Permian marine ecosystem collapse began in deeper waters: evidence from brachiopod diversity and body size changes. Geobiol [doi: 10.1111/gbi.12119]

Hodges MS, Stanley GD Jr (2015) North American coral recovery after the end-Triassic mass extinction, New York Canyon, Nevada, USA. GSA Today 25:4–9

Hoegh-Guldberg O (2014) Coral reefs in the Anthropocene: persistence or the end of the line? Waters CN, Zalasiewicz JA, Williams M, Ellis MA, Snelling AM (eds) A stratigraphical basis for the Anthropocene. Geol Soc London Spec Publ 395:167–183

Hoegh Guldberg O, Mumby PJ, Hooten AJ, Steneck RS, Greenfield P, Gomez E, Harvell CD, Sale F, Edwards AJ, Caldeira K., Knowlton N, Eakin CM, Iglesias-Prieto R, Muthiga N, Bradbury HR, Dubi A, Hatziolos ME (2007) Coral reefs under rapid climate change and ocean acidification. Science 318:1737–1742

Hönisch B, Ridgwell A, Schmidt DN, Thomas E, Gibbs SJ, Sluijs A, Zeebe R, Kump L, Martindale RC, Greene SE, Kiessling W, Ries J, Zachos JC, Royer DL, Barker S, Marchitto Jr TM, Moyer R, Pelejero C, Ziveri P, Foster GL,Williams B (2012) The geological record of ocean acidification. Science 335:1058–1063

Hopley D, Smithers SG, Parnell KG (2007) The geomorphology of the Great Barrier Reef: development, diversity, and change. Cambridge Univ Press, Cambridge, 532 p

Hubbard DK (2009) Depth-related and species-related patterns of Holocene reef accretion in the Caribbean and western Atlantic: a critical assessment of existing models. Int. Assoc. Sedimentol. Spec Publ 41:1–18

Hubbard DK (2014) Holocene accretion rates and styles for Caribbean coral reefs: lessons for the past and future. In: Harris M (ed) Facies architecture of shelf-edge carbonates. SEPM Special Publication 105:264–281

Hubbard DK, Gill IP, Burke RB (2013) Holocene reef building on eastern St. Croix, US Virgin Islands: Lang Bank revisited. Coral Reefs 32:653–669

Hueerkamp C, Glynn PW, D'Croz L, Maté L, Colley SB (2001) Bleaching and recovery of five eastern Pacific corals in an El Niño-related temperature experiment. Bull Mar Sci 69:215–236

Hughes TP, Baird AH, Bellwood DR, Card M, Connolly SR, Folke C, Grosberg R, Hoegh-Guldberg O, Jackson JBC, Kleypas J, Lough JM, Marshall P, Nyström M, Palumbi SR, Pandolfi JM, Rosen B, Roughgarden J (2003) Climate change, human impacts, and the resilience of coral reefs. Science 301:929–933

Insalaco E, Hallam A, Rosen B (2008) Oxfordian (Upper Jurassic) coral reefs in Western Europe: reef types and conceptual depositional model. Sedimentol 44:707–734

IPCC (2014) Summary for policymakers. In: Field, CB, Barros VR, Dokken DJ, Mach KJ, Mastrandrea MD, Bilir TE, Chatterjee M, Ebi KL, Estrada YO, Genova RC, Girma B, Kissel ES, Levy AN, MacCracken S, Mastrandrea PR, White LL (eds) Climate Change 2014: Impacts, Adaptation, and Vulnerability. Part A: Global and sectoral Aspects. Contribution of Working Group II to the Fifth Assessment Report of the Intergovernmental Panel on Climate Change. Cambridge University Press, Cambridge, pp. 1–32

Isozaki Y, Aljinović D (2009) End-Guadalupian extinction of the Permian gigantic bivalve Alatoconchidae: end of gigantism in tropical seas by cooling. Palaeogeogr Palaeoclimatol Palaeoecol 284:1–11

Joachimski MM, Lai XL, Shen SZ, Jiang HS, Luo GM, Chen B, Chen J, Sun YD (2012) Climate warming in the latest Permian and the Permian–Triassic mass extinction. Geol 40:195–198

Johnson CC (2002) The rise and fall of rudist reefs. Amer Sci 90:148–153

Johnson CC, Kauffman EG (2001) Cretaceous evolution of reef ecosystems: a regional synthesis of the Caribbean tropics. In: Stanley GD Jr. (ed) The history and sedimentology of ancient reef systems. Kluwer Academic/Plenum Pub, New York, pp. 311–349

Johnson KG, Jackson JBC, Budd AF (2008) Caribbean reef development was independent of coral diversity over 28 million years. Science 319:1521–1523

Jourdan F, Féraud G, Bertrand H, Watkeys MK, Renne PR (2008) The 40Ar/39Ar ages of the sill complex of Karoo large igneous province: implications for the Pliensbachian-Toarcian climate change. Geochem Geophys Geosyst 9:Q06009, doi:10.1029/2008GC001994

Karhu J, Epstein S (1986) The implication of the oxygen isotope records in coexisting cherts and phosphates. Geochem Cosmochem Acta 30:1745–1756

Karlson RH, Cornell HV, Hughes TP (2004) Coral communities are regionally enriched along an oceanic biodiversity gradient. Nature 429:867–870

Kauffman EG, Johnson CC (1988) The morphological and ecological evolution of Middle and Upper Cretaceous reef-building rudistids. Palaios 3:194–216

Kiessling W (2002) Secular variations in the Phanerozoic reef ecosystem In Kiessling, W, Flügel, E and Golonka, J (eds) Phanerozoic reef patterns. Tulsa, SEPM Spec Pub 72:391–463

Kiessling W (2009) Geologic and biologic controls on the evolution of reefs. Ann Rev Ecol Evol Syst 40:173–192

Kiessling W (2010) Reef expansion during the Triassic: spread of photosymbiosis balancing climatic cooling. Palaeogeogr Palaeoclimatol Palaeoecol 290:11–19

Kiessling W (2011) On the potential for ocean acidification to be a general cause of ancient reef crises. Glob Change Biol 17:56–67

Kiessling W, Baron-Szabo RC (2004) Extinction and recovery patterns of scleractinian corals at the Cretaceous-Tertiary boundary. Palaeogeogr Palaeoclimatol Palaeoecol 214:195–223

Kiessling W, Kocsis AT (2015) Biodiversity dynamics and environmental occupancy of fossil azooxanthellate and zooxanthellate scleractinian corals. Paleobiol 41:402–414

Kiessling W, Simpson C, Beck B, Mewis H, Pandolfi JM (2012) Equatorial decline of reef corals during the last Pleistocene interglacial. Proc Nat Acad Sci 109:21378–21383

Kiessling W, Simpson C, Foote M (2010) Reefs as cradles of evolution and sources of biodiversity in the Phanerozoic. Science 327:196–198

Kleppel GS, Dodge RE, Reese CJ (1989) Changes in pigmentation associated with the bleaching of stony corals. Limnol and Oceanogr 34:1331–1335

Kleypas JA, Yates KK (2009) Coral reefs and ocean acidification. Oceanogr 22:108–117

Knowlton N, Brainard RE, Fishcer R, Moews M, Plaisance L, Caley MJ (2010) Coral reef diversity. In: MacIntyre AD (ed) Life in the world's oceans: diversity, distribution, and abundance. Wiley-Blackwell, Oxford, pp. 65–77

Knowlton N, Jackson JBC (2008) Shifting baselines, local impacts, and global change on coral reefs. PLoS Biol 6:215–220

Knowlton N, Jackson JBC (2011) Evolutionary diversity and ecological complexity of coral reefs. Paleontol Soc Pap 17:111–120

Köhler P, Fischer H, Schmitt J (2010) Atmospheric $\delta^{13}CO_2$ and its relation to pCO_2 and deep ocean $\delta^{13}C$ during the late Pleistocene. Paleoceanogr 25 doi:10.1029/2008PA001703

Ladner T, Barshis DJ, Palumbi, SR (2012) Protein evolution in two co-occurring types of *Symbiodinium*: an exploration into the genetic basis of thermal tolerance in *Symbiodinium* clade D. Evol Biol. 12:217 http://www.biomedcentral.com/1471-2148/12/217

Landing E, Geyer G, Brasier MD, Bowring, SA (2013a) Cambrian evolutionary radiation: context, correlation, and chronostratigraphy—overcoming deficiencies of the first appearance datum (FAD) concept. Earth-Sci Rev 123:133–172

Landing E, Westrop SR, Bowring SA (2013b) Reconstructing the Avalonia palaeocontinent in the Cambrian: a 519 Ma caliche in South Wales and transcontinental middle Terreneuvian sandstones. Geol Mag 150:1022–1046

Lathuilière B, Marchal D (2009) Extinction, survival and recovery of corals from the Triassic to Middle Jurassic time: Terra Nova, v. 21, no. 1, p. 57–66.

Lees A (1988) Waulsortian reefs the history of a concept. Mem l'Inst Geol l'Universite Louvain 34:43–55

Leigh EG, O'Dea A, Vermeij GJ (2014) Historical biogeography of the Isthmus of Panama. Biol Rev 89:148–172

Leinfelder RR (2001) Jurassic reef ecosystems. In: Stanley GD Jr. (ed) The history and sedimentology of ancient reef systems. Kluwer Academic/Plenum Pub, New York pp 251–309

Lessios HA (2008) The great American schism: divergence of marine organisms after the rise of the central American isthmus. Ann Rev Ecol Evol Syst 39:63–91

Lipps JH (1970) Plankton evolution. Evol 24:1–22

Lipps JH (2006) Major features of protistan evolution: Controversies, problems and a few answers. Anu Inst Geociên–UFRJ 29:55–80

Lipps JH (2011) Reef restoration—the good and the bad, a paleobiologic perspective. Paleontol Soc Papers 17:139–152

Little AF, Van Oppen JH, Willis BL (2004) Flexibility in algal endosymbioses shapes growth in reef corals. Science 304:1492–1494

McCauley DJ, Pinsky ML, Palumbi SR, Estes JA, Joyce FH, Warner RR (2015) Marine defaunation: animal loss in the global ocean. Science 347(6219) doi: 10.1126/science.1255641

Medina M, Collins A, Takaoka TL, Kuehl J, Boore JL (2006) Naked corals: skeleton loss in Scleratinia. Proc Nat Acad Sci 103:9096–9100

Melchin MJ, Mitchell CE, Holmden, C, Štorch P (2013) Environmental changes in the Late Ordovician–early Silurian: review and new insights from black shales and nitrogen isotopes. Geol Soc Amer Bull 125:1635–1670

Meyer CP, Geller JB, Paulay G (2005) Fine scale endemism on coral reefs: archipelagic differentiation in turbinid gastropods. Evol 59:113–125

Miller J, Muller E, Rogers C, Waara R, Atkinson A, Whelan KRT, Patterson M, Witcher B (2009) Coral disease following massive bleaching in 2005 causes 60% decline in coral cover on reefs in the US Virgin Islands. Coral Reefs 28:925–937

Moffitt SE, Hill TM, Roopnarine PD, Kennett JP (2015) Response of seafloor ecosystems to abrupt global climate change. Proc Nat Acad Sci 112:4684–4689

Montaggioni LF (2005) History of Indo-Pacific coral reef systems since the last glaciation: development patterns and controlling factors. Earth-Sci Rev 71:1–75

Muscatine L, Goiran C, Land L, Jaubert J, Cuif J-P, Allemand D (2005) Stable isotopes (^{13}C and ^{15}N) of organic matrix from coral skeleton. Proc Nat Acad Sci 102:1525–1530

Muscatine L. (1990) The role of symbiotic algae in carbon and energy flux in reef corals. In: Dubinsky Z (ed) Ecosystems of the world: Coral reefs. Elsevier, Amsterdam, pp. 75–87

Newell, ND (1971) An outline history of tropical organic reefs. American Mus Novitiates 2465:1–37

Newell ND, Rigby JK, Fischer AG, Whiteman AJ, Hicks LE, Bradley J. (1953) The Permian Reef Complex of the Guadalupe Mountains region, Texas and New Mexico: a study in paleoecology. WH Freeman, San Francisco. 236 p.

Oliver WA Jr (1996) Origins and relationships of Paleozoic coral groups and the origin of the Scleractinia. Paleontol Soc Papers 1:107–134

Oreskes N, Conway EM (2010) Merchants of doubt—how a handful of scientists obscured the truth on issues from tobacco smoke to global warming. Bloomsbury Press, New York, 345 p.

Pandolfi JM (1996) Limited membership in Pleistocene reef coral assemblages from the Huon Peninsula, Papua New Guinea: constancy during global change. Paleobiol 22:152–176

Pandolfi JM (2002) Coral community dynamics at multiple scales. Coral Reefs 21:13–23

Pandolfi JM, Jackson JBC, Baron N, Bradbury RH, Guzman HM, Hughes TP, Kappel CV, Micheli F, Ogden JC, Possingham HP, Sala E (2005) Are U.S. Coral Reefs on the slippery slope to slime? Science 307:1725–1726

Paulay G (1997) Diversity and distribution of reef organisms. In: Birkeland C (ed) Life and death of coral reefs. Chapman Hall, New York, pp. 298–353

Payne JL, Clapham ME (2012) End-Permian mass extinction in the oceans: an ancient analog for the Twenty-First Century? Ann Rev Earth Planet Sci 40:89–111

Perrin C (2002) Tertiary: the emergence of modern reef ecosystems. SEPM Spec Publ 71:587–621

Perrin C, Kiessling W (2010) Latitudinal trends in Cenozoic reef patterns and their relationship to climate. In: Piller W, Mutti M, Betzler C (eds) Carbonate systems during the Oligocene–Miocene climatic transition. Internat Assoc Sedimentol 42:17–34

Pilkey OH, Pilkey KC (2011) Global climate change—a primer. Duke Univ Press, Durham, 142 p.

Planavsky NJ, Reinhart CT, Wang X, Thomson D, McGoldrick P, Rainbird RH, Johnson T, Fischer WW, Lyons TW (2014) Low Mid-Proterozoic atmospheric oxygen levels and the delayed rise of animals. Science 346: 635–638

Pochon X, Montoya-Burgos JI, Stadelmann, B, Pawlowski JP (2006) Molecular phylogeny, evolutionary rates, and divergence timing of the symbiotic dinoflagellate genus *Symbiodinium*. Mol Phylogen Evol 38:20–30

Pope KO (2002) Impact dust not the cause of the Cretaceous-Tertiary mass extinction. Geol 30:99–102

Pruss SB, Bottjer DJ (2005) The reorganization of reef communities following the end-Permian mass extinction. Compt Rend Palevol 4:553–568

Reaka-Kudla ML (1997) The global diversity of coral reefs: a comparison with rain forests. In: Reaka-Kudla ML, Wilson DE, Wilson EO (eds) Biodiversity II: Understanding and protecting our biological resources. Joseph Henry Press, Washington, DC, pp. 83–108

Richards MA, Alvarez W, Self S, Karlstrom L, Renne PR, Manga M, Sprain CJ, Smit, J, Vanderkluysen L, Gibson SA (2015) Triggering of the largest Deccan eruptions by the Chicxulub impact. Geol Soc Amer Bull doi: 10.1130/B31167.1

Robertson DS, Lewis WM, Sheehan PM, Toon OB (2013) K-Pg extinction: reevaluation of the heat-fire hypothesis. J Geophys Res: Biogeosci 118:329–336

Romano SL, Palumbi SR (1996) Evolution of scleractinian corals inferred from molecular systematics. Science 271:640–642

Roniewicz E (2010) Uniform habit spectrum vs. taxonomic discrepancy between two succeeding Triassic coral faunas: a proof of the intra-Norian faunal turnover: Palaeoworld 19:410–413

Rosen BR (2000) Algal symbiosis, and the collapse and recovery of reef communities: Lazarus corals across the K-T boundary. In: Culver SJ, Rawson PA (eds) Biotic response to global change: the last 145 million years. Cambridge Univ Press, Cambridge, pp. 164–180

Rosen BR, Turnšek D (1989) Extinction patterns and biogeography of scleractinian corals across the Cretaceous/Tertiary boundary. In: Jell PA, Pickett JW (eds) Fossil Cnidaria 5 Mem Assoc Australas Paleontol 8:355–370

Rowan R (2004) Coral bleaching: thermal adaptation in reef coral symbionts. Nature 430:742

Rowland SM (2001) Archaeocyaths—a history of phylogenetic interpretation. J Paleontol 75:1065–1078

Rowland SM, Hicks M (2004) The Early Cambrian experiment in reef-building by metazoans. In: Lipps JH,Waggoner BM (eds) Neoproterozoic—Cambrian biological revolutions. Paleontol Soc Papers 10:107–130

Rowland SM, Shapiro SH (2002) Reef patterns and environmental influences in the Cambrian and earliest Ordovician. SEPM Spec Publ 72:95–128

Salvat B (ed) (1987) Human impacts on coral reefs: facts and recommendations. Antenne Mus EPHE, French Polynesia, 253 p

Salvat B, Aubanel A, Adjeroud M, Bouisset P, Calmet D, Chancerelle Y, Cochennec N, Davies N, Fougerousse A, Galzin R, Lagouy R, Lo C, Monier C, Ponsonnet C, Remoissenet G, Schneider D, Stein A, Tatarata M, Villiers L (2008) Monitoring of French Polynesia coral reefs and their recent development. Rev Ecol-La Terre La Vie 63:145–177

Schlager W (1981) The paradox of drowned reefs and carbonate platforms. Bull.Geol. Soc Amer 92:197–211

Schoene B, Samperton KM, Eddy MP, Keller G, Adatte, T, Bowring SA, Khadri SFR, Gertsch B (2015) U-Pb geochronology of the Deccan Traps and relation to the end-Cretaceous mass extinction. Science 347:182–184

Schopf JW (1992) Paleobiology of the Archean. In: Schopf JW, Klein C (eds) The Proterozoic biosphere. Cambridge Univ Press, Cambridge, pp. 25–39

Scott RW (1988) Evolution of Late Jurassic and Early Cretaceous reef biotas. Palaios 3:184–193

Scott RW, Fernández-Mendiola PA, Gili E, Simó A (1990) Persistence of coral-rudist reefs into the Late Cretaceous. Palaios 5:98–110

Sella B, Ovtcharova M, Guex J, Bartolini A, Jourdand F, Spangenberg JE, Vicente J-C, Schaltegger U (2014) Evaluating the temporal link between the Karoo LIP and climatic–biologic events of the Toarcian Stage with high-precision U–Pb geochronology. Earth Planet Sci Let 408:48–56

Shen S-Z, Bowring SA (2014) The end-Permian mass extinction: a still unexplained catastrophe. Nat Sci Rev 1: 492–495

Signor PW, Lipps JH (1982) Sampling bias, gradual extinction patterns and catastrophes in the fossil record. Geol Soc Amer Spec Papers 190:291–296

Simon SL, Robinson WL (1997) A compilation of nuclear weapons test detonation data for U.S. Pacific Ocean tests. Health Phys 73:258–264

Simpson C, Kiessling W, Mewis H, Baron-Szabo RC, Müller J (2011) Evolutionary diversification of reef corals: a comparison of the molecular and fossil records. Evol 65:3274–3284

Spalding MD, Brown BE (2015) Warm-water coral reefs and climate change. Science 350:769–772

Stanley GD Jr (1981) Early history of scleractinian corals and its geological consequences. Geology 9:507–511

Stanley GD Jr (1988) The history of early Mesozoic reef communities: a three-step process. Palaios 3:170–183

Stanley GD Jr (1991) Tropical reef ecosystems. In Nierenberg WA (ed) Encyclopedia of earth system science. Academic Press, New York 4:375–388

Stanley GD Jr (2001) Introduction to reef ecosystems and their evolution. In: Stanley GD Jr (ed) The history and sedimentology of ancient reef systems. Kluwer Academic/Plenum Publ, New York, pp. 1–39

Stanley GD Jr (2003) The evolution of modern corals and their early history. Earth Sci Rev 60:195–225

Stanley GD Jr (2005) Coral microatolls from the Triassic of Nevada: oldest scleractinian examples. Coral Reefs 24:247

Stanley GD Jr (2006) Photosymbiosis and the evolution of modern coral reefs. Science 312: 857–860

Stanley GD Jr (2011) The naked Lazarus effect and the recovery of corals after the end-Permian mass extinction. Kölner Forum Geol Paläontol 19:164–165

Stanley GD Jr (2015) Geologic history of reefs. In: McGraw-Hill Yearbook of Science and Technology 2015, McGraw-Hill publishers, New York, pp. 124–126.

Stanley GD Jr, Fautin DF (2001) The origins of modern corals. Science 291:1913–1914

Stanley GD Jr, Helmle KP (2010) Middle Triassic coral growth bands and their implication for photosymbiosis. Palaios 25:754–763

Stanley GD Jr, Lipps JH (2011) Photosymbiosis: the driving force for reef success and failure. Paleontol Soc Papers 17:33–60

Stanley GD Jr, Swart PK (1995) Evolution of the coral-zooxanthellae symbiosis during the Triassic: a geochemical approach. Paleobiol 21:179–199

Stanley SM, Hardie A (1998) Secular oscillations in the carbonate mineralogy of reef-building and sediment-producing organisms driven by tectonically forced shifts in seawater chemistry. Palaeogeogr Palaeoclimatol Palaeoecol 144:3–19

Stanton RJ (2006) Nutrient models for the development and location of ancient reefs. Geo. Alps 3:191–206

Stanton RJ, Flügel E (1987) Paleoecology of Upper Triassic reefs in the Northern Calcareous Alps: reef communities. Facies 16:157–186

Steuber T, Mitchell SF, Buhl D, Gunter G, Kasper HU (2002) Catastrophic extinction of Caribbean rudist bivalves at the Cretaceous-Tertiary boundary. Geology 30:999–1002

Stigall AL (2012) Speciation collapse and invasive species dynamics during the Late Devonian "Mass Extinction". GSA Today 22:4–8

Sun Y, Joachimski MM, Wignell PB, Yan C, Chen Y, Jiang H, Wang L, Lai X (2012) Lethally hot temperatures during the Early Triassic greenhouse. Science 338:366–370

Svensen H, Planke S, Chevallier L, Malthe-Sørenssen A, Corfu F, Jamtveit B (2007) Hydrothermal venting of greenhouse gases triggering Early Jurassic global warming. Earth Planet Sci Let 256:554–566

Svensen H, Planke S, Polozov AG, Schmidbauer N, Corfu F, Podladchikov YY, Jamtveit B (2009) Siberian gas venting and the end-Permian environmental crisis. Earth Planet Sci Letters 277:490–500

Talent JA (1988) Organic reef-building: episodes of extinction and symbiosis? Lethaea 69:315–368

Talge HK, Hallock P (2003) Ultrastructural responses in field-bleached and experimentally stressed *Amphistegina gibbosa* (Class Foraminifera). J Eukary Microbiol 50:324–333

Toon OB, Zahnle K, Morrison D, Turco RP, Covey C (1997) Environmental perturbations caused by the impacts of asteroids and comets. Rev Geophys 35: 41–78

Tripati AK, Roberts CD, Eagle RA (2009) Coupling of CO_2 and ice sheet stability over major climate transitions of the last 20 million years. Science 326:1394–1397

Van Geldern R, Joachimski MM, Day J, Jansen U, Alvarez F, Yolkin EA, Ma XP (2006) Carbon, oxygen and strontium isotope records of Devonian brachiopod shell calcite. Palaeogeogr Palaeoclim Palaeoecol 240:47–67

Van Oppen MJH, Lough JM (eds) (2009) Coral bleaching: patterns, processes, causes, and consequences. Springer-Verlag Berlin, Heidelberg

Veron JEN (2008) Mass extinctions and ocean acidification: biological constraints on geological dilemmas. Coral Reefs 27:459–472

Veron JEN (2011) Ocean acidification and coral reefs: an emerging big picture. Diversity 3:262–274

Walter MR (1983) Archean stromatolites: evidence of the Earth's earliest benthos. In: Schopf JW (ed) Earth's earliest biosphere. Princeton Univ Press, Princeton NJ, pp. 187–213

Webby BD (2002) Patterns of Ordovician reef development. SEPM Spec Publ 72:129–179

Weidlich O (2002) Middle and Late Permian reefs—distributional patterns and reservoir potential. SEPM Spec Publ 72:339–390

West RR (1988) Temporal changes in Carboniferous reef mound communities. Palaios 3:152–169

Wignall PB, Twitchett RJ (1996) Oceanic anoxia and the end Permian mass extinction. Science 272:1155–1158

Wood R (1993) Nutrients, predation and the history of reef-building. Palaios 8:526–543

Wood R (1999) Reef evolution. Oxford Univ Press, Oxford, 414 p

Wood R, Curtis A (2014) Extensive metazoan reefs from the Ediacaran Nama Group, Namibia: the rise of benthic suspension feeding. Geobiol doi: 10.1111/gbi.12122

Woods AD (2014) Assessing Early Triassic paleoceanographic conditions via unusual sedimentary fabrics and features. Earth-Sci Rev 137:6–18

Zachos J, Pagani M, Sloan L, Thomas E, Billups K (2001) Trends, rhythms, and aberrations in global climate 65 ma to present. Science 292:686–693

Zachos JC, Schouten S, Bohaty SM, Quattlebaum T, Sluijs A, Brinkhuis H, Gibbs SJ, Bralower TJ (2006) Extreme warming of mid-latitude coastal ocean during the Paleocene-Eocene Thermal maximum: inferences from TEX86 and isotope data. Geol 34:737–740

Zamagni J, Mutti M, Košir A (2012) The evolution of mid Paleocene-early Eocene coral communities: how to survive during rapid global warming. Palaeogeogr Palaeoclimatol Palaeoecol 317–318:48–65

Zeebe EE, Wolf-Gladrow DA (2001) CO_2 in seawater equilibrium, kinetics and isotopes. Elsevier Oceanography Series No 65, Elsevier B.V., 346

Zhang Y, Payne JL (2012) Size-frequency distributions along a latitudinal gradient in Middle Permian fusulinoideans. PLoS One 7(6): e38603. doi:10.1371/journal.pone.0038603

Zhuravlev AY (2001) Paleoecology of Cambrian reef ecosystems. In: Stanley GD Jr (ed) The history and sedimentology of ancienkenneth Johnson Caribbean Acroporat reef systems. Kluwer Academic/Plenum Publishers, New York, p 121–157

Climate Change, Ocean Chemistry, and the Evolution of Reefs Through Time

9

Christopher P. Jury and Paul L. Jokiel

Abstract

Reef building has responded to changes in climate, ocean chemistry, and a variety of other physical and biological factors during the geologic past, as have the taxa involved. Many of the data revealed by the geologic record are also relevant to human impacts on coral reefs today and their success moving forward. This chapter reviews the responses of reefs and reef builders to environmental changes over Earth's history and relates this information to projected changes due to anthropogenic activities going forward. These changes include increasing temperature, ocean acidification, more intense storms, sea-level rise, nutrification, and sedimentation. Past events provide some insights, but are somewhat limited proxies of future impacts, largely because of the perhaps unprecedented current rate of CO_2 release today. Present-day rates of climate change and ocean acidification may be higher than at any point in the geologic past, and may exceed the capacity for corals and other reef builders to tolerate or adapt to the changing environment.

Keywords

Coral • Reef • Biomineralization • Calcification • Ocean acidification • PETM • Calcite • Aragonite • Climate change

9.1 Introduction

Environmental changes can impact the capacity for corals and other calcifying organisms to build reefs. Many of the changes that are projected to occur in response to anthropogenic influences have occurred in the geologic past. This chapter complements the discussion of calcification by modern corals in Chap. 2 and the geologic history of photosymbiosis in Chap. 3 by considering pre-anthropogenic variations in the physical, chemical and biological regime of the world ocean and their effects on the evolution of reefs and reef building organisms through time. It summarizes what is known about how organisms responded to past climate, ocean chemistry, and other factors and provides insight into how future reefs may react to environmental changes driven by human activities including increasing temperature, ocean acidification (OA), more intense storms, sea-level rise, nutrification, and sedimentation. Relevant reviews include: Stanley (2003), on the evolution of modern scleractinians; Kiessling (2009), on geologic and biologic controls of reef evolution; Pandolfi et al. (2011) on modern aspects of coral reef paleoecology within a historical framework; Hönisch et al. (2012), on geological evidence of OA and it effects over the past ~300 million years; Hansen et al. (2013) on what past climate change might tell us about future impacts to reefs and on the humans that depend on them.

This chapter builds on these discussions and the chronicling of reef history in Chap. 8, tying past events more closely to changing ocean conditions. It also distinguishes between the evolving suite of carbonate-

C.P. Jury (✉) • P.L. Jokiel
Hawaii Institute of Marine Biology, University of Hawaii, 1346Kaneohe, HI 96744, USA
e-mail: jurycp@hawaii.edu; jokiel@hawaii.edu

D.K. Hubbard et al. (eds.), *Coral Reefs at the Crossroads*, Coral Reefs of the World 6, DOI 10.1007/978-94-017-7567-0_9

producing organisms and the development of the structures they built and inhabited. Chapter 6 points out that, while the two are inextricably linked, reef building involves more than just calcification. Looking to the future, accelerating sea-level rise will challenge the ability of reefs to keep up; the overview in the following pages attempts to put this into the context of reef evolution in the geologic past as well as the possible fate of reefs in the twenty-first century and beyond.

We begin with an overview of the processes that are most intimately tied to coral growth and reef building. We then chronicle the evolution of reefs since the Archean Eon in response to changes in physical and chemical conditions and ecological interactions. Finally, these lessons are applied to possible scenarios in the twenty-first century to provide a contrast between pre- and post-anthropogenic factors related to reef building.

9.2 Physical, Chemical, and Biological Drivers of Reef Building

Reef accretion in tropical and subtropical waters is driven by high rates of calcification, primarily among photosymbiotic corals and calcifying algae, which exceed local rates of reef erosion and dissolution. Not all shallow-water corals are symbiotic, and not all symbiotic corals grow fast enough to build reefs. Conversely, some non-photosynthetic species of primarily deep-water corals can still produce reefs. They tend to be slower-growing than their shallow-water counterparts, but live in environments with lower disturbance, allowing for the accumulation of reef carbonates over time. The ability of shallow- and deep-water corals to create reefs is constrained by environmental and biological factors. Changes in these parameters modulate rates of reef building across space and time.

Organisms cope with changes in their environment through three major mechanisms: migration, acclimatization, and adaptation. In some situations organisms can cope with changing environmental conditions by simply moving from one place to another to find suitable habitat. Sessile species such as adult corals cannot migrate, but their planktonic larvae can disperse, potentially over great distances. A variety of physical (e.g., current velocity) and biological (e.g., planktonic larval duration) factors constrain how fast and far marine organisms can move from one place to another. Successful migration is only possible if suitable habitat exists at the new location. Under global change, there may be nowhere to run, requiring organisms to engage in other responses to survive.

One possibility is acclimatization (also referred to as phenotypic plasticity). This occurs through physiological or behavioral changes within a *single organism* based on previous conditioning, and occurs *within a single generation*. Another possible option is adaptation (i.e., evolution by natural selection). This occurs due to selective pressure acting on heritable variation within a *population*, and occurs over *one or more generations*. On an evolutionary scale, acclimatization is relatively fast-acting, but is inherently limited in the scope of variation it can generate. Adaptation may require long periods of time to achieve major changes, but it is less constrained in the degree of variation it can generate.

By way of example, a bird with a broken wing which is unable to fly would lose mass in its flight muscles due to a lack of exercise, but would regain that mass after the bone heals and the muscles can be exercised again. That process of losing and regaining muscle mass is acclimatization—the flight muscles undergo physiological changes according to usage. In contrast, turning the forearm of a dinosaur into the wing of a bird is an exquisite adaptation that requires evolution. The rate of evolution is often slow but it can increase dramatically in a rapidly changing environment, given strong selective pressure.

Sometimes the rate of environmental change exceeds the rate at which organisms can migrate, acclimatize, or adapt leaving only one option—to die. Results can range from the death of individuals to the collapse of entire populations, leading to the extinction of species. These events are more common than we might imagine; as Carl Sagan quipped, "Extinction is the Rule. Survival is the exception." And, even where species are able to survive under environmental change, their ecological roles may shift under the new conditions. At many times in the geologic past, for example, corals survived major environmental perturbations even as coral-reef building ground to a halt.

Whether we are looking into the deep past or the immediate future, it is important to understand the factors which fundamentally impact growth rates of organisms and the reefs that they build. Many of these factors have undergone changes in the past, and can help us to understand natural changes in reefs. However, environmental conditions are changing rapidly due to climate change and other human influences such that migration, acclimatization or adaptation may not be viable options for some species. The following pages will focus on how environmental and biological factors interact to drive changes in reef building. We then examine the rise and fall of reefs and reef biota throughout geologic time, focusing in particular on ocean and climate processes that are most relevant to the present-day situation. Finally, we use these past responses to consider the fate of future reefs as the rate of change in key environmental factors continues to accelerate.

9.2.1 Light, Turbidity, and Sedimentation

Calcification in shallow-water, reef building corals is fundamentally a light-driven process and some scleractinian corals have been photosymbiotic since at least the Triassic (Muscatine et al. 2005; Stanley and Helmle 2010; see also Chaps. 3 and 8). The fact that calcification rates of symbiotic corals tend to be higher in the light than in the dark has been known for decades (Kawaguti and Sakumoto 1948; Muscatine and Cernichiari 1969). A variety of possible mechanisms for "light-enhanced calcification" have been offered (see Chap. 2), though the best-supported hypothesis to date involves photosynthesis providing the raw materials needed for calcification, particularly energy-rich photosynthetic products and oxygen (Colombo-Pallotta et al. 2010; Holcomb et al. 2014). Photosynthesis also provides a portion of the precursors needed to synthesize organic matrix molecules, which are critical for the nucleation and growth of calcium carbonate crystals by calcifying organisms (Allemand et al. 1998; Muscatine et al. 2005).

Any factor that clouds the water, such as elevated turbidity from sediment runoff or an increase in plankton, can negatively impact reef building by reducing the amount of light available to drive calcification. Reef building corals growing in clear, oceanic waters may not reach their minimum light requirements until 80 m depth (Lesser et al. 2010), whereas on a turbid reef, similar limits of irradiance can be reached at less than 12 m (Dunne and Brown 1996). In a highly turbid estuary, irradiance may fall below the limits needed to support symbiotic corals within centimeters of the surface, preventing the development of reefs. Sedimentation is a natural process on reefs (Hubbard 1986; Rogers 1990; Anthony and Fabricius 2000), but human activities have led to increased levels of turbidity and sedimentation in some reef areas due to enhanced sediment loading from land-based sources, or dredging (McCulloch et al. 2003; Brodie et al. 2012; Cramer et al. 2012; Pollock et al. 2014). Reduced light penetration due to increased turbidity combined with either direct or indirect effects of excessive sedimentation (e.g., impacts on disease, coral-algal competition, recruitment, etc.) can negatively impact reef building (Hubbard 1986; Cramer et al. 2012; Jokiel et al. 2014; Perez et al. 2014; Pollock et al. 2014). Some coral species are well-adapted to high sedimentation, however, and can continue reef building even under sediment loads that would devastate others (Anthony and Fabricius 2000; Perry et al. 2012).

With increasing water depth, shallow-water coral species become light-limited, but at greater depths their role as reef builders is sometimes taken over by other species of mesophotic corals (30–165 m; Maragos and Jokiel 1986; Locker et al. 2010; Sherman et al. 2010; Weinstein et al. 2014). Mesophotic reef communities are often distinctive compared to shallow-water assemblages, but many of the coral species are photosynthetic reef builders, as are calcifying algae (reviewed by Kahng et al. 2010). A suite of adaptations are needed for these organisms to thrive under such low irradiance, though these adaptations are not well understood. A few coral species can be found from shallow to mesophotic depths and shift from higher contributions of autotrophy to more heterotrophy along this depth gradient (Lesser et al. 2010). Most corals are more restricted in distribution, either to shallow or deeper waters, presumably because their light tolerances are more limited than in widely distributed species. Descending further still into the water column, beyond the euphotic zone, one eventually reaches cold and dark water where aposymbiotic (i.e., non-photosynthetic) corals build deep-water reefs (reviewed by Roberts et al. 2003; Freiwald et al. 2004; Roberts et al. 2006). These corals rely on light only insofar as it drives the growth of plankton in shallower waters that eventually serves as food for the coral communities in deep water. Even though deep-water corals tend to grow much more slowly than their shallow-water counterparts, they can nonetheless build structural reefs given enough time.

9.2.2 Temperature

On most contemporary coral reefs, normal seasonal maximum temperatures range from ~27 to 30 °C, though some reefs fall well outside these limits (Kleypas et al. 1999; Jokiel and Brown 2004). On the cool end of the range, the reefs around Lord Howe Island, Australia, reach a normal seasonal maximum temperature of only 24 °C. In contrast, the reefs of the Arabian Gulf reach normal seasonal maximum temperatures of 33–34 °C—a full 9–10 °C warmer (Hughes et al. 2003). This regional variation correlates with differences in coral thermal limits. Corals tend to undergo paling or bleaching (the loss of symbiotic algae or algal pigments) at temperatures 1–2 °C above the normal maximum to which they are accustomed, even at the extremes of 24 °C and 34 °C (Coles et al. 1976; Jokiel and Brown 2004), and both adaptation and acclimatization appear to play roles in setting these limits (Maynard et al. 2008; Barshis et al. 2010; Guest et al. 2012; Barshis et al. 2013). Some equatorial reefs rarely experience temperatures below 27–28 °C whereas high latitude reefs can experience temperatures as low as 18–19 °C for up to several months (Jokiel and Coles 1977; Kleypas et al. 1999; Jokiel and Brown 2004). Some subtropical corals tolerate temperatures as low as 15–16 °C for short periods of time, and exceptional corals in the Arabian Gulf even tolerate brief exposure to temperatures as low as 11 °C (Jokiel and Coles 1977; Coles and Fadlallah 1991). Perhaps the most extreme example of low temperature tolerance among symbiotic corals is exhibited by *Oulastrea crispata* which is known to survive brief exposure to temperatures as low as 0 °C (Yajima et al. 1986).

While some reefbuilding corals can survive exposure to quite cool temperatures, reef development is exceptionally rare in areas where the temperature drops below 18 °C for more than a few weeks (Jokiel and Coles 1977; Veron 1995). The total annual coral growth rates on cooler, high-latitude reefs may simply be too low to support reef building, though some subtropical corals grow as fast as tropical species (Ross et al. 2015). Other covariates (e.g., nutrient availability, irradiance, herbivory, etc.) probably reinforce the 18 °C isotherm as the lower temperature limit for shallow-water reef building. No such upper thermal limit appears to occur in modern oceans, though anthropogenic climate change may induce a transient upper thermal limit to reef building in the future. Hence, upper thermal limits for reef formation vary from region to region and appear to be flexible over evolutionary timescales, but the lower thermal limit for reef building of 18 °C seems to be more fixed (see Sect. 9.4 below for a discussion of reef building under climate change).

The temperature tolerances of deep-water corals are not as well studied as those of shallow-water corals, but appear to provide equally strong constraints on deep-water reef development. *Lophelia* reefs tend to be associated with water masses that have a temperature of 4–12 °C (reviewed by Roberts et al. 2003; Freiwald et al. 2004; Roberts et al. 2006).

9.2.3 Nutrients, Herbivory, and Bioerosion

Shallow-water coral reefs are restricted to oligotrophic and mesotrophic waters where the concentrations of dissolved inorganic nutrients (e.g, ammonium, nitrate, and phosphate) are relatively low. The absolute nutrient concentrations and the availability of particulate organic nutrient sources, however, vary widely on shallow-water reefs, and some communities are subject to anthropogenic nutrient enrichment (reviewed by Kleypas et al. 1999; Szmant 2002; Atkinson 2011). Reefs are also characterized by high biomass and high rates of primary production (reviewed by Atkinson 2011). The role of nutrients in coral-reef building can be summarized as follows: corals and other calcifiers require a supply of nutrients to maintain normal physiological function, but excessive nutrient supply (either due to natural processes or human influence) inhibits carbonate accretion through both direct and indirect mechanisms. Reef organisms are adapted to low and relatively constant supplies of nutrients, but also rapidly take up nutrients during infrequent periods of elevated supply. While this is an efficient use of available resources, it also makes them sensitive to small increases in nutrients that would not negatively impact many other natural communities (Atkinson 2011).

Under some circumstances, elevated nutrient concentrations have direct, negative effects on coral health and growth (Muscatine et al. 1989; Stimson and Kinzie 1991; Falkowski et al. 1993; Jokiel et al. 1994; Marubini and Atkinson 1999; Ferrier-Pagès et al. 2000; Koop et al. 2001; Loya et al. 2004). Elevated nutrient supply may also interact with other stressors, such as temperature or light stress, leading to negative outcomes for corals (Vega Thurber et al. 2013; Wiedenmann et al. 2013).

In aquarium experiments, where the corals are removed from confounding stressors like adverse light and temperature levels or excessive algae, coral growth can be faster under conditions of modest nutrient enrichment than without it (Godinot et al. 2011; Dunn et al. 2012). DeCarlo et al. (2015) have made similar observations in nature. However, at some point a threshold is reached where higher nutrient supply drives ecological changes (e.g., algal proliferation) that are adverse to coral growth.

In contrast with direct impacts on corals, excessive nutrients inhibit reef building largely through indirect pathways. Nutrient enrichment stimulates the growth and virulence of some coral pathogens, increasing rates of coral loss due to disease (Bruno et al. 2003; Voss and Richardson 2006; Vega Thurber et al. 2013). Elevated nutrients also stimulate the growth of algae, including phytoplankton and benthic macroalgae and microalgae. The latter then may overgrow and kill corals and inhibit coral recruitment. There has been much debate over the relative importance of bottom-up (i.e., nutrient supply) vs. top-down (i.e., herbivory) controls on reef-algae abundance (Hughes 1994; Lapointe 1997; Lirman 2001; Thacker et al. 2001; Szmant 2002; Hughes et al. 2007), but it is clear that both processes are important. When algae have more nutrients they grow faster, and may overgrow and kill corals. If there is insufficient herbivory algae can overgrow and damage corals, and block coral recruitment. With too many nutrients, too little herbivory, or a combination, reefs become choked with algae, and reef building grinds to a halt.

Perhaps most insidious are the effects of nutrient enrichment on bioerosion. The overall rate of reef building depends not just on the constructional processes (e.g., coral growth rates), but equally on the erosional ones (e.g., bioerosion—see Chap. 4). Grazing organisms (primarily parrotfish and urchins) grind away substrate as they consume algae. A diverse array of organisms make a living by boring into the reef itself, and many of these feed on plankton, detritus, or benthic algae. Nutrient enrichment increases the food supply for bioeroders, dramatically enhancing substrate loss in many cases, but not universally (reviewed by Glynn 1997; Chazottes 2002; DeCarlo et al. 2015; Glynn and Manzello 2015). Higher planktonic food availability often enhances coral growth rates (Ferrier-Pagès et al. 2003; Edmunds 2011; Forsman et al. 2012; DeCarlo et al. 2015), but not as much as it enhances bioerosion. Small increases in nutrient supply are often tolerable or even beneficial for reef builders, and corals may grow fastest under slightly mesotrophic conditions. However, elevated nutrients can still be detrimental to the geological process of reef building due to a larger relative increase in bioerosion (Perry et al. 2014; DeCarlo et al. 2015).

9.2.4 Water Motion and Storm Damage

Water motion is a major factor controlling the exchange of nutrients, oxygen, and other vital compounds between the reef benthos and the overlying water column (Shashar et al. 1993; Kühl et al. 1995; Gardella and Edmunds 1999; Finelli et al. 2006; Atkinson 2011). Flushing of the water overlying reefs also moderates fluctuations in temperature, chemistry, and other parameters that result from local physical and biogeochemical fluxes. As a negative factor, water motion can increase sedimentation rates by re-suspending sediment, but as a positive factor it helps to remove sediment from corals and other surfaces. Water motion is, therefore, a critically important parameter for the healthy function of corals and other benthic reef organisms.

Different coral species are adapted to different flow regimes, with delicate-branching, foliaceous, or fleshy forms (such as *Euphyllia* spp.) well-adapted to gentle flow, whereas many mounding, encrusting, or thick-branching species can tolerate (or even require) much stronger water motion. Coral species are capable of acclimatizing to a range of mean and maximum water flow velocities, but the breadth of this range varies. Jokiel (1978) found that the coral *Pocillopora meandrina* is restricted to turbulent environments in Kāneʻohe Bay, Hawaiʻi whereas the congener *Pocillopora damicornis* is most abundant on semi-protected reefs. These differences in habitat preference parallel different capacities to tolerate high and low extremes in water motion by each species.

Tropical storms can deliver huge amounts of wave energy onto reefs, reducing reef building by physically damaging and killing corals, and by washing away reef carbonates (Chap. 6). Periodic removal of sand and silt, however, benefits reef building by flushing away sediments which would otherwise accumulate and eventually smother the corals (Hubbard, 1986; Hubbard and Miller 1990; Scoffin, 1993; Jokiel 2006). The role of tropical storms in structuring reef communities and the impacts of storms on reef building depend largely on the frequency and intensity of disturbance, which vary substantially over space and time (Hubbard et al. 2008; Hubbard 2011). Many of the Caribbean islands, the Bahamas, and South Florida, for example, experience frequent hurricanes whereas reefs in the southernmost Caribbean (from roughly Nicaragua to Venezuela and the Dutch Antilles) are only rarely struck by major storms (National Hurricane Center 2014). Branching corals and others with less robust growth forms tend to be more severely affected by hurricanes than are mounding corals, due to their greater propensity to be broken and damaged. Some of the coral mortality associated with storm damage is also driven by secondary disease outbreaks, with the large-scale tissue damage facilitating infection (Knowlton et al., 1981; Miller et al. 2009).

Under anthropogenic climate change, the strength of tropical storms (and possibly their frequency) is expected to increase (Webster et al. 2005; IPCC 2013). Greater storm intensity is a straightforward consequence of higher ocean and atmospheric temperatures. Therefore, storm damage on coral reefs is likely to increase in the future, though the relative importance of this increase will undoubtedly vary. Conversely, tropical storms can also cool shallow waters by mixing them with deeper waters, through evaporative cooling, and by reducing solar input, all of which reduce thermal stress on reefs. The beneficial cooling effects of tropical storms extend up to several hundred kilometers away from the center of the storm, whereas the negative effects from storm damage typically reach <100 km (Manzello et al. 2007). Storm damage is likely to increase in the future, yet tropical storms may also provide reefs with a temporary respite from higher seawater temperatures.

9.2.5 Carbonate Chemistry

Human activities, particularly the burning of fossil fuels, deforestation, and cement production, are releasing CO_2 to the environment. Roughly one-quarter of these emissions are absorbed by the ocean, leading to changes in seawater chemistry (Caldeira and Wickett 2003), a process known as ocean acidification (OA). Interest in understanding the impacts of OA on marine organisms and ecosystems has grown over the last decade. Chapter 2 discusses the effects of carbonate chemistry on modern coral reefs. The effects of carbonate chemistry (including OA) on reef building over geologic time are considered below.

9.2.6 Sea-Level Rise

High sea levels lead to the inundation of continents (i.e., marine transgressions) and the generation of shallow, inland seas with resulting formation of huge areas of suitable habitat for shallow-water reef building. Some of the highest rates of reef building and the greatest radiations of marine species coincided with sea-level highstands in the geologic past (Droser 2003; Kiessling 2006; Servais et al. 2008; Kiessling 2009; Munnecke et al. 2010; see also Chap. 8). The effects of sea-level *rise* on reef building, however, depend largely on how fast the rise is occurring. Under low or moderate rates of sea-level rise, vertical reef accretion can keep pace with or even catch up to the rising ocean. As the reef reaches the surface of the water, it is unable to build any higher, but a moderate rate of sea-level rise allows reefs to continue accreting vertically (Neumann and Macintyre 1985). Once the reef has filled all available accommodation space, it will either build seaward (e.g., in the Caribbean: Chap. 6: Fig. 6.1) or landward in the form of widening reef flats such as those that developed along many Indo-Pacific and GBR reefs after sea level reached its maximum between 8000 and 6000 years ago (Davies and Hopley 1983; Davies et al. 1985; Hopley et al. 2007).

Under high rates of sea-level rise, even rapid reef accretion may be unable to keep up, ultimately resulting in reefs "drowning" below the waves. Tropical oceans around the world are littered with the submerged remains of old reefs. Some of these are the remnants of reefs that were unable to keep up with the rapid sea-level rise that occurred between the last glacial period and the current interglacial. Sea-level rise continues to accelerate with sea level expected to increase by at least 0.5 m, and perhaps as much as 2 m by the end of this century (Jevrejeva et al. 2009, 2012; IPCC 2013). The present rate of sea-level rise (~3.3 mm yr^{-1}: Church and White 2006, 2011) is in excess of the rate at which many coral reefs can accrete vertically, even under optimal conditions (perhaps more than half of the world's reefs: see Chap. 6) and conditions are far from optimal on many reefs today due to a variety of anthropogenic stressors. Without limiting climate change, sea-level rise will continue for centuries or millennia and could easily drown most of the world's present-day coral reefs.

9.3 The Rise and Fall of Reefs Through Time

Any rocky promontory can support diverse and productive biological communities, but coral reef organisms build the structures on which they flourish. The sheer abundance and ecological complexity of biogenic reefs during the Phanerozoic Eon, the last 542 million years (Ma), have established their prominence over both space and time.

The history of reefs over geologic time is one of boom and bust, characterized by major episodes of biological turnover and several reef crises (see Chap. 8). Some of these changes correlate with major extinction events, but others have been difficult to explain. Reefs have been successful and repressed during both greenhouse and icehouse climates; they have both flourished and floundered during times of (likely) high atmospheric CO_2. The general evolutionary patterns of reefs and reef builders through geologic time have already been described in Chaps. 8 and 3 stresses the importance of the evolving photosymbiotic association between reef autotrophs and heterotrophs. The following overview focuses on reef responses to past changes in the climatic, oceanographic, and the ecological drivers of reef building detailed above.

9.3.1 The First Reefs

Stromatolites are layered, carbonate structures that formed through the cementation of sediments by mineral precipitation and first appear in the geologic record during the Archean, some 3.5 billion years ago (Ba). Some stromatolites may have formed through strictly abiotic processes, but many were formed through the action of microbial assemblages (Allwood et al. 2007; Lepot et al. 2008). In particular, mats of cyanobacteria and eventually primordial eukaryotes were responsible for the formation of many stromatolites and thrombolites. Such mats and biofilms readily trap sediments. During the daytime the mats consume CO_2 through photosynthesis, raising the local pH and facilitating the cementation of sediment into a hard structure. Because of the light-dependent nature of this process, stromatolites tend to form in shallow water and to accrete vertically. The oldest stromatolites of confirmed microbial origin date to 2.74 Ba (Lepot et al. 2008), though there is evidence of microbial stromatolites formed 3.43 billion years ago—among the oldest evidence of life on Earth (Allwood et al. 2007).

Following a long run of increasing stromatolite abundance and diversity, they started to decline around 1.25 Ba (Allwood et al. 2007), coincident with the proliferation of metazoan burrowers and grazers. Structures potentially formed by early metazoan burrowers have been reported from 1.1 Ba (Seilacher et al. 1998), though subsequent work has suggested that these were abiotic in origin (Budd and Jensen 2000; Jensen 2003). By the Ediacaran period (~565 Ma), horizontal burrows formed directly under the microbial mats reflecting clear biotic activity, perhaps to avoid predators (Dzik 2007). Vertical burrowing in the Early Cambrian (~542 Ma) was patchy and developed at different rates in different places (Dornbos et al. 2004). During the Ordovician (~485–443 Ma), metazoan animals underwent a major evolutionary radiation and stromatolites declined further. This pattern continued afterward, with the exception of brief intervals following the end-Ordovician and end-Permian mass extinction events, until metazoans re-established (Sheehan and Harris 2004). Stromatolites still form today, such as in Shark Bay, Australia and the Bahamas, but are restricted to hypersaline lagoons that discourage metazoans (Playford and Cockbain 1976) or areas where high currents and intermittent burial similarly restrict grazing (Dravis 1983; Dill et al. 1986).

9.3.2 The Paleozoic Rise of Metazoan Reefs

The process of cementation associated with early stromatolites was related to a photosynthetically driven increase in pH (i.e., "biologically induced biomineralization"). The earliest evidence of "biologically controlled biomineralization", in which organisms directly invested energy and resources into biominerals comes from protists as early as 800 Ma (Cohen et al. 2011). These unicellular organisms produced phosphatic biominerals. Throughout the remainder of this chapter, "biomineralization" refers to this latter, direct production of skeletal material.

Calcareous, sponge-like organisms were present by ~650 Ma, but biomineralization was relatively rare through most of the Neoproterozoic. During the latter Ediacaran (~580–542 Ma) the size and complexity of metazoans began to increase notably, as did the presence of biominerals. However, it was during the Early Cambrian (~542–510 Ma) that biomineralizing metazoans first became widespread and abundant, allowing them to build reefs. Whether or not this proliferation of calcifying species was a response to increasing predation, it was facilitated by favorable carbonate chemistry conditions (reviewed by Stanley 2006; Ries 2010) and increasingly proved to be advantageous as grazing organisms became more important. Though biomineralization appears to have evolved independently in numerous lineages, some of the genes and many of the basic mechanisms involved appear to be conserved among diverse groups of organisms (e.g., corals, mollusks, vertebrates, etc.). Some of these genes appear to be co-opted from those involved in fundamental metabolic processes, such as calcium and inorganic carbon transport and regulation of intracellular chemistry (Zoccola et al. 2004; Moya et al. 2008; Tambutté et al. 2011; Drake et al. 2013).

The first recognizable coral reefs were built by tabulate and rugose corals, not the scleractinians that build modern reefs. While it's been suggested that scleractinian corals may have evolved from one or more groups of the Rugosa, this origin is unlikely (see Chap. 8). Both septal development and symmetry are dramatically different in the two coral types. Also, rugose and tabulate corals produced calcitic skeletons, in contrast to aragonitic scleractinians. Hence, rugose, tabulate, and scleractinian corals most likely had different origins and simply show convergence in their roles as reef builders. Recent isotopic data suggests that reef building tabulate corals were photosymbiotic, but the type of algae involved and the nature of the symbiosis are unknown (Zapalski 2013; see also Chap. 3).

The tabulate and rugosan corals from the Ordovician through the Devonian (~485–359 Ma) may have built more abundant and widely distributed reefs than those constructed by modern corals (Kiessling 2006; Fig. 9.1). However, they declined severely following the late-Devonian mass extinction. The contributions of these corals to reef building was greatly curtailed through the Carboniferous and the Permian (~359–252 Ma: Kiessling 2009). Both tabulate and rugose corals disappeared during the Permian-Triassic extinction event.

While tabulate and rugose corals were the major framework builders of many Paleozoic reefs, other organisms contributed significantly to carbonate production and cementation. Microbial cementation was important, but generally decreased after the Cambrian (Kiessling 2009). Various bryozoans produced heavily calcified calcite skeletons, particularly from Ordovician to Devonian time (Boardman and Cheetham 1987). Stromatoporid sponges played an important cementing role on Paleozoic reefs, though there is controversy as to their primary skeletal mineralogy (Stanley 2006).

During the latter Paleozoic, particularly from Carboniferous to Permian time, various groups of algae contributed

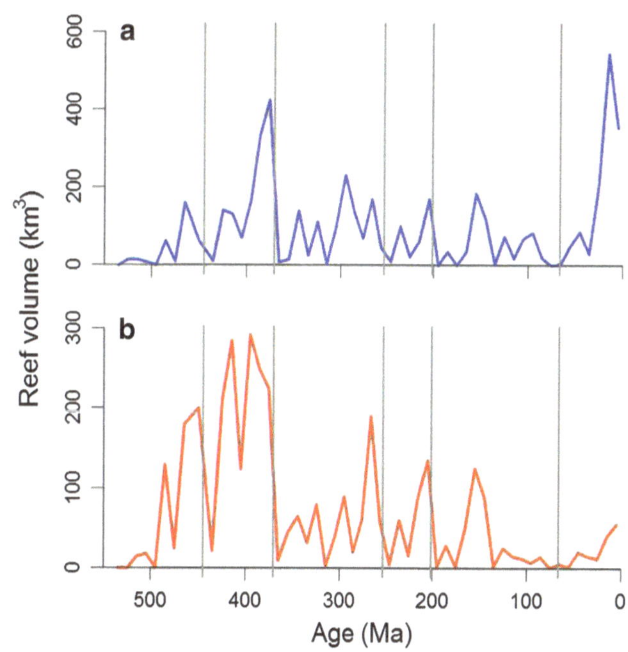

Fig. 9.1 Reef volume (km^3) produced during each interval of the Phanerozoic, plotted in 10 My bins. (**a**) shows the original series based on preserved, fossil reefs, whereas (**b**) shows the data transformed to account for biases in the fossil record, particularly reduced preservation of ancient reefs. Note that in both panels the timing of relative peaks and valleys in reef building are maintained, but accounting for biases in the fossil record in panel (**b**) alters their relative magnitudes. Vertical, grey lines show mass extinctions (Modified from Kiessling 2006)

significantly to reef building. Phylloid algae built substantial aragonitic mounds, similar to deposits formed by codiacean algae (e.g., *Halimeda, Udotea*) on modern reefs (Kirkland et al. 1993; Martín et al. 1997; Stanley 2006). Likewise, various sponges contributed significantly to reef building during the latter Paleozoic, with most having primary aragonitic mineralogy (Stanley 2006). Algae with high-Mg calcite skeletons also contributed to building Paleozoic reefs, and some may have been progenitors of modern coralline algae (Stanley 2006). The increased prominence of organisms which produced aragonite and high-Mg calcite skeletons and reduced importance of low-Mg calcitic corals as reef builders during the latter Paleozoic may be related to long-term changes in seawater chemistry, as discussed in Sect. 9.5 below. Bioeroding taxa also experienced major radiations during the Ordovician, which were unmatched until the Jurassic (Taylor and Wilson 2003; Wilson and Palmer 2006).

9.3.3 Origination and Diversification of the Scleractinia

Scleractinian corals have been a component of the primary framework builders of reefs since the Middle Triassic (~245 Ma), but their origins are debated. Scleractinian corals were already diverse and differentiated at the level of suborder by the time they were able build significant reefs (Wells 1956). Thus, either an earlier origination and/or a major evolutionary radiation is necessary to explain their sudden rise as reef builders in the geologic record.

Romano and Palumbi (1996) used genetic evidence to place the origins of the Scleractinia further back in the geologic past. From these data they inferred that modern scleractinians are represented by two major groups: the "Complex" and "Robust" clades. Based on molecular clocks, their divergence was estimated to have occurred on the order of 300 million years ago—well into the Paleozoic. More recent molecular evidence has revealed a wider diversity of extant scleractinians than previously known. Kitahara et al. (2010) and Stolarski et al. (2011) confirmed the basic correctness of distinctive Complex and Robust clades which diverged in the Paleozoic, and identified additional scleractinian clades. The "Basal" scleractinian clade, represented by the solitary, deep-sea families Gardineriidae and Micrabaciidae, was estimated to have diverged from the Complexa and Robusta by ~425 Ma, whereas divergence of the Complexa and Robusta was estimated at ~415 Ma. A Paleozoic origination of scleractinians helps to reconcile the geologically sudden appearance of diverse suborders in the Middle Triassic as well as the occurrence of "scleractiniamorph" fossils in the Paleozoic (Stolarski et al. 2011). The oldest scleratianiamorphs are known from the Ordovician, ~450 Ma. They are indistinguishable from later scleractinians in terms of septal insertion and are morphologically similar to Basal scleractinians (Stolarski et al. 2011). Thus, it appears that scleractinians originated and began calcifying in the Paleozoic. Further, scleractiniamorphs are bona fide Paleozoic scleractinians that apparently did not engage in reef building until the Middle Triassic. Unfortunately, Paleozoic scleractians (or scleractianiamorphs) may have left a poor fossil record of their early evolution.

The "Naked Coral Hypothesis" proposes that the geologically sudden reappearances of coral skeletons and subsequent gaps in the fossil record may have been the result of the temporary loss of skeletons during periods when environmental conditions were unfavorable for calcification. These so-called "naked corals" would not have left a fossil record until a resumption of calcification when conditions improved (Stanley and Fautin 2001; see also Chap. 8). Medina et al. (2006) found that the non-calcifying Corallimorpharia form a genetic clade nested within the Scleractinia, supporting this idea. They estimated that corallimorphs diverged from scleractinians during the mid-to Late Cretaceous (~132–110 Ma), when coral-reef building was depressed globally. However, new genetic data from Kitahara et al. (2010) and Stolarski et al. (2011) suggest that the Corallimorpharia and Scleractinia represent distinct, monophyletic clades from closely related sister groups. Further study is needed to determine whether these two are closely related but distinct sister groups, or if corallimorphs are instead scleractinians that lost the capacity to calcify at some point in the geologic past (Kitahara et al. 2014), making them truly "naked corals." Whatever the outcome it is possible that periodic "reef gaps" favored reduced skeletons during times when environmental conditions were unfavorable for calcification.

9.3.4 Reef Building Through the Mesozoic

Scleractinians were extremely successful reef builders from the Middle to the Late Triassic (~245–201 Ma: Stanley 2003). This may have been facilitated by the establishment of photosymbiosis. Stable isotope data and growth bands suggest that at least some scleractinians were photosymbiotic by the Middle Triassic (Muscatine et al. 2005; Stanley and Helmle 2010; see also Chap. 3). The coral species that dominated Triassic reefs were distinctive from those on modern reefs. They were primarily phaceloid-dendroid, uniserial (i.e., one polyp per branch), and had non-porous skeletons (Stanley 2003). In contrast, many dominant corals on modern reefs have complex, branching morphologies, are multiserial (i.e., many polyps per branch), and have porous skeletons.

Following their success during the Triassic, scleractinians suffered severe extinction coincident with the broader Triassic-Jurassic mass extinction. Reef building remained relatively low during the Early Jurassic, but expanded substantially during the Late Jurassic to earliest Cretaceous (~160–140 Ma; Stanley 2003; Stanley 2006; Kiessling 2009). During this time, they likely exceeded modern rates of reef formation (Kiessling 2006; Fig. 9.1). However, coral reef building was relatively low during much of the Cretaceous, and became significantly depressed during the latest Cretaceous (~80–66 Ma) when coral extinction rates increased dramatically. This Late Cretaceous extinction was second only to the Triassic-Jurassic mass extinction event in terms of scleractinian extinction rates and faunal turnover (Kiessling and Baron-Szabo 2004).

The decline of corals during the Cretaceous was coincident with a proliferation of coccolithophores (Stanley 2006) and the rapid rise to prominence of rudist bivalves at a time when tropical temperatures were significantly elevated. Rudists secreted a combination of aragonite and low-Mg calcite and they came to dominate many tropical reef environments during much of the Cretaceous. While it has been argued that rudists competitively excluded scleractinian corals on reefs during this interval (Kauffman and Johnson 1988), competitive exclusion seems unlikely for several reasons. First, rudists and corals appear to have had distinct (though overlapping) paleoenvironmental preferences. Rudists were gregarious settlers on soft sediments, similar to many bivalves and tubeworms, whereas most corals required stable, hard substrata (Gili et al. 1995). Second, coral generic diversity remained high throughout most of the Cretaceous, inconsistent with their being out-competed by rudists (Stanley 2003; Kiessling 2009). Finally, reef building by rudists collapsed at the end of the Cretaceous, more than 1 million years before the Cretaceous-Paleogene extinction event, yet corals did not return to dominance. Indeed, extensive coral-reef building did not occur until much later in the Cenozoic, long after the extinction of the rudists.

The Cretaceous-Paleogene mass extinction resulted in heavy losses among many groups, notably the (non-avian) dinosaurs, but was comparatively mild in its effects on scleractinians. Kiessling and Baron-Szabo (2004) suggest that the high rates of extinction experienced by corals during the latest Cretaceous (~70 Ma) may have left remaining species more tolerant of adverse environmental conditions, and more likely to survive through the K-Pg boundary.

As in the Paleozoic, various calcifying sponges were major contributors to reef building during much of the Mesozoic (Stanley 2003; Stanley 2006; Kiessling 2009). Calcifying algae of various types were also important contributors to Mesozoic reefs, though to a much lesser extent than on Carboniferous and Permian reefs, or later Cenozoic reefs (Kiessling 2009). Microbes contributed significantly to reef building during the Triassic and Late Jurassic, especially at mesophotic depths (Stanley 2003), but became far less important from the Cretaceous onwards (Kiessling 2009).

9.3.5 Coral-Reef Building Through the Cenozoic

Compared to the latest Cretaceous, coral-reef building increased noticeably during the Paleogene, but still remained at rather modest levels (Kiessling and Baron-Szabo 2004). Over recent years, a great deal of interest has been placed on understanding the causes and consequences of the Paleocene-Eocene Thermal Maximum (PETM, ~56 Ma) as a possible analogue to modern climate change and ocean acidification from human activities (e.g., Zachos et al. 2005). Reef building decreased dramatically at the PETM, whereas coral extinction rates increased only modestly (Kiessling and Baron-Szabo 2004). By the Late Eocene (~35 Ma), a larger increase in coral extinction rates roughly coincided with the onset of Antarctic glaciation and rapid climate cooling (Lisiecki and Raymo 2005; Zanazzi et al. 2007). However, coral-reef building was little affected (Kiessling and Baron-Szabo 2004) and increased rapidly during the late Oligocene (~25 Ma). While reef building has continued at relatively high levels over the last ~20 Ma, it never returned to the levels seen in the Triassic and Late Jurassic (Kiessling 2006; Fig. 9.1).

Most recently, a modest increase in coral extinction rates occurred during the Pliocene and Pleistocene (~5–1 Ma), coinciding with the closure of the Isthmus of Panama and the initiation of Pleistocene glacial-interglacial cycles (see Chap. 6). This extinction was especially profound in the Caribbean and Western Tropical Atlantic where more than 40 % of the coral genera (18 of 43) and perhaps 60 % of the coral species went extinct (Johnson et al. 2008; van Woesik et al. 2012). Of the 18 genera that disappeared at this time, six experienced global extinction while 12 are still extant in the Indo-Pacific (*Alveopora, Caulastraea, Galaxea, Gardinoseris, Goniopora, Isopora, Pavona, Pocillopora, Psammocora, Stylophora* and *Trachyphyllia*). It is particularly noteworthy that reef building shows no relationship with coral-species diversity in the Caribbean over the last 28 Ma (Johnson et al. 2008). Reef building was relatively widespread in the Caribbean during the late Oligocene when coral species diversity is estimated to have been on the order of ~70 species. From the late Oligocene to the Early Pliocene coral species diversity roughly doubled to ~130 species, though reef building became rare and highly localized across the region. Approximately 60 % of these coral species went extinct during the Pliocene and Pleistocene, and the regional coral diversity was reduced to ~50 species. Subsequently, reef building again became widespread across the Tropical Western Atlantic. Both the regional coral extinctions and low rates of reef building may have been driven largely by changes in regional upwelling (Edinger and Risk 1994) and associated fluctuations in temperature, nutrient supply, and seawater chemistry. Extensive reef building resumed only after upwelling diminished.

Unlike in earlier reef buildups, microbes and calcifying sponges contributed very little to the structure of Cenozoic reefs. In contrast, calcifying algae (and especially corallines) made major contributions to reef structure, especially since the Eocene. Indeed, modern reefs are often referred to as coral-algal constructions. A variety of other calcifiers including forams, molluscs, echinoderms, etc. also contributed to reef building over the Cenozoic, but corals and coralline algae have been by far the most important contributors to the reef framework and the carbonate incorporated into the reef structure.

Perhaps the most important development in the Quaternary (2.6 Ma to present) was the rise to prominence of the coral genus *Acropora*. The earliest known fossil *Acropora* comes from the Late Paleocene (~54 Ma), in Somalia (Carbone et al. 1994). However, it is not until the Late Oligocene (~25 Ma) that the genus is first recorded as dominating a reef community, in Greece (Schuster 2002). Some have argued that the genus has undergone a recent evolutionary radiation with all modern species having descended from a single ancestral form over the last ~5 Ma (Veron 1995; Fukami et al. 2000). However, recent fossil evidence presented by Wallace and Rosen (2006) contradicts this view. Veron and Wallace (1984) divided the genus *Acropora* into 20 'species groups' based on shared skeletal characters and morphological phylogeny. Representatives of at least one-third of these species groups make their first known fossil appearances in English and French deposits from the Eocene (~49–34 Ma), suggesting that the genus has deeper roots than previously recognized and that its modern diversity has developed more slowly over time.

Acropora became established in the Western Atlantic and Caribbean region by the Late Eocene but appears to have been absent across most of the Indo-Pacific region until the Late Oligocene (Wallace and Rosen 2006). The genus disappeared from the European region during the mid-Miocene, following the closure of the Tethys seaway between the Indian Ocean and modern-day Mediterranean Sea. This extinction was coincident with rapid global cooling, Antarctic glaciation, and the extinction of all tropical and reefal organisms from the region (Wallace and Rosen 2006). By this time the genus was well-established across the Indo-Pacific, perhaps as offshoots from earlier Western Indian Ocean and European lineages. Since then, it has become a primary contributor to the carbonate budget of modern reefs and, until recently, was the dominant shallow-water reef builder in the western Atlantic.

9.4 Climate Change and Reef Building in the Past

Individual corals are extremely sensitive to temperatures outside the range to which they are adapted, experiencing bleaching and eventually death at sustained temperatures as little as 1–2 °C above or below normal (reviewed by Jokiel and Coles 1990). For example, when colonies of *Acropora millepora* were reciprocally transplanted between the Central and Southern Great Barrier Reef, corals from the cooler, southern population bleached and died during summer at the northern site, while corals from the warmer, northern population bleached and died during winter at the southern site (Howells et al. 2011). Corals and coral reefs, however, clearly have the capacity to adjust to different thermal regimes. Upper thermal limits for corals span at least 9–10 °C in the modern ocean (Hughes et al. 2003). Given enough time to acclimatize or adapt, corals are capable of surviving and building reefs over an even broader temperature range.

The climate has changed dramatically over the last several hundred million years, and across a variety of temporal and spatial scales. Coral reefs have tolerated many of these changes, and have likely benefitted from some of them, but some have also been adversely affected. Understanding the responses of coral species and coral reefs to climate change in the past before significant human impacts is particularly important today, as the world faces the possibility of large-scale, extremely rapid climate change due to human activities. In this section we will discuss climate change in the past and the responses of corals and reefs to these changes within a modern, physio-ecological framework.

9.4.1 Climate Change Over the Phanerozoic

Over the last 542 million years (My) global climate has experienced a general pattern of four quasi-periodic fluctuations between hotter, greenhouse climates and colder, icehouse climates (Royer et al. 2004; Fig. 9.2). The period of these fluctuations has been on the order of 100–200 My. This long-term pattern of warming and cooling is punctuated by numerous shorter-term climate fluctuations, many of which have been quite substantial. A number of factors have contributed to these patterns, but changes in the climatic forcing provided by atmospheric CO_2 appears to be among the most important drivers of climate change over both long and short timescales (Royer et al. 2004; Royer 2006; Came et al. 2007). A critical piece to understand here is that

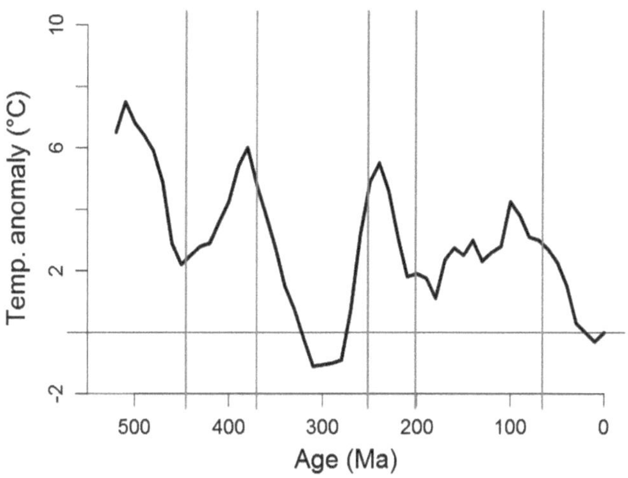

Fig. 9.2 Long-term, secular changes in tropical seawater temperature over the Phanerozoic, shown at 10 My resolution. Vertical, grey lines show mass extinctions (Modified from Royer et al. 2004)

atmospheric CO_2 is not a thermostat, *per se*, and a given atmospheric CO_2 concentration does not imply a specific global mean temperature. Rather, CO_2 provides a particular *forcing* on the climate system, and the equilibrium climate response to that forcing also depends on the other drivers and feedbacks within the system. In particular, solar output has slowly increased over the last several billion years (the sun was at about 70 % of current output four billion years ago) and it will continue to increase in the future (Gough 1981). Reconstructed atmospheric CO_2 concentrations were extremely high on the early Earth, as were some other greenhouse gases, helping to resolve the "Faint Early Sun Paradox" of a relatively warm planet with liquid water under low solar irradiance (Sagan and Mullen 1972; Pavlov et al. 2000; Royer et al. 2004; Ueno et al. 2009). Uncertainty over atmospheric CO_2 concentration increases substantially the further back we look into the geologic past. Other factors (such as the positions of the continents) also had important effects on the long-term evolution of the climate (see Chap. 6), but variation in atmospheric CO_2 is a primary driver of climate change.

The coldest climates over the Phanerozoic are represented by glacial maxima during the Pleistocene glaciations (2.6–0.12 Ma; reviewed by IPCC (2013). During these glacial maxima, global climate cooled to 4–6 °C below the late-twentieth century average. The Pleistocene glaciations were punctuated by interglacial periods of a few thousand years when climate warmed relatively rapidly by several degrees Celsius. Pleistocene interglacials have ranged from slightly cooler than the late-twentieth century mean global temperature to perhaps as much as 2–3 °C warmer than our recent climate during the peak warmth of the Eemian interglacial (~130,000–115,000 years ago).

The Pleistocene was preceded by a general global cooling trend over the previous 50 My from the very warm climates of the Paleocene and early Eocene, with glaciation taking hold in Antarctica by ~30 Ma. This cooling trend was driven by the drawdown of atmospheric CO_2 and the migration of Antarctica toward the South Pole. Southern migration of Antarctica facilitated its glaciation, increasing the Earth's albedo and providing a positive feedback that enhanced the cooling trend. A period which was likely as cold or colder than the Pleistocene had occurred during the Late Carboniferous to Early Permian (~360–260 Ma), triggering extensive glaciation. As in the Pleistocene, cooling and glaciation were associated with the draw-down of atmospheric CO_2, perhaps due to the expansion of terrestrial forests. Another glacial period occurred during the Late Ordovician to Early Silurian (~450–420 Ma), though this time appears to have been somewhat warmer than during the Pleistocene glaciations, even during the height of the event at the end of the Ordovician (~443 Ma). Another relative cool period occurred during the late Jurassic to early Cretaceous (~160–90 Ma), though this interval was still warmer than the present-day climate.

Between these long-lasting cool periods the planet experienced warmer greenhouse climates. The late Cambrian to early Ordovician (520–480 Ma), Devonian (~400–360 Ma), Permian to early Triassic (290–240 Ma), and late Cretaceous to early Cenozoic (~120–50 Ma) climates were all much warmer than our present-day climate. At their peaks, these periods may have reached global temperatures as much as 10–15 °C higher than the late-twentieth century average. These were all periods of strong, greenhouse gas forcing, and ice was rare or entirely absent globally. Because continental glaciers were minimal, changes in sea level were driven primarily by the expansion and contraction of seawater and of the ocean basins themselves, as there was so little water exchanged between the limited continental ice sheets and the world ocean.

When climate changes, absolute temperature changes tend to be greatest over land and toward the poles and smallest in the ocean and toward the equator. Hence, these major climate fluctuations have resulted in more modest changes in temperature of the tropical oceans. Relative to the last 10 My, the tropical oceans averaged 1–2 °C cooler during the Carboniferous glaciations (Royer et al. 2004), but probably reached even cooler temperatures during the height of glaciation. In contrast, during the Cambrian-Ordovician, Devonian, and Permian-Triassic hothouse intervals, tropical ocean temperatures averaged 6–8 °C warmer than the average over the last 10 My, and perhaps 4–6 °C warmer during the Cretaceous-Cenozoic hothouse. Hence, the long-term average for tropical ocean temperature has varied over a range of about 10 °C during the Phanerozoic, though this range was certainly a bit larger when considering climatic fluctuations over shorter timescales.

9.4.2 Past Effects of Secular Changes in Climate on Reef Building

Any time-series can be decomposed into periodic as well as secular (i.e., long-term, non-periodic) variation. The coming and going of the seasons represents a periodic change in temperature. Gradual warming or cooling of the climate represents secular change. In this section we consider the effects of longer-term secular changes in climate on reef building—the meandering between hothouse and icehouse climates that Earth has experienced over the Phanerozoic.

Rugose and tabulate coral reefs flourished during the Silurian, under relatively cooler climatic conditions, and also thrived during the Devonian under an intense hothouse regime. These reefs became extremely repressed during the Carboniferous and Permian, under climates that ranged from intense icehouse to intense hothouse conditions. Rugose and tabulate corals succumbed to the end-Permian mass extinction and scleractinan corals replaced their ecological role as primary reef builders in the Middle Triassic. Scleractinian reefs were extremely successful throughout the Triassic, under intense hothouse conditions, but were devastated by the end-Triassic extinction event. They again rose to major prominence as reef builders during the late Jurassic, under a relatively cooler climate (but still warmer than present-day), but floundered during the late Cretaceous and early Cenozoic hot climates. Scleractinians again rose to dominate reef building during the cooling climate of the Neogene, and have been very successful in our recent icehouse climate. These variations in response suggest that temperature alone has not had a consistent control on geologic reef building.

An upper thermal limit for metazoan life is thought to be 45–47 °C (Pörtner 2001). The tropical ocean has rarely if ever reached such a limit during the Phanerozoic, though equatorial ocean temperatures have soared into the upper 30's and perhaps even the lower 40's°C under the most intense hothouse climates (Sun et al. 2012). Lethally high seawater temperatures beyond which reef builders could not adapt may have occurred in semi-enclosed, shallow seas. Nevertheless, it appears unlikely that the climate has been so hot as to widely prevent reef building during the Phanerozoic, given appropriate evolutionary timescales needed for reef builders to adapt. In contrast, in the modern ocean, coral reefs (except for deep, cold-water reefs) give way to non-reefal communities where the seasonal minimum seawater temperature drops below 18 °C for significant periods of time. Adaptation and/or acclimatization are also required to allow builders to tolerate these low thermal limits. Thus, at least for modern reefs, the lower temperature limit of roughly 18 °C appears to be more akin to a fixed limit. Ancient reefs, such as those built by rugose and tabulate corals, likewise were restricted to tropical and subtropical climates, and a lower temperature limit for shallow-water reef building of somewhere around 18 °C has likely operated for hundreds of millions of years. Unlike the upper thermal limit, which has rarely if ever constrained reef building over long timescales, the lower thermal limit for reef building has frequently limited the expansion of reefs to higher latitudes.

Over long, steady-state conditions where reef organisms have plenty of time to adapt to temperature change, all else being equal, warmer climates should be more favorable to reef building than cooler climates simply because the latitudinal extent and total area open to colonization increases under a warm climate as compared to a cool one. Shallow-water reefs have thrived under hothouse climates, but a warm climate by itself is not sufficient to predict a period of successful reef building. Reefs were less abundant during the Carboniferous icehouse, yet thrived during the late Jurassic cool period and our more recent, Neogene, icehouse climate. More than anything, this pattern demonstrates that reef organisms can indeed adapt to climate change over sufficiently long time scales, and that factors other than just climate are also critical for determining the relative success of reef building organisms.

9.4.3 Coral Reefs Under Rapid Climate Change

The climate can be relatively stable for long periods of time; however, it can also change rapidly, especially after a major perturbation to the carbon cycle. Rapid climate change may exceed the rate at which reef builders are able to migrate, acclimatize, or adapt, leading to the extinction of reef organisms and reduced reef building. Kiessling and Simpson (2011) identified five major reef crises during the Phanerozoic. All five appear to be associated with rapid climate change—one with rapid cooling, and the other four with rapid warming. During the late Devonian the climate cooled rapidly, associated with the drawdown of atmospheric of CO_2 due to the expansion of vascular land plants (Algeo and Scheckler 1998). Rapid cooling and the associated drop in sea level may have contributed to the collapse of reef building during this time, though low data resolution makes it difficult to test this hypothesis. The evolution and proliferation of land plants also led to higher terrestrial weathering rates and increased supply of micronutrients to the ocean that possibly resulted in eutrophication (Algeo and Scheckler 1998). Increased oceanic productivity may have contributed to lower reef building at this time through negative effects on calcifiers adapted to oligotrophic conditions and positive effects on bioeroding organisms. The degree to which the latter would have been an important factor depends on the extent to which the reef-bioeroder community had proliferated. Cooling that began in the mid-Devonian intensified throughout the Carboniferous and into early Permian time, and undoubtedly contributed to a reduction in reef building during this interval. Lower contributions by corals to reef building during the latter Paleozic, however, were reinforced by changes in ocean chemistry (see below). A minor, coral/sponge-reef crisis occurred during the Middle-Late Permian (~260 Ma), but there is not evidence for rapid climate change at this time (Kiessling and Simpson 2011).

A clearer role for rapid climate change in suppressing reef building comes from the Permian-Triassic extinction event (~251 Ma) that continued through the Early Triassic. The climate warmed very rapidly at the end of the Permian and the Early Triassic was characterized by a great deal of upheaval and climatic instability (Kiessling and Simpson 2011; Sun et al. 2012). As a result, reef building was substantially suppressed during this interval. Once the climate began to stabilize during the Middle Triassic, the first scleractinian reefs rapidly burst onto the scene and were extremely successful throughout the Triassic. Reefs were again devastated by the Triassic-Jurassic extinction events (~200 Ma) and rapid global warming is similarly implicated as tropical taxa were more severely affected than were non-tropical ones (Kiessling and Aberhan 2007). Throughout the earliest Jurassic, reefs and corals were concentrated toward the mid-latitudes, suggesting an escape from heat (Flügel and Kiessling 2002; Lathuilière and Marchal 2009; Kiessling and Simpson 2011).

Additional reef crises occurred during the Early Jurassic (~183 Ma) and the Paleocene-Eocene (~56 Ma). Temperature changes during the latter were among the most rapid in the geologic record (Kiessling and Simpson 2011) and some have proposed this event as an analogue to twenty-first century climate change. Not every variation in the climate was associated with declines in reef building, nor has every period of relatively rapid climate change brought about a reef crisis. The question then becomes, how much climate change can reef builders cope with via migration, acclimatization, and adaptation, and at what point does it exceed these capabilities, leading to a reef crisis?

An instructive example comes from the geologically recent past. During the Eemian interglacial period (~130,000–115,000 years ago) the climate was warmer than during our present interglacial (the Holocene), by perhaps 2–3 °C. This warmth was likely driven by differences in orbital forcing, particularly at high latitudes. Annual mean temperatures at low latitudes appear to have been only slightly warmer than present-day temperatures—perhaps 0.7 °C warmer (McKay et al. 2011)—though seasonality was higher due to changes in orbital forcing (Winter 2003). Sea level was ~6 m higher than today, largely due to the retreat of the Greenland ice sheet, but coral reefs flourished as they have during the Holocene. Indeed, reefs and tropical coral species spread to higher latitudes than during the present interglacial (Greenstein and Pandolfi 2008). What is perhaps most striking about the Eemian is that not only did coral reefs persist, but coral species distributions changed rapidly. Many tropical and subtropical corals spread further toward the poles than during the Holocene, but some taxa became rare or disappeared in the equatorial zones (Kiessling et al. 2012). Hence, the reef response to the Eemian interglacial was one of continued growth, facilitated by adaptation/acclimatization *in situ* as well as migration to higher latitudes by species that were unable to cope with the warm, equatorial temperatures.

This response was more pronounced in the Northern Hemisphere, consistent with both modeling and proxy data showing higher levels of warming in the Northern as compared to the Southern Hemisphere, but temperatures were higher world-wide during the Eemian (Kiessling et al. 2012). Proxy data from the Western Pacific Warm Pool show seawater temperature during the Holocene of about 29–30 °C, similar to present-day seasonal maxima. In contrast, during the Eemian the same proxy data suggest seawater temperatures >30 °C, and up to 31.5 °C at times (Visser et al. 2003; Leduc et al. 2010; Nikolova et al. 2013). Similarly, proxy data shows Eemian seawater temperatures ~1–1.5 °C higher in the Caribbean (Herbert and Schuffert 2000) and up to 3 °C warmer in the Eastern Equatorial Pacific than present-day temperatures (Lea et al. 2006; Nikolova et al. 2013). These temperature increases occurred over not more than a few centuries, and followed on the heels of warming from the previous glacial period. A temperature increase of 1–3 °C in coral reef waters over the course of a few centuries during the Eemian implies a maximum warming rate on the order of about 0.5 °C per century. This rate of warming was not detrimental to reef building, however, there was rapid reorganization of the coral community structure, especially at low latitudes and may be near the upper limit of what many coral species can cope with. Without a doubt, the rate of warming is important, and perhaps even more so than the absolute magnitude of warming. For example, a 1–2 °C increase in seawater temperature over a single year would likely have greater effects than the same magnitude of warming spread over a few centuries. A longer timescale simply allows more time for corals to migrate, acclimatize, or adapt, leading to a slow, adaptive turnover in the reef community, rather than the abrupt loss of corals and recolonization by the few remaining survivors.

As described earlier, thermal tolerance is related to the long-term regional temperatures to which corals have adapted and acclimatized rather than to any absolute global threshold. On the reefs of the Arabian Gulf, corals tolerate temperatures of at least 34 °C, but bleach when the temperature reaches 35–36 °C for prolonged periods. This is at least 4–6 °C higher than the bleaching threshold for other corals in the Indian Ocean (~30–31 °C), from which Arabian corals are derived (Riegl et al. 2011). Arabian corals appear to have gained this exceptional thermal tolerance within the last 6000 years, which would imply a minimum rate of adaptation/acclimatization on the order of 1 °C per millennium (i.e., 6 °C increase in 6000 years), or 0.1 °C per century. While the actual rates of increase in thermal tolerance are not well constrained, 6000 years can be regarded as an upper limit on the timescale. Thus, some reef building corals are capable of tolerating a warming rate of least 0.1 °C per century, and 4–6 °C of total warming.

Field observations of contemporary, repeat-bleaching events as well as manipulative experiments help to further define the potential for corals and coral reefs to tolerate rapid climate change. The mass-bleaching event of 1997/1998 struck coral reefs around the world, including the Central Great Barrier Reef (GBR). Here bleaching was extensive, though mortality was relatively modest. Bleaching impacted the GBR again in 2002. In spite of more severe thermal stress (and higher irradiance, which significantly exacerbates thermal stress), the bleaching observed on the Central GBR was less severe in 2002 than expected based on prior observations (Maynard et al. 2008). Indeed, applying the bleaching thresholds observed in 1998 substantially over-predicted actual rates of bleaching in 2002. Similar findings come from Southeast Asia between the 1998 and 2010 bleaching events (Guest et al. 2012). Here rates of mortality were more severe than on the GBR. Not only were corals more thermally tolerant in 2010 as compared to 1998, but genera such as *Acropora* and *Pocillopora* which usually are among the most thermally sensitive also showed the greatest increases in resistance. Overall, these observations suggest an increase in the thermal tolerance of corals on the order of 0.5–1 °C after the single bleaching event. The changes were apparently due to acclimatization of the resident corals, or adaptation within the changing coral populations. Hence, even rapid increases in seawater temperature of <1 °C may be tolerable for some (but not necessarily all) species of coral. Experimental and modeling studies suggest that larger increases in thermal tolerance, on the order of 1–2 °C, are also possible for at least some corals over timescales ranging from weeks to a century. These changes may occur through physiological acclimatization, switching to more thermally tolerant algal symbionts by the host corals, and/or adaptation (Middlebrook et al. 2008; Baskett et al. 2009; Howells et al. 2011; Logan et al. 2013).

The emerging picture from these various lines of evidence is that some coral species can achieve small increases in thermal tolerance (<1 °C) over timescales ranging from weeks to a few years. Larger increases on the order of 1–2 °C appear to be possible over short timescales for some corals, but most corals likely require decades to centuries (i.e., multiple generations) to achieve these higher tolerances. The timescales required for increases in tolerance >2 °C have not yet been clearly demonstrated. A maximum warming rate on the order of 0.5 °C per century, such as during the Eemian, appears to be near the upper limit of what many corals can tolerate. The oceans are expected to warm by 1–3 °C this century due to anthropogenic climate change (IPCC 2013). This higher warming rate may exceed what many corals can cope with, resulting in major reductions in reef building and ultimately extinctions during the twenty-first century and beyond.

9.5 Reefs Under Changing Ocean Chemistry

Changes in seawater carbonate chemistry can significantly impact calcification rates of corals and other reef builders (reviewed by Pandolfi et al. 2011). The carbonate system in seawater involves the interaction of several, dissolved species: Ca^{2+}, $CO_{2(aq)}$, HCO_3^-, CO_3^{2-}, H^+ (or its surrogate, pH) as well as solid phases of $CaCO_3$ (aragonite vs. calcite). The reader is referred to Chap. 2 for a discussion of carbonate chemistry as it relates to coral calcification. Additional parameters important to the carbonate system include the aragonite or calcite saturation state ($\Omega_{arag/calc}$) and the magnesium to calcium ratio (Mg/Ca). Dramatic changes over the Phanerozoic have influenced reef building and may provide useful analogs to the immediate future. We review these below.

9.5.1 Calcite and Aragonite Seas

Calcifying organisms primarily produce one of three polymorphs of $CaCO_3$, although some species produce shells and skeletons of mixed composition. These three polymorphs are low-Mg calcite (or simply, calcite), aragonite, and hi-Mg calcite (Ca,$MgCO_3$). Low-Mg calcite and hi-Mg calcite differ in the quantity of Mg incorporated into the $CaCO_3$ crystal lattice. Low-Mg calcite contains <4 mol % Mg, whereas hi-Mg calcite contains >4 mol% Mg, and usually in the range of 11–20 mol% Mg in modern seawater. Differences in lattice structure and chemical composition among polymorphs lead to differences in solubility and other physical properties of the minerals, with low-Mg calcite being the least soluble, followed by aragonite, and finally with hi-Mg calcite being the most soluble (reviewed by Zeebe and Wolf-Gladrow 2001). Some coralline algae also produce dolomite, which is 50 mol% Mg, and is highly resistant to dissolution (Nash et al. 2013; Diaz-Pulido et al. 2014).

Seawater interacts with basalt in areas of seafloor spreading such that the Mg^{2+} concentration decreases and the Ca^{2+} concentration increases. Hence, during periods of rapid seafloor spreading the Mg/Ca ratio decreases whereas during periods of slower seafloor spreading the Mg/Ca ratio increases (Stanley and Hardie 1998). Variations in the rate of seafloor spreading and chemical weathering on land over the Phanerozoic have led to shifts in the seawater Mg/Ca ratio from ~0.9 to 5.2. Under conditions where the Mg/Ca ratio is low (<2), the precipitation of low-Mg calcite is thermodynamically favored over aragonite or hi-Mg calcite. In contrast, at a high Mg/Ca ratio (>2) the precipitation of aragonite and hi-Mg calcite are favored over low-Mg calcite.

As their names imply, calcite seas are characterized by high representation of low-Mg calcite among abiotic precipitates, such as hardgrounds, and to some extent among calcifying taxa (Fig. 9.3). In contrast, aragonite seas are characterized by higher representation of aragonite and hi-Mg calcite in abiotic precipitates and to some extent among calcifiers. These fluctuations in seawater chemistry are also associated with variation in hypersaline precipitates, with more KCl formed during calcite seas and more $MgSO_4$ formed during aragonite seas (Stanley and Hardie 1998). During the Phanerozoic there have been two major periods of calcite seas (the Cambrian through the Early Carboniferous and the late Jurassic through the early Neogene) and two periods of aragonite seas (the middle Carboniferous through the late Jurassic and the early Neogene to present). This had a significant impact on the success of particular calcifying species as the seawater chemistry changed. Many carbonate-producing taxa preferentially adopted a skeletal mineralogy that was more favorable within the chemical regime at the

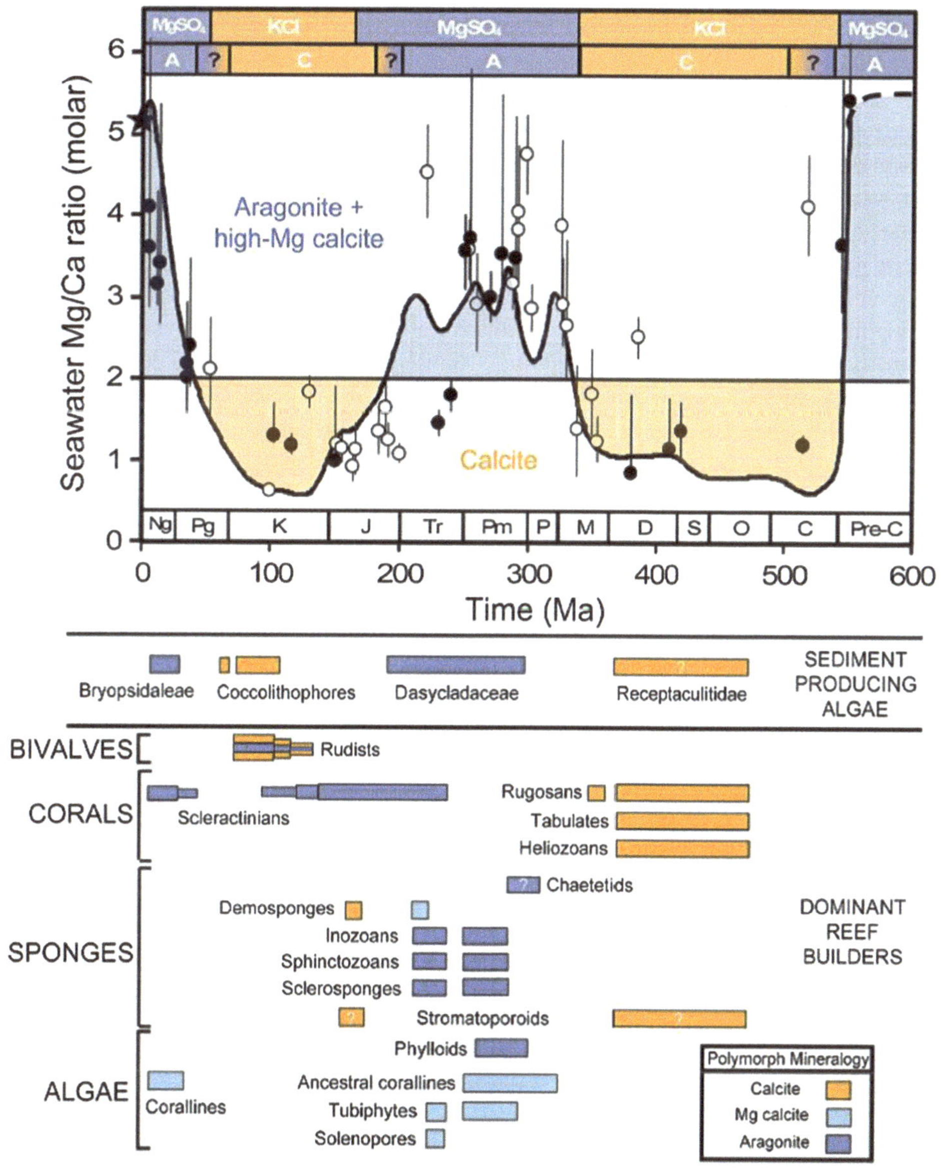

Fig. 9.3 Changes in the Mg/Ca ratio of sea water and of the major calcifying taxa over the Phanerozoic. Curve shows modeled Mg/Ca ratio in comparison to proxy data from halite inclusions (*closed circles*) and fossil echinoids (*open circles*). Top bars show periods of calcite and aragonite seas, and associated KCl or $MgSO_4$ evaporite deposits. Horizontal line divides aragonite and calcite nucleation fields. Lower bars show relative changes in abundance of dominant calcifiers along with their associated mineralogies (From Ries 2010)

time they were first evolving, though a few taxa deviate from the pattern (Porter 2010). As a result, changes in ocean chemistry can potentially provide important constraints on skeletonization that might reinforce the thermodynamically favored carbonate polymorph.

For example, rugose and tabulate corals evolved in the calcite seas of the Ordovician (Fig. 9.3). They produced low-Mg calcite skeletons and thrived in the calcite seas of the Ordovician, Silurian, and Devonian oceans. However, they declined dramatically as ocean chemistry shifted in the late Devonian and Early Carboniferous. Rugosans and tabulates showed some recovery in the Late Carboniferous, and the late Permian, but they never again achieved the dominance they had during earlier times. The aragonite seas that characterized this interval may have contributed to their reduced roles as reef builders.

However, this relationship was not straightforward. Scleractinian corals, which principally produce aragonite, first began building reefs during the Middle Triassic under aragonite-sea conditions. However, recent evidence suggests that they first evolved skeletons in the calcite seas of the Paleozoic (Stolarski et al. 2011). They were extremely successful during the Triassic, until the Triassic-Jurassic extinction event, generally coincident with a shift toward a calcitic ocean. As the Mg/Ca ratio in seawater fell further into the calcite nucleation field (the chemical parameter space where the precipitation of calcite is thermodynamically favored) during the late Jurassic, however, scleractinians experienced a major resurgence in reef building. Subsequently, scleractinian reefs declined throughout the Cretaceous, during a period when the Mg/Ca was squarely in the calcite nucleation field (~1). At the same time, rudist bivalves, which produced skeletons consisting of a combination of aragonite and low-Mg calcite, were highly successful, as were calcite-producing coccolithophores, and this success was likely facilitated by seawater chemistry which favored their skeletal mineralogies. Scleractinian reefs increased in abundance following the Cretaceous-Paleogene extinction event (though rudists had already declined long before). They were knocked back again by the Paleocene-Eocene Thermal Maximum, a response more to temperature than seawater chemistry (see below). Finally, they expanded dramatically during the early Neogene, coinciding with a shift of the Mg/Ca ratio back into the aragonite nucleation field.

Inconsistent responses of scleractinians to variation in the Mg/Ca ratio show that it alone is not a master control over reef building, though it does appear to influence the relative success of calcifiers during a given interval. In particular, scleractinians were highly successful during the calcite seas of Late Jurassic, but their success dropped off as the Mg/Ca fell even lower during the Cretaceous. Thus, reef accretion can still occur under somewhat unfavorable seawater chemistry, but there are also limits to what organisms can tolerate. It appears that the threshold for scleractinian reefs was crossed during the Cretaceous, when the Mg/Ca ratio fell to ~1, but not during the Jurassic at a Mg/Ca ratio of ~2. This conclusion contrasts with previous suggestions that corals were out-competed by rudists during the Cretaceous. Instead, the floundering of corals and the flourishing of rudists during Cretaceous time were both likely driven by changes in ocean chemistry and the relative success of each group was probably independent of the other.

Experimental evidence from modern calcifiers (including scleractinian corals) shows that, while they tend to calcify at much higher rates in seawater that favors the precipitation of their $CaCO_3$ polymorph, they are less tied to their original mineralogy than has generally been assumed (reviewed by Stanley 2006; Ries 2010). Some organisms which produce hi-Mg calcite in modern seawater with a high Mg/Ca ratio will switch to precipitation of low-Mg calcite when the Mg/Ca ratio is lowered experimentally (Stanley 2006; Ries 2010). Likewise, scleractinian corals produce moderate quantities of calcite under experimental conditions as the Mg/Ca ratio dips into the calcite nucleation field (Ries et al. 2006; Higuchi et al. 2014). Hence, skeletal composition is not under absolute control by the organisms—at least when seawater chemistry changes rapidly. Over geologic time, however, most taxa maintain the same skeletal mineralogy in spite of shifts in the Mg/Ca ratio of sea water (Porter 2010). It is not known why calcifiers tend to produce one $CaCO_3$ polymorph vs. another, but the ability to precipitate the favored polymorph appears to be an important determinant of their calcification rates. Consistent mineralogy within taxa over geologic time may be an evolutionary holdover. That is, the costs of evolving skeletons with a new mineralogy may exceed the costs of calcifying in a less favorable nucleation field, which would explain why switching from one mineralogy to another is rare in the geologic record, when the rate of chemical change is slow relative to reasonable evolutionary responses. Nevertheless, exceptions to the rule do occur (Porter 2010). At least one species of scleractinian coral appears to have switched to calcite as its primary skeletal mineralogy during the calcite seas of the Cretaceous (Stolarski et al. 2007). How this species was able to achieve the switch, and why other species did not follow suit, are unknown.

Regardless of the variability described here, rugose, tabulate, and scleractinian corals as well as other calcifying taxa have generally been more productive when seawater chemistry favored their preferred carbonate mineralogies, but an unfavorable nucleation field is not an insurmountable obstacle (Kiessling et al. 2008). In particular, scleractinians were extremely successful in the late Jurassic despite a relatively low Mg/Ca ratio. Coccolithophores were most successful during the calcite seas of the Cretaceous, but are still prolific today under a geologically high Mg/Ca ratio (5.2). In the end, the seawater Mg/Ca ratio appears to influence the relative success of calcifying taxa, but is not the master control of either reef-building taxa or the rate of reef building.

9.5.2 Ocean Acidification

Ocean acidification (OA) describes the reduction in seawater pH and associated changes in seawater chemistry (i.e., reduced carbonate concentration, $[CO_3^{2-}]$, reduced aragonite and calcite saturation states, ($\Omega_{arag/calc}$, etc.) due to the addition of acid to the ocean). The terminology used with OA can sometimes be confusing. The process of adding an acid to a solution, ultimately reducing that solution's pH, is referred to as *acidification* and makes the solution *more acidic*. However, a solution only becomes *acidic* (as opposed to basic) when the pH drops below neutral (pH ~7). Hence, OA is correctly referred to as acidification, and makes the oceans more acidic, but bulk seawater pH will never drop below neutral, and the oceans will never become "acidic" by accepted chemical standards. Having made this distinction, carbonate minerals will begin to dissolve well above a pH of 7 in normal seawater, and OA makes seawater more corrosive to carbonates.

At local or regional scales, the introduction of mineral acids (e.g., sulfuric acid through volcanic activity) can lead to short-lived OA, but global scale OA is generally only achievable through the release of large amounts of CO_2. When CO_2 dissolves into seawater, it reacts to produce carbonic acid (Caldeira and Wickett 2003; Orr et al. 2005). Over geologic timescales, chemical weathering on land and carbonate dissolution in the ocean can neutralize excess CO_2, releasing alkalinity and cations (Ca^{2+}, Mg^{2+}, etc.), and maintain relatively stable $\Omega_{arag/calc}$, though not necessarily stable pH (Archer and Brovkin 2008; Hönisch et al. 2012). Hence, OA is inherently an ephemeral process on geological timescales, restricted to a few tens of thousands of years at most.

OA is occurring today due to human activities, but geological evidence suggests that it has also occurred in the past, causing negative consequences for calcifying taxa and for reef building. The current rate of OA is potentially unprecedented for at least the last 300 Ma of Earth's history, and could push organisms and ecosystems into unknown territory (Hönisch et al. 2012).

The Paleocene-Eocene Thermal Maximum (PETM: ~56 M) has been proposed as an analogous event and may provide a window into the future absent aggressive reductions in anthropogenic CO_2 emissions (Kiessling and Simpson 2011; Hönisch et al. 2012). The rate of acidification during the PETM is not clear, but Wright and Schaller (2013) argue, based on high-resolution proxy data, that the injection of carbon into the atmosphere during the PETM may have occurred within as little as a year, and surface ocean acidification could have occurred in as little as 13 years. This conclusion is hotly debated (Pearson and Nicholas 2014; Stassen et al. 2014; Wright and Schaller 2014; Zeebe et al. 2014). In contrast, most studies have suggested that the carbon release and resultant OA during the PETM occurred over a timescale of several millennia (e.g., Penman et al. 2014; Zeebe et al. 2014). Hence, depending on the timescale of acidification during the PETM (estimates ranging from ~10 to 10,000 years), the event proceeded at a rate between an order of magnitude faster and two orders of magnitude slower than anthropogenic OA. Additional proxy data and modeling studies are clearly needed to resolve this issue. Nevertheless, the PETM provides our best proxy for possible impacts following rapid CO_2 release.

The PETM was associated with a 0.25–0.45 pH unit reduction, and an average Ω_{arag} reduction from ~3 to ~1.5 in the surface, mixed layer of the ocean within a period of less than 10,000 years (Zachos et al. 2010; Penman et al. 2014). These changes in seawater chemistry are similar in magnitude to those associated with medium to high CO_2 emissions from human activity during this century (IPCC 2013). The critical factor would be the timeframe over which these changes occurred. Based on a timescale of a few thousand years, the rate of acidification during the PETM would have been an order of magnitude slower than modern OA due to human activities and any twenty-first century scenario would be more dramatic.

As mentioned previously, the PETM was mild in its effects on coral extinction rates, though it was associated with a mass extinction event among benthic formainifera (Kiessling and Baron-Szabo 2004; Hönisch et al. 2012). While corals did not experience a major increase in extinction rates during the PETM, reef building was substantially reduced. In fact, the PETM constitutes one of the five major reef crises during the Phanerozoic recognized by Kiessling and Simpson (2011). Therefore, even though changes in coral diversity may have been modest, the impact of OA on reef building was not. Chapter 6 argues that, although related, coral growth and reef building should not be confused. Evidence from the PETM suggests that reef-wide calcification (and, therefore, accretion) may be compromised well in advance of community structure breaking down. That is, reef growth may plummet even when coral cover is high and the ecosystem visually appears to be healthy. This has implications for our ability to understand reef building in the past using taxonomic information alone. Looking to the future, it suggests that lowered calcification and other biological (bioerosion), physical (storms), and chemical factors (acidification) may be as important as coral abundance or diversity in predicting which reefs will keep up with rising sea level and which ones will lag behind.

After the PETM, perhaps the next most similar event to modern OA for which there are also good, geochemical constraints is the last Pleistocene glacial-interglacial transition. During this interval, atmospheric CO_2 increased from ~189 to ~265 μatm, resulting in a ~0.15 decrease in surface ocean pH. This level of pH change is at the low end of the magnitude of anthropogenic OA (0.15 pH units vs. 0.15–0.45 pH units) proposed by the end of the century, depending on anthropogenic CO_2 emissions. However, it is worth nothing that this occurred more than an order of magnitude more slowly (~6000 vs. <200 years). Hence, Pleistocene glacial-interglacial transitions are of limited utility in understanding the impacts of rapid OA on reef building. Nevertheless, these transitions were associated with reduced shell weight of foraminifera (Barker and Elderfield 2002) and coccolithophores (Beaufort et al. 2011) even though there was no obvious impacts on reef building (Hönisch et al. 2012). How relevant this is to reef accretion in the immediate future lay in the importance of an order of magnitude faster increase going forward.

The Triassic-Jurassic mass extinction event (~200 Ma) may have also been associated with substantial OA, and could provide another useful proxy for recent events, though the geologic record is much poorer this far back in time. As discussed previously, this extinction event devastated scleractinians and reef building collapsed—another of the five reef crises identified by Kiessling and Simpson (2011) over the Phanerozoic. However, while there is evidence for a rapid increase in atmospheric CO_2 at the time of this mass extinction event, few data are available with the resolution required to unambiguously implicate OA.

Other potential OA events have been identified in the fossil record, including the Permian-Triassic mass extinction event (~252 Ma) and an Early Jurassic ocean anoxia event (Toarcian OAE, ~183 Ma). Both of these correspond to reef crises (Kiessling and Simpson 2011). The Toarcian event led to a modest extinction among corals and calcifying sponges, but had little effect on extinction rates for other organisms. In contrast, the Permian-Triassic event corresponds to the most severe global mass extinction and reef crisis in the fossil record. Two Cretaceous ocean anoxia events (OAE 1a, ~120 Ma; OAE 2, ~93 Ma) may have also involved OA, though the evidence for acidification during these events is not as clear as in those discussed above.

The four reef crises mentioned above all appear to coincide with OA events. In contrast, neither the Late Devonian mass extinction event and reef crisis (~375 Ma) nor the minor coral-sponge extinction event of the Late Permian (~260 Ma) appear to have been associated with a severe drop in pH. Similarly, neither the Late Ordovician (~446–444 Ma) nor the Cretaceous-Paleogene (~66 Ma) mass extinction events appear to be associated with a decline in pH, nor were there global reef crises during these intervals. Hence, there is not a perfect correspondence among OA events, reef crises, and mass extinction events. Nonetheless, there is compelling geologic evidence for the occurrence of at least four major OA events over the Phanerozic, and these correspond to four of the five major reef crises over geologic time. The conditions associated with OA are not the only way to hinder reef building, but they appear to be a highly effective mechanism.

9.5.3 Ocean Acidification vs. Rapid Climate Change

The geologically rapid release of large amounts of CO_2 to the biosphere results in OA, but also leads to rapid climate change. The four major reef crises in the fossil record that appear to be associated with OA events were also characterized by rapid global warming, making it challenging to decipher which factor (acidification vs. warming) was most responsible for the observed biological and geochemical responses. In fact, rapid warming may have been the more important factor in driving reef decline and elevated extinction rates in at least some of these events. During the Triassic-Jurassic reef crisis, tropical taxa were more severely affected than were non-tropical taxa, and reef building in the earliest Jurassic was concentrated primarily in mid-latitudes and deeper waters (Flügel and Kiessling 2002; Kiessling and Aberhan 2007; Kiessling et al. 2009; Lathuilière and Marchal 2009). Likewise, during the PETM the loss of reefs propagated from tropical to subtropical locations (Scheibner and Speijer 2008). These observations suggest that an escape from heat due to rapid global warming rather than ocean chemistry may have been the more important proximate factor driving the loss of reefs. OA, however, certainly could have reinforced the pattern or acted synergistically with global warming to enhance organisms' sensitivity to heat stress, as has been shown for some contemporary corals and coralline algae (Anthony et al. 2008).

Acidification and rapid warming in combination seem to act as an effective "kill switch" for reef building. Corals are capable of adapting or acclimatizing to a range of temperatures given enough time, as are other reef builders. The extent to which reef builders could adapt or acclimatize to OA conditions is not yet clear, though data showing adaptive responses to OA is beginning to emerge for other calcifiers (Lohbeck et al. 2012; Pespeni et al. 2013; Schlüter et al. 2014; Suckling et al. 2014). OA favors carbonate dissolution rather than accumulation and preservation, however, and even if reef builders are able to tolerate novel environmental conditions during OA/climate change events, OA could still imperil the geologic process of reef building by reducing the calcification rate of reef biota, increasing both bioerosion and dissolution, and decreasing abiotic cementation (Manzello et al. 2008; Eyre et al. 2014). The latter process may become increasingly important as erosion and breakage due to more intense storm activity remove loose sediment and rubble (Webster et al. 2005).

Reduced preservation of reef carbonates under OA may be especially important to deep-water coral reefs, such as those built by *Lophelia pertusa*, because they naturally occur closer to the chemical conditions which favor carbonate dissolution. In laboratory experiments, calcification by *Lophelia* appeard to be quite resistant to OA (Maier et al. 2011, 2013), yet *Lophelia* reefs tend to disappear where natural processes make seawater too acidic. OA may substantially reduce deep-water reef building, even if the corals themselves are able to tolerate the new conditions (Turley et al. 2007; Davies et al. 2008). The combined impacts of OA and rapid climate change have led to the global collapse of reef building in the geologic past, and a future collapse of reef building is a real possibility under anthropogenic OA and climate change today.

9.5.4 The Long Tail of CO_2

Following the geologically rapid release of large amounts of CO_2, tens of thousands of years are required for climate and ocean chemistry to return to levels roughly similar to those prior to the event. This response is driven in part by the quasi-exponential decay in atmospheric CO_2 concentration, leading to a long right tail in the anomaly distribution and a slow decrease in the future. Under anthropogenic OA and climate change, global mean temperature will remain near the peak level of warming even 1000 years after anthropogenic CO_2 emissions cease, and more than half of the peak level of warming may remain after 10,000 years (Archer and Brovkin 2008; Solomon et al. 2009; Meissner et al. 2012).

The time needed for elevated temperatures to return to initial levels is on the order of 20,000–100,000 years, depending on the ultimate size of the CO_2 perturbation and feedbacks within the Earth system (Archer and Brovkin 2008; Meissner et al. 2012). Hansen et al. (2013) note that geologic precedent points to a higher long-term climate sensitivity and potentially much greater warming of the climate over millennial timescales. Hence, rather than a decay in the level of warming after 10,000 years, the Earth system may continue to warm, leading to almost twice the rise in temperature expected from the medium-term equilibrium climate response. The recovery of seawater chemistry is likewise a slow process, requiring tens of thousands of years, but will occur somewhat more rapidly than the recovery of climate (Archer and Brovkin 2008; Meissner et al. 2012).

These are extremely slow processes on human timescales, but such long periods are supported by both modeling and geologic precedent. About 100,000 years was required for oceanic processes, such as carbonate deposition, to return to normal following the PETM, for example (Zachos et al. 2005). Thus, over timescales relevant to human society, climate change and OA are both "irreversible" (*sensu* Solomon et al. 2009), but they are not unstoppable. The magnitude of the perturbations from anthropogenic CO_2 to the Earth system and to reef building depends heavily on the quantity of CO_2 released by human activities over coming decades. The most severe impacts of OA and climate change can be avoided if human-induced CO_2 emissions are rapidly curtailed.

9.5.5 Reef Building Under Persistent, Low pH

Proxy data and modeling studies show that seawater pH and carbonate ion concentration $[CO_3^{2-}]$ were much lower and pCO_2 was much higher during most of the Phanerozoic as compared to the modern ocean. Given the negative impacts of OA on reef building, the success of reefs throughout much of geologic time at first seems counterintuitive. This paradox can be solved by recognizing that reef building does not depend simply on seawater pH, $[CO_3^{2-}]$, or pCO_2 *per se*, but that these parameters are part of a more complex set of chemical drivers of reef building.

Over geologically short timescales, $[Ca^{2+}]$ is essentially constant in sea water, but varies over long timescales due to changes in the rate of chemical weathering and seafloor spreading, as discussed above. Over the Phanerozoic, seawater $[Ca^{2+}]$ has varied from a minimum of ~10.3 mM in modern sea water to a maximum of perhaps ~30–40 mM during the Cretaceous (Hardie 1996; Lowenstein et al. 2001; Demicco et al. 2003; 2005). Carbonate ion concentration depends on seawater pH and carbonate alkalinity (at a given temperature, pressure, and salinity), which are themselves determined by the interaction of rates of chemical weathering, $CaCO_3$ production, and preservation. During much of the Phanerozoic, variation in oceanic $[CO_3^{2-}]$ (as well as pH) and $[Ca^{2+}]$ have tended to be opposite in phase (Ridgwell 2005). The overall effect of these changes has been relatively stable $\Omega_{arag/calc}$ when averaged over long timescales, though the individual parameters of the carbonate system have varied dramatically over geologic time (Ridgwell 2005; Fig. 9.4). This long-term stability does not include short-term perturbations to the carbonate system, such as those associated with an OA event.

The rate of reef building depends on the net outcome of the opposing processes of carbonate production and carbonate erosion/dissolution. Post-depositional carbonate preservation depends heavily on seawater $\Omega_{arag/calc}$ (Eyre et al. 2014), though other factors such as nutrient availability also affect reduced preservation via bioerosion. The relative

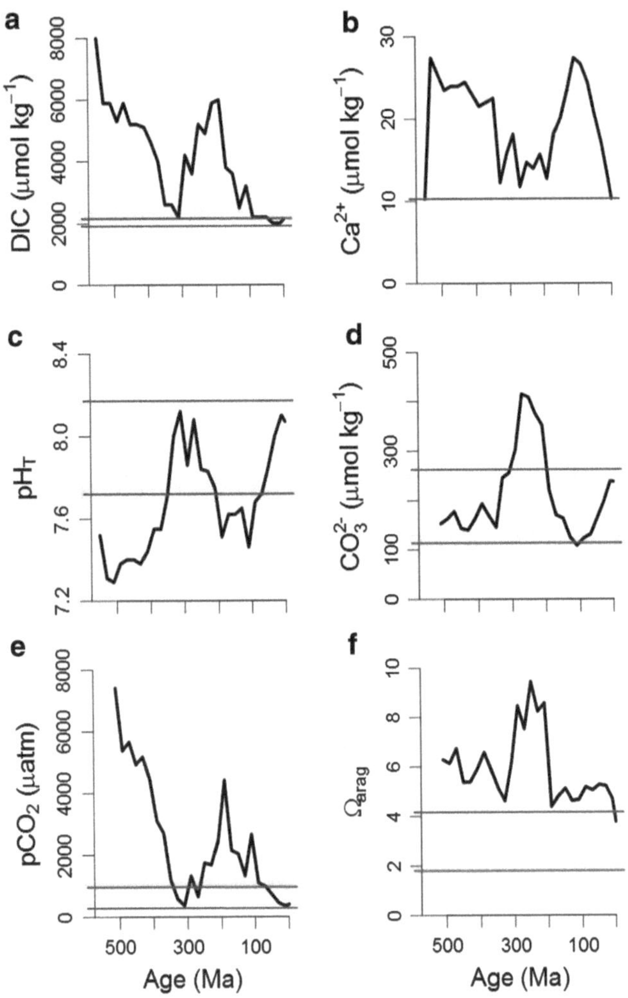

Fig. 9.4 Secular variation in seawater carbonate chemistry over the Phanerozoic. *Black curves* are best estimates of (**a**) dissolved inorganic carbon, (**b**) calcium concentration, (**c**) pH, (**d**) carbonate concentration, (**e**) carbon dioxide partial pressure, and (**f**) aragonite saturation state for the tropical, surface ocean derived from modeling and constrained by geochemical data. Values shown as 20 My averages; error estimates omitted for clarity. Horizontal lines in each plot show modern values for the preindustrial (*blue*) and the end-of-the-century conditions under business-as-usual CO_2 emissions (*red*). Preindustrial and end-of-the-century lines are overlapping in (**b**) (Modified from Ridgwell 2005)

stability of oceanic $\Omega_{arag/calc}$ over long timescales suggests that major changes in reef building generally were not driven by thermodynamically controlled changes in carbonate preservation. This is in contrast to ephemeral OA events, where reduced carbonate preservation and increased dissolution were important factors in the observed reductions in reef-building capacity. Instead, major changes in reef accretion over long timescales were likely driven more by variability in carbonate production, the relative success of reef-building taxa, and the importance of bioeriosion, cementation, and encrustation of loose detritus within a given interval.

Much research to date has considered the effects of OA or changes in carbonate chemistry on calcifiers in the context of changing $\Omega_{arag/calc}$, but physiological data show that such an approach is an oversimplification of the true chemical drivers of organismal calcification (see Chap. 2). While average oceanic $\Omega_{arag/calc}$ has been relatively stable over much of geologic time, the other components of the carbonate system have not, and changes in these other parameters may help to explain variation in reef building. Regardless, the apparent paradox of reef building under steady-state low pH, low $[CO_3^{2-}]$, and high pCO_2 is solved by the compensating effects of higher seawater $[Ca^{2+}]$ and carbonate alkalinity on carbonate production and preservation. However, these mechanisms only apply under steady-state conditions, where chemical weathering compensates for high atmospheric CO_2, and are probably not directly informative of the impacts of OA on reef building on shorter, human timescales.

9.6 Coral Reefs in a Brave New Ocean

Human activities today are assaulting coral reefs on multiple fronts. Over the last few decades these pressures have resulted in the loss of at least half the coral cover across the wider Caribbean and Indo-Pacific, though the decline has been highly variable in space and time (Gardner et al. 2003; Bruno and Selig 2007; Jackson et al. 2014). Rates of reef building have likely been reduced by at least these magnitudes, if not more due to reduced growth rates of the surviving corals and community shifts toward less productive species (De'ath et al. 2009; Perry et al. 2013, 2015). The bright spot in this gloomy picture is that relatively healthy reefs still exist, and many reefs which have lost substantial portions of their coral cover could still recover over the decadal timescales relevant to human society. Indeed, without chronic, local, human stressors, some reefs can recover from catastrophic disturbance in as little as 12–20 years (Gilmour et al. 2013; Bahr et al. 2015). The fate of coral reefs over the next 10,000–100,000 years depends heavily on human decisions within the next few decades. With aggressive societal action, many reefs can be put on the path to recovery before the end of this century; without it, many reefs will continue on a downward spiral, culminating in their collapse.

The only scenarios which could save a significant portion of the world's coral reefs are ones where society substantially reduces emissions of greenhouse gases (particularly CO_2) and improves local reef management (i.e., reductions in sedimentation, pollution, overfishing, nutrification, etc.) within the next few decades (Kennedy et al. 2013). Without both global reductions in CO_2 emissions and effective local management, coral reefs will become increasingly degraded as bleaching events, acidification impacts, outbreaks of disease, infestations of Crown-of-Thorns starfish, more powerful storms, sedimentation, pollution, and a host of other stressors kill corals and erode the reef structures they have built. Under increasing stress, reef building will collapse in most places before the end of the century, present-day coral reefs will drown below the rising sea, and reefs will remain rare long into the future, as the geologic history of reef building clearly illustrates. The tools needed to prevent such a decline already exist, but they must be implemented quickly if society wishes to prevent the global collapse of coral reefs.

9.7 Conclusions

The geologic history of reef building is one of immense change. Reefs were not always built by corals, and both the dominant taxa and geologic processes of reef building have waxed and waned over time (see Chap. 8). Cyanobacteria and other algal assemblages built the first reefs billions of years ago, but their stromatolites and thrombolites became rare as animal life began to influence what had previously been a microbial world. Some early metazoans and their progenitors began to build mineralized shells and skeletons long ago, becoming obvious by the latter Ediacaran, but first became widespread during the Early Cambrian. Rugose and tabulate corals as well as hypercalcifying sponges came to dominate reef building throughout much of the Paleozoic. These coral-sponge reefs were especially successful during the Ordovician, Silurian, and Devonian, and much less dominant during the Carboniferous and Permian. Changes in the relative success of these calcitic reef builders may be explained in part by changes in the Mg/Ca ratio of seawater, which led to calcite seas during the earlier part of the Paleozoic, but aragonite seas during the latter portion. Rugose and tabulate corals succumbed to the Permian-Triassic mass extinction event.

Scleractinian corals appear to have originated and began calcifying in the calcite seas of the Paleozoic but did not build reefs until the Mesozoic. Their reefs burst onto the scene during the Middle Triassic and scleractinians dominated reef building until the Triassic-Jurassic mass extinction event, which devastated them. They again rose to prominence during the Late Jurassic to Early Cretaceous, but experienced a much-reduced role as reef builders during much of the Cretaceous, perhaps driven in part by a decline of the seawater Mg/Ca ratio deep into the calcite nucleation field. In contrast, calcitic taxa such as some rudist bivalves and planktonic coccolithophores experienced a great deal of success in the calcite seas of the Cretaceous. Rudists went extinct by the end of the Cretaceous and scleractinians survived the Cretaceous-Paleogene mass extinction event. Scleractinian reef building remained at relatively modest levels until the end of the Paleogene, when the seawater Mg/Ca ratio rapidly rose into the aragonite nucleation field.

Slow, secular changes in climate have influenced the latitudinal extent of reefs over time, with reefs expanding toward the poles under warmer climates and retreating toward the equator during cooler periods. Prevailing climate alone, however, is not a good predictor of the relative success of reef building over time. All else being equal, warmer climates expand the potentially habitable area for tropical reef builders, but other factors can override favorable climatic conditions. Similarly, slow, secular changes in seawater pH and CO_2 concentration have been balanced by the release of alkalinity and calcium to the ocean over geologic timescales, maintaining suitable carbonate chemistry for reef building across a range of pH and CO_2 levels.

In contrast to slow, secular, changes in climate and ocean chemistry, the geologically rapid release of CO_2 to the biosphere leads to rapid climate change and OA. It is not entirely clear which of these factors, or their combination has the most severe effects on reef building, but episodes of rapid climate change and OA are associated with reef crises in the geologic past that may be realistic models for the near future. The fate of coral reefs during this century and beyond depends heavily on decisions regarding global CO_2 emissions and local reef management over the next few decades. Optimistically, CO_2 emissions and local human impacts can be reduced sufficiently to allow coral reefs to begin to recover from the assaults of recent decades. Less optimistically, human actions will lead to the next major coral reef crisis, leaving an indelible mark on the geologic history of reef building.

Acknowledgements This chapter is funded in part by a grant/cooperative agreement from the National Oceanic and Atmospheric Administration, Project R/IR-23, which is sponsored by the University of Hawaii Sea Grant College Program, SOEST, under Institutional Grant No. NA09OAR4170060 from NOAA Office of Sea Grant, Department of Commerce. The views expressed herein are those of the author (s) and do not necessarily reflect the views of NOAA or any of its subagencies. UNIHI-SEAGRANT- BC-13-07. Additional funding is provided by the National Science Foundation (OA#1416889), and NOAA Grant "Research in Support of the NWHI Coral Reef Ecosystem Reserve and the Pacific Island Climate Change Cooperative (PICCC)".

References

Algeo T, Scheckler S (1998) Terrestrial-marine teleconnections in the Devonian: links between the evolution of land plants, weathering processes, and marine anoxic events. Philos Trans R Soc B Biol Sci 353:113–130

Allemand D, Tambutté E, Girard J, Jaubert J (1998) Organic matrix synthesis in the scleractinian coral *Stylophora pistillata*: role in biomineralization and potential target of the organotin tributyltin. J Exp Biol 201:2001–2009

Allwood A, Walter M, Burch I, Kamber B (2007) 3.43 billion-year-old stromatolite reef from the Pilbara Craton of Western Australia: ecosystem-scale insights to early life on Earth. Precambrian Res 158:198–227
Anthony K, Fabricius K (2000) Shifting roles of heterotrophy and autotrophy in coral energetics under varying turbidity. J Exp Mar Biol Ecol 252:221–253
Anthony K, Kline D, Diaz-Pulido G, Dove S, Hoegh-Guldberg O (2008) Ocean acidification causes bleaching and productivity loss in coral reef builders. Proc Natl Acad Sci 105:17442–17446
Archer D, Brovkin V (2008) The millennial atmospheric lifetime of anthropogenic CO_2. Clim Change 90:283–297
Atkinson M (2011) Biogeochemistry of nutrients. In Dubinsky Z, Stambler N (eds) Coral reefs: an ecosystem in transition. Springer, New York, pp 199–206
Bahr K, Jokiel P, Toonen R (2015) The unnatural history of Kāne'ohe Bay: coral reef resilience in the face of centuries of anthropogenic impacts. PeerJ 3:e950
Barker S, Elderfield H (2002) Foraminiferal calcification response to glacial-interglacial changes in atmospheric CO_2. Science 297:833–836
Barshis D, Ladner J, Oliver T, Seneca F, Traylor-Knowles N, Palumbi S (2013) Genomic basis for coral resilience to climate change. Proc Natl Acad Sci 110:1387–1392
Barshis D, Stillman J, Gates R, Toonen R, Smith L, Birkeland C (2010) Protein expression and genetic structure of the coral *Porites lobata* in an environmentally extreme Samoan back reef: does host genotype limit phenotypic plasticity? Mol Ecol 19:1705–1720
Baskett M, Gaines S, Nisbet R (2009) Symbiont diversity may help coral reefs survive moderate climate change. Ecol Appl 19:3–17
Beaufort L, Probert I, de Garidel-Thoron T, Bendif E, Ruiz-Pino D, Metzl N, Goyet C, Buchet N, Coupel P, Grelaud M, Rost B, Rickaby R, de Vargas C (2011) Sensitivity of coccolithophores to carbonate chemistry and ocean acidification. Nature 476:80–83
Boardman R, Cheetham A (1987) Phylum Bryozoa. In: Boardman R, Cheetham, A, Rowell A (eds) Fossil invertebrates. Blackwell, Palo Alto, pp 497–549
Brodie J, Wolanski E, Lewis S, Bainbridge Z (2012) An assessment of residence times of land-sourced contaminants in the Great Barrier Reef lagoon and the implications for management and reef recovery. Mar Pollut Bull 65:267–279
Bruno J, Petes L, Harvell D, Hettinger A (2003) Nutrient enrichment can increase the severity of coral diseases. Ecol Lett 6:1056–1061
Bruno J, Selig E (2007) Regional decline of coral cover in the Indo-Pacific: timing, extent, and subregional comparisons. PLoS ONE 2: e711
Budd G, Jensen S (2000) A critical reappraisal of the fossil record of the bilaterian phyla. Biol Rev 75:253–295
Caldeira K, Wickett M (2003) Oceanography: anthropogenic carbon and ocean pH. Nature 425:365–365
Came R, Eiler J, Veizer J, Azmy K, Brand U, Weidman C (2007) Coupling of surface temperatures and atmospheric CO_2 concentrations during the Palaeozoic era. Nature 449:198–201
Carbone F, Matteucci R, Pignatti J, Russo A (1994) Facies analysis and biostratigraphy of the Auradu limestone formation in the Berbera-Sheikh area, north-western Somalia. Geol Romana 29:213–235
Chazottes V (2002) The effects of eutrophication-related alterations to coral reef communities on agents and rates of bioerosion (Reunion Island, Indian Ocean). Coral Reefs 21:375–390
Church J, White N (2006) A 20th century acceleration in global sea-level rise. Geophys Res Lett 33: L01602 4 p. doi:10.1029/2005GL024826
Church J, White N (2011) Sea-level rise from the late 19th to the early 21st Century. Surv Geophys 32:585–602
Cohen P, Schopf J, Butterfield N, Kudryavtsev A, Macdonald F (2011) Phosphate biomineralization in mid-Neoproterozoic protists. Geology 39:539–542
Coles S, Fadlallah Y (1991) Reef coral survival and mortality at low temperatures in the Arabian Gulf: new species-specific lower temperature limits. Coral Reefs 9:231–237
Coles SL, Jokiel PL, Lewis CR (1976) Thermal tolerance in tropical versus subtropical Pacific reef corals. Pac Sci 30:155–166
Colombo-Pallotta M, Rodríguez-Román A, Iglesias-Prieto R (2010) Calcification in bleached and unbleached *Montastraea faveolata*: evaluating the role of oxygen and glycerol. Coral Reefs 29:899–907
Cramer K, Jackson J, Angioletti C, Leonard-Pingel J, Guilderson T (2012) Anthropogenic mortality on coral reefs in Caribbean Panama predates coral disease and bleaching. Ecol Lett 15:561–567
Davies A, Wisshak M, Orr J, Murray Roberts J (2008) Predicting suitable habitat for the cold-water coral *Lophelia pertusa* (Scleractinia). Deep Sea Research Part I: Oceanographic Research Papers 55:1048–1062
Davies P, Hopley D (1983) Growth fabrics and growth rates of Holocene reefs in the Great Barrier Reef. BMR J of Australian Geol and Geophys 8: 237–251
Davies P, Marshall J, Hopley D (1985) Relationship between reef growth and sea level in the Great Barrier Reef. Proc 5th Int Coral Reef Symp 3: 95–103
De'ath G, Lough J, Fabricius K (2009) Declining coral calcification on the Great Barrier Reef. Science 323:116–119
DeCarlo T, Cohen A, Barkley H, Cobban Q, Young C, Shamberger K, Brainard R, Golbuu Y (2015) Coral macrobioerosion is accelerated by ocean acidification and nutrients. Geology 43:7–10
Demicco R, Lowenstein T, Hardie L (2003) Atmospheric pCO2 since 60 Ma from records of seawater pH, calcium, and primary carbonate mineralogy. Geology 31:793
Demicco R, Lowenstein T, Hardie L, Spencer R (2005) Model of seawater composition for the Phanerozoic. Geology 33:877
Diaz-Pulido G, Nash M, Anthony K, Bender D, Opdyke B, Reyes-Nivia C, Troitzsch U (2014) Greenhouse conditions induce mineralogical changes and dolomite accumulation in coralline algae on tropical reefs. Nature Communications 5: Article 3310. Doi:10.1038/ncomms4310
Dill R, Shinn E, Jones A, Kelly K, Steinen R (1986) Giant subtidal stromatolites forming in normal salinity waters. Nature 324:55–58
Dornbos S, Bottjer D, Chen J (2004) Evidence for seafloor microbial mats and associated metazoan lifestyles in Lower Cambrian phosphorites of Southwest China. Lethaia 37:127–137
Drake J, Mass T, Haramaty L, Zelzion E, Bhattacharya D, Falkowski P (2013) Proteomic analysis of skeletal organic matrix from the stony coral *Stylophora pistillata*. Proc Natl Acad Sci 110:3788–3793
Dravis J (1983) Hardened subtidal stromatolites, Bahamas. Science 219:385–386
Droser M (2003) The Ordovician radiation: a follow-up to the Cambrian explosion? Integr Comp Biol 43:178–184
Dunn J, Sammarco P, LaFleur G (2012) Effects of phosphate on growth and skeletal density in the scleractinian coral *Acropora muricata*: A controlled experimental approach. J Exp Mar Biol Ecol 411:34–44
Dunne R, Brown B (1996) Penetration of solar UVB radiation in shallow tropical waters and its potential biological effects on coral reefs; results from the central Indian Ocean and Andaman Sea. Mar Ecol Prog Ser 144:109–118
Dzik J (2007) The Verdun Syndrome: simultaneous origin of protective armour and infaunal shelters at the Precambrian Cambrian transition. Geol Soc Spec Publ 286:405–414
Edinger E, Risk M (1994) Oligocene-Miocene extinction and geographic restriction of Caribbean corals: roles of turbidity, temperature, and nutrients. Palaios 9:576

Edmunds P (2011) Zooplanktivory ameliorates the effects of ocean acidification on the reef coral *Porites* spp. Limnol Oceanogr 56:2402–2410

Eyre B, Andersson A, Cyronak T (2014) Benthic coral reef calcium carbonate dissolution in an acidifying ocean. Nat Clim Chang 4:969–976

Falkowski P, Dubinsky Z, Muscatine L, McCloskey L (1993) Population control in symbiotic corals. BioScience 43:606–611

Ferrier-Pagès C, Gattuso J, Dallot S, Jaubert J (2000) Effect of nutrient enrichment on growth and photosynthesis of the zooxanthellate coral *Stylophora pistillata*. Coral Reefs 19:103–113

Ferrier-Pagès C, Witting J, Tambutté E, Sebens K (2003) Effect of natural zooplankton feeding on the tissue and skeletal growth of the scleractinian coral *Stylophora pistillata*. Coral Reefs 22:229–240

Finelli C, Helmuth B, Pentcheff N, Wethey D (2006) Water flow influences oxygen transport and photosynthetic efficiency in corals. Coral Reefs 25:47–57

Flügel E, Kiessling W (2002) Patterns of Phanerozoic reef crises. In: Kiessling W, Flügel E, Golonka J (eds) Phanerozoic Reef Patterns. SEPM Special Publication 72, Society for Sedimentary Geology, Tulsa, OK, USA, pp 691–733

Forsman Z, Kimokeo B, Bird C, Hunter C, Toonen R (2012) Coral farming: effects of light, water motion and artificial foods. J Mar Biol Assoc UK 92:721–729

Freiwald A, Fosså J, Grehan A, Koslow T, Roberts J (2004) Cold-water coral reefs. UNEP-WCMC, Cambridge, UK

Fukami H, Omori M, Hatta M (2000) Phylogenetic relationships in the coral family Acroporidae, reassessed by inference from mitochondrial genes. Zoolog Sci 17:689

Gardella D, Edmunds P (1999) The oxygen microenvironment adjacent to the tissue of the scleractinian *Dichocoenia stokesii* and its effects on symbiont metabolism. Mar Biol 135:289–295

Gardner T, Côté I, Gill J, Grant A, Watkinson A (2003) Long-term region-wide declines in Caribbean corals. Science 301:958–960

Gili E, Masse J, Skelton P (1995) Rudists as gregarious sediment-dwellers, not reef-builders, on Cretaceous carbonate platforms. Palaeogeography, Palaeoclimatology, Palaeoecology 118:245–267

Gilmour J, Smith L, Heyward A, Baird A, Pratchett M (2013) Recovery of an isolated coral reef system following severe disturbance. Science 340:69–71

Glynn P (1997) Bioerosion and coral reef growth: a dynamic balance. In Birkeland C (ed) Life and death of coral reefs, Chapman and Hall, New York, pp 68–95

Glynn P, Manzello D (2015) Bioerosion and coral reef growth: a dynamic balance, In: Birkeland C (Ed) Coral reefs in the Anthropocene, Springer, 67–98

Godinot C, Grover R, Allemand D, Ferrier-Pagès C (2011) High phosphate uptake requirements of the scleractinian coral *Stylophora pistillata*. J Exp Biol 214:2749–2754

Gough D (1981) Solar interior structure and luminosity variations. Solar Physics 74:21–34

Greenstein B, Pandolfi J (2008) Escaping the heat: range shifts of reef coral taxa in coastal Western Australia. Glob Chang Biol 14:513–528

Guest J, Baird A, Maynard J, Muttaqin E, Edwards A, Campbell S, Yewdall K, Affendi Y, Chou L (2012) Contrasting patterns of coral bleaching susceptibility in 2010 suggest an adaptive response to thermal stress. PLoS ONE 7:e33353

Hansen J, Sato M, Russell G, Kharecha P (2013) Climate sensitivity, sea level and atmospheric carbon dioxide. Philo Trans R Soc A Math Phys Eng Sci 371:20120294

Hardie L (1996) Secular variation in seawater chemistry: An explanation for the coupled secular variation in the mineralogies of marine limestones and potash evaporites over the past 600 m.y. Geology 24:279

Herbert T, Schuffert J (2000) Alkenone unsaturation estimates of sea-surface temperatures at DP Site 1002 over a full glacial cycle. Proc Ocean Drill Program Sci Results 165:239–247

Higuchi T, Fujimura H, Yuyama I, Harii S, Agostini S, Oomori T (2014) Biotic control of skeletal growth by scleractinian corals in aragonite–calcite seas. PLoS ONE 9:e91021

Holcomb M, Tambutté E, Allemand D, Tambutté S (2014) Light enhanced calcification in *Stylophora pistillata*: effects of glucose, glycerol and oxygen. PeerJ 2:e375

Hönisch B, Ridgwell A, Schmidt D, Thomas E, Gibbs S, Sluijs A, Zeebe R, Kump L, Martindale R, Greene S, Kiessling W, Ries J, Zachos J, Royer D, Barker S, Marchitto T, Moyer R, Pelejero C, Ziveri P, Foster G, Williams B (2012) The geological record of ocean acidification. Science 335:1058–1063

Hopley D, Smithers S, Parnell K (2007) The geomorphology of the Great Barrier Reef: development, diversity and change. Cambridge University Press, p532

Howells E, Beltran V, Larsen N, Bay L, Willis B, van Oppen M (2011) Coral thermal tolerance shaped by local adaptation of photosymbionts. Nat Clim Chang 2:116–120

Hubbard D (1986) Sedimentation as a control of reef development: St. Croix, USVI. Coral Reefs 5:117–125

Hubbard DK (2011) Eastern Caribbean reefs. In: Hopley D, Encyclopedia of modern coral reefs, Springer, the Netherlands, p 338-348

Hubbard D, Miller A (1990) Production and cycling of calcium carbonate in a shelf-edge reef system (St. Croix, U.S. Virgin Islands): applications to the nature of reef systems in the fossil record. J Sediment Res 60:335–360

Hubbard DK, Burke RB, Gill IP, Ramirez W, Sherman C (2008) Coral-reef geology: Puerto Rico and the U.S. Virgin Islands. In: Riegl, B. M., and Dodge, R.E., (eds) Coral Reefs of the USA, Springer-Verlag, p 263–302

Hughes T (1994) Catastrophes, phase shifts, and large-scale degradation of a Caribbean coral reef. Science 265:1547–1551

Hughes T, Baird A, Bellwood D, Card M, Connolly S, Folke C, Grosberg R, Hoegh-Guldberg O, Jackson J, Kleypas J, Lough J, Marshall P, Nyström M, Palumbi S, Pandolfi J, Rosen B, Roughgarden J (2003) Climate change, human impacts, and the resilience of coral reefs. Science 301:929–933

Hughes T, Rodrigues M, Bellwood D, Ceccarelli D, Hoegh-Guldberg O, McCook L, Moltschaniwskyj N, Pratchett M, Steneck R, Willis B (2007) Phase shifts, herbivory, and the resilience of coral reefs to climate change. Curr Biol 17:360–365

IPCC (2013) Climate Change 2013: The physical science basis. Contribution of Working Group I to the Fifth Assessment Report of the Intergovernmental Panel on Climate Change. Stocker T, Qin D, Plattner G, Tignor M, Allen S, Boschung J, Nauels A, Xia Y, Bex V, Midgley P (eds) Cambridge University Press, Cambridge, United Kingdom and New York, NY, USA, 1535 pp, doi:10.1017/CBO9781107415324

Jackson J, Donovan M, Cramer K, Lam V (eds) (2014) Status and trends of Caribbean coral reefs: 1970–2012. Global Coral Reef Monitoring Network, IUCN, Gland Switzerland, p304

Jensen S (2003) The Proterozoic and earliest Cambrian trace fossil record: patterns, problems and perspectives. Integr Comp Biol 43:219–228

Jevrejeva S, Grinsted A, Moore J (2009) Anthropogenic forcing dominates sea level rise since 1850. Geophys Res Lett 36:doi:10.1029/2009GL040216

Jevrejeva S, Moore J, Grinsted A (2012) Sea level projections to AD2500 with a new generation of climate change scenarios. Glob Planet Change 80–81:14–20

Johnson K, Jackson J, Budd A (2008) Caribbean reef development was independent of coral diversity over 28 million years. Science 319:1521–1523

Jokiel P (1978) Effects of water motion on reef corals. J Exp Mar Biol Ecol 35:87–97
Jokiel P (2006) Impact of storm waves and storm floods on Hawaiian reefs. Proc 10th Int Coral Reef Symp 1:390–398
Jokiel P, Brown E (2004) Global warming, regional trends and inshore environmental conditions influence coral bleaching in Hawaii. Glob Chang Biol 10:1627–1641
Jokiel P, Coles S (1977) Effects of temperature on the mortality and growth of Hawaiian reef corals. Mar Biol 43:201–208
Jokiel P, Coles S (1990) Response of Hawaiian and other Indo-Pacific reef corals to elevated temperature. Coral Reefs 8:155–162
Jokiel P, Dubinsky Z, Stambler N (1994) Results of the United States-Israel workship, "Nutrient limitation in the symbiotic associated between zooxanthellae and reef-building coral". Pac Sci 48:215–218
Jokiel P, Rodgers K, Storlazzi C, Field M, Lager C, Lager D (2014) Response of reef corals on a fringing reef flat to elevated suspended-sediment concentrations: Moloka'i, Hawai'i. PeerJ 2:e699
Kahng S, Garcia-Sais J, Spalding H, Brokovich E, Wagner D, Weil E, Hinderstein L, Toonen R (2010) Community ecology of mesophotic coral reef ecosystems. Coral Reefs 29:255–275
Kauffman E, Johnson C (1988) The morphological and ecological evolution of Middle and Upper Cretaceous reef-building rudistids. Palaios 3:194
Kawaguti S, Sakumoto D (1948) The effect of light on the calcium deposition of corals. Bull Oceanogr Inst 4:65–70
Kennedy E, Perry C, Halloran P, Iglesias-Prieto R, Schönberg C, Wisshak M, Form A, Carricart-Ganivet J, Fine M, Eakin C, Mumby P (2013) Avoiding coral reef functional collapse requires local and global action. Curr Biol 23:912–918
Kiessling W (2006) Towards an unbiased estimate of fluctuations in reef abundance and volume during the Phanerozoic. Biogeosciences 3:15–27
Kiessling W (2009) Geologic and biologic controls on the evolution of reefs. Annu Rev Ecol Evol Syst 40:173–192
Kiessling W, Aberhan M (2007) Environmental determinants of marine benthic biodiversity dynamics through Triassic–Jurassic time. Paleobiology 33:414–434
Kiessling W, Aberhan M, Villier L (2008) Phanerozoic trends in skeletal mineralogy driven by mass extinctions. Nat Geosci 1:527–530
Kiessling W, Roniewicz E, Villier L, Leonide P, Struck U (2009) An early Hettangian coral reef in southern France: implications for the end-Triassic reef crisis. Palaios, 24:657–671
Kiessling W, Baron-Szabo R (2004) Extinction and recovery patterns of scleractinian corals at the Cretaceous-Tertiary boundary. Palaeogeography, Palaeoclimatology, Palaeoecology 214:195–223
Kiessling W, Simpson C (2011) On the potential for ocean acidification to be a general cause of ancient reef crises. Glol Chang Biol 17:56–67
Kiessling W, Simpson C, Beck B, Mewis H, Pandolfi J (2012) Equatorial decline of reef corals during the last Pleistocene interglacial. Proc Natl Acad Sci 109:21378–21383
Kirkland B, Moore C, Dickson J (1993) Well preserved, aragonitic phylloid Algae (Eugonophyllum, Udoteaceae) from the Pennsylvanian Holder Formation, Sacramento Mountains, New Mexico. Palaios 8:111
Kitahara M, Cairns S, Stolarski J, Blair D, Miller D (2010) A comprehensive phylogenetic analysis of the Scleractinia (Cnidaria, Anthozoa) based on mitochondrial CO1 sequence data. PLoS ONE 5:e11490
Kitahara M, Lin M, Forêt S, Huttley G, Miller D, Chen C (2014) The "Naked Coral" Hypothesis revisited – evidence for and against scleractinian monophyly. PLoS ONE 9:e94774
Kleypas J, McManus J, Meñez L (1999) Environmental limits to coral reef development: where do we draw the line? Am Zool 39:146–159
Knowlton N, Lang J, Christine Rooney M, Clifford P (1981) Evidence for delayed mortality in hurricane-damaged Jamaican staghorn corals. Nature 294:251–252
Koop K, Booth D, Broadbent A, Brodie J, Bucher D, Capone D, Coll J, Dennison W, Erdmann M, Harrison P, Hoegh-Guldberg O, Hutchings P, Jones G, Larkum A, O'Neil J, Steven A, Tentori E, Ward S, Williamson J, Yellowlees D (2001) ENCORE: the effect of nutrient enrichment on coral reefs. Synthesis of results and conclusions. Mar Pollut Bull 42:91–120
Kühl M, Cohen Y, Dalsgaard T, Jorgensen B, Revsbech N (1995) Microenvironment and photosynthesis of zooxanthellae in scleractinian corals studied with microsensors for O_2, pH and light. Mar Ecol Prog Ser 117:159–172
Lapointe B (1997) Nutrient thresholds for bottom-up control of macroalgal blooms on coral reefs in Jamaica and southeast Florida. Limnol Oceangr 42:1119–1131
Lathuilière B, Marchal D (2009) Extinction, survival and recovery of corals from the Triassic to Middle Jurassic time. Terra Nova 21:57–66
Lea D, Pak D, Belanger C, Spero H, Hall M, Shackleton N (2006) Paleoclimate history of Galápagos surface waters over the last 135,000 yr. Quat Sci Rev 25:1152–1167
Leduc G, Schneider R, Kim J, Lohmann G (2010) Holocene and Eemian sea surface temperature trends as revealed by alkenone and Mg/Ca paleothermometry. Quat Sci Rev 29:989–1004
Lepot K, Benzerara K, Brown G, Philippot P (2008) Microbially influenced formation of 2,724-million-year-old stromatolites. Nat Geosci 1:118–121
Lesser M, Slattery M, Stat M, Ojimi M, Gates R, Grottoli A (2010) Photoacclimatization by the coral *Montastraea cavernosa* in the mesophotic zone: light, food, and genetics. Ecology 91:990–1003
Lirman D (2001) Competition between macroalgae and corals: effects of herbivore exclusion and increased algal biomass on coral survivorship and growth. Coral Reefs 19:392–399
Lisiecki L, Raymo M (2005) A Pliocene-Pleistocene stack of 57 globally distributed benthic $\delta^{18}O$ records. Paleoceanography 20: PA1003.doi:10.1029/2009PA001071
Locker S, Armstrong R, Battista T, Rooney J, Sherman C, Zawada D (2010) Geomorphology of mesophotic coral ecosystems: current perspectives on morphology, distribution, and mapping strategies. Coral Reefs 29:329–345
Logan C, Dunne J, Eakin C, Donner S (2013) Incorporating adaptive responses into future projections of coral bleaching. Glob Chang Biol 20:125–139
Lohbeck K, Riebesell U, Reusch T (2012) Adaptive evolution of a key phytoplankton species to ocean acidification. Nat Geosci 5:917–917
Lowenstein T, Tomfeeff M, Brennan S, Hardie L, Demicco R (2001) Oscillations in Phanerozoic seawater chemistry: evidence from fluid inclusions. Science 294:1086–1088
Loya Y, Lubinevsky H, Rosenfeld M, Kramarsky-Winter E (2004) Nutrient enrichment caused by *in situ* fish farms at Eilat, Red Sea is detrimental to coral reproduction. Mar Pollut Bull 49:344–353
Maier C, Bils F, Weinbauer M, Watremez P, Peck M, Gattuso J (2013) Respiration of Mediterranean cold-water corals is not affected by ocean acidification as projected for the end of the century. Biogeosciences Discussions 10:7617–7640
Maier C, Watremez P, Taviani M, Weinbauer M, Gattuso J (2011) Calcification rates and the effect of ocean acidification on Mediterranean cold-water corals. Proc Royal Soc B Biol Sci 279:1716–1723
Manzello D, Brandt M, Smith T, Lirman D, Hendee J, Nemeth R (2007) Hurricanes benefit bleached corals. Proc Natl Acad Sci 104:12035–12039
Manzello D, Kleypas J, Budd D, Eakin C, Glynn P, Langdon C (2008) Poorly cemented coral reefs of the eastern tropical Pacific: Possible

insights into reef development in a high-CO2 world. Proc Natl Acad Sci 105:10450–10455

Maragos J, Jokiel P (1986) Reef corals of Johnston Atoll: one of the world's most isolated reefs. Coral Reefs 4:141–150

Martín J, Braga J, Riding R (1997) Late Miocene Halimeda alga-microbial segment reefs in the marginal Mediterranean Sorbas Basin, Spain. Sedimentology 44:441–456

Marubini F, Atkinson M (1999) Effects of lowered pH and elevated nitrate on coral calcification. Mar Ecol Prog Ser 188:117–121

Maynard J, Anthony K, Marshall P, Masiri I (2008) Major bleaching events can lead to increased thermal tolerance in corals. Mar Biol 155:173–182

McCulloch M, Fallon S, Wyndham T, Hendy E, Lough J, Barnes D (2003) Coral record of increased sediment flux to the inner Great Barrier Reef since European settlement. Nature 421:727–730

McKay N, Overpeck J, Otto-Bliesner B (2011) The role of ocean thermal expansion in Last Interglacial sea level rise. Geophys Res Lett 38: 10.1029/2011GL048280

Medina M, Collins A, Takaoka T, Kuehl J, Boore J (2006) Naked corals: Skeleton loss in Scleractinia. Proc Natl Acad Sci 103:9096–9100

Meissner K, McNeil B, Eby M, Wiebe E (2012) The importance of the terrestrial weathering feedback for multimillennial coral reef habitat recovery. Global Biogeochem Cycles 26:10.1029/2011GB004098

Middlebrook R, Hoegh-Guldberg O, Leggat W (2008) The effect of thermal history on the susceptibility of reef-building corals to thermal stress. J Exp Biol 211:1050–1056

Miller J, Muller E, Rogers C, Waara R, Atkinson A, Whelan K, Patterson M, Witcher B (2009) Coral disease following massive bleaching in 2005 causes 60 % decline in coral cover on reefs in the US Virgin Islands. Coral Reefs 28:925–937

Moya A, Tambutté S, Bertucci A, Tambutté E, Lotto S, Vullo D, Supuran C, Allemand D, Zoccola D (2008) Carbonic anhydrase in the scleractinian coral *Stylophora pistillata*: characterization, localization, and role in biomineralization. J Biol Chem 283:25475–25484

Munnecke A, Calner M, Harper D, Servais T (2010) Ordovician and Silurian sea–water chemistry, sea level, and climate: A synopsis. Palaeogeography, Palaeoclimatology, Palaeoecology 296:389–413

Muscatine L, Cernichiari E (1969) Assimilation of photosynthetic products of zooxanthellae by a reef coral. Biol Bull 137:506

Muscatine L, Falkowski P, Dubinsky Z, Cook P, McCloskey L (1989) The effect of external nutrient resources on the population dynamics of zooxanthellae in a reef coral. Proc R Soc London B Biol 236:311–324

Muscatine L, Goiran C, Land L, Jaubert J, Cuif J, Allemand D (2005) Stable isotopes (^{13}C and ^{15}N) of organic matrix from coral skeleton. Proc Natl Acad Sci 102:1525–1530

Nash M, Opdyke B, Troitzsch U, Russell B, Adey W, Kato A, Diaz-Pulido G, Brent C, Gardner M, Prichard J, Kline D (2013) Dolomite-rich coralline algae in reefs resist dissolution in acidified conditions. Nat Clim Chang 3:268–272

National Hurricane Center (2014) Tropical cyclone climatology. http://www.nhc.noaa.gov/climo/

Neumann A, Macintyre I (1985) Reef responses to sea level rise: keep-up, catch-up or give-up. Proc 5th Intl Coral Reef Symp 3:105–110

Nikolova I, Yin Q, Berger A, Singh U, Karami M (2013) The last interglacial (Eemian) climate simulated by LOVECLIM and CCSM3. Clim Past Discuss 8:5293–5340

Orr J, Fabry V, Aumont O, Bopp L, Doney S, Feely R, Gnanadesikan A, Gruber N, Ishida A, Joos F, Key R, Lindsay K, Maier-Reimer E, Matear R, Monfray P, Mouchet A, Najjar R, Plattner G, Rodgers K, Sabine C, Sarmiento J, Schlitzer R, Slater R, Totterdell I, Weirig M, Yamanaka Yool A (2005) Anthropogenic ocean acidification over the twenty-first century and its impact on calcifying organisms. Nature 437:681–686

Pandolfi J, Connolly S, Marshall D, Cohen A (2011) Projecting coral reef futures under global warming and ocean acidification. Science 333:418–422

Pavlov A, Kasting J, Brown L, Rages K, Freedman R (2000) Greenhouse warming by CH_4 in the atmosphere of early Earth. J Geophys Res 105:11981

Pearson P, Nicholas C (2014) Layering in the Paleocene/Eocene boundary of the Millville core is drilling disturbance. Proc Natl Acad Sci 111:E1064–E1065

Penman D, Hönisch B, Zeebe R, Thomas E, Zachos J (2014) Rapid and sustained surface ocean acidification during the Paleocene-Eocene Thermal Maximum. Paleoceanography 29:357–369

Perez K, Rodgers K, Jokiel P, Lager C, Lager D (2014) Effects of terrigenous sediment on settlement and survival of the reef coral *Pocillopora damicornis*. PeerJ 2:e387

Perry C, Murphy G, Kench P, Smithers S, Edinger E, Steneck R, Mumby P (2013) Caribbean-wide decline in carbonate production threatens coral reef growth. Nat Commun 4:1402

Perry C, Smithers S, Gulliver P, Browne N (2012) Evidence of very rapid reef accretion and reef growth under high turbidity and terrigenous sedimentation. Geology 40:719–722

Perry C, Steneck R, Murphy G, Kench P, Edinger E, Smithers S, Mumby P (2014) Regional-scale dominance of non-framework building corals on Caribbean reefs affects carbonate production and future reef growth. Glob Chang Biol 21:1153–1164

Perry CT, Murphy GN, Kench PS, Edinger EN, Smithers SG, Steneck RS, Mumby PJ (2015) Changing dynamics of Caribbean reef carbonate budgets: emergence of reef bioeroders as critical controls on present and future reef growth potential. Proc R Soc London B 281: doi:10.1098/rspb.2014.2018

Pespeni M, Sanford E, Gaylord B, Hill T, Hosfelt J, Jaris H, LaVigne M, Lenz E, Russell A, Young M, Palumbi S (2013) Evolutionary change during experimental ocean acidification. Proc Natl Acad Sci 110:6937–6942

Playford P, Cockbain A (1976) Modern algal stromatolites at Hamelin Pool, a hypersaline barred basin in Shark Bay, Western Australia. In: Walter, M (ed) Stromatolites: developments in sedimentology. Elsevier, Amsterdam, pp 398–411

Pollock F, Lamb J, Field S, Heron S, Schaffelke B, Shedrawi G, Bourne D, Willis B (2014) Sediment and turbidity associated with offshore dredging increase coral disease prevalence on nearby reefs. PLoS ONE 9:e102498

Porter S (2010) Calcite and aragonite seas and the *de novo* acquisition of carbonate skeletons. Geobiology 8:256–277

Pörtner H (2001) Climate change and temperature-dependent biogeography: oxygen limitation of thermal tolerance in animals. Naturwiss 88:137–146

Ridgwell A (2005) A Mid Mesozoic revolution in the regulation of ocean chemistry. Mar Geol 217:339–357

Riegl B, Purkis S, Al-Cibahy A, Abdel-Moati M, Hoegh-Guldberg O (2011) Present limits to heat-adaptability in corals and population-level responses to climate extremes. PLoS ONE 6:e24802

Ries J (2010) Review: geological and experimental evidence for secular variation in seawater Mg/Ca (calcite-aragonite seas) and its effects on marine biological calcification. Biogeosciences 7:2795–2849

Ries J, Stanley SHardie L (2006) Scleractinian corals produce calcite, and grow more slowly, in artificial Cretaceous seawater. Geology 34:525

Roberts J, Long D, Wilson J, Mortensen P, Gage J (2003) The cold-water coral *Lophelia pertusa* (Scleractinia) and enigmatic seabed mounds along the north-east Atlantic margin: are they related? Mar Pollut Bull 46:7–20

Roberts J, Wheeler A, Friewald A (2006) Reefs of the deep: the biology and geology of cold-water coral ecosystems. Science 312:543–547

Rogers C (1990) Responses of coral reefs and reef organisms to sedimentation. Mar Ecol Prog Ser 62:185–202

Romano S, Palumbi S (1996) Evolution of scleractinian corals inferred from molecular systematics. Science 271:640–642

Ross C, Falter J, Schoepf V, McCulloch M (2015) Perennial growth of hermatypic corals at Rottnest Island, Western Australia (32°S). PeerJ 3:e781

Royer D (2006) CO_2-forced climate thresholds during the Phanerozoic. Geochim Cosmochim Acta 70:5665–5675

Royer D, Berner R, Montañez I, Tabor N, Beerling D (2004) CO_2 as a primary driver of Phanerozoic climate. GSA Today 14:4

Sagan C, Mullen G (1972) Earth and Mars: evolution of atmospheres and surface temperatures. Science 177:52–56

Scheibner C, Speijer R (2008) Decline of coral reefs during late Paleocene to early Eocene global warming. eEarth 3:19–26

Schlüter L, Lohbeck K, Gutowska M, Gröger J, Riebesell U, Reusch T (2014) Adaptation of a globally important coccolithophore to ocean warming and acidification. Nat Clim Chang 4:1024–1030

Schuster F (2002) Oligocene and Miocene examples of *Acropora* dominated palaeo-environments: Mesohellenic Basin (NW Greece) and northern Gulf of Suez (Egypt). Proc 9th Int Coral Reef Symp 1:199–204

Scoffin T (1993) The geological effects of hurricanes on coral reefs and the interpretation of storm deposits. Coral Reefs 12:203–221

Seilacher A, Bose P, Pflüger F (1998) Triploblastic animals more than 1 billion years ago: Trace fossil evidence from India. Science 282:80–83

Servais T, Lehnert O, Li J, Mullins G, Munnecke A, Nützel A, Vecoli M (2008) The Ordovician biodiversification: revolution in the oceanic trophic chain. Lethaia 41:99–109

Shashar N, Cohen Y, Loya Y (1993) Extreme diel fluctuations of oxygen in diffusive boundary layers surrounding stony corals. Biol Bull 185:455

Sheehan P, Harris M (2004) Microbialite resurgence after the Late Ordovician extinction. Nature 430:75–78

Sherman C, Nemeth M, Ruíz H, Bejarano I, Appeldoorn R, Pagán F, Schärer M, Weil E (2010) Geomorphology and benthic cover of mesophotic coral ecosystems of the upper insular slope of southwest Puerto Rico. Coral Reefs 29:347–360

Solomon S, Plattner G, Knutti R, Friedlingstein P (2009) Irreversible climate change due to carbon dioxide emissions. Proc Natl Acad Sci 106:1704–1709

Stanley G (2003) The evolution of modern corals and their early history. Earth-Science Reviews 60:195–225

Stanley G, Fautin D (2001) The origin of modern corals. Science 291:1913–1914

Stanley G, Helmle K (2010) Middle Triassic coral growth bands and their implication for photosymbiosis. Palaios 25:754–763

Stanley S (2006) Influence of seawater chemistry on biomineralization throughout phanerozoic time: paleontological and experimental evidence. Palaeogeography, Palaeoclimatology, Palaeoecology 232:214–236

Stanley S, Hardie L (1998) Secular oscillations in the carbonate mineralogy of reef-building and sediment-producing organisms driven by tectonically forced shifts in seawater chemistry. Palaeogeography, Palaeoclimatology, Palaeoecology 144:3–19

Stassen P, Speijer R, Thomas E (2014) Unsettled puzzle of the Marlboro clays. Proc Natl Acad Sci 111:E1066–E1067

Stimson J, Kinzie R (1991) The temporal pattern and rate of release of zooxanthellae from the reef coral *Pocillopora damicornis* (Linnaeus) under nitrogen-enrichment and control conditions. J Exp Mar Biol Ecol 153:63–74

Stolarski J, Kitahara M, Miller D, Cairns S, Mazur Meibom A (2011) The ancient evolutionary origins of Scleractinia revealed by azooxanthellate corals. BMC Evol Biol 11:316

Stolarski J, Melbom A, Przeniosto R, Mazur M (2007) A Cretaceous coral with a calcitic skeleton. Science 318:92–94

Suckling C, Clark M, Beveridge C, Brunner L, Hughes A, Harper E, Cook E, Davies A, Peck L (2014) Experimental influence of pH on the early life-stages of sea urchins II: increasing parental exposure times gives rise to different responses. Invertebr Reprod Dev 58:161–175

Sun Y, Joachimski M, Wignall P, Yan C, Chen Y, Jiang H, Wang Lai X (2012) Lethally hot temperatures during the Early Triassic greenhouse. Science 338:366–370

Szmant A (2002) Nutrient enrichment on coral reefs: Is it a major cause of coral reef decline? Estuaries 25:743–766

Tambutté S, Holcomb M, Ferrier-Pagès C, Reynaud S, Tambutté É, Zoccola D, Allemand D (2011) Coral biomineralization: from the gene to the environment. J Exp Mar Biol Ecol 408:58–78

Taylor P, Wilson M (2003) Palaeoecology and evolution of marine hard substrate communities. Earth-Science Reviews 62:1–103

Thacker R, Ginsburg D, Paul V (2001) Effects of herbivore exclusion and nutrient enrichment on coral reef macroalgae and cyanobacteria. Coral Reefs 19:318–329

Turley C, Roberts J, Guinotte J (2007) Corals in deep-water: will the unseen hand of ocean acidification destroy cold-water ecosystems? Coral Reefs 26:445–448

Ueno Y, Johnson M, Danielache S, Eskebjerg C, Pandey A, Yoshida N (2009) Geological sulfur isotopes indicate elevated OCS in the Archean atmosphere, solving faint young sun paradox. Proc Natl Acad Sci 106:14784–14789

van Woesik R, Franklin E, O'Leary J, McClanahan T, Klaus J, Budd A (2012) Hosts of the Plio-Pleistocene past reflect modern-day coral vulnerability. Proc R Soc London B Biol 279:2448–2456

Vega Thurber R, Burkepile D, Fuchs C, Shantz A, McMinds R, Zaneveld J (2013) Chronic nutrient enrichment increases prevalence and severity of coral disease and bleaching. Glob Chang Biol 20:544–554

Veron J (1995) Corals in space and time. Comstock/Cornell, Ithaca

Veron J, Wallace C (1984) Scleractinia of Eastern Australia. Part V. Family Acroporidae. Australian Institute of Marine Science Monograph Series, Townsville, Australia

Visser K, Thunell R, Stott L (2003) Magnitude and timing of temperature change in the Indo-Pacific warm pool during deglaciation. Nature 421:152–155

Voss J, Richardson L (2006) Nutrient enrichment enhances black band disease progression in corals. Coral Reefs 25:569–576

Wallace C, Rosen B (2006) Diverse staghorn corals (*Acropora*) in high-latitude Eocene assemblages: implications for the evolution of modern diversity patterns of reef corals. Proc R Soc London B Biol 273:975–982

Webster P, Holland G, Curry J, Chang H (2005) Changes in tropical cyclone number, duration, and intensity in a warming environment. Science 309:1844–1846

Weinstein D, Smith T, Klaus J (2014) Mesophotic bioerosion: variability and structural impact on US Virgin Island deep reefs. Geomorphology 222:14–24

Wells J (1956) Scleractinia. In: Moore R (ed) Treatise on invertebrate paleontology. Geological Society of America and University of Kansas Press, Lawrence, pp 328–444

Wiedenmann J, D'Angelo C, Smith E, Hunt A, Legiret F, Postle A, Achterberg E (2013) Nutrient enrichment can increase the susceptibility of reef corals to bleaching. Nat Clim Chang 3:160–164

Wilson M, Palmer T (2006) Patterns and processes in the Ordovician bioerosion revolution. Ichnos 13:109–112

Winter A (2003) Orbital control of low-latitude seasonality during the Eemian. Geophys Res Lett 30:doi:10.1029/2002GL016275

Wright J, Schaller M (2013) Evidence for a rapid release of carbon at the Paleocene-Eocene thermal maximum. Proc Natl Acad Sci 110:15908–15913

Wright J, Schaller M (2014) Reply to Pearson and Nicholas, Stassen et al., and Zeebe et al.: Teasing out the missing piece of the PETM puzzle. Proc Natl Acad Sci 111:E1068–E1071

Yajima T, Sano O, Okamoto T, Shirai Y, Shinya T, Matada M (1986) Ecological distribution of the reef coral, *Oulastrea crispata* (Lamarck) at the shore region in the vicinity of Tukumo Bay. Bull Japan Sea Res Inst 18:21–36

Zachos J, McCarren H, Murphy B, Röhl U, Westerhold T (2010) Tempo and scale of late Paleocene and early Eocene carbon isotope cycles: implications for the origin of hyperthermals. Earth Planet Sci Lett 299:242–249

Zachos J, Röhl U, Schellenberg S, Sluijs A, Hodell D, Kelly D, Thomas E, Nicolo M, Raffi I, Lourens L, McCarren H, Kroon D (2005) Rapid acidification of the ocean during the Paleocene-Eocene thermal maximum. Science 308:1611–1615

Zanazzi A, Kohn M, MacFadden B, Terry D (2007) Large temperature drop across the Eocene–Oligocene transition in central North America. Nature 445:639–642

Zapalski M (2013) Evidence of photosymbiosis in Palaeozoic tabulate corals. Proc Royal Soc London B Biol 281:doi:10.1098/rspb.2013.2663

Zeebe R, Dickens G, Ridgwell A, Sluijs A, Thomas E (2014) Onset of carbon isotope excursion at the Paleocene-Eocene thermal maximum took millennia, not 13 years. Proc Natl Acad Sci 111:E1062–E1063

Zeebe R, Wolf-Gladrow D (2001) CO_2 in seawater: equilibrium, kinetics, isotopes. Elsevier Oceanography Series, Amsterdam, The Netherlands, p346

Zoccola D, Tambutté E, Kulhanek E, Puverel S, Scimeca J, Allemand D, Tambutté S (2004) Molecular cloning and localization of a PMCA P-type calcium ATPase from the coral Stylophora pistillata. Biochimica et Biophysica Acta (BBA) - Biomembranes 1663:117–126

Living and Mineral Reefs: Are they Comparable and What Ecological Information Is Preserved? 10

Halard Lescinsky

Abstract

Reefs are complex ecosystems on many scales and a host of simplifications, assumptions, and limitations are inherent in surveying and characterizing them. This chapter examines the techniques used to collect data on living and mineral reefs and asks what potential biases may arise from equating the two very different types of "reefs" (i.e., a living community vs. its mineral remains). Data from each of the two has its own limitations. Although fossil assemblages are famous for lost detail, vast amounts of ecological information are also lost when surveying living reefs. Despite inherent shortcomings, living and mineral reefs both provide important ecological context that is needed to address many of todays most relevant reef questions. While data on living organisms are the foundation for documenting the status of modern reefs, mineral reef deposits provide the baseline needed to put that information in a broader context.

Keywords

Fossil reefs • Pre-human baseline • Taphonomy • Time averaging • Preservation • Pleistocene • Holocene

10.1 Introduction – What Do Living and Mineral Reefs Represent?

In graduate school, I dove with a fellow student who would go on to become a fish ecologist. Comparing notes after dives, we realized that we rarely saw the same things. I tended to look down at the benthos and he looked out at the fish. Our different perspectives led us to different views of what was there, not unlike the differences that can develop between geologists who focus on the mineral reef, and biologists who see only the living one. The sheer aesthetic pleasure of swimming over a living reef provides, for most people, a vivid mental image of what a reef should look like. If they consider the mineral reef at all, they assume it is a simple extrapolation of the outer living surface. However, reef interiors run the gamut from "garbage piles" of storm rubble to intact interlocking biological remains. This chapter identifies and explores the potential limitations and biases inherent in comparing mineral and living reefs, and then illustrates that, despite these shortcomings, mineral reefs are some of the most ecologically intact fossil deposits known. They provide invaluable information on past reef ecology and they constitute the basis for developing the pre-human baseline that we need to evaluate the fate of modern and future reefs in the face of global change.

10.1.1 What Does the Living Reef Represent?

An observer on a living reef sees the "reef community"- the veneer of organisms that are alive at or near the reef's surface (Sheppard et al. 2009). The underlying skeletal buildup that is the hallmark of geological reef definitions

H. Lescinsky (✉)
Department of Biology and Earth Science, Otterbein University, Westerville, OH 43081, USA
e-mail: hlescinsky@otterbein.edu

D.K. Hubbard et al. (eds.), *Coral Reefs at the Crossroads*, Coral Reefs of the World 6, DOI 10.1007/978-94-017-7567-0_10

(e.g., Fagerstrom 1987; Wood 1999) is generally not observed. Modern "reef" monitoring is therefore best considered "community" monitoring. To address the condition of the "reef" itself, geological techniques such as carbonate budget analysis (e.g., Stearn and Scoffin 1977; Hubbard et al. 1990; Perry et al. 2008, 2012, 2013) and coring (e.g., Macintyre and Glynn 1976; Smithers et al. 2006; Roff et al. 2013) are needed.

Characterizing living reefs is also complicated because the reef community does not occupy a simple planar area. Reef organisms inhabit not only the complex three-dimensional reef surface, but also the many holes and crypts within the reef framework. As a result, total reef surface area can be up to 8 times greater than the planar surface projection (Richter and Wunsch 1999; Scheffers et al. 2004). Hidden surfaces are sites of extensive nutrient cycling and phytoplankton consumption (Richter and Wunsch 1999), hold numerous species and guilds of mobile invertebrates (Moreno-Forero et al. 1998), and harbor up to 40 % of the fish species and 50 % of the fish individuals (Ackerman and Bellwood 2000). In this sense, modern reef monitoring is similar to assessing the health of the Amazon Rainforest from aerial photographs. Gross deforestation patterns are obvious and important to quantify but other potentially important ecological data remain hidden beneath the living canopy.

Characterization of living reefs is also limited by short time span. Benthic surveys of living reefs yield instantaneous snapshots of the exposed part of the reef community, and long-term monitoring programs have broadened the perspective to include multi-year or decadal patterns. Still, these intervals are very short compared to relevant ecological time scales. Reefs are classically structured by infrequent disturbances (Connell 1978) that introduce large changes on decadal timescales. Reef accretion proceeds at much slower rates still (e.g., 3 mm/y average in the Caribbean: Hubbard 2009). The discrepancies in time frame between biologists and geologists are well illustrated by some Pacific reefs that are now dominated by macroalgae. Vroom et al. (2006) argued that the reefs were "healthy", albeit naturally lacking in coral. In contrast, Pandolfi and Jackson (2007) countered that the very edifice of carbonate that underlies the macroalgae attests to millennia of coral accretion and that macroalgal dominance could not be considered the norm at an appropriate timescale.

10.1.2 What Does the Mineral Reef Represent?

While the living veneer of reefs is dominated by biological growth, the mineral reef is the result of the related but more physically driven process of accretion (see Chap. 6). Field insights into the mineral interiors of modern reefs are rarely possible and our knowledge of them comes primarily from geologic outcrops and reef cores. Referring to these types of exposures as "mineral reefs", rather than "fossil reefs", in the following pages separates the process of reef accretion from issues of antiquity (age) and fossilization (taphonomy). It may seem a semantic point, but Holocene reef cores lack the antiquity to be considered fossils by most definitions, and the widely used phrase "reef taphonomy" (e.g., Scoffin 1992; Greenstein 2006) is misleading because it blurs processes that actively form the reef (accretion) and the post-mortem processes that alter the skeletons of dead organisms (taphonomy). Reefs are not actually alive, even if their constituent organisms are, and thus the reef itself cannot have a true post-mortem history. Along any given reef surface some organisms are alive at the same time that others are dead. For example, corals thriving on the reef fronts of many fringing reefs on the Great Barrier Reef are living directly on 5000-year old corals (Smithers et al. 2006), and thus cavity dwelling organisms are boring into, encrusting on, and living within very old and dead corals. The taphonomy of one is the life of another, and reef accretion is a different, and only partially related process.

A potential intermediate between the living and mineral reef is the "death assemblage". Although a reef's death assemblage should in principle include all dead corals (both in-situ and unattached) in a given reef area, in practice studies of coral death assemblages have followed protocols developed for studies of mollusk death assemblages and have included primarily smaller pieces of loose rubble (Greenstein 2006). Defined in this way, there is no tight correspondence between the rubble of the death assemblage and the living reef community (Pandolfi and Minchin 1995; Greenstein and Pandolfi 1997; Edinger et al. 2001; Greenstein 2006). Differences probably arise because the rubble portion of the death assemblage is skewed towards fragile and faster growing morphologies (see Sect. 10.5.2), and large massive colonies are not included. Mineral reefs are therefore often characterized by corals that are much better preserved and more similar in composition to living reefs than to those in the smaller detritus of the death assemblage (e.g., Greenstein et al. 1998, Greenstein 2006).

The surface topography of the reef acts to trap or baffle sediment within the reef. As Fagerstrom (1988) pointed out, erect organisms such as gorgonians and sponges (his "baffler guild") act to slow down and retain suspended sediment, which then accumulates in reef crypts. Trapped sediment and coral rubble can be the majority of the bulk volume of a reef (Hubbard et al. 1990, 1998). The mineral reef thus includes well-preserved corals that are buried as the upward growth of the reef organisms is balanced by the in-filling of the base of the reef. The trapped sediment subsequently protects the dead coral surfaces from the deep bioerosion and encrustation that occurs on most exposed dead reef surfaces.

10.1.3 The Need for Comparative Data from Living and Mineral Reefs

Data from both types of reefs lack certain ecological information that was either unavailable or not collected. This data, albeit imperfect, is all that we have to address increasingly important questions about the continued existence of reef communities. Data on living reefs are needed to map short-term community changes, and data from mineral reefs provide the only information available to address the increasingly important questions of change over spans of more than a few decades. To facilitate the comparison of data from the two reef types, this chapter examines the similarities and differences in how that quantitative information is collected and then reviews the considerations necessary to extract the best paleo-ecological information from mineral reefs. Finally, it reviews the wealth of ecological and evolutionary information that has already been successfully extracted from mineral reefs. While mineral reefs will never be suitable for certain types of questions, when biases in preservation and the process of reef accretion are understood, geologic data can provide information relevant to the many increasingly important reef-ecology questions.

10.2 How Do We Sample/Monitor Living and Mineral Reefs?

10.2.1 Living Reefs

Many specific protocols are used to measure living reef communities around the world (reviewed by Hil and Wilkinson 2004). Most monitoring systems are based on quantitative counting along a known distance or area and the subsequent derivation of "percent cover" metrics.

The standard method recommended by the Global Coral Reef Monitoring Network (GCRMN) is the line-intercept transect (LIT) in which a tape measure is strung taut between two pegs and the endpoints of each organism or type of substrate directly below the tape is recorded. Point-intercept transects (PITs) are a variation where the substrate is identified at set increments along a tape (e.g., every 50 cm in the "Reef Check" protocol). Increasingly popular are belt transects in which a known width or belt on either side of the tape is examined usually via underwater photographs or video. Percent cover is then calculated by superimposing random points or a grid on the still image and counting the benthos beneath the points.

The length of transects used to monitor recent reefs varies from 10 m (e.g., Virgin Islands National Park) to 50 m (GBR), with 20 m recommended by the GCRMN. There may also be replicate transects (often 5) per reef zone, and transects are generally arranged parallel to the reef front so each follows a consistent depth. Although monitoring is done carefully to maximize precision, recent studies suggest that simpler and less precise techniques could yield comparable results for many questions (e.g., Dumas et al. 2009; Jimenez et al. 2010; Pante and Dustan 2012).

Another type of monitoring on living reefs is a targeted search in which a trained observer creates a list of observed organisms over a large area. This "roving diver" method is useful for rapid surveys, the observation of rare occurrences, and for mobile organisms such as fish and invertebrates (Hil and Wilkinson 2004). Abundance is quantified per unit of search effort (e.g., a timed interval of swimming). Chapter 12 discusses the evolution of many sampling strategies in detail.

10.2.2 Mineral Reefs

Historically, paleontologists have characterized fossil deposits primarily by species lists, but more recent efforts use quantitative methods designed to be similar to those used on modern reefs. Line-intercept transects are conducted along a vertical outcrop by anchoring a tape measure every few meters along the face. The tape is laid out parallel to the perceived reef surface (horizontal if the face is parallel to paleodepth; slightly dipping if it follows some presumed slope) so that the entire transect approximates a single time slice. Given adequate exposures, continuous transects many meters long are possible. For example, 25-m transects were used to characterize the Holocene reefs of the Huon Peninsula (Pandolfi et al. 2006), 40 m for the Pleistocene of Curacao (Pandolfi and Jackson 2001), and up to 125 m for the Holocene of the Dominican Republic (Lescinsky et al. 2012).

Belt transects and quadrats are used less commonly for mineral reefs because they require relatively extensive horizontal exposures. One spectacular example is the large bedding plane of a Devonian reef that is exposed along the sides of Ohio River (Greb et al. 1993). The geological equivalent of roving diver sampling, a timed and targeted search, is also widely used for mineral reefs. As with living reefs, this approach is useful for finding rare and mobile species (e.g., mollusc shells).

Bulk sampling is another method commonly applied to mineral reefs, though uncommon for living reefs. For bulk sampling, data from a known volume of substrate is systematically collected and examined later in detail, generally focusing on smaller shelled species such as molluscs and foraminifera. On modern reefs, bulk sampling is used in lagoon sediments (Miller 1988; Parsons-Hubbard 2003; Ferguson and Miller 2007), but is generally too destructive to be applied to zones with reef framework and its crypts.

Cores are another widely used sampling technique for mineral reefs, and it is through cores that we have learned the vast majority of what we know about the development of Holocene reefs. The library of cores from reefs around the world is now in the hundreds from the Indo-Pacific (Montaggioni and Braithwaite 2009), the Great Barrier Reef (Hopley et al. 2007), and the Caribbean (Hubbard 2009). In addition, short "low tech" cores, in which a PVC or aluminum pipe is manually driven into the reef substrate are becoming increasingly popular (e.g., Aronson et al. 1998; Roche et al. 2011; Roff et al. 2013).

Drill cores provide a unique "outcrop" shape that makes them difficult to interpret ecologically. They are very narrow (generally <100 mm in diameter, and as little as 25 mm), and very tall (most Holocene reefs have accreted <20 m, though a few are thicker than 50 m: Montaggioni and Braithwaite 2009). This shape precludes ecological analyses at a specific time unless many cores are taken along a transect. Cores intervals are therefore logged primarily with respect to reef zone or the dominant fauna (e.g., lagoon, branching coral zone, massive coral zone, rhodoliths; sediment/rubble) and are used most often to investigate changes in water depth, environmental conditions, and accretion rates over thousands of years. Only rarely have core data been used, probably too optimistically, to determine and compare reef-community composition at different times in the past (e.g., Roche et al. 2011).

10.2.3 How Congruent Are Mineral and Living Reef Samples?

Sampling of mineral reefs is constrained by available outcrop exposures, but the best examples of mineral reefs are laterally continuous outcrops that can be sampled in much the same way as living reefs using multiple quadrats or line-intercept transects of comparable length. In addition, both types of reefs rely on targeted searches for data on rare and mobile taxa. The different nature of living and mineral reefs will of course lead to important subtle and not so subtle differences that must be addressed when comparing data from the two. These differences are discussed below as biases that are inherent in the rock record and fossil preservation. Nevertheless, numerous studies have shown that reef monitoring data are robust across sampling methodologies (Dumas et al. 2009; Jimenez et al. 2010; Pante and Dustan 2012) and this provides encouragement that mineral- and living-reef data are, at least in principle, broadly comparable when sufficient care is taken.

10.3 Distribution of Pleistocene to Recent Mineral Reef Exposures

The distribution of modern reefs is broad and well known (Darwin 1842; Spalding et al. 2001) and cores through many of these reefs have provided a chronicle of their histories. In a very few cases, cross sections through mineral reefs have also been exposed in underwater excavations made during the dredging and cutting of boat channels (e.g., Adey et al. 1977; Lighty et al. 1978).

Outcrop exposures of uplifted mineral reefs are a second source of data (Table 10.1). Episodic tectonic uplift (earthquakes), usually associated with subduction zones, is responsible for the most spectacular examples of Holocene and Pleistocene mineral reefs. On rare occasions, such as in the Persian Gulf, reefs can also be uplifted by salt diapirs (Bruthans et al. 2006; Samimi-Namin and Riegl 2012).

Two well-documented modern examples of reef emergence are the 1.5-m uplift of reefs along the south coast of Simeulue Island, Sumatra, during the 2004 Indian Ocean earthquake and tsunami (Meltzner et al. 2006) and 7-m uplift of reefs in Urvina Bay, Isabela Island, Galapagos in 1954 (Malmquist 1991). The cumulative effect of on-going tectonic uplift and sea-level variation over time may lead to a series of spectacular reef terraces (Cabioch 2011) with older ones perched above younger ones, in apparent contradiction to the stratigraphic law of superposition (Fig. 10.1a, b).

An interesting variant on uplifted reefs occurs in the Dominican Republic where an entire Holocene reef complex (6000+ ybp) has become exposed not by uplift, but by evaporation once the Enriquillo embayment was cut off from the ocean (Mann et al. 1984; see Fig. 10.1c, this chapter). The mineral reefs remain in their original position but water level in the now-hypersaline lake has dropped by 42 m leaving the reefs sitting subaerially along the lakeshore.

The third main pathway to exposure of mineral reefs is stranding following past intervals of higher sea level. For example, between 130,000 and 110,000 years ago (MIS-5e), sea level was 5-8 m higher than it is today and reef deposits, radiometrically dated at ca. 125 kypb, occur just above sea-level in many parts of the world including western Australia (Greenstein and Pandolfi 2008), the Red Sea (Dullo 1990), and the Caribbean (Greenstein and Moffat

Table 10.1 Examples of Quaternary emergent reefs of the world

	References
Caribbean	
Holocene	
Dominican Republic	Mann et al. (1984)
Pleistocene	
Bahamas	Greenstein et al. (1998), Meyer et al. (2003)
Barbados	Mesolella (1967)
Columbia	Pandolfi and Jackson (2006)
Cuba	Iturralde Vinent (1995), Toscano et al. (1999)
Curacao	Pandolfi and Jackson (2001)
Dominican Republic	Geister (1982), Klaus and Budd (2003)
Florida	Stanley (1966)
Guadeloupe	Feuillet et al. (2004)
Haiti	Dodge et al. (1983)
Pacific	
Holocene	
Galapagos	Malmquist (1991)
Indonesia, Java	Azmy et al. (2010)
Japan	Sugihara et al. (2003), Hamanaka et al. (2012)
New Britain	Riker-Coleman et al. (2006)
Papua New Guinea	Pandolfi et al. (2006)
Taiwan	Inoue et al. (2011)
Tonga	Taylor and Bloom (1977)
Pleistocene	
Fiji	Nunn et al. (2002)
Indonesia, Alor	Hantoro et al. (1994)
Indonesia, Atauro, Timor	Chappell and Veeh (1978)
Indonesia, Sumba	Pirazzoli et al. (1991)
Indonesia, Sulawesi	Crabbe et al. (2006)
Japan	Nakamori et al. (1995), Humblet et al. (2009)
New Hebrides/ Vanuatu	Taylor et al. (1987)
Papua New Guinea	Chappell (1974), Pandolfi (1996)
Pitcairn Group, Henderson	Pandolfi (1995)
Solomon Islands	Taylor et al. (2005)
Indian Ocean	
Holocene	
Red Sea	Gvirtzman et al. (1992)
Pleistocene	
Eastern Africa	Accordi et al. (2010), Crame (1986)
Mauritius	Montaggioni (1982)
Persian Gulf, Red Sea	Dullo (1990), El-Sorogy (2008), Preusser et al. (2003)

1996; Budd and Pandolfi 2004). Similarly, in the Pacific, there was a Mid-Holocene (4 ky) sea-level peak that was 1–2 + m above present that left exposed Holocene Reefs above modern Pacific shorelines (Taylor and Bloom 1977; Dickinson 2009).

10.4 Biases Related to Sampling and Monitoring

Reef data may include four general types of biases. First are the effects of sampling decisions such as replicate design and how to account for topography along a transect. Second are traits inherent to reef organisms; do they have skeletons and are they sessile or mobile? A third type of bias relates to time resolution: how close are preserved reef deposits to ecological "snapshots". The fourth type operates over longer time scales and broader geographic ranges, referred to as "megabiases" (see Behrensmeyer et al. 2000). Each of these is described in detail below, before discussing the most important processes that can make modern-ancient reef comparisons problematic.

10.4.1 Sampling Strategy: Replicates, and Randomness

Large-scale monitoring programs that are designed by statisticians consistently include elements of sample design: randomness and replication. The starting points of transects can be randomly assigned once a reef is selected and outlined based on objective criteria (Rogers et al. 2000), and the major regional monitoring programs reviewed by Hil and Wilkinson (2004) each include five replicate transects per reef zone, with replicates within 10 s of meters of each other.

Sampling of mineral reefs is much more opportunistic. Reef outcrops depend on adequate exposure and accessibility, and transect orientation is dictated by the outcrop. Sea cliffs (e.g., the Huon Peninsula; SE Barbados) may facilitate reef-parallel transects similar in orientation to those used in monitoring programs. However, outcrops carved by streams usually provide transects along the reef slope exposing several reef zones and not constraining paleo-depth.

Replicate sampling, similar to that done on modern reefs, is also possible for many well-exposed mineral reefs. For example, Pandolfi and Jackson (2001) surveyed the Pleistocene reefs of Curacao, using 40-m line-intercept transects. They examined seven transects in each of three reef zones (windward crest, windward back reef and leeward reef crest) at each of two sites. Replicate transects at each site were constrained by the location of erosional breaks in the reef terraces ("bokas") and thus were located further apart than ideal for modern reef monitoring (100 s to a 1000 m vs 10 s of meters for modern reef replicates). Similar restrictions on the proximity of replicate transects are found in other well-exposed mineral reefs such as those on the Huon terraces (Pandolfi 1996; Pandolfi et al. 2006; Fig. 10.2) and in the Dominican Republic (Hubbard et al. 2008; Lescinsky et al. 2012).

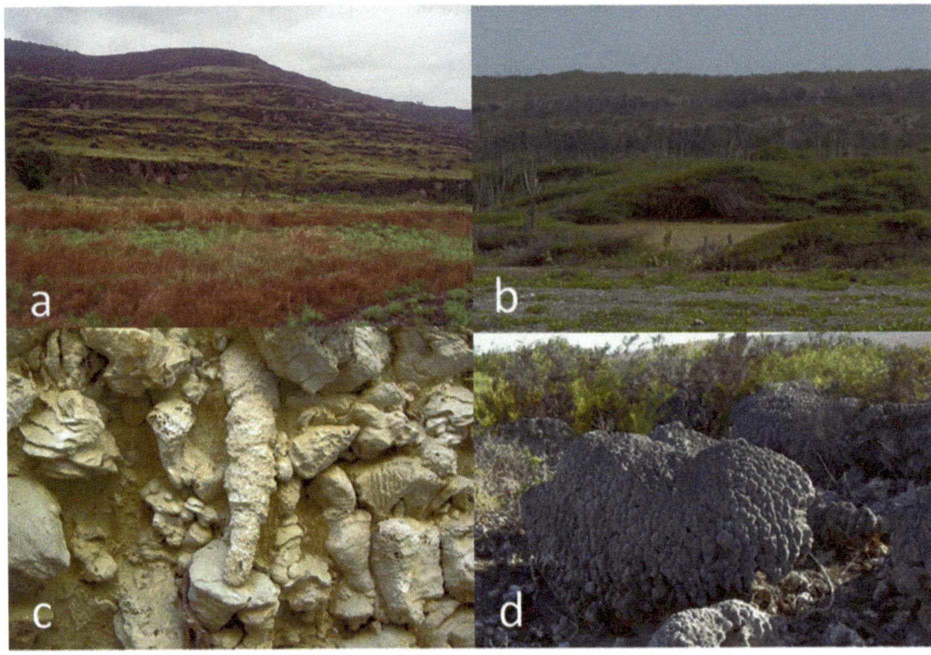

Fig. 10.1 Examples of mineral reef exposures. (**a**) Huon Peninsula, Papua New Guinea, numerous Pleistocene reef terraces sit on the Holocene terrace that rises from the shore. (**b**) Curacao Pleistocene reef terraces, photo from the top of the youngest terrace (125 kya). (**c**) Enriquillo Valley, Dominican Republic. Holocene reefs are exposed by intermittent stream cuts below sea level in an evaporated basin. (**d**) Urvina Bay, Galapagos: reef was uplifted about 5 m in a 1954 earthquake, photograph 50 years later

A different type of replication that is possible in mineral (but not living) reefs is temporal replication. A reef outcrop might be many meters high and a series of transects, one above the other, can be used to compare the reef community at various times in the past. Successive transects differ in ecological time, but can accurately characterize the preserved reef community at that site at a time scale much longer than our oldest monitoring records (Pandolfi 1996, Pandolfi and Jackson 2006).

10.4.2 Reef-Surface Topography

Surfaces of reefs are far from flat, yet most survey methods treat them as such. An exception is a chain-intercept transect (CIT: reviewed in Hil and Wilkinson 2004) in which a chain is draped loosely over the reef so that it can conform to the reef topography. The resulting line-intercept data collected along the length of the chain includes the three-dimensional surface of the reef. However, CITs are rarely used because they are time intensive and may damage living coral. Ignoring reef topography, as most monitoring systems of living reefs do, leads to potential bias in at least two important ways. First, the degree of rugosity or topography of the reef surface is a large part of the overall reef complexity and this is critical to the distribution and abundance of fish and mobile invertebrates (e.g., Chabanet et al. 1997; Gratwicke and Speight 2005). Another concern is that the true length of the reef surface (surface area for belt transects) is far greater than the linear distance that is measured along a tightly strung transect tape. The result is that the extra distance must be somehow factored out of the data collection, via decisions on what surface is counted and what surface is omitted (Fig. 10.3). Video and photographs taken straight down ("planar transects") will record the uppermost organism (i.e., a plate coral in Fig. 10.3b), while omitting lower organisms. Goatley and Bellwood (2011) examined the impact of this "canopy effect" and found that coral cover dropped by almost half and turf algae cover increased by more than two thirds when lower, rather than upper surfaces, were counted ("benthic transects", Fig. 10.3c). In addition, the canopy-forming corals or other organisms that project laterally are the most likely to be broken and lost during storms suggesting that planar transects will record greater impacts from storms than if benthic transects are used.

In mineral reefs, the three-dimensional nature of the reef surface is even more difficult to account for. Although the internal fabric may preserve the general reef slope, spaces between living coral heads are filled by coral growth, new recruits and sediment over time, thus erasing the reef's original profile and rugosity. Only in rare instances, such as during rapid burial, is the reef surface preserved in an identifiable way. For example, in the Holocene reef of Enriquillo Valley, Dominican Republic (Hubbard et al. 2008; Lescinsky et al. 2012) layers of detrital carbonate, presumably stirred up and re-suspended during a storm, drape over and preserve a 150-m long reef surface that existed approximately 8500 years ago. Paleoecologists often exploit these large burial horizons ("obrution deposits", Brett and Seilacher 1991) because they yield excellent preservation and provide an "ecological snapshot" of the community at a particular time. Obrution deposits are most common in shallow continental seas such as those that formed during very high sea level in the lower to

Fig. 10.2 Holocene mineral reef, Huon Peninsula. (**a**) Photo of sea cliff with scaffolding constructed to facilitate sampling of 25 m horizontal transects every 1 m up the vertical face of the cliff. (**b**) High resilience in the section occurs as branching corals are preserved growing directly on top of a debris flow. (**c**) Close-up of well-preserved mineral reef showing molluscs that retain original color

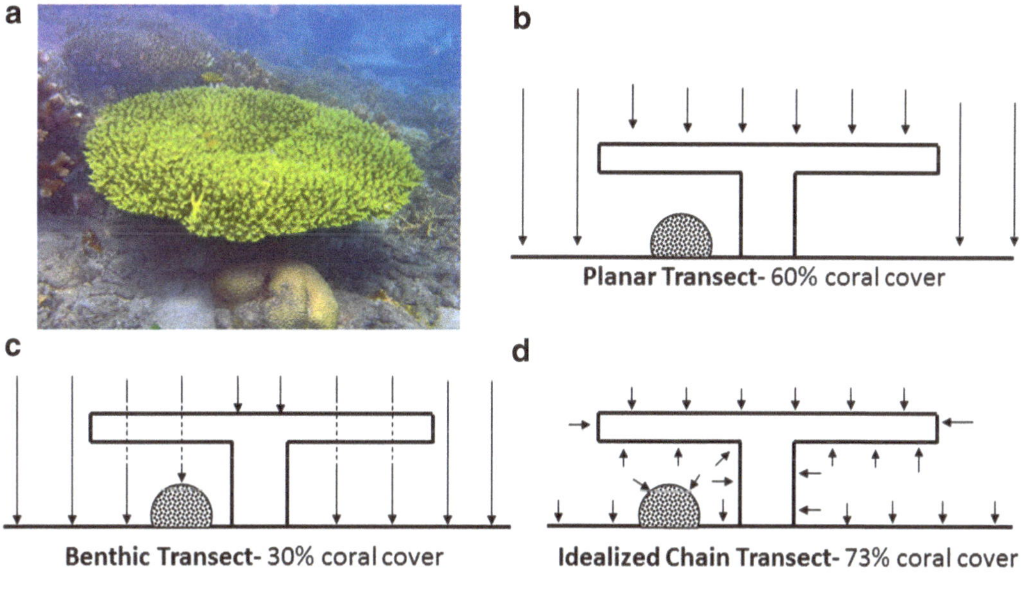

Fig. 10.3 How reef surface topography affects measurement of coral cover with different sampling systems. (**a**) Living coral with a small massive colony under it, (**b**) "Planar transect" of the reef counts only the uppermost surface and would record 60 % coral cover. It would not include the lower coral, (**c**) "Benthic transect" (lowest surface) would record 30 % coral cover. d. Idealized "chain transect" that traces the entire surface topography would record 73 % coral cover. Mineral reefs sampling is most similar to benthic sampling suggesting that mineral reef transects may underestimate coral cover as compared to living (planar) transects (Modified from Goatley and Bellwood (2011))

mid-Paleozoic and late Mesozoic, but these settings are rare today (Edinger et al. 2002).

A further difference between living- and mineral-reef transects is that, while "planar" transects are generally used in living-reef monitoring, mineral-reef transects are closer to "benthic transects" (Fig. 10.3c; Goatley and Bellwood 2011). Benthic transects underestimate coral cover and overestimate the cryptic benthos compared to planar transects, suggesting that mineral reefs are probably more biased toward lower total coral cover than a comparable transect on a living reef. This underestimate is at least partially offset by the loss of soft-bodied organisms which would be counted as live cover on a living reef, but might not be preserved along a mineral-reef transect. Another consequence of this difference is that pockets of sand and rubble, which are hidden and uncounted below the coral canopy in planar transects of living reefs, are much more visible in outcrop. This difference probably accounts for the discrepancy between living reef transects which find unconsolidated sand to form only a few percent of the reef surface, while geological studies routinely find that half or more of the reef volume is composed of loose sand and rubble rather than skeletal framework (Hubbard et al. 1998).

10.4.3 Site Selection

The "Crown Jewel" living reefs that make up National Parks and marine reserves are obviously not a random sampling of reefs in general (Chap. 12), nor are the locations of marine labs and dive sites random with respect to overall reef quality or initial live coral cover. However, for obvious reasons, these are the locations where much reef monitoring is focused. Likewise, the establishment of 'citizen" monitoring programs such as Reef Check in the Pacific and AGRRA and REEF in the Caribbean foster the monitoring and measuring of reefs that were not randomly chosen and were perhaps perceived as "healthy" reefs. The danger is that regional and global compilations and meta-analyses (e.g., Gardner et al. 2003; Bruno and Selig 2007) may be biased toward reefs that had unusually high coral cover to start with (Sweatman et al. 2011). While this cannot explain away recent reef decline, it nevertheless taints the data on which perceptions are based and decisions are made.

Mineral reefs are much more limited in number, so site selection is primarily constrained by outcrop area, preservation, and accessibility rather than aesthetic considerations, though in principle similar bias could exist. Mineral reefs may also be potentially biased by geologic setting. Preserved Pleistocene reefs often come from uplifted terraces in tectonically active areas where the reef profile is steep, leading to disproportionately high coral cover. In contrast, most Paleozoic mineral reefs come from shallow turbid continental seas, a very different environment than is typical today or near most Pleistocene reefs.

10.4.4 Depth Bias

Most data on both living and mineral reefs are biased toward relatively shallow water. Living-reef transects are often placed at 10 m or less (Hil and Wilkinson 2004), even though deeper reef sites are easily accessible on SCUBA. As a result, reefs below recreational SCUBA depths are much less surveyed (Menza et al. 2008; Lesser et al. 2009) even though they can have higher coral cover, serve as refugia (Bongaerts et al. 2010) and have important conservation implications (Hinderstein et al. 2010).

Mineral reefs are likewise often sampled primarily in the shallowest reef zones. Although deeper reef zones are probably more widely preserved than shallow zones, they are less likely to be exposed by geological processes. For example, only the upper few meters of reefs built during Pleistocene sea-level highstands sit above present sea level, and so only the shallowest reef zones are accessible for study. Deeper zones are hidden underwater beneath the existing Holocene reef. Likewise, tectonic uplift rates associated with reef-terrace formation are usually too slow or too small to produce the 20+ m of offset necessary to lift deeper reef zones out of the water where they can be observed. Even dredge excavations to create port channels are dominated by the uppermost shallow facies (e.g., Adey et al. 1977; Lighty et al. 1978). The continental reefs that dominate the ancient reef record existed in relatively flat-bottomed, shallow seas and rarely extended to significant depths. Cores of living reefs might in principle sample deeper reef zones (e.g., Hubbard et al. 1986, 1997, 2005; Hubbard 2011) but, in practice, drill-site selection can still result in reef cores preserving thick sequences of primarily shallow branching coral zones (Blanchon and Blakeway 2003).

10.5 Biases Related to Organism Attributes

10.5.1 Mineralization of Taxa

Unmineralized organisms are generally absent from the rock record except in rare cases where rapid burial and low oxygen lead to the preservation of soft tissues like the outline of fish in deposits called "Lagerstatten". Reefs generally occur in energetic areas that are unsuited for this type of preservation. High circulation, oxygenation, and scavenging by predators cause the rapid decay and breakup of soft bodied organisms and only very rarely are soft-bodied reef organisms preserved and then only in areas adjacent to the reef proper (see section on reef fish below). Evidence of soft-bodied organisms can be left behind, however, by bioimmuration in which they are preserved in negative relief by the overgrowth of an adjacent skeletonized organism (Taylor 1990). For example, mutualisms between sabellid annelids (Garberoglio and Lazo 2011) and small crustaceans (Abelson et al. 1991) have been inferred from bioimmuration and gall formation in their coral hosts.

Mineral-reef deposits thus preserve primarily the mineralized organisms and the percent of actual species, individuals, biomass, or cover of non-skeletal organisms that this represents is poorly known. Schopf (1978) suggested that about only a third of the benthic species in sand, mud and rock communities in temperate environments had mineralized skeletons that might be preserved. In Caribbean reef crypts, the loss of soft taxa such as sponges, algae, and ascidians could result in the loss of 85 % of the living cover (Rasmussen and Brett 1985).

The impact of these factors for typical reef environments has not been quantified, probably because total diversity of a given reef community is difficult to measure, and counting the numbers of individuals or the biomass of the various species is even more difficult. The result is that surveys of living- and mineral-reef communities generally ignore the small, mobile, and cryptic taxa in favor of the sessile benthos. For example, 2 kg of dead *Acropora palmata* branches from Columbia were shown to contain 7574 associated individuals from over 80 taxa including many polychaetes and sipunculids that were only identified to family (Moreno-Forero et al. 1998). This type of biodiversity is generally omitted from both living- and mineral-reef survey that focus on the abundance or larger species.

The focus on sessile benthic communities for living reefs facilitates comparisons with mineral reefs since reef surfaces have been dominated by taxa with skeletons throughout geologic time. In addition, since the organisms are preserved mostly in original orientation and relative abundance, reef deposits rank among the fossil assemblages with the best potential for unlocking ecological information and characterizing evolutionary changes in community structure.

The ecological fidelity of individual mineral-reef deposits will vary with original composition and environment of deposition. Some living reefs may have near 100 % stony-coral cover, but most include soft organisms such as brown algae, sponges, anemones, gorgonians and alcyonarians. These organisms may contribute only a few percent of the cover on some living reefs, but on others such as macroalgae-rich degraded reefs, the removal of soft-bodied organisms from the record would result in the loss of information about a significant percentage of the reef community. Notable exceptions include alcyonarians that have internal spicules that can be preserved in reef sediments, and fused spicules from near the base of a few genera such as *Sinularia* (Konishi 1981). For example, in southern Taiwan, alcyonarian "spiculites" constitute up to 30–40 % of the living reef community and 40 % of an adjacent Holocene mineral reef (Jeng et al. 2011).

10.5.2 Growth Form and Robustness of Taxa

Reef organisms vary greatly in durability, and fragile species are generally removed from the record in a greater proportion than more robust taxa. This bias has been detected in the shells of reef molluscs where radiocarbon dating revealed that large, thick shells survived 2–3 times longer than small and thin ones in a Great Barrier Reef lagoon (Kosnik et al. 2009). Similar biases also occur in skeletal sands where robust grains outlast more fragile ones (e.g., *Halimeda*: Ford and Kench 2012). The same undoubtedly holds for at least the most brittle corals. For example, experimentally killed plate corals (*Acropora hyacinthus*) lost about 50 % of their area in 20 weeks due to external breakage, while more robust branching acroporids (*A. gemmifera* and *A. cuneata*) had little loss (Musso 1992). Another study compared living corals in the Madang Lagoon, Papua New Guinea with adjacent dead coral pieces (Pandolfi and Minchin 1995) and found that the life and death assemblages varied primarily in the loss of the lagoon's delicate living growth forms in the death assemblage.

While there is an obvious bias against the retention of the most fragile corals into the mineral reef, the effect does not appear to be important for most other corals (Greenstein and Pandolfi 2003; Greenstein 2006). The composition of living reefs and nearby fossil reefs in Papua New Guinea (Edinger et al. 2001), Florida (Greenstein et al. 1998), and the Bahamas (Greenstein and Moffat 1996) found remarkable similarity in taxonomic composition without growth-form bias.

It is perhaps counter intuitive, but the rapid destruction of very fragile forms and the general persistence of most other forms leads to generally better overall surface preservation of fragile corals when they are found (Greenstein 2006). This is because in deeper- and lower-energy zones, biological alteration outpaces mechanical destruction. If wave energy is high or if corals are very delicate, corals may be transported away from the reef and buried prior to undergoing extensive biological alteration. This provides a cautionary note to not always put too much weight on a specimen's degree of external degradation (taphograde). Well-preserved coral or mollusc specimens will reflect freshness and/or rapid burial (Lescinsky et al. 2002) but specimens with lesser preservation are difficult to interpret. Post-mortem degradation can accumulate through different pathways in different local settings (Powell et al. 2011), and even adjacent shells with similar appearances can differ in age by 100 s to 1000 s of years (Meldahl et al. 1997; Kosnik et al. 2009).

10.5.3 Growth Rate

A potential bias for interpreting the relative abundance of organisms in the fossil-reef record is related to their life spans (Kidwell and Rothfus 2010). Shorter-lived species have faster turnover and, therefore, their skeletons will accumulate at a faster rate than for long-lived species even if their abundances at any one time are similar (Van Valen 1964; Vermeij and Herbert 2004). In mineral reefs, the faster

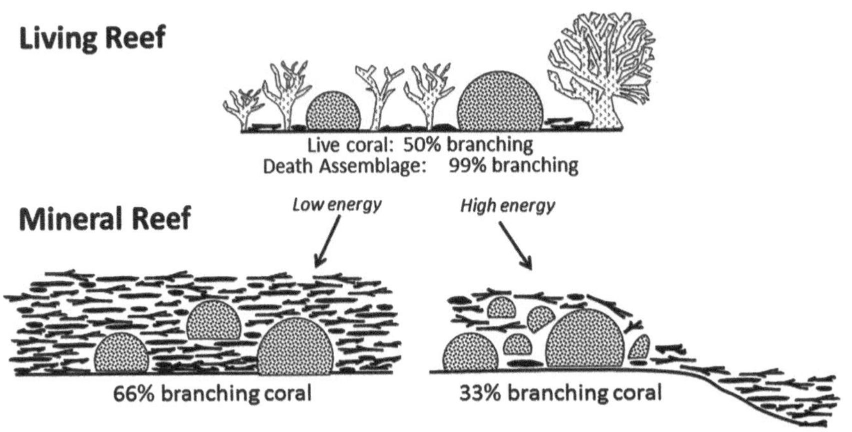

Fig. 10.4 Potential growth-rate bias in mineral reef deposits. (**a**) Fast growing and easily broken branching corals compose only half the life assemblage, but make up most of the death assemblage. How this translates into the mineral reef is dependent on local transport. (**b**) In quiet lagoons, the fast growth rate of branching corals may bias the mineral reef towards branching coral. Where storm activity is greater, waves may export branching rubble to deeper water and the mineral reef may be biased away from branching coral (Modified and expanded from Edinger et al. (2001))

growth rates in branching corals potentially cause them to be over-represented as compared to massive forms (Pandolfi and Minchin 1995; Edinger et al. 2001). The magnitude of this effect can be great, particularly in death assemblages from lagoons where export is low (Fig. 10.4). For example, Edinger et al. (2001) found that about 95 % of the coral in a Papua New Guinea death assemblage was branching (primarily rapidly growing *Acropora* sp. with high turnover rates) while the living community had a higher cover of longer-lived massive corals (e.g., Faviidae).

The overall impact of life span/growth rate may actually be small in most cases. A recent review found that for mollusc shells, bias related to life span was generally minimal in a variety of environments (Kidwell and Rothfus 2010). For reefs, the higher production rate of branching corals is at least partially offset by elevated bioerosion and export potential. Branching coral rubble can, therefore, be relatively ephemeral and export in storms may effectively reset the death assemblage, resulting in large accumulations of branching coral rubble in adjacent reef zones (Shinn et al. 2003).

10.6 Time Resolution Bias

Modern reef surveys focus on data at a precise moment in ecological time, but they may also record important temporal information. Some monitoring programs intentionally include categories such as recently dead and/or dying corals to highlight instability. In addition, the measurement of total coral cover (live plus dead coral) approximates the reef community over a much longer time period, and might be more similar to that of a comparable mineral reef. Thus, even surveys of modern reefs include an element of time averaging when total (i.e., live and dead) coral cover is tallied.

Few line-transect surveys from mineral reefs are as close to "ecological snapshots" as those from living reefs because it is difficult to determine whether adjacent corals in outcrop were alive at the same time. Nevertheless, in a few cases, comparable time horizons can be identified using geological clues. Reef surfaces may be buried by storm sedimentation (Lescinsky et al. 2012) or volcanic ash fall (Pandolfi et al. 2006), and an earthquake may cause a rapid depth shift resulting in a distinct preservational horizon (Strasser and Strohmenger 1997). Likewise, corals growing on reef flats at sea level may form flat topped "microatolls" that mark a true reef surface defined by sea level (Woodroffe and Mclean 1990; Azmy et al. 2010).

As just described, most mineral-reef horizons (horizontal transects) will include some degree of time averaging because not all corals along a transect will have been alive at the same time. In the most comprehensive study of time averaging in a mineral reef, Edinger et al. (2007) radiometrically dated 15–18 adjacent corals along horizontal transects at each of three sites on a Holocene reef in Papua, New Guinea. Coral ages along a single transect varied by up to 1000 years (Fig. 10.5), reflecting a combination of dating imprecision (from the choice of which growth interval in the coral was dated) and time averaging attributable to the three-dimensional topography of the original reef surface. Some of the oldest corals were also the largest, suggesting that smaller corals subsequently filled in depressions between them. Still, because the variation in mean coral age along the transect was comparable to the lifespan of the colonies, it is likely that "most laterally adjacent whole upright corals in fossil reefs probably co-existed in life" (Edinger et al. 2007, p. 45).

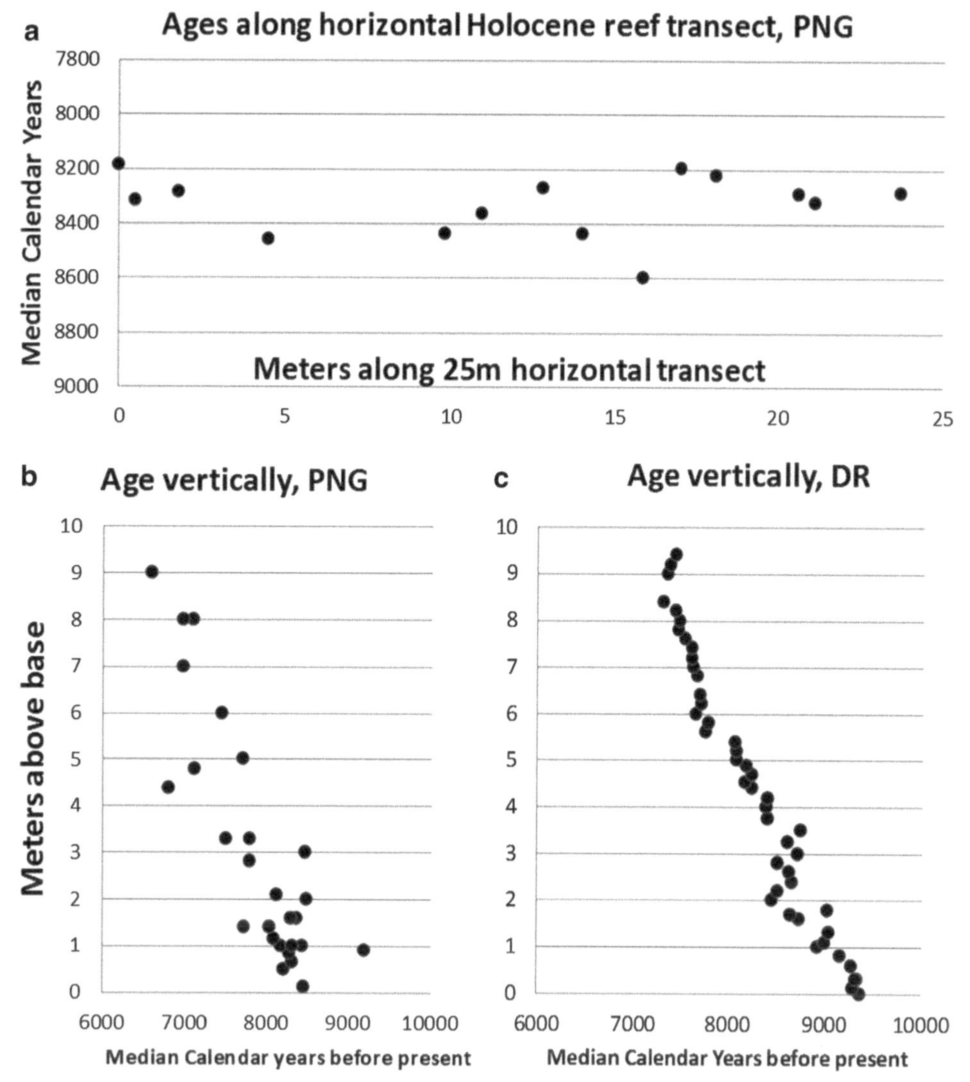

Fig. 10.5 Time resolution in mineral reefs from carbon dated Holocene corals. (**a**) Adjacent corals along a horizontal transect in a PNG Holocene reef, vary in age by several centuries, probably reflecting the topography of the paleo-reef surface and the long life spans of individual colonies. (**b** and **c**) Ages of successive corals in vertical outcrops of **b.** PNG massive corals and **c.** Dominican Republic *Acropora cervicornis*. Dates in both vertical sections are generally linear with stratigraphic position suggesting broadly constant accretion rates. However, in detail, both sections have examples of older corals resting on top of younger corals suggesting that reworking, may at times be difficult to recognize in the field (PNG data redrawn from Edinger et al. (2007); Kilasairo NW and DR data redrawn from Greer et al. (2009))

A related approach to examining time resolution in coral communities is to examine the age of successive corals that are stacked vertically in a section. Two studies (Edinger et al. 2007; Greer et al. 2009) determined ^{14}C dates for successive corals in vertical sections through fossil Holocene reefs. Both found the expected relationship between height within the section and age, but they also found examples of age reversals (older corals on top of younger corals) and centuries for which no corals were preserved. Field observations did not identify evidence of reworking within the reef deposits, though it is not hard to imagine, especially within the near mono-specific assemblage of Holocene *Acropora cervicornis* from the Dominican Republic (Greer et al. 2009). These results suggest that mixing in reef deposits may at times be hard to identify in outcrop though further studies along such tightly sampled transects are needed to fully explore the origin of such time discrepancies.

The degree to which time averaging introduces bias into paleoecological interpretations of mineral-reef transects can also be investigated by examining a preserved mineral-reef surface and comparing this to a horizontal transect from the same locale. In the Dominican Republic, storm rubble instantaneously buried the reef, leaving an easily discerned time surface. Live coral cover along this surface was reflected in pristine surface textures of corals relative to the altered character of colonies that were dead prior to burial. Based on these criteria, the abundance of live coral was nearly 80 % despite this reef's location in a protected embayment with high sedimentation levels. Adjacent horizontal transects in the same reef zone were similar in species composition but had somewhat lower estimates of live coral cover (ca. 60 %: Lescinsky et al. 2012). This study suggests that the inclusion of corals of slightly different ages (horizontal transects) had little influence on coral community composition, but did lower coral cover estimates that were based on the true irregular reef surface.

In contrast with the reef framework, reef sediments have more pervasive time averaging because deposit-feeding organisms continually stir up and homogenize the skeletal grains. For example, calianassid shrimp may rework sediments and homogenize the molluscan assemblages up to 4 m deep in lagoons (Tudhope and Scoffin 1984; Ziebis et al. 1996; Parsons-Hubbard et al. 2014) and mollusk shells differing in age by hundreds to thousands of years may occur adjacent to each other in reef lagoon sediments (Kosnik et al. 2009). Long-term time averaging such as this will accentuate preservation biases because each shell type has a sedimentary half-life that is proportional to its durability (controlled by mineralogy, structure, size and other factors).

Despite the potential for the loss of much of the ecological/abundance signal for smaller organisms like some molluscs and foraminifera, a review of numerous studies of live and dead species suggests that abundance data for death assemblages of molluscs can be quite consistent with those of the living community (Kidwell 2001). It is likely that the pervasive time averaging found in lagoon sediments is much diminished in framework reefs where sediment accumulates in smaller pockets and crypts and bioturbation are less extensive, but this has not been experimentally shown.

10.7 Megabiasess

10.7.1 Is the Reef Fabric Primarily Rubble?

Connell (1978) argued that episodic hurricanes and their widespread destruction are fundamental to maintaining high diversity and the structure of reefs in general. The geological correlate of this observation is that hurricanes profoundly influence the mineral reef fabric. Hurricane-impacted reefs may be comprised primarily of storm rubble intermixed with occasional in-place corals (Hubbard et al. 1990; Blanchon et al. 1997; Hubbard et al. 1998), rather than in place coral "framework" as was traditionally assumed (Lowenstam 1950; Newell et al. 1953; Fagerstrom 1987).

While many mineral reefs are comprised primarily of storm-generated rubble and these only approximate the living reef in a rough sense, there are also many mineral reefs that retain the original fabric of the biological reef community. These are suitable for more detailed ecological reconstructions. Variation in the importance of storm impacts to reefs is related to a number of variables, including the reef's proximity to storm tracks and the equator. Cyclonic storms require the apparent deflection of winds via the Coriolis Effect and thus storms are rare near the equator. The impact of hurricanes on reef structure is well illustrated in Caribbean Pleistocene Reefs. Curacao, in the southern Caribbean, experiences few hurricanes and the reefs there retain about 90 % of the fossil corals in life position, while Bahamian reefs of San Salvador experience significantly more hurricanes and have only about 40 % of the corals still in life position (Meyer et al. 2003). Likewise, the famous and outstanding fossil reefs of the Huon Peninsula in Papua New Guinea are adjacent to the equator, experience few hurricanes, and retain corals primarily in life position.

Colony orientation (i.e., is a colony toppled, see Sect. 10.8.1 below) is a good proxy for ecological integrity and it can be easily applied to mineral reef deposits to ensure that they are adequate ecological analogues to living reefs. Similar analysis of ecological integrity in reef cores is more problematic because in-situ orientation is difficult to identify given the limited lateral perspective of the core. Nonetheless, coral skeletons in cores are often assumed to be in life position, and their presence has been widely used to reconstruct paleo-water depth and sea-level curves (Fairbanks 1989; Camoin et al. 2011). While it may be possible to ascertain life orientation of corals in cores (Macintyre and Glynn 1976), this is problematic because even within obvious coral rubble deposits, pieces of flat-branched corals such as *Acropora palmata* tend to be stacked in approximate life orientations (Hubbard 2009).

10.7.2 Diagenesis

Diagenesis refers to the physical changes such as compaction, cementation, and recrystallization that occur in sediments and biological remains. During typical fossilization, diagenetic processes occur primarily after the burial and removal of skeletons from the taphonomically active zone (TAZ). In reefs, diagenesis is conceptually different because diagenetic processes such as cementation and recrystallization occur during the physical process of reef accretion while various reef organisms are still alive and interacting. Reef cementation, although no longer regarded as pervasive as it once was (Macintyre and Marshall 1988), is still common near the surface in many reef zones. The result is that diagenetic processes are integral to the stability and habitat of the reef, and are occurring during the ecologic time in which organisms are living on, in, and around the reef.

From the point of view of preservational bias, many diagenetic processes are readily recognizable, and are therefore unlikely to skew ecological interpretations. However, recrystallization is a rampant and important process on reefs that has many side effects. Most reef organisms secrete skeletons made of one of three carbonate minerals: aragonite, high-magnesium calcite and low-magnesium calcite. These minerals differ in their chemical reactivity. Aragonite is the most reactive and, therefore, tends to convert to the more stable low-magnesium calcite through time.

Modern, scleractinian corals and green siphonous algae such as *Halimeda* and *Penicillus* are ubiquitous around reefs and generate much of the sediment yet they have aragonite skeletons that are highly reactive. Snails and many bivalves are also aragonite. Magnesium-calcite excreting organisms include many red algae, echinoderms, and foraminifera, while organisms that secrete low-magnesium calcite include some bivalves (e.g., oysters, scallops), bryozoans, and barnacles. Paleozoic corals were also made of calcite. Because the majority of modern reef-producing taxa have aragonite skeletons, they recrystallize relatively rapidly, especially during low sea level when subaerial reef deposits are exposed to rain and groundwater. For example, the upper 10–20 cm of aragonite sands in the Bahamas have lithified after less than 10 years of exposure (Dravis 1996). Reefs formed during Pleistocene high stands have now been exposed for at least 100 ky and have experienced intense digenesis resulting in the easily recognizable contact in cores between indurated Pleistocene reefs and their un-indurated overlying Holocene counterparts.

Rapid recrystallization of reef carbonates has several important taphonomic implications. For geochemical investigations it requires that extreme care be taken to ensure that corals or other skeletal elements have not been recrystallized prior to analysis. For ecological investigations it means that taxa may be over or under represented based on their original mineralogy. This mineralogical bias is well known in Paleozoic fossil deposits (Cherns and Wright 2000; Wright et al. 2003) where amateur fossil hunters collect dozens of well-preserved brachiopods (calcite), but overlook the nearly as numerous bivalve fossils that have lost their original aragonite shell and are preserved only as the mud casts of their interior. A similar effect has been documented in uplifted reef terraces along the Red Sea where there is a decrease in most aragonitic molluscs while calcitic taxa, such as *Tridacna*, remain well preserved (Montaggioni and Braithwaite 2009).

Perhaps the biggest impact of rapid recrystallization and lithification of carbonate skeletons in mineral reefs is that the texture and composition of exposed reef deposits can be rapidly lost, making quantitative analyses of many outcrops difficult. For example, mineral reefs are often preserved as a series of uplifted terraces and upper terraces are often highly recrystallized and difficult to survey. Surficial weathering compounds the problem, particularly in arid environments. Coastal reef limestones, sometimes called "ironstones", develop sharp dissolution features known as "karren" which are treacherous to walk on and which obscure the outlines of most fossils (Fig. 10.6). Many fossil reef deposits, therefore, cannot be surveyed except in fortuitous fresh outcroppings such as road cuts, quarries, and streams.

The loss of surface detail means that species with the most distinct identifiable morphologies may be over-represented in transects with poor preservation. This bias may be particularly important in inter-ocean comparisons. Caribbean reefs have fewer coral species and common rubble from the breakdown of staghorn corals can be assigned to a specific taxon (*Acropora cervicornis*) while similar rubble in the Pacific could be from several *Acropora* species (Greenstein and Pandolfi 1997). Analyses based on growth form rather than taxon can still be performed (e.g., Pandolfi and Greenstein 1997), but this clearly results in the loss of significant ecological information.

Fig. 10.6 Weathering surface of the top of the lower terrace, Curacao. Dissolution and the formation of sharp karren make study of many Pleistocene mineral reef deposits difficult

10.7.3 Uneven Distribution of Mineral Reefs in Time

Mineral reefs are not evenly distributed throughout the geologic past. While some of this results from changing environmental conditions that precluded reef formation, some of these "gaps" are the result of differential preservation or vagaries of sea level. For example, we are missing the reef record for most of the last million + years. All living reefs today are geologically young (mostly <9000 years old, Hopley et al. 2007; Montaggioni and Braithwaite 2009) because prior to that time sea level was much lower (100 + m) and the locations of today's living reefs were dry land. Because glacial cycles have occurred around every 100,000 years during the Late Pleistocene/Holocene, interglacial periods with sea level on par with modern sea level have been comparatively rare, corresponding to only 10–15 % of the last 450 ky (Hopley et al. 2007). On a first approximation then, most reef building, and the majority of the time during which reefs have existed over the last few 100,000 years, is now located up to 100+ m below the modern sea surface at disphotic depths. We now have evidence of submerged reefs adjacent to many modern reefs (e.g., GBR, Abbey et al. 2011; Hawaii, Fletcher et al. 2008; Marquesas, Cabioch et al. 2008; Barbados, Fairbanks 1989; Tahiti, Camoin et al. 2011) as well as along coasts where reefs are poorly developed today. For example, during low sea level a 1300-km long barrier reef with substantial topographic relief formed along the west coast of India (Vora et al. 1996), and numerous reefs dotted the Gulf coast of Florida (Hine et al. 2008). Unfortunately, these lowstand reefs which account for 85–90 % of Pleistocene time are known almost entirely from topographic profiles captured on side scan sonar. Only recently do we have any ecological information on these "lowstand reefs" (e.g., Tager et al. 2010).

Reef accretion during sea-level change introduces another potential bias since our record of mineral reefs is largely limited to reefs that formed during sea-level rise. Transgression provided reefs with accommodation space (the depth of water above the reef in which it is possible to accumulate sediments and framework) and this allowed upward accretion of all reef zones. Once sea level stabilized, as it had until the recent anthropogenic resurgence, reefs sitting near sea level were limited to accretion on their deeper flanks or along reef flats. As a result, the overall pattern of reef building was lateral accretion. The comparison of transgressive mineral reefs with laterally accreting high-stand living reefs introduces a potential bias that has not yet been explored.

10.7.4 Summary of Biases

The biases outlined above affect transect data from all reefs in varying ways, and thus no two transects will be exactly comparable, particularly between mineral and living reefs. How problematic this is when comparing modern and mineral reefs depends on the nature of the surveys and the questions being asked. Several lines of evidence suggest that, despite inherent biases, mineral-reef data are broadly comparable to those collected from living reefs if appropriate criteria have been met to assure ecological fidelity (discussed below). Analyses of modern monitoring methods have found that patterns of benthic composition are very robust to differences in sampling intensity and technique (e.g., Dumas et al. 2009; Jimenez et al. 2010; Pante and Dustan 2012; Chap. 12). In addition, radiocarbon dating suggests that the effect of time averaging along transects may be low since age variation is within the life span of individual coral colonies (Edinger et al. 2007). The stability of local coral-community composition through time (Sect. 10.9.10) also suggests low temporal bias. These lines of evidence suggest that with careful selection of appropriate localities and samples (Sect. 10.8), detailed paleoecological information can be unlocked from ancient reefs. Section 10.9 reviews the many types ecological information that have been obtained for Pleistocene and Holocene reefs and how these data have be used to provide a baseline for judging recent anthropomorphically induced trends in reef ecology.

10.8 Criteria for Assessing Ecologic Fidelity in Mineral Reefs

Ecological information can be extracted from mineral reefs if they have appropriate "ecological fidelity". Brett and Baird (1986) reviewed various fidelity proxies for fossil assemblages and suggested that the most useful criteria in approximate order of increasing physical disruption and decreasing biological signal were: re-orientation, disarticulation, fragmentation, and corrosion/bioerosion (loss of surface texture).

10.8.1 Reorientation

Re-orientation and transport of specimens from life position are common initial post-mortem effects for unattached organisms such as epifaunal molluscs. However, on reefs, many organisms such as corals are attached to the reef framework and, therefore, will be reoriented only when their attachment is broken. This can be a result of storms (see discussion of fragmentation below) or more benign toppling of substrates weakened by bioerosion. Even corals that are not attached to reef framework are often massive and have a stable morphology (such as a hemispherical dome) that is resistant to toppling and transport. Thus, compared to most typical fossil communities, mineral-reef deposits can preserve orientation and spatial relations between organisms with high fidelity.

Corals are of course toppled and moved during storms, and an analysis of the percent of corals in life position (e.g., Meyer et al. 2003; Lescinsky 2008) is an important criterion for examining the ecological fidelity of the mineral reef and the impact of storms on the coral assemblage. Toppling, however, does not always lead to coral death and corals can continue to grow after re-orientation. Toppling of larger corals is also unlikely to appreciably move them or mix corals between ecological zones.

10.8.2 Disarticulation

Skeletons with multiple mineralized pieces may disarticulate after death and the decomposition of the connective tissues. This includes the two shells of a bivalve, the many ossicles of echinoderms, the bones and teeth of fish, and the segments of marine algae such as *Halimeda*. Since disarticulation results primarily from biological decomposition rather than physical fragmentation during storms (see below) live articulated organisms may be moved and left in storm deposits without significant disarticulation. Whether they ultimately enter the mineral reef record still intact depends on how rapidly they are buried below the TAZ. On reefs with high energy and deep bioturbation (up to 3.0 m deep: Tudhope and Scoffin 1984; Ziebis et al. 1996; Parsons-Hubbard et al. 2014) disarticulation is the rule within most reef sediments. In a few cases, coral framework and rubble may hinder bioturbation by deposit feeders. Likewise, articulated infaunal organisms such as bivalves and burrowing echinoids are sometimes found in life position among coral skeletons.

10.8.3 Fragmentation

Skeleton fragmentation on reefs occurs via both biological and physical processes, though it is dominated by physical mechanisms such as storm waves. Biological processes on reefs include predation, such as the feeding bouts of sting rays and stomatopods, and a few special cases such as green algae (i.e., *Halimeda*), that intentionally shed skeletal parts to reduce fouling (Littler and Littler 1999). Bioerosion is an important contributing factor to fragmentation because it weakens the skeletons and the strength of their attachment.

Interestingly, natural selection has dictated that most organisms occupying wave-scoured environments, including reef corals, are extremely "over built" to withstand dislodgement and fragmentation in the waves (Denny 2006). Thus, the high velocities associated with storm waves may create drag and lift on massive corals, but these forces are easily compensated by relatively small areas of firm attachment. As Massel and Done (1993, p. 153) state "analysis of shear, compression and tension forces generated by waves indicate that corals firmly attached to solid substratum, even if only over a small proportion of their base, can resist all waves, regardless of colony size or shape, cyclone intensity or region".

Of course hurricane force waves do overturn and fragment corals, but these instances are generally limited to extreme events. For example, a survey of coral damage from cyclones on the Great Barrier Reef (Fabricius et al. 2008) found that little skeletal damage occurred until a high threshold of wave energy and/or duration was reached. At inshore sites with branching and foliose corals the wind-speed threshold for coral fragmentation was about 33 m/s and a duration of >12 h. By comparison, pervasive damage to offshore reefs with more robust morphologies only occurred after wind speeds surpassed 40 m/s. These thresholds approximate the definition of category 2 to 3 storms. In the Atlantic where there are fewer corals, especially finely branched ones, higher wind speeds (>50 m/s) are required for significant coral breakage (Gardner et al. 2005).

The relevance of these studies for reef geologists is that coral fragments result primarily from one of two processes. Rare severe storms will break even large corals and the entrained pieces will be rolled and abraded and perhaps deposited in a mineral reef deposit as rubble (e.g., Hubbard et al. 1998; Rasser and Riegl 2002). Branching corals are at a disadvantage due to both their shape and the lower density of their skeletons (Shinn et al. 2003) though fragmentation can also serve as a dispersal adaptation (Highsmith 1982) that ultimately increases their abundance.

However, the degree of fragmentation is not necessarily controlled only by storm duration or intensity. The threshold for breakage can be measurably reduced by bioerosion such that colonies break at significantly lower wave energies than normally associated with Category 3 storms. Even on healthy reefs the bases of corals may still be dead and easily broken. For example, Hudson (1977) reported that across the Florida reef tract virtually all of the living *Montastraea annularis* colonies exhibited extensively bored bases while still alive. Increased boring will thus facilitate breakage during less powerful storms. As a result, heavily bioeroded corals may actually represent pieces that broke in smaller storms and were deposited close to where they lived.

10.8.4 Loss of Original Surface Texture

Once dead, corals and other skeletons soon begin to lose the fine detail of their surface texture. This degradation can be used as the basis for semi-quantitative scales of degradation known as taphonomic grade or "taphogrades" (Greenstein and Moffat 1996). Surface degradation occurs via a number of physical (dissolution, abrasion, corrosion) and biological (boring and encrustation) processes. On reefs, biological processes typically predominate. Waters surrounding reefs are supersaturated with respect to aragonite and therefore dissolution is low, and corals that remain in place are not subjected to abrasion from transport. On the other hand, biological post-mortem alteration is rapid and pervasive.

The first organisms to attack skeletons are microborers, a guild of small (<1 mm in diameter) organisms (primarily green algae, fungi and bacteria) that dissolve carbonate substrates. Vogel et al. (2000) deployed experimental substrates at a variety of depths and sub-environments in the Bahamas and along the Great Barrier Reef. Within 2 weeks, all samples in less than 30 m of water were colonized by microborers, and by 3 months up to 90 % of the skeletal surfaces were significantly impacted.

Although microborers are individually indistinguishable to the naked eye, their collective effect is to degrade the surface texture enough that specimens can be assigned taphogrades that highly correlate with microboring intensity (Fig. 10.7; Lescinsky et al. 2012). Thus, in mineral reefs, surface texture and taphogrades are robust indicators of how long a surface was exposed prior to burial. With extended exposure, surface texture rapidly declines primarily via encrustation by coralline algae, foraminiferans, and others, and by scraping bioeroders (grazers) which can remove the original surface texture and topography of the corallite. Thus, well-preserved surface texture is typically correlated with rapid burial.

10.8.5 Summary of Taphonomic Indicators in Mineral Reefs

Reef corals are not nearly as susceptible to the processes that remove ecological information in other fossil assemblages such as marine invertebrate communities on muddy bottoms. Coral skeletons are largely locked in place and resistant to physical degradation, except during large storms. Corals in life position are, therefore, probably a good indicator of the biological community, even if they are heavily bored. Nevertheless, it cannot be assumed that all corals within a reef were alive or dead at the same time.

The principle processes that degrade recently killed corals are biological and start with a brief period of predominately microboring, followed by increasing encrustation and surficial bioerosion by grazers that may erase original surface texture. In-place coral fossils with well-preserved surface texture suggest that they were alive at the time of burial and may have been killed by re-suspended storm sediment that can blanket a reef at infrequent intervals. Corals with moderate surface texture loss are probably buried more gradually by suspended sediment traveling across the reef that is slowed down by organisms on the reef (baffled, Fagerstrom 1988) and settles into crypts and low spots. Regardless of the precise burial history, sediment and displaced corals can make up the majority of the reef volume (Hubbard et al. 1990).

10.9 Lessons from Mineral Reefs: Results of Ecological Studies

Much attention has been paid to the patterns of reef change over recent decades and the factors that might be responsible. We have increasingly broadened our interests from local top-down or bottom-up interactions to include regional to global changes in temperature, pollution, acidity and general human impact. With this shift, reef managers have started to think about not only factors acting on individual reefs but also those that compromise connectivity between reefs. In order to address these questions, a historical baseline is needed to interpret possible changes. Although, as outlined above, care is needed when unlocking ecological information from mineral reefs, these reefs are our only models for "natural systems" before human intervention. This section reviews the various types of information that have been learned so far from mineral reefs, and how this information has increased out understanding of the ecological patterns of modern reefs.

10.9.1 Baselines for Coral Community Composition

Presence/absence data are the most commonly used metric for studying fossil communities and estimates of relative abundance are initially viewed with suspicion due to time averaging, as well as biases such as transport, winnowing, and variable rates of accumulation. However, field studies comparing live communities and their dead assemblages have found that for hard-shelled invertebrates (e.g., bivalve molluscs) the rank abundances of common species are quite accurately preserved (Kidwell 2001). For reefs, the mineral record may be better. The main space occupiers on reefs have mineral skeletons that are attached and have a higher possibility of being preserved either in place or at least close to where they lived. This allows for a better characterization of absolute abundance in ancient reef

Taphograde			Microboring intensity with depth into skeleton
1	Original ornamentation on septa, corallite topography preserved		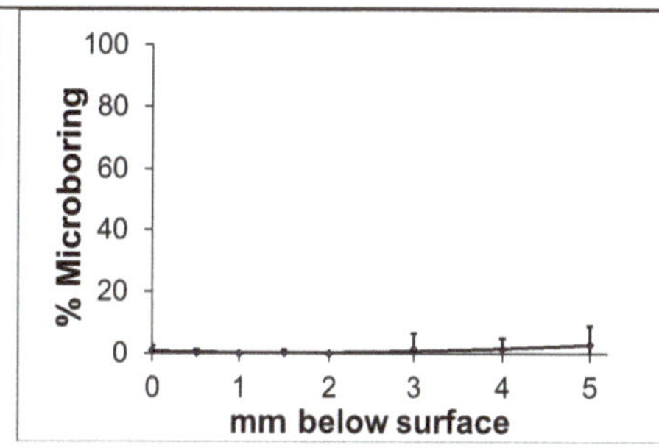
2	3-D topography of corallite and, septal ridges preserved, some original detail lost		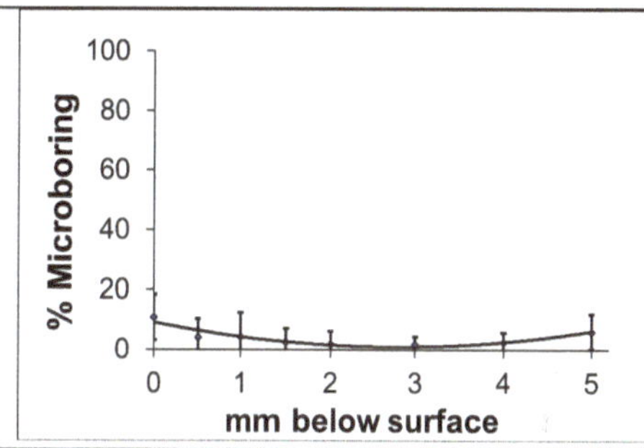
3	3D topography of corallites and septa lost, may have minor pitting/encrusting		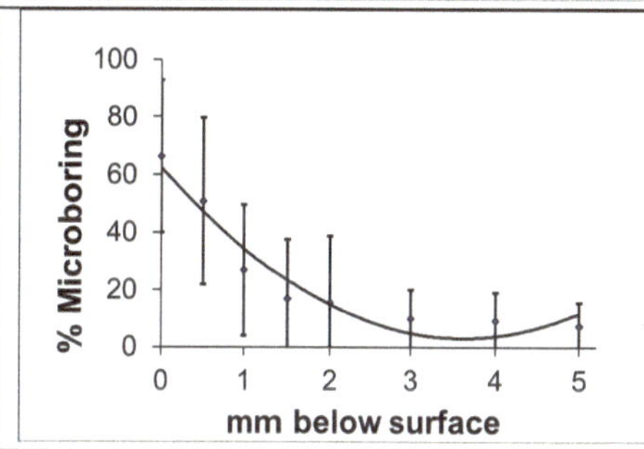
4	Surface irregular, and fine corallite detail obscured, can ID in field		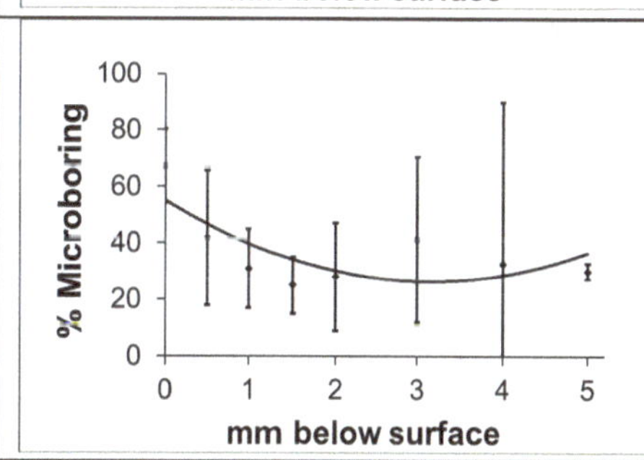
5	Rubble, pitting and encrustation common, unidentifiable in field	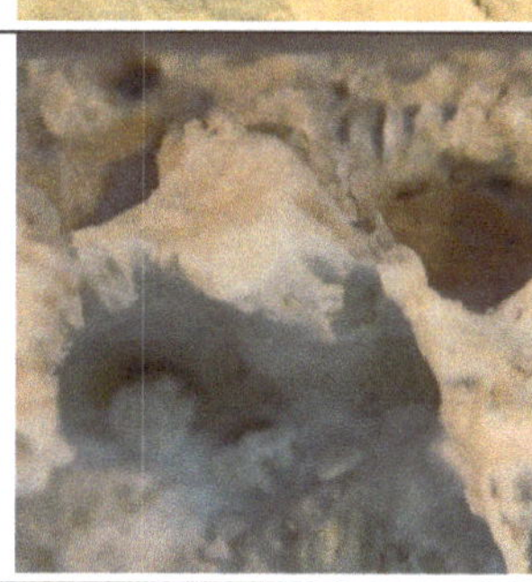	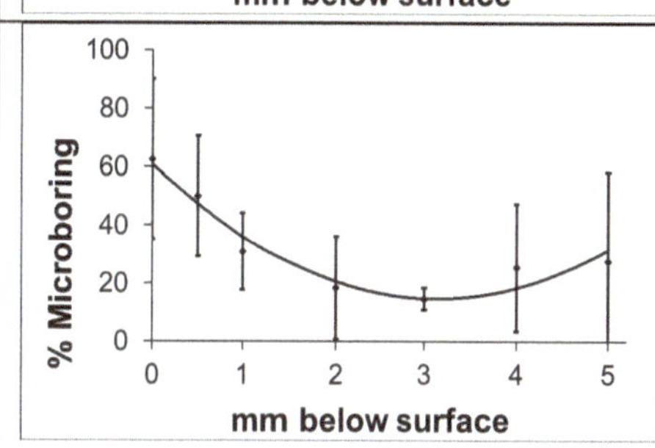

Fig. 10.7 Semi-quantitative taphograde analysis of the Holocene Dominican Republic mineral reef. Surface texture of corals correlates strongly with microboring intensity (Data from Lescinsky et al. (2012))

communities (e.g., Pandolfi and Jackson 2006; Pandolfi et al. 2006).

Quantitative characterization of Pleistocene and Holocene reefs provides a pre-human baseline that can be used to evaluate stability and change in coral communities. In general, there has been little change in species composition between the late Pleistocene and now. Notable exceptions include the Caribbean loss of *Pocillopora*, and organ-pipe *Orbicella* after the Pleistocene (Pandolfi et al. 2002). Ancient mineral-reef communities are also often similar to those of adjacent living reefs (e.g., Pandolfi 1996, Pandolfi and Jackson 2006) suggesting that ancient physical environments were quite similar.

Where differences exist between Pleistocene and living reef communities, the possibility that they are unrelated to recent anthropogenic causes should be considered. For example, the extensive windward reef-crest community that existed along the north east coast of Curacao during the Pleistocene (Pandolfi and Jackson 2001; Meyer et al. 2003) did not return during the Holocene and is still absent today. Similarly, nearly a third of the well preserved Pleistocene coral fauna of Henderson Island, at the southern edge of reef development in the Pacific are not found on contemporary reefs in the area for reasons unrelated to human influence (Paulay and Spencer 1988). The possibility of such discrepancies calls for examining changes in physical variables such as tectonic uplift, paleo-oceanographic variations and other natural factors that might be responsible for differences between past and present coral communities.

10.9.2 Baselines for Reef Fish Communities

The recent loss of grazers and other top-end predators has been widely cited as an important factor in the decline of reef corals (Jackson 1997; Jackson et al. 2014). However, mineral-reef deposits contain no fossil fish, save for an occasional tooth or vertebrae and are thus not suitable to provide a geological baseline. Reef environments are highly energetic, their sediments are intensely bioturbated, and high predation rapidly consumes fish carcasses. Knowledge of ancient reef fish comes from fossil deposits in other environments where the species and families of fish that characterize coral reefs also live (Bellwood 1996, 1998). A few lagerstatten (areas of exceptional preservation) in originally anoxic lagoon muds provide most of what we know about ancient reef fish (Goatley et al. 2010). Quantitative fish data must be very cautiously interpreted. To date, the main insights provided by fossil reef fish concern evolutionary trends in diversity, niche specialization, herbivory (Bellwood 2003), and nocturnal feeding (via eye size, Goatley et al. 2010).

10.9.3 Baselines for Shelled Invertebrates

Unlike fish, many shelled invertebrates are readily preserved in reef sediments, and can be relatively easy to count and assess for abundance in mineral reefs. It may even be argued that representative surveys of some shelled invertebrates (e.g., molluscs) are easier and more reliable in mineral reefs than for living communities. For example, reef snails and bivalves can be collected in volume from fossil deposits (e.g., Crame 1986; Cerridwen and Jones 1991; Jackson et al. 1999; Gardiner 2001). Yet modern reef-mollusc communities typically have low density, high variability, and high diversity making quantitative sampling difficult (McClanahan 1989). This is related to the fact that the shelly fauna in sediments accumulated over a longer time interval. As a result, mobile species have a much higher probability of having died at a particular point over that time span than being alive at that particular spot today (Miller 1988).

Likewise, modern foraminifera assemblages are usually surveyed via sediment grabs and it is difficult to separate living organisms from dead tests. Several studies (e.g., Glenn-Sullivan and Evans 2001; Langer and Lipps 2003) have shown that the total assemblage (live and dead tests together) provides a more robust and less variable indicator of reef environment than samples restricted to live foraminifera.

The record of other shelled organisms such as echinoderms is very poor. Echinoderms such as sea-urchins and sea stars (which have many calcified ossicles) disarticulate rapidly after death, with some such as *Diadema* losing most spines in the first day and having the central corona collapse into pieces within a week (Greenstein 1991). Disarticulated echinoid ossicles are porous, easily abraded and, therefore, uncommon in reef sediments (Donovan 2005; Dynowski 2012). Infaunal echinoids (e.g., sea biscuits) may have a somewhat better fossil record since they can be preserved in life position within the sediment.

The low preservation potential of epifaunal echinoderms limits attempts to document either a rapid increase (e.g., Crown of Thorn outbreaks in the GBR) or decrease (e.g., *Diadema* die-off) of echinoderms using the mineral-reef record or sediments in the modern reef. For example, Greenstein (1989) found no recognizable signal in the sediment record in the US Virgin Islands for the Caribbean-wide die-off of *Diadema* in 1983–84. When putative patterns are identified, interpretation remains problematic. Peaks in the number of echinoderm ossicles at a particular horizon within reef sediments have been interpreted alternatively as resulting from either a dramatic increase (Walbran et al. 1989) or decrease (Gischler 2010) in the number of echinoderms alive at that time. If peaks are based on relatively few ossicles (10–24) from much larger sediment samples and do not correlate between nearby cores (e.g.,

Gischler 2010), they are probably best interpreted as an artifact of coarser grains that were concentrated during bioturbation by callianassid shrimp, or as the result of chance sampling of one or more individuals whose ossicles were clumped in the sediment (Pandolfi 1992; Greenstein et al. 1995).

10.9.4 Patterns of Reef Zonation

Qualitative and quantitative data from well-exposed ancient reef tracts suggest that Pleistocene (125 k) reefs of the Caribbean had zonation patterns similar to those of modern reefs (Fig. 10.8; Mesolella 1967; Geister 1981). This zonation also generally reflects the same wave-energy gradient that is seen today (Geister 1977). For example, until recently the crest of moderately high-energy Caribbean reefs today were dominated by branching *Acropora palmata* and nearby Pleistocene communities are also dominated by this species. This suggests that many Pleistocene reefs shared physical variables, such as wave and storm tracks, with their modern counterparts. These insights allow geologists to extrapolate ancient depth and wave-energy patterns for fossil outcrops, and to reconstruct the development and accretion histories of reefs in cores. For example, an upward transition from massive corals to branching corals in cores is interpreted as a shallowing-upward record, indicating that accretion was faster than sea level rise (the "catch-up" reef scenario of Neumann and Macintyre 1985). This is the most common pattern observed in cores (Montaggioni and Braithwaite 2009), although cores indicating increasing depth (reef drowning) or constant depth (accretion rate equals sea level rise) are also found (the "give-up" and "keep-up" scenarios of Neumann and MacIntyre 1985, respectively). For a more detailed discussion of this, see Chap. 6.

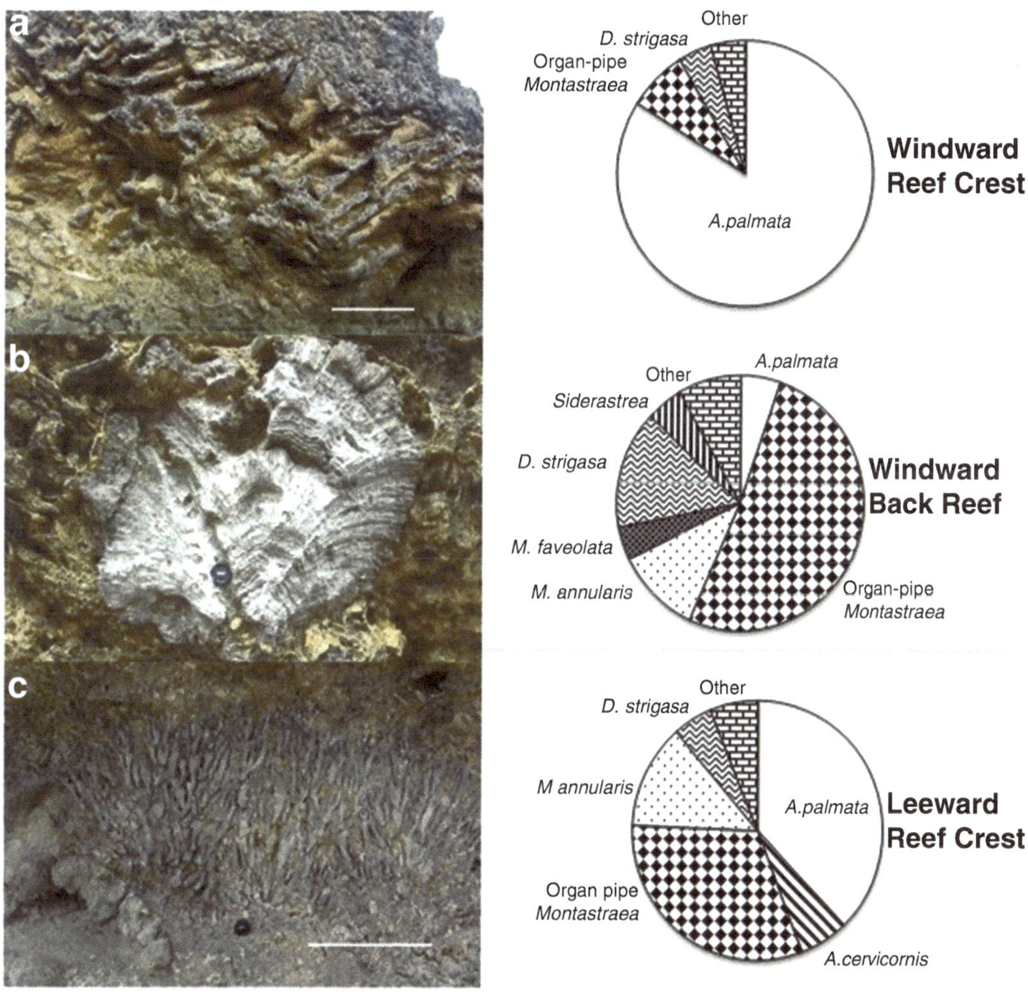

Fig. 10.8 Faunal composition and zonation of a Pleistocene Reef, Curacao. Taphonomic indicators such as re-orientation and fragmentation suggest high ecological fidelity. Reef was replicate sampled using 40 m linear intercept transects in three reef zones: (**a**) windward reef crest (**b**) windward backreef and (**c**) leeward reef crest (Data from Pandolfi and Jackson (2001))

10.9.5 Geographic Range and Climate Change

Fossil reef deposits provide information about the geographic range of corals at various times in the past and recent interest has focused on coral-range extensions during warmer climates. Living corals have begun to expand their ranges poleward in response to a warming climate (Precht and Aronson 2004; Yamano et al. 2011) and similar range extensions have been documented for the mid-Holocene when temperatures were higher. Coral ranges extended further north in the Caribbean (along the Florida coast: Lighty et al. 1978; Precht and Aronson 2004), and Pacific (Ryukyu Islands of Japan: Veron 1992; Webster et al. 2004; Hongo 2012), and further south around Australia (Lord Howe Island: Woodroffe et al. 2010). Other range extensions in corals have been tied to warming associated with changes of ocean currents during the Pleistocene (Kennedy et al. 2007; Greenstein and Pandolfi 2008). The importance of climate in influencing coral ranges is perhaps best detailed in studies of Holocene reef terraces in the Ryukyu Islands (Abram et al. 2001; Hamanaka et al. 2012). Coral species composition, here at the northern limit of some Indo-Pacific corals, has varied closely with the rise and fall of sea surface temperature over the last 4000 years.

10.9.6 Limited Evidence of Ecological Interactions

Mineral reefs in general preserve little record of ecological interactions between specific reef organisms. Although a surprisingly long list of complex interactions has been documented in the fossil record (Boucot and Poinar 2010), evidence on reefs is limited to a few special cases. Evidence for predation can come from the damage to shells of reef invertebrates. For example, molluscs are commonly drilled by naticid and muricid gastropods, and the characteristic drill holes are easy to identify in fossil deposits (Kelley and Hansen 2003). Mollusc shells are also crushed during predation by crabs and some fish and, therefore, broken shells can also provide a predation metric (Vermeij 1989; Stafford and Leighton 2011). Rays have characteristic feeding pits in some lagoonal environments, and these have been identified in the fossil record to provide evidence for ancient ray feeding behavior (Howard et al. 1977). Another type of feeding that can be identified in mineral reefs is damselfish "farming". Many damselfish nip at coral to create dead spots where algae can grow and be farmed. The coral continues to grow up around the algae forming a distinctive "collar" that is recognizable in the fossil record, particularly on staghorn acroporids (Kaufman 1981).

Intense spatial competition on reefs occurs via overgrowths among coral and other modular organisms (e.g., coralline algae, sponges) that occupy reef surfaces (Lang and Chornesky 1990), and overgrowth patterns between mineralized taxa can be examined (reviewed in West et al. 2011). To date, there has been little investigation of competitive interactions in near-recent mineral reefs, but ancient overgrowth patterns have been important for reconstructing the ecology of older reefs that were very different from their modern counterparts (e.g., Cambrian, Zhuravlev 2001), and analyzing evolutionary trends in spatial competition over time (McKinney and Jackson 1989).

Ecological information on symbiotic associations between reef dwellers is available in some instances. For example, a few reef organisms with skeletons grow with other organisms preserving a record of their symbiosis. Christmas tree worms (*Spirobranchus*) are obligate commensals with living coral and settle on the coral, which then grows up and around the worm's calcareous tube; similar examples have been recorded in the fossil record (Garberoglio and Lazo 2011). Other obligate symbiotic relationships that would be preserved in the mineral reef would include records of gall crabs (Abelson et al. 1991) and gammaridean amphipods (Bergsma and Martinez 2011), both of which distort coral growth in a distinctive way that benefits the coral and would be easily recognizable in the rock record.

Relationships between encrusters using dead substrates and discarded mollusc shells provide other examples of ecological relationships that are similar to symbioses. Encrusting organisms may settle on live or dead substrates and hermit crabs occupy shells of dead molluscs. In the mineral reef record, hermited shells can be recognized by taphonomic criteria including wear and encrustation patterns (Walker 1992).

10.9.7 Can We Estimate Live-Coral Cover from Mineral Reefs?

Percent live-coral cover is the metric most often used to quantify reef changes and reef "health" today (e.g., Gardner et al. 2003; Bruno and Selig 2007), but an equivalent metric for mineral reefs is problematic. On a living reef, the condition of two adjacent corals, one alive and the other dead, is easy to discern and they would be counted differently. In outcrop all corals are dead making it difficult or impossible to tell which were contemporaneous with one another. However, live coral cover can be estimated in mineral reefs if strict conditions are met. Rapid burial of a reef surface will preserve the original three-dimensional reef profile below the TAZ effectively "freezing" the pre-burial surface in time. The surface texture of corals along such a profile can be examined to quantify the percent of corals that were alive when the reef was buried. A burial horizon created about

8500 ybp was identified by Hubbard et al. 2008 and traced for 125 m in the Holocene reef complex of the Enriquillo Valley, Dominican Republic. Lescinsky et al. (2012) used the taphonomic condition of the buried corals to quantify the percentage of live and dead corals on this section of the reef at the time. Corals devoid of microboring were taken as having been buried alive and live coral cover on various parts of the Holocene reef was estimated to be between 59 and 80 % (Fig. 10.9), suggesting that low coral cover found on most Caribbean reefs today differs markedly from what was present at this site throughout most of the Holocene.

10.9.8 Coral Reef "Health" in the Past

With the incidence of coral bleaching and diseases rising today, mineral reefs provide a potential baseline for comparison. Two approaches have been used to address coral disease in mineral reefs. The first is time-series analysis to look for gaps in the occurrence of particular corals that might indicate a disease outbreak that eliminated that species for a period of time. The simplest application of this approach involves large coral heads with growth histories preserved as annual skeletal-growth bands. Continuous growth implies centuries of conditions amenable to that coral.

Similar reasoning has been used by comparing successive colonies over longer time periods. For example, prior to the 1980s die-off of Caribbean *Acropora* due from White Band Disease, *Acropora* had dominated most sites in the Caribbean in the geologic and historic past (Pandolfi 2002; Fig. 10.10). Cores through the last 3000 years of patch-reef development in Belize (Aronson et al. 1998) and continuous outcrop sampling of 6–9000 year old Holocene reef deposits in the Dominican Republic (Greer et al. 2009) found no

Fig. 10.9 Live coral cover in the fossil record, Enriquillo Valley, Dominican Republic. (**a**) an event bed (above *white dashed line*) buried a 8500+ kya fossil reef surface. (**b**) Detailed analysis of surface texture and microborings along the 125 m event bed transect, and non-event bed transects in various zones of the paleo reef suggest high (60–80 %) live coral cover in this Holocene reef (From Lescinsky et al. (2012))

evidence of any interruption to the deposition of *Acropora cervicornis* beds. However, the synthesis of many core records from throughout the Caribbean identifies several possible gaps in *Acropora palmata* (Hubbard 2009) and *A. cervicornis* (Shinn et al. 2003; Shinn 2004) that require further study.

A second approach for estimating the condition of corals in ancient reefs is the use of partial mortality as a "health" proxy, similar to how it has been applied to living reefs (e.g., Ginsburg et al. 2001; Garzon-Ferreira et al. 2005). In mineral reefs partial mortality is preserved as growth irregularities within a coral cross section. The dead portion of a colony will be bored and encrusted and, if corals subsequently regrow over that surface ("re-sheeting"), evidence of the partial mortality will be preserved. Identification of growth disturbances within corals have been used to examine a variety of past reef disturbance including Crown of Thorns outbreaks (DeVantier and Done 2007), coral bleaching (Halley and Hudson 2007), disease (Lescinsky 2012), and other stresses (Hudson 1977, 1981). In coral X-radiograms, these horizons are referred to as "stress bands".

This approach has been used to investigate the presence of widespread White Band Disease in ancient *A. palmata*. Many colonies that died from White Band Disease in the 1980s remain standing and some are now being recolonized by new *A. palmata* recruits that re-sheet over the dead, encrusted and bored skeleton (Jordan-Dahlgren 1992). The re-sheeted colony appears outwardly to be healthy and of great age, but is actually a new colony that preserves an internal bioeroded horizon (Bonito and Grober-Dunsmore 2005). A search in Pleistocene reefs of Curacao of 1450

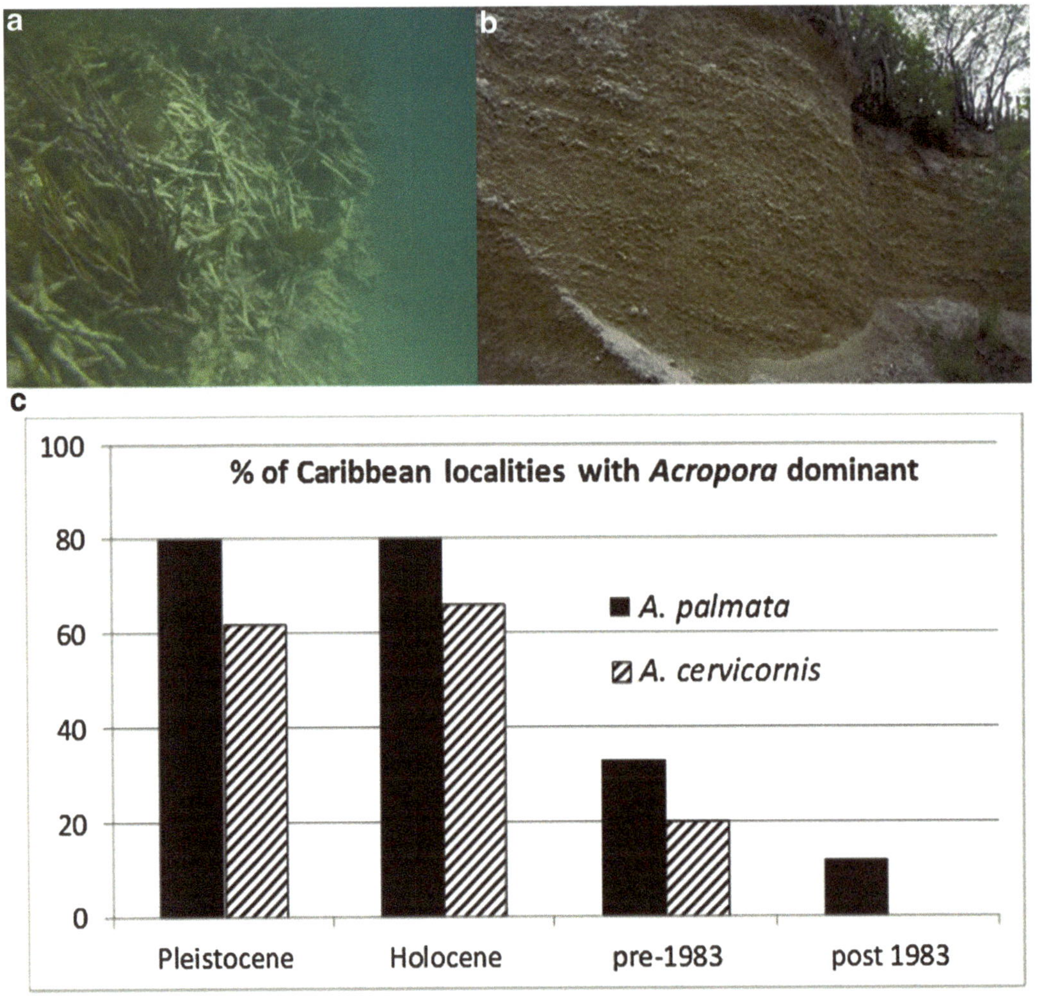

Fig. 10.10 Mineral reef evidence for long-term *Acropora* dominance in the Caribbean. (**a**) Cross section through (former) *A. cervicornis* patch reef exposed in recent underwater earthquake scarp, central Belize coast. Cores in the area by Aronson et al. (1998) found a 3000 year uninterrupted record of *A. cervicornis*. (**b**) Field exposure of continuous 6–9000 kya record of *A. cervicornis* in the Dominican Republic (see Greer et al. 2009). (**c**) Regional compilation suggests that *Acropora* has dominated most sites in the Caribbean from the Pleistocene until the 1980s die off suggesting that the recent decline is anomalous (Redrawn from Pandolfi 2002)

large *A. palmata* skeletons from three time intervals and three reef environments found <1 % of the corals had any evidence of partial mortality, and none had evidence of extensive regrowth (Lescinsky 2012; Fig. 10.11). This, and a low incidence of growth irregularities in the cross sections of large (>1 m diameter) coral heads in the same outcrops suggest that disease was rare on these ancient reefs.

10.9.9 Disturbance and Resilience

Paleoecological data can provide an important baseline for the frequency of ancient community disturbance and subsequent resilience (e.g., Willis et al. 2010; Garcia Massini and Jacobs 2011; Reuter and Piller 2011). For reefs, storms are a principle agent of physical disturbance and, although hurricane debris is often washed to deeper water or piled on shore (Kobluk and Lysenko 1992; Scoffin 1993; Blanchon and Jones 1997), it may also be incorporated into the reef structure (Blanchon et al. 1997; Hubbard et al. 1998). Layers of storm debris can, therefore, provide a record of reef disturbance, and if there is also evidence of in-situ coral growing on top of hurricane deposits, it is also possible to evaluate community resilience. For example, storm frequency was estimated for Holocene reefs in the South China Sea based on dating of large *Porites* heads that accumulated in shallow water following strong storms with return frequencies of ca. 160 years (Yu et al. 2004). For the Pleistocene reefs of the Barbados, Perry (2001) identified storm layers based on cycles of photophilic and cryptic encrusting organisms. The storm deposits suggested a high frequency of disturbance, but the low coral diversity (primarily *A. palmata*) and an absence of in-situ corals prevented a detailed analysis of community resilience. Nevertheless, the study concluded that the patterns of encrustation associated with windward and leeward colonies were similar in adjacent modern and Pleistocene reefs.

Disturbance and resilience have been studied in Holocene reefs of the Huon Peninsula of Papua New Guinea (Pandolfi et al. 2006; Reymond et al. 2011). This equatorial reef

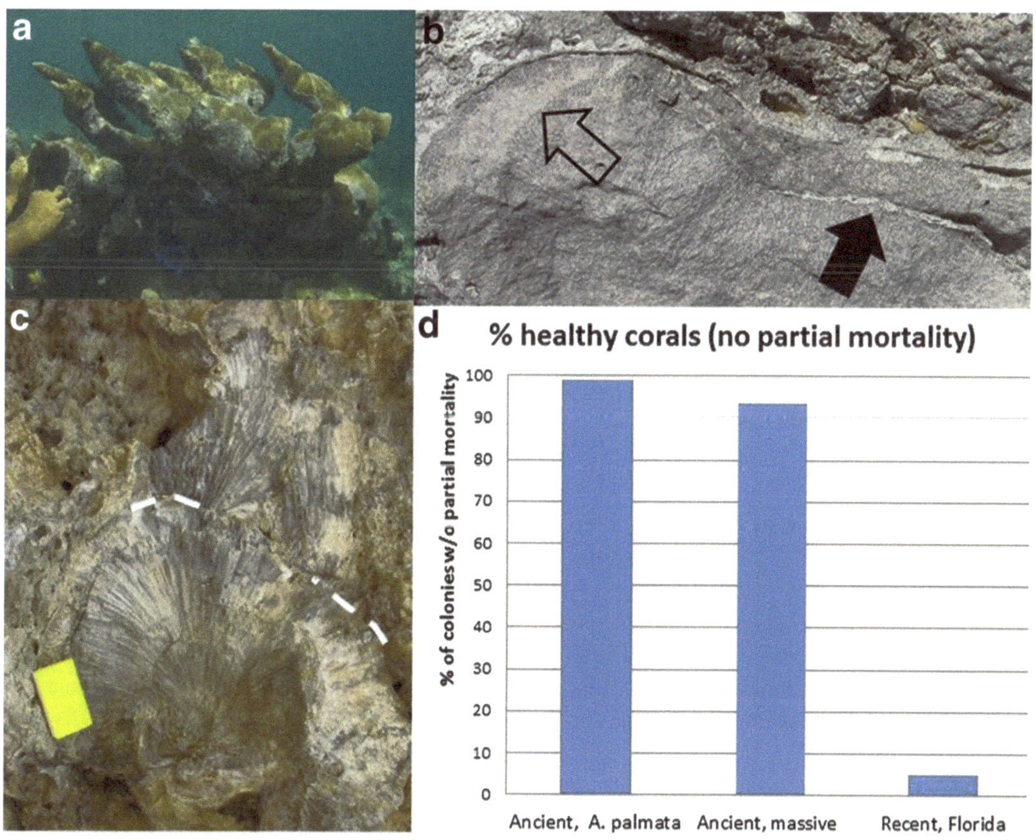

Fig. 10.11 Health of coral colonies in Pleistocene reefs as indicated by partial mortality. (**a**) Dead skeleton of *A. palmata* is resheeted by new recruits of *A. palmata* that will preserve an encruster/bioerosion horizon within the skeleton. (**b**) Cross section of a Pleistocene *A. palmata* shows re-growth of healthy colony (*hollow arrow*) over dead portion of same colony (*black arrow*). (**c**) 1 m *Colpophyllia* underwent partial mortality (*white dashes*) and regrowth. (**d**) Data from many colonies in the Pleistocene of Curacao suggest very low rates (1–7 %) of partial mortality and regrowth or resheeting in the past (Data from Lescinsky 2012). By comparison, Ginsburg et al. (2001) found that 95 % large modern *Montastraea* colonies in Florida have partial mortality of at least 10 % of their surface

system had no identified hurricane deposits, but natural disturbances resulted from volcanic eruptions and debris flows at a frequency of about 1500 years. Disturbances were characterized by thick overgrowths of coralline algae followed by a rapid return to a coral-rich community. The recurrence of similar corals and foraminiferans suggests overall high reef resilience.

In the absence of disturbance horizons, another measure of hurricane disruption in fossil reefs is the percent of larger coral heads that are preserved in life position (Meyer et al. 2003). Hurricane disturbance varies geographically and differences in hurricane frequency were easily recognizable in Caribbean Pleistocene reefs. Curacao, with few hurricanes, had 93 % of coral colonies upright, while Inagua in the Bahamas, with intermediate storm frequency had 79 % upright, and San Salvador, Bahamas, with the highest storm frequency, had only 38 % of Pleistocene colonies in life position.

Another disturbance proxy that has been used in living reefs is the relative abundance of large and small coral colonies (Bak and Meesters 1998; Smith et al. 2005; Cooper et al. 2009). Larger average coral size suggests longer intervals between major disturbances, and the presence of abundant small age classes of corals may represent recruitment/resilience following a disturbance (Crabbe 2009). Alternatively, few small corals on a reef may suggest low recruitment due to chronic disturbance such as frequent flood plumes (Smith et al. 2005) or anthropogenic pollution (Meesters et al. 2001). Coral-population structure as a proxy for disturbance should also be widely applicable to mineral reefs, although to date, few if any mineral reef studies have applied this approach.

10.9.10 Stability of Reef Communities

Questions of the stability of reef communities over time intervals longer than the last few decades can only be addressed with the temporal data preserved in mineral reefs. Ecologists have long noted recurrent patterns in coral distribution that are related to depth and wave energy (e.g., Goreau 1959; Adey and Burke 1977; Geister 1977), and paleo-ecologists have found similar zonation in Holocene and Pleistocene mineral reefs (Mesolella 1967; Geister 1981). More recently, data from corals exposed on successive terraces have been used to examine coral-community composition through time at particular sites and between sites at various scales (e.g., Jackson 1992; Pandolfi 1996; Pandolfi and Jackson 2001, 2006). Although successive reef communities represent "new" reefs formed during successive sea-level cycles, the reefs generally had a very stable composition with the same few species dominating.

Recurring coral assemblages undoubtedly reflect the physical requirements of the constituent species (see Chap. 7), but they may also reflect biological processes that maintain a coherent community. Pandolfi and Jackson (2006) review six categories of biological mechanisms that might structure communities at broad temporal and spatial scales. They suggest that short-term ecological processes such as recruitment and dispersal limitations and ecological saturation are less important to the maintenance of community stability over the longer term and that their data are consistent with structuring from biological interactions and evolutionary adaptations as well as the possible interplay of dispersal and competition (i.e., metapopulation dynamics) at longer time scales.

The pattern of Pleistocene community stability is corroborated in Holocene reefs of Papua New Guinea. Both coral (Pandolfi 1996) and foraminifera communities (Reymond et al. 2011) are remarkably stable through most of the 2200-year record despite episodic volcanic disturbances. Reef-community composition only began to shift high in the section as sea-level rise slowed and water depth decreased as the reef had an opportunity to "catch up".

10.9.11 Speciation and Evolution

Fossil corals provide an important morphological record of coral evolution, though limitations to the utility of this record have been identified by recent molecular phylogenetic studies. For example, traditional morphological analyses have failed to discriminate between many sibling species (e.g., Knowlton and Jackson 1994) and similar looking, but not closely related Atlantic and Pacific coral taxa (Fukami et al. 2004). Nevertheless, recent morphological analyses based primarily on microstructural characters of the skeleton still permit the recognition of formerly cryptic species and their application to the fossil record (Budd and Johnson 1996; Pandolfi and Budd 2008).

Insights gained from the fossil record of corals include (a) the overall pattern and timing of major events in the evolution of scleractinians (see reviews in Budd et al. 2010; Simpson et al. 2011), (b) the identification of reefs as the site of origination for many evolutionary novelties (e.g., corals: Kiessling et al. 2010; molluscs: Johnson et al. 2007; Vermeij 2012) and (c) the low turnover of coral species since the Pliocene (Johnson et al. 1995). Within the well-studied *Montastraea* species complex, fossil specimens have been used to document the competitive release of morphology following Pleistocene extinction (Pandolfi et al. 2002) and the geographic distribution of novel traits in space and time (Budd and Pandolfi 2010).

The fossil record has also been critical in highlighting the importance of extinction in shaping modern coral diversity.

Although the Pacific currently is much more species rich than the Atlantic, the Caribbean fauna of the past held many of the genera that we take to be characteristic of the Indo-Pacific today. For example, *Isopora*, *Goniopora*, *Stylophora*, *Galaxea*, *Leptoseris*, *Pavona*, *Pocillopora* and *Psammocora* were all found in the Caribbean five million years ago (Budd and Wallace 2008), suggesting that the modern discrepancy in diversity between the Pacific and Atlantic results more from extinction than origination patterns.

Extinction during the Phanerozoic has also been examined to suggest how modern coral species may react to ocean acidification (Kiessling and Simpson 2011). Best studied are the reefs of the Paleocene/Eocene boundary (55 mya) where a dramatic increase in paleotemperature has been documented. Rising sea-level, warming and acidity, possibly caused by the release of methane hydrates correlate strongly with the loss of coral/algal reefs and their replacement by large foraminifera across Africa and Europe (Scheibner and Speijer 2008). Interestingly, the loss of the extensive Paleocene reef tract may have occurred without a significant drop in coral diversity (Zamagni et al. 2012). If this is true, it suggests that, while modern ocean acidification might decrease calcification and lead to the loss of reef ecosystems, it may not lead directly to coral extinction because corals could continue to live in non-reef assemblages.

10.9.12 Reef Accretion: Coral Growth and Bioerosion

Dated horizons within mineral-reef cores provide the necessary data to determine reef-accretion rates. One principle finding of these studies is that the physical process of reef accretion is largely decoupled from the biological process of coral growth (see Chap. 6). For example, in the Caribbean, deeper water reef zones have slower growing corals but accrete at the same pace as shallower reef zones comprised largely of fast-growing branching corals (Gischler 2008; Hubbard 2009). Many fringing reefs on the Great Barrier Reef have no net accretion, despite high coral cover (Smithers et al. 2006). Accretion and coral-growth rates can actually be inversely correlated. For example, on some polluted Indonesian reefs rapidly feeding corals have high growth rates, yet the reef as whole has negative net accretion due to low coral cover and high bioerosion (Edinger et al. 2000).

Reef accretion rates are also of interest in light of recent anthropogenic sea-level rise (Hubbard 2011, 2014; Hubbard et al. 2014). During the Holocene, typical accretion rates were about half what they were originally assumed (3.5–5 m/ky vs. 10–14 m/ky) and many extant Caribbean reefs are accreting at an insufficient rate to keep pace with accelerating sea-level rise (Hubbard 2009; Chap. 6).

Coral-growth rates are easily calculated from coral cores by identifying annual density bands, which are preserved much like tree rings (Knutson et al. 1972). Growth rates can then be correlated with a variety of environmental and paleoenvironmental proxies to provide information on factors that affect growth rates such as depth, temperature, turbidity, location along an inshore offshore gradient, and others (reviewed in Lough and Cooper 2011). In addition, large coral heads provide centuries of data (up to 400 years: Dunbar et al. 1994) that can be used to relate growth rates to long-term climate cycles (Bessat and Buigues 2001), pollution (Hudson et al. 1994), and recent climate change (De'ath et al. 2009; Helmle et al. 2011; Cooper et al. 2012). These studies have found that, while pollution and recent trends in bleaching and acidification have decreased growth rate in many corals, some from high latitudes actually have increased in growth rate in response to higher sea temperatures (Cooper et al. 2012).

Similar analyses are possible using fossil corals. For example, the identification of annual growth bands in Middle Triassic scleractinian corals and their similarity to modern corals strongly suggest that the Triassic corals were zooxanthellate (Stanley and Swart 1995; Stanley and Helmle 2010) and overall, coral growth rates have not changed significantly over the last 30 my (Brachert et al. 2006; Johnson and Perez 2006). In a more modern example, growth rates of living corals in Sulawesi are similar to Holocene fossil corals preserved in nearby terraces (Crabbe et al. 2006).

Mineral reefs are also well suited to preserve evidence of bioerosion, which proceeds via grazing organisms that scrape outer surfaces and borers that excavate cavities within the substrate. Geological studies have focused almost exclusively on infaunal bioerosion (e.g., Pleydell and Jones 1988; Palmer and Plewes 1993; Perry 1996, 2000) because borings are preserved as discrete voids and are easy to quantify. Studies of fossil reefs (e.g., Perry 2000) suggest that borer type and intensity varies with environment as has been found in modern reefs (Risk et al. 1995; Holmes et al. 2000).

Many modern experiments have suggested that bioerosion by grazing may be an order of magnitude greater than infaunal bioerosion in the Pacific (Kiene and Hutchings 1994; Chazottes et al. 1995; Pari et al. 1998). However, long-term bioerosion studies in the Caribbean suggest this may not be a general pattern (Lescinsky et al. 2008). Quantifying bioerosion by grazing in mineral reefs is difficult because it involves loss of the skeletal surface. However, one technique that would be applicable would be noting the removal of annual growth increments between two parts of a single coral colony (Hudson 1977). Fossil corals with irregular, "pitted" surfaces probably reflect the scars of grazing bioerosion but it is difficult to derive rates from these. Conversely, the preservation of detailed corallite topography (i.e., the elevated collars of *Acropora*) provides evidence for low grazing. Applying these principles across three zones in the Pleistocene reefs of Curacao, Lescinsky (2008) determined that grazing bioerosion (irregular pitting) occurred in <20 % of corals with <10 % of the surface of those coral affected and the Holocene reefs of the Dominican Republic had even lower rates of grazing bioerosion across four reef zones (Lescinsky et al. 2012). Boring bioerosion was also low in both studies suggesting a low incidence of grazing and boring bioerosion in ancient reefs with high coral cover.

10.10 Coral Skeletons as Archives of Environmental Information

Corals also contain paleoenvironmental information in the form of isotopes locked within their $CaCO_3$ skeleton, in the organic matrix that is preserved in the skeleton, and in the many inclusions of inorganic and organic molecules that are incorporated into the skeleton in small amounts. These data can be derived from living corals to yield a precise temporal record during their lifetime. Longer but less precise records have been derived from reef cores and fossil outcrops to hindcast environmental conditions for thousands of years.

10.10.1 Physical Conditions of Past Reef Environments

Most coral cores are taken from large living corals and are primarily used to reconstruct temperature and other physical variables over the last few centuries (see reviews by Grottoli and Eakin 2007; Jones et al. 2009). The continuous climate record is generally limited by the age of the oldest extant coral heads, around 400 years (Dunbar et al. 1994), though in a few cases a longer record has been assembled by including overlapping cores from radiocarbon dated corals sampled from old mega-storm deposits on adjacent beaches or in lagoons (so called "fossil corals" of Cobb et al. 2003).

Geochemical markers retrieved from coral skeletons can, with care, be used as proxies for a variety of physical parameters including temperature (Sr/Ca, δO^{18}), salinity (discrepancies between Sr/Ca and δO^{18}: Correge 2006), cloud cover and turbidity (δC^{13}: Grottoli and Wellington 1999), pH (δB^{11}: Hönisch et al. 2004), upwelling (Mn/Ca, Ba/Ca: Matthews et al. 2008), nutrient levels (P/Ca: LaVigne et al. 2010; δN^{1}: Yamazaki et al. 2011), and rain and river outflow (luminescence: Isdale 1984; Susic et al. 1991; Lough 2011). In addition, markers of anthropogenic pollution are also recorded in the geochemical record of corals including petroleum (Pb/Ca: Fallon et al. 2002), heavy metals (Al-Rousan et al. 2007), and nitrogen fertilizers (Marion et al. 2005).

While these data are crucial in providing baseline information for understanding climate and land-use change in reefs over the last decades to centuries, many of the proxies can also be employed examine geologic-scale changes, and to understand the ecology of ancient reefs. For example, estimates of seasonal variations in temperature and solar radiation in 300+ kya Pacific corals suggest seasonal ranges that are similar to those experienced by corals today (Kilbourne et al. 2004; Ayling et al. 2006). Temperature variation within ancient colonies has also been used to study ancient El Niños. Cores of Holocene and Pleistocene *Porites* from the uplifted terraces of Papua New Guinea preserved evidence for ancient El Niños, of lower frequency and intensity than those of the last few decades (Tudhope et al. 2001).

Delineating the physical parameters of ancient environments allows us to use the past as a backdrop to present and future reef change. For example, the Pliocene tropics were likely similar in important ways to environments anticipated in twenty-first century seas (Haywood et al. 2009). Future reefs may begin to resemble Pliocene reefs by becoming increasingly patchy, having lowered topographic relief, and favoring corals with large polyps as occurred under the more mesophotic conditions of the Pliocene (Klaus et al. 2011). Also, the coral genera lost from the Atlantic during the late Pliocene extinction are the same as many of the modern Pacific taxa that have been identified as most susceptible to local stresses (van Woesik et al. 2012). These findings validate aspects of current models of coral vulnerability and also suggest that broad geographic range and high abundance typical of extirpated Caribbean genera are not necessarily sufficient to protect corals from future extinction events.

10.10.2 Biological Signals in Reef Cores

Reef cores may also preserve information that can be used to reconstruct the coral's original metabolic and biologic conditions. For example, both coral carbonate (Stanley and Swart 1995) and the organic matrix preserved within the carbonate (Muscatine et al. 2005) may retain isotopic signatures that can be used to separate zooxanthellate and azooxanthellate corals. These isotopic signatures are preserved in older corals and have been used to suggest that scleractinian corals, as far back as in the Triassic, had adopted zooxanthellae (Stanley and Swart 1995; Muscatine et al. 2005; also see Chaps. 3, 8 and 9).

Of particular interest to many authors is whether or not evidence of stress (e.g., bleaching) is preserved in ancient corals. One approach has used the drop in δ^{13} C associated with bleaching to identify intervals of high and low level stress in the past (Porter et al. 1989; Suzuki et al. 2003). Slowed or halted growth associated with bleaching is also visible in coral skeletons as dense "stress bands" and these should be readily identifiable in the past. For example, Yu et al. (2010) identified stress bands in Holocene (6000+ y) corals that suggest several intervals where growth was slowed or halted. In this case, the growth hiatuses were unlikely to be caused by a widespread bleaching because horizons of different colonies occurred during different seasons, based on temperature proxies.

Another approach to identifying stress intervals in cores is looking for bands of intense microboring from endolithic algae (Carilli et al. 2010; Hartmann et al. 2010). The alga *Ostreobium* lives in low numbers within the skeleton below the living coral polyp where it excavates chambers. In response to the elevated light levels that it occasionally experiences when the overlying coral polyp expels some of its symbionts, the algae bloom and form discrete "green bands" in the coral skeleton. Severe bleaching does not appear to cause blooms due to photo-inhibition, and so green bands are suggested as a proxy for the partial loss of zooxanthellae ("paling" of Carilli et al. 2010). This proxy provides a potential, though still unutilized tool for investigating zooxanthellae densities in ancient corals.

A final approach to estimating coral stress from cores is the examination of organic matter that is preserved in the coral skeleton. Ingalls et al. (2003) showed that a variety of organic compounds, including some that might be related to coral stress are preserved within coral skeletons for at least centuries, providing the potential to identify new proxies for ancient coral stress. In a similar approach, the ratio of aspartic acid to other amino acids preserved in the skeleton may vary directly with coral growth/biomineralization and may provide a proxy for disturbance intervals that is independent of environmental variables (Gupta et al. 2006, 2007).

10.11 Summary and Conclusions

To extract ecological and environmental information from both living and mineral reefs, a number of simplifications are routinely made. For example, in monitoring living reefs, the complex topography of the reef surface is generally reduced to a plane for analysis, and even though reef communities extend to great depth, transects are typically established in shallow water. Reef communities are famous for their diversity, yet only a subset of this biodiversity is usually monitored (i.e., sessile benthic cover). Many of the same simplifying assumptions that are used for studying living reefs are inherent in the mineral-reef record, suggesting that, at least in principle, mineral and living reefs are broadly comparable for certain types of data.

Although the ecological information preserved in mineral reefs may pale in comparison to that observable in living reefs, when compared to other paleontological assemblages, mineral reefs preserve some of the most detailed and accurate paleoecological information. In reef environments, the principle landscape architects have always had relatively large and robust mineralized skeletons that are difficult to overturn, move or destroy. Life-position and abundance data are, therefore, frequently preserved and detailed quantitative reconstructions are possible.

A principal limiting factor for reef preservation is the shallow, high-energy environments that reefs characteristically occupy. Large storms in the tropics may periodically turn most corals into rubble, potentially destroying the ecological record of the reef community. On the other hand, reefs outside of storm tracts, such as equatorial reefs, are rarely impacted by such extreme disturbances and may retain primarily intact coral framework. In addition, the complex three-dimensional topography of the reef slows down ("baffles") sediment and continuously traps it within the structure, providing a source of burial even within highly turbulent zones.

Not all reef deposits meet the detailed criteria that taphonomists have developed to distinguish between reworked assemblages (i.e., storm accumulations) and intact ancient communities, but where mineral reefs do meet these criteria, they can yield tremendously detailed ecological information on many questions. While it is doubtful that appropriate fossil deposits will ever be found to yield, for example, quantitative data on reef fish size and abundance in Pleistocene reefs, similar questions about coral abundance, distribution, health and resilience have all been carefully addressed using select fossil deposits. Only time will tell what additional types of ecological information can be unlocked from mineral reefs and used as comparative information to help us better understand modern and future reefs.

References

Abbey E, Webster JM, Beaman RJ (2011) Geomorphology of submerged reefs on the shelf edge of the Great Barrier Reef: The influence of oscillating Pleistocene sea-levels. Mar Geo 288:61–78

Abelson A, Galil BS, Loya Y (1991) Skeletal modifications in stony corals caused by indwelling crabs: hydrodynamical advantages for crab feeding. Symbiosis 10:233–248

Abram NJ, Webster JM, Davies PJ, Dullo WC (2001) Biological response of coral reefs to sea surface temperature variation: evidence from the raised Holocene reefs of Kikai-jima (Ryukyu Islands, Japan) Coral Reefs 20:221–234

Accordi G, Brilli M, Carbone F, Voltaggio M (2010) The raised coral reef complex of the Kenyan coast: Tridacna gigas U-series dates and geological implications. J Afr Earth Sci 58:97–114

Ackerman JL, Bellwood DR (2000) Reef fish assemblages: a re-evaluation using enclosed rotenone stations. Mar Ecol Prog Ser 206:227–237

Adey W, Macintyre IG, Stuckenrath R (1977) Relict barrier reef system off St. Croix: its implications with respect to Late Cenozoic coral reef development in the western Atlantic. Proc 3rd Int Coral Reef Symp 2:15–21

Adey WH, Burke RB (1977) Holocene bioherms of Lesser Antilles - geologic control of development. In: Frost SH, Weiss MP, Saunders JB (eds) Reefs and related carbonates. AAPG Studies in Geology 4: 67–81

Al-Rousan SA, Al-Shioul RN, Al-Horani FA, Abu-Hilal AH (2007) Heavy metal contents in growth bands of Porites corals: record of anthropogenic and human developments from the Jordanian Gulf of Aqaba. Mar Poll Bull 54:1912–1922

Aronson RB, Precht WF, Macintyre IG (1998) Extrinsic control of species replacement on a Holocene reef in Belize: the role of coral disease. Coral Reefs 17:223–230

Ayling BF, McCulloch MT, Gagan MK, Stirling CH, Andersen MB, Blake SG (2006) Sr/Ca and delta O-18 seasonality in a Porites coral from the MIS 9 (339–303 ka) interglacial. Earth Planet Sc Lett 248:462–475

Azmy K, Edinger E, Lundberg J, Diegor W, (2010) Sea level and paleotemperature records from a mid-Holocene reef on the north coast of Java, Indonesia. Int J Earth Sci 99:231–244

Bak RPM, Meesters EH (1998) Coral population structure: the hidden information of colony size-frequency distributions. Mar Ecol Prog Ser 162:301–306

Behrensmeyer AK, Kidwell SM, Gastaldo RA (2000) Taphonomy and paleobiology. Paleobiology 26, Supplement 103–147

Bellwood DR (1996) The Eocene fishes of Monte Bolca: the earliest coral reef fish assemblage. Coral Reefs 15:11–19

Bellwood DR (1998) What are reef fishes? – Comment on the report by D. R. Robertson: do coral-reef fish faunas have a distinctive taxonomic structure? Coral Reefs 17:187–189

Bellwood DR (2003) Origins and escalation of herbivory in fishes: a functional perspective. Paleobiology 29:71–83

Bergsma GS, Martinez CM (2011) Mutualist-induced morphological changes enhance growth and survival of corals. Mar Biol 158:2267–2277

Bessat F, Buigues D (2001) Two centuries of variation in coral growth in a massive *Porites* colony from Moorea (French Polynesia): a response of ocean–atmosphere variability from south central Pacific. Palaeogeogr Palaeoclimatol Palaeoecol 175:381–392

Blanchon P, Blakeway D (2003) Are catch-up reefs an artefact of coring? Sedimentology 50:1271–1282

Blanchon P, Jones B (1997) Hurricane control on shelf edge reef architecture around Grand Cayman. Sedimentology 44:479–506

Blanchon P, Jones B, Kalbfleisch W (1997) Anatomy of a fringing reef around Grand Cayman: storm rubble, not coral framework. J Sed Res 67:1–16

Bongaerts P T, Ridgway T, Sampayo EM, Hoegh-Guldberg O (2010) Assessing the 'deep reef refugia' hypothesis: focus on Caribbean reefs. Coral Reefs 29:309–327

Bonito V, Grober-Dunsmore R (2005) Resheeting of relict *Acropora palmata* framework may promote fast growth, but does it compromise the structural integrity of the colony? Coral Reefs 25:46

Boucot AJ, Poinar GO (2010) Fossil Behavior Compendium. CRC Press, Boca Raton

Brachert TC; Reuter M, Kroeger KF, Lough JM (2006) Coral growth bands: a new and easy to use paleothermometer in paleoenvironment analysis and paleoceanography (late Miocene, Greece) Paleoceanography 21:1–12

Brett CE, Baird GC (1986) Comparative taphonomy: a key to paleoenvironmental interpretation based on fossil preservation. Palaios 1:207–227

Brett CE, Seilacher A (1991) Fossil lagerstatten: a taphonomic consequence of event sedimentation. In: Einsele E (ed) Cycles and events in stratigraphy. Springer, Berlin pp 283–297

Bruno JF, Selig ER (2007) Regional decline of coral cover in the Indo-Pacific: timing, extent, and subregional comparisons. PLoS ONE 2 (8): e711. doi:10.1371/journal.pone.0000711

Bruthans J, Filippi M, Gerl M, Zare M, Melkova J, Pazdur A, Bosak P (2006) Holocene marine terraces on two salt diapirs in the Persian Gulf, Iran: age, depositional history and uplift rates. J Quat Sci 21:843–857

Budd AF, Johnson KG (1996) Recognizing species of Late Cenozoic Scleractinia and their evolutionary patterns. Paleontol Soc Papers 1:59–79

Budd AF, Pandolfi JM (2004) Overlapping species boundaries within the *Montastraea "annularis"* species complex in the late Pleistocene of the Bahamas Islands. Paleobiology 30:396–425

Budd AF, Pandolfi JM (2010) Evolutionary novelty is concentrated at the edge of coral species distributions. Science 328:1558–1561

Budd AF, Romano SL, Smith ND, Barbeitos MS (2010) Rethinking the phylogeny of Scleractinian corals: a review of morphological and molecular data. Integr Comp Biol 50: 411–427

Budd AF, Wallace CC (2008) First record of the Indo-Pacific reef coral genus *Isopora* in the Caribbean region: two new species from the Neogene of Curacao, Netherlands Antilles. Palaeontology 51:1387–1401

Cabioch G (2011) Emerged reefs. In Hopley D (ed) Encyclopedia of modern coral reefs: structure, form and process. Springer, New York pp 373–380

Cabioch G, Montaggioni L, Frank N, Seard C, Sallé E, Payri C, Pelletier B, Paterne M (2008) Successive reef depositional events along the Marquesas foreslopes (French Polynesia) since 26 ka. Mar Geol 254:18–34

Camoin GF, Seard C, Deschamps P, Webster JM, Abbey E, Braga JC, Iryu Y, Durand N, Bard E, Hamelin B, Yokoyama Y, Thomas AL, Henderson GM, Dussouillez P (2011) Reef response to sea-level and environmental change during the last deglaciation: integrated ocean drilling program expedition 310, Tahiti sea level. Geology 40: 643–646

Carilli JE, Godfrey J, Norris RD, Sandin SA, Smith JE (2010) Periodic endolithic algal blooms in *Montastraea faveolata* corals may represent periods of low-level stress. Bull Mar Sci 86:709–718

Cerridwen SA, Jones B (1991) Distribution of bivalves and gastropods in the Pleistocene Ironshore Formation, Grand Cayman. Caribbean J Sci 27:97–116

Chabanet P, Ralambondrainy H, Amanieu M, Faure G, Galzin R (1997) Relationships between coral reef substrata and fish. Coral Reefs 16:93–102

Chappell J (1974) Geology of coral terraces, Huon Peninsula, New Guinea: a study of Quaternary tectonic movements and sea-level changes. Geol Soc Amer Bull 85:553–570

Chappell J, Veeh HH (1978) Late Quaternary tectonic movements and sea-level changes at Timor and Atauro Island. Geol Soc Amer Bull 89:356–367

Chazottes V, LeCampion-Alsumard T, Peyrot-Clausade M (1995) Bioerosion rates on coral reefs: interactions between macroborers, microborers, and grazers (Moorea, French Polynesia). Palaeogeogr Palaeoclimatol Palaeoecol 113:189–198

Cherns L, Wright VP (2000) Missing molluscs as evidence of large-scale, early skeletal aragonite dissolution in a Silurian sea. Geology 28:791–794

Cobb KM, Charles CD, Cheng H, Edwards RL (2003) El Nino/Southern Oscillation and tropical Pacific climate during the last millennium. Nature 424:271–276

Connell JH (1978) Diversity in tropical rain forests and coral reefs. Science 199:1302–1310

Cooper TF, Gilmour JP, Fabricius KE (2009) Bioindicators of changes in water quality on coral reefs: review and recommendations for monitoring programmes. Coral Reefs 28:589–606

Cooper TF, O'Leary RA, Lough JM (2012) Growth of Western Australian corals in the Anthropocene. Science 335:593–596

Correge T (2006) Sea surface temperature and salinity reconstruction from coral geochemical tracers. Palaeogeogr Palaeoclimatol Palaeocecol 232:408–428

Crabbe MJC (2009). Scleractinian coral population size structures and growth rates indicate coral resilience on the fringing reefs of North Jamaica. Mar Environ Res 67:189–198

Crabbe MJC, Wilson MEJ, Smith DJ (2006) Quaternary corals from reefs in the Wakatobi Marine National Park, SE Sulawesi, Indonesia, show similar growth rates to modern corals from the same area. J Quat Sci 21:803–809

Crame JA (1986) Late Pleistocene molluscan assemblages from the coral reefs of the Kenya coast. Coral Reefs 4:183–196

Darwin CR (1842) The structure and distribution of coral reefs. Smith Elder and Co, London

De'ath G, Lough JM, Fabricius KE (2009) Declining coral calcification on the Great Barrier Reef. Science 323:116–119

Denny MW (2006) Ocean waves, nearshore ecology, and natural selection. Aqua Ecol 40:439–461

DeVantier LM, Done TJ (2007) Inferring past outbreaks of the crown-of-thorns seastar from scar patterns on coral heads. In: Aronson RB (ed) Geological approaches to coral reef ecology. Springer, New York pp 85–125

Dickinson WR (2009) Pacific atoll living: how long already and until when? GSA Today 19:4–10

Dodge RE, Fairbanks RG, Benninger LK, Maurrasse F (1983) Pleistocene sea levels from raised coral reefs of Haiti. Science 219:1423–1425

Donovan SK (2005) The fossil record of *Diadema* in the Caribbean. Coral Reefs 24:603–605

Dravis JJ (1996) Rapidity of freshwater calcite cementation: implications for carbonate diagenesis and sequence stratigraphy. Sediment Geol 107:1–10

Dullo W-C (1990) Facies, fossil record, and age of Pleistocene reefs from the Red Sea (Saudi Arabia). Facies 22:1–46

Dumas P, Bertaud A, Peignon C, Léopold M, Pelletier D (2009) A "quick and clean" photographic method for the description of coral reef habitats. J Exp Mar Biol Ecol 368:161–168

Dunbar R, Wellington GM, Colgan MW, Glynn PW (1994) Eastern Pacific sea surface temperature since 1600 AD: the delta "SUP 18" O record of climate variability in Galapagos corals. Paleocean 9:291–315

Dynowski JF (2012) Echinoderm remains in shallow-water carbonates at Fernandez Bay, San Salvador Island, Bahamas. Palaios 27:183–191

Edinger EN, Burr GS, Pandolfi JM, Ortiz JC (2007) Age accuracy and resolution of Quaternary corals used as proxies for sea level. Earth Planet Sci Letters 253:37–49

Edinger EN, Copper P, Risk MJ, Atmoyo W (2002) Oceanography and reefs of recent and Paleozoic tropical epeiric seas. Facies 47:127–150

Edinger EN, Limmon GV, Jompa J, Widjatmoko W, Heikoop JM, Risk MJ (2000) Normal coral growth rates on dying reefs: are coral growth rates good indicators of reef health? Mar Pollut Bull 5:404–425

Edinger EN, Pandolfi JM, Kelley RA (2001) Community structure of Quaternary coral reefs compared with Recent life and death assemblages. Paleobiol 27:669–694

El-Sorogy, AS (2008) Contributions to the Pleistocene coral reefs of the Red Sea coast, Egypt. Arab Gulf J Sci Res 26:63–85

Fabricius KE, De'ath G, Puotinen ML, Done T, Cooper TF, Burgess SC (2008) Disturbance gradients on inshore and offshore coral reefs caused by a severe tropical cyclone. Limnol Oceanogr 53:690–704

Fagerstrom JA (1987) The evolution of reef communities. John Wiley & Sons, New York

Fagerstrom JA (1988) A structural model for reef communities. Palaios 3:217–220

Fairbanks RG (1989) A 17,000-year glacio-eustatic sea level record: influence of glacial melting rates on the Younger Dryas event and deep-ocean circulation. Nature 342:637–642

Fallon SJ, White JC, McCulloch MT (2002) *Porites* corals as recorders of mining and environmental impacts: Misima Island, Papua New Guinea. Geochim Cosmochim Acta 66:45–62

Ferguson CA, Miller AI (2007) A sea change in Smuggler's Cove? Detection of decadal-scale compositional transitions in the subfossil record. Paleogeo Paleoclim 254: 418–429

Feuillet N, Tapponnier P, Manighetti I (2004) Differential uplift and tilt of Pleistocene reef platforms and Quaternary slip rate on the Morne-Piton normal fault (Guadeloupe, French West Indies) J Geophysical R 109:B02404

Fletcher CH, Bochicchio C, Conger CL, Engels MS, Feirstein EJ, Frazer N, Glenn CR, Grigg RW, Grossman EE, Harney JN, Isoun E, Murray-Wallace CV, Rooney JJ, Rubin KH, Sherman CE, Vitousek S (2008) Geology of Hawaii Reefs. In Riegl BM, Dodge RE (eds) Coral reefs of the USA. Springer, New York pp 435–488

Ford MR, Kench PS (2012) The durability of bioclastic sediments and implications for coral reef deposit formation. Sedimentology 59:830–842

Fukami H, Budd AF, Paulay G, Sole-Cava A, Chen CA, Iwao K, Knowlton N (2004) Conventional taxonomy obscures deep divergence between Pacific and Atlantic corals. Nature 427: 832–835

Garberoglio RM, Lazo DG (2011) Post-mortem and symbiotic sabellid and serpulid-coral associations from the Lower Cretaceous of Argentina. Rev Bras Paleontol 14:215–228

Garcia Massini JL; Jacobs BF (2011) The effects of volcanism on Oligocene-age plant communities from the Ethiopian Plateau, and implications for vegetational resilience in a heterogeneous landscape Rev Palaeobot Palynol 164:211–222

Gardiner L (2001) Stability of Late Pleistocene reef mollusks from San Salvador Island, Bahamas: Palaios 16:372–386

Gardner TA, Cote IM, Gill JA, Grant A, Watkinson AR (2003) Long-term region-wide declines in Caribbean corals. Science 301:958–960

Gardner TA, Coté IM, Gill JA, Grant A, Watkinson AR (2005) Hurricanes and Caribbean coral reefs: Impacts, recovery patterns, and role in long-term decline. Ecology 86:174–184

Garzon-Ferreira J, Zea S, Diaz JM (2005) Incidence of partial mortality and other health indicators in hard-coral communities of four southwestern Caribbean atolls. Bull Mar Sci 76:105–122

Geister J (1977) The influence of wave exposure on the ecological zonation of Caribbean coral reefs. Proc. 3rd Int Coral Reef Symp 1:23–29

Geister J (1981) Calm-water reefs and rough-water reefs of the Caribbean Pleistocene. Acta Palaeo Polonica 25:541–556

Geister J (1982) Pleistocene reef terraces and coral environments at Santo Domingo and near Boca Chica, southern coast of the Dominican Republic. Trans Ninth Caribbean Geol Conf 2:689–703

Ginsburg RN, Gischler E, Kiene WE (2001) Partial mortality of massive reef-building corals: an index of patch reef condition, Florida Reef Tract. Bull Mar Sci 69:1149–1173

Gischler E (2008) Accretion patterns in Holocene tropical coral reefs: do massive coral reefs in deeper water with slowly growing corals accrete faster than shallower branched coral reefs with rapidly growing corals? Int J Earth Sci (Geol Rundsch) 97:851–859

Gischler E (2010) Possible fossil echinoid mass mortality detected in Holocene lagoons, Belize. Palaios 25:260–268
Glenn-Sullivan EC, Evans I (2001) The effects of time-averaging and taphonomy on the identification of reefal sub-environments using larger foraminifera: Apo Reef, Mindoro, Philippines. Palaios 16: 399–408
Goatley CHR, Bellwood DR (2011) The roles of dimensionality, canopies and complexity in ecosystem monitoring. PLoS ONE 6 (11): e27307
Goatley CHR, Bellwood DR, Bellwood O (2010) Fishes on coral reefs: changing roles over the past 240 million years. Paleobiol 36:415–427
Goreau, TF (1959) The ecology of Jamaica reefs. I. Species composition and zonation. Ecology 70: 275–279
Gratwicke B, Speight MR (2005) The relationship between fish species richness, abundance and habitat complexity in a range of shallow tropical marine habitats. J Fish Biol 66:650–667
Greb SF, Hendricks RT, Chesnut DR (1993) Fossil beds of the falls of the Ohio, Kentucky Geological Survey, Series XI, Special Publication 19
Greenstein BJ (1989) Mass mortality of the West-Indian echinoid *Diadema antillarum* (Echinodermata: Echinoidea): A natural experiment in taphonomy: Palaios 4: 487–492
Greenstein BJ (1991) An integrated study of echinoid taphonomy: predictions for the fossil record of four echinoid families. Palaios 6:519–540
Greenstein BJ (2006) Taphonomy: detecting critical events in fossil reef-coral assemblages. In: Aronson RB (ed) Geological approaches to coral reef ecology. Springer, New York pp 31–60
Greenstein BJ, Harris LA, Curran HA (1998) Comparison of recent coral life and death assemblages to Pleistocene reef communities: implications for rapid faunal replacement observed on modern reefs. Carbonate Evaporite 13:23–31
Greenstein BJ, Moffat HA (1996) Comparative taphonomy of modern and Pleistocene corals, San Salvador, Bahamas. Palaios 96:57–63
Greenstein BJ, Pandolfi JM (1997) Preservation of community structure in modern reef coral life and death assemblages of the Florida Keys: implications for the Quaternary record of coral reefs. Bull Mar Sci 19:39–59
Greenstein, BJ, Pandolfi JM (2003) Taphonomic alteration of reef corals: effects of reef environment and coral growth form. II: The Florida Keys. Palaios 18:495–509
Greenstein BJ, Pandolfi JM (2008) Escaping the heat: range shifts of reef coral taxa in coastal, Western Australia. Global Change Biology 14:1–16
Greenstein BJ, Pandolfi JM, Moran PJ (1995) Taphonomy of crown-of-thorn starfish: implications for recognizing ancient population outbreaks. Coral Reefs 14:91–97
Greer L, Jackson J, Curran HA, Guilderson T, Teneva T (2009) How vulnerable is *Acropora cervicornis* to environmental change? Lessons from the Early to Middle Holocene. Geology 37:263–266
Grottoli AG, Eakin CM (2007) A review of modern coral delta O-18 and delta C-14 proxy records. Earth Sci Rev 81:67–91
Grottoli AG, Wellington GM (1999) Effect of light and zooplankton on skeletal δ13C values in the eastern Pacific corals *Pavona clavus* and *Pavona gigantean*. Coral Reefs 18:29–41
Gupta LP, Suzuki A, Kawahata H (2006) Aspartic acid concentrations in coral skeletons as recorders of past disturbances of metabolic rates. Coral reefs 25:599–606
Gupta LP, Suzuki A, Kawahata H (2007) Endolithic aspartic acid as a proxy of fluctuations in coral growth. J Geophys Res-Biogeosci 112 G01001
Gvirtzman G, Kronfeld J, Buchbinder B (1992) Dated coral reefs of southern Sinai (Red Sea) and their implication to late Quaternary sea levels. Mar Geol 108:29–37
Halley RB, Hudson JH (2007) Fidelity of annual growth in *Montastraea faveolata* and the recentness of coral bleaching in Florida. In Aronson RB (ed) Geological approaches to coral reef ecology. Springer, New York pp. 161–177
Hamanaka N, Kan H, Yokoyama Y, Okamoto T, Nakashima Y, Kawana T (2012) Disturbances with hiatuses in high-latitude coral reef growth during the Holocene: correlation with millennial-scale global climate change. Global Planet Change 81:21–35
Hantoro WS, Pirazzoli PA, Jouaunic C, Faure H, Hoang CT, Radtke U, Causse C, Borel Best M, Lafont R, Bieda S Lambeck K (1994) Quaternary uplifted coral reef terraces on Alor Island, East Indonesia. Coral Reefs 13:215–223
Hartmann AC, Carilli JE, Norris RD, Charles CD, Deheynlli DD (2010) Stable isotopic records of bleaching and endolithic algae blooms in the skeleton of the boulder forming coral *Montastraea faveolata*. Coral Reefs 29:1079–1089
Haywood AM, Dowsett HJ, Valdes PJ, Lunt DJ, Francis JE, Sellwood BW (2009) Introduction. Pliocene climate, processes and problems. Roy Soc London Phil Trans Ser A 367:3–17
Helmle KP, Dodge RE, Swart PK, Gledhill DK, Eakin CM (2011) Growth rates of Florida corals from 1937 to 1996 and their response to climate change. Nature Comm. 2:215 doi:10.1038/ncomms1222
Highsmith RC (1982) Reproduction by fragmentation in corals. Mar Ecol Progress Ser 7:207–226
Hil J, Wilkinson C (2004) Methods for ecological monitoring of coral reefs: a resource for managers. Australian Institute of Marine Science (AIMS), Townsville, Australia
Hinderstein LM, Marr JCA, Martinez FA, Dowgiallo MJ, Puglise KA, Pyle RL, Zawada DG, Appeldoorn R (2010) Mesophotic coral ecosystems: characterization, ecology, and management. Coral Reefs 29:247–251
Hine AC, Halley RB, Locker SD, Jarrett BD, Jaap WC, Mallinson DJ, Ciembronowicz KT, Ogden NB, Donahue BT, Naar DF (2008) Coral reefs, present and past, on the west Florida shelf and platform margin. In: Riegl BM, Dodge RE (eds) Coral Reefs of the USA. Springer, New York pp 127–173
Holmes KE, Edinger EN, Hariyadi H, Limmon GV, Risk MJ (2000) Bioerosion of live massive corals and branching coral rubble on Indonesian coral reefs. Mar Pollut Bull 40:606–617
Hongo, C (2012) Holocene key coral species in the Northwest Pacific: indicators of reef formation and reef ecosystem responses to global climate change and anthropogenic stresses in the near future. Quaternary Sci Rev 35:82–99
Hönisch BN, Hemming G, Grottoli AG, Amat A, Hanson GN, Bijma J (2004) Assessing scleractinian corals as recorders for paleo-pH: empirical calibration and vital effects. Geochim Cosmochim Acta 68:3675–3685
Hopley D, Smithers SG, Parnell KE (2007) The geomorphology of the Great Barrier Reef. Cambridge University Press, Cambridge
Howard JD, Mayou TV, Heard RW (1977) Biogenic sedimentary structures formed by rays. J Sediment Petrol 47:339–346
Hubbard D, Gischler E, Davies P, Montaggioni L, Camoin G, Dullo WC, Storlazzi C, Field M, Fletcher C, Grossman E, Sheppard C, Lescinsky H, Fenner D, McManus J, Scheffers S (2014) Island outlook "warm and swampy". Science 345: 1461
Hubbard DK (2009) Depth-related and species-related patterns of Holocene reef accretion in the Caribbean and western Atlantic: a critical assessment of existing models. Int Assoc Sedim Spec Publ 41:1–18
Hubbard DK (2011) Reefs through the looking glass. In: Stanley G (ed) Corals and coral reefs: crises, collapse and change. Paleontological Society Papers 14:95–110
Hubbard DK (2014) Holocene accretion rates and styles for Caribbean coral reefs: lessons for the past and future. In: Harris PM (ed) Facies architecture of shelf-edge carbonates. SEPM Spec Pub 105:264–281

Hubbard, DK, Burke RB Gill IP (1986) Styles of reef accretion along a steep, shelf-edge reef, St. Croix, U.S. Virgin Islands J Sedim Petrol. 56: 848–861

Hubbard DK, Burke RB, Gill IP (1998) Where's the reef: the role of framework in the Holocene. Carbonates and Evaporites 13: 3–9

Hubbard, DK, Gill IP, Burke RB Morelock J (1997) Holocene reef backstepping – southwestern Puerto Rico Shelf. Proc 8th Int Coral Reef Sym 1: 1779–84

Hubbard DK, Miller AI, Scaturo D (1990) Production and cycling of calcium carbonate in a shelf-edge reef system (St. Croix, U.S. Virgin Islands); applications to the nature of reef systems in the fossil record. J Sedim Petrol 60:335–360

Hubbard DK, Ramirez W, Cuevas D, Erickson T, Estep A (2008) Holocene reef accretion along the north side of Bahia Enriquillo (western Dominican Republic): unique insights into patterns of reef development in response to sea-level rise. Proc 11th Inter Coral Reef Symp 1:43–47

Hubbard DK, Zankl H, Van Heerden I, Gill IP (2005) Holocene reef development along the northeastern St. Croix shelf, Buck Island, U.S. Virgin Islands. J Sed Res 75: 97–113

Hudson JH (1977) Long term bioerosion rates on a Florida Reef: a new method. Proc 3rd Inter Coral Reef Symp 1:491–497

Hudson JH (1981) Growth rates in *Montastrea annularis*: a record of environmental change in Key Largo Reef Marine Sanctuary, Florida. Bull Mar Sci 31:444–459

Hudson JH, Hanson KJ, Halley RB, Kindinger JL (1994) Environmental implications of growth-rate changes in *Montastrea annularis*, Biscayne National Park, Florida. Bull Mar Sci 54:647–669

Humblet M, Iryu Y, Nakamori T (2009) Variations in Pleistocene coral assemblages in space and time in southern and northern Central Ryukyu Islands, Japan. Mar Geol 259:1–20

Ingalls A, Lee C, Druffel ERM (2003) Preservation of organic matter in mound-forming coral skeletons. Geochim Cosmochim Acta 67:2827–2841

Inoue S, Kayanne H, Matta N, Chen WS, Ikeda Y (2011) Holocene uplifted coral reefs in Lanyu and Lutao Islands to the southeast of Taiwan. Coral Reefs 30:581–592

Isdale P (1984) Fluorescent bands in massive corals record centuries of coastal rainfall. Nature 310: 578–579

Iturralde Vinent M (1995) Sedimentary geology of western Cuba, SEPM Field Guide, 1st SEPM Congress on Sedimentary Geology, St. Pete Beach, Florida

Jackson JBC (1992) Pleistocene perspectives on coral reef community structure. Am Zool 32:719–731

Jackson JBC (1997) Reefs since Columbus. Coral Reefs 16 Suppl:S23-S32

Jackson JBC, Donovan MK, Cramer KL, Lam VV (2014) Status and trends of Caribbean coral reefs: 1970–2012. Global Coral Reef Monitoring Network, IUCN, Gland, Switzerland

Jackson JBC, Todd JA, Fortunato H, Jung P (1999) Diversity and assemblages of Neogene Caribbean mollusca of lower Central America. In: Collins LS, Coates AG (eds) A paleobiotic survey of the Caribbean faunas from the Neogene of the Isthmus of Panama. Bull Amer Paleon 357:193–230

Jeng MS, Huang HD, Dai CF, Hsiao YC, Benayahu Y (2011) Sclerite calcification and reef-building in the fleshy octocoral genus *Sinularia* (Octocorallia: Alcyonacea). Coral Reefs 30:925–933

Jimenez H, Dumas P, Bigot L, Amouroux JM, Ferraris J (2010) Taxonomic resolution needed to describe invertebrate assemblages and to detect harvesting effects on coral reef ecosystems. Mar Ecol Prog Ser 406:211–222

Johnson KG, Budd AE, Stemann TA (1995) Extinction selectivity and ecology of Neogene Caribbean corals. Paleobiol 21:52–73

Johnson KG, Perez ME (2006) Skeletal extension rates of Cenozoic Caribbean reef corals. Palaios 21:262–271

Johnson KG, Todd JA, Jackson JBC (2007) Coral reef development drives molluscan diversity increase at local and regional scales in the late Neogene and Quaternary of the southwestern Caribbean. Paleobiol 33:24–52

Jones PD, Briffa KR, Osborn TJ, Lough JM, van Ommen TD, Vinther BM, Luterbacher J, Wahl ER, Zwiers FW, Mann ME, Schmidt GA, Ammann CM, Buckley BM, Cobb KM, Esper J, Goosse H, Graham N, Jansen E, Kierer T, Kull C, Kuttel M, Mosley-Thompson E, Overpeck JT, Riedwyl N, Schulz M, Tudhope AW, Villalba R, Wanner H, Wolff E, Xoplaki E (2009) High-resolution palaeoclimatology of the last millennium: a review of current status and future prospects. The Holocene 19:3–49

Jordan-Dahlgren E (1992) Recolonization patterns of *Acropora palmata* in a marginal environment. Bull Mar Sci 51:104–117

Kaufman L (1981) There was biological disturbance on Pleistocene coral reefs. Paleobiol 7:527–532

Kelley PH, Hansen TA (2003) The fossil record of drilling predation on bivalves and gastropods in predator prey interactions in the fossil record In: Kelley PH, Kowalewski M, Hansen TA (eds) Predator–prey interactions in the fossil record. Kluwer Academic/Plenum Publishers, New York pp 113–139

Kennedy DM, Woodroffe CD, Zhao JX (2007) Reef development at high-latitudes during multiple interglacial cycles: new evidence from Lord Howe Island, southwestern Pacific. Carbonate Evaporite 22:23–32

Kidwell SM (2001) Preservation of species abundance in marine death assemblages. Science 294:1191–1194

Kidwell SM, Rothfus TA (2010) The living, the dead, and the expected dead: variation in life span yields little bias of proportional abundances in bivalve death assemblages. Paleobiol 36:615–640

Kiene WE, Hutchings PA (1994) Bioerosion experiments at Lizard Island, Great Barrier Reef. Coral Reefs 13:91–98

Kiessling W, Simpson C (2011) On the potential for ocean acidification to be a general cause of ancient reef crises. Global Change Biology 17:56–67

Kiessling W, Simpson C, Michael Foote M (2010) Reefs as cradles of evolution and sources of biodiversity in the Phanerozoic. Science 327:196–198

Kilbourne KH, Quinn TM, Taylor FW (2004) A fossil coral perspective on western tropical Pacific climate 350 ka. Paleoceanography 19: PA1019

Klaus JS, Budd AF (2003) Comparison of Caribbean coral reef communities before and after the Plio-Pleistocene faunal turnover: analyses of two Dominican Republic reef sequences. Palaios 18:3–21

Klaus JS, Lutz BP, McNeill DF, Budd AF, Johnson KG, Ishman SE (2011) Rise and fall of Pliocene free-living corals in the Caribbean. Geology 39:375–378

Knowlton N, Jackson JBC (1994) New taxonomy and niche partitioning on coral reefs – Jack of all trades or master of some. Trends Ecol Evol 9:7–9

Knutson DW, Buddemeier RW, Smith SV (1972) Coral chronometers: seasonal growth bands in reef corals. Science 177:270–272

Kobluk DR, Lysenko MA (1992) Storm features on a southern Caribbean fringing coral reef. Palaios 7:213–221

Konishi K (1981) Alcyonarian spiculite: limestone of soft corals. Proc 4th Int Coral Reef Symp 1:643–649

Kosnik MA, Hua Q, Kaufman DS, Wust RA (2009) Taphonomic bias and time-averaging in tropical molluscan death assemblages: differential shell half-lives in Great Barrier Reef sediment. Paleobiol 35:565–586

Lang JC, Chornesky A (1990) Competition between scleractinian reef corals-a review of mechanisms and effects. In: Dubinsky Z (ed) Ecosystems of the world 25, coral reefs. Elsevier, Amsterdam 209–252

Langer MR, Lipps JH (2003) Foraminiferal distribution and diversity, Madang Reef and Lagoon, Papua New Guinea. Coral Reefs 22:143–154

LaVigne M, Matthews KA, Grottoli AG, Cobb KM, Anagnostou E, Cabioch G, Sherrell RM (2010) Coral skeleton P/Ca proxy for seawater phosphate: multi-colony calibration with a contemporaneous seawater phosphate record. Geochim Cosmochim Acta 74:1282–1293

Lescinsky HL (2008) Bioerosion and encrustation on Curacao Pleistocene reefs: evaluating grazing in the fossil record. In: Park LE, Freile D (eds) 13th Symposium on the Geology of the Bahamas and other Carbonate Regions. San Salvador Bahamian Field Station, pp 197–210

Lescinsky HL 2012 Partial mortality and re-sheeting in fossil corals: a disease proxy. Proc 12th Inter Coral Reef Symp 2B

Lescinsky HL, Edinger E, Risk M (2002) Mollusc shell encrustation and bioerosion rates in a modern epeiric sea: taphonomy experiments in the Java Sea, Indonesia. Palaios 17:171–191

Lescinsky HL, Hill M, Hoedt BA (2008) Results of long-term bioerosion study: Belize patch reefs. Proc 11th Inter Coral Reef Symp 1:26–30

Lescinsky HL, Titus B, Hubbard DK (2012) Live coral cover in the fossil record: an example from Holocene reefs of the Dominican Republic. Coral Reefs 31:335–346

Lesser MP, Slattery M, Leichter JJ (2009) Ecology of mesophotic coral reefs. J Exper Mar Biol Ecol 375:1–8

Lighty RG, Macintyre IG, Stuckenrath R (1978) Submerged early Holocene barrier reef south-east Florida shelf. Nature 276: 59–60

Littler MM, Littler DS (1999) Blade abandonment/proliferation: a novel mechanism for rapid epiphyte control in marine macrophytes. Ecology 80:1736–1746

Lough JM (2011) Measured coral luminescence as a freshwater proxy: comparison with visual indices and a potential age artifact. Coral Reefs 30:169–182

Lough JM, Cooper TF (2011) New insights from coral growth band studies in an era of rapid environmental change. Earth Sci Rev 108:170–184

Lowenstam HA (1950) Niagaran reefs of the Great Lakes area. GSA Mem. 67: 215–248

Macintyre IG, Glynn PW (1976) Evolution of modern Caribbean fringing reef, Galeta Point, Panama. AAPG Bull 60:1054–1072

Macintyre IG, Marshall JF (1988) Submarine lithification in coral reefs: some facts and misconceptions. Proc 6th Inter Coral Reef Symp 1:263–272

Malmquist DL (1991) The past as a key to the present: taphonomy and paleoecology of the Urvina Bay uplift. In: James MJ (Ed) Galapagos marine invertebrates. Plenum Press, New York pp 393–421

Mann P, Taylor FW, Burke K, Kulstad R (1984) Subaerially exposed Holocene coral reef, Enriquillo Valley, Dominican Republic. Geol Soc Amer Bull 95:1084–1092

Marion GS, Dunbar RB, Mucciarone DA, Kremer JN, Lansing JS, Arthawiguna A (2005) Coral skeletal d15N reveals isotopic traces of an agricultural revolution. Marine Pollut Bull 50:931–944

Massel SR, Done TJ (1993) Effects of cyclone waves on massive coral assemblages on the Great Barrier Reef: meteorology, hydrodynamics and demography. Coral Reefs 12:153–166

Matthews KA, Grottoli AG, McDonough WF, Palardy JE (2008) Upwelling, species, and depth effects on the natural variability of coral skeletal cadmium-to-calcium ratios (Cd/Ca). Geochim Cosmochim Acta 72: 4537–4550

McClanahan TR (1989) Kenyan coral reef-associated gastropod fauna: a comparison between protected and unprotected reefs. Mar Ecol Prog Ser 53:11–20

McKinney FK, Jackson JBC (1989) Bryozoan Evolution, University of Chicago Press, Chicago

Meesters EH, Hilterman M, Kardinaal E, Keetman M, de Vries M, Bak RPM (2001) Colony size–frequency distributions of scleractinian coral populations: spatial and interspecific variation. Mar Ecol Prog Ser 209:43–54

Meldahl KH, Flessa KW, Cutler AH (1997) Time averaging and postmortem skeletal survival in benthic fossil assemblages: quantitative comparisons among Holocene environments. Paleobiology 23:207–229

Meltzner AJ, Sieh K, Abrams M, Agnew DC, Hudnut KW, Avouac JP, Natawidjaja DH (2006) Uplift and subsidence associated with the great Aceh-Andaman earthquake of 2004. J Geophys Res 111: B02407

Menza C, Kendall M, Hile S (2008) The deeper we go the less we know. Rev Biol Trop 56 (Suppl. 1):11–24

Mesolella KJ (1967) Zonation of uplifted Pleistocene coral reefs on Barbados, West Indies. Science 156:638–640

Meyer DL, Bries JM, Greenstein BJ, Debrot AO (2003) Preservation of in situ reef framework in regions of low hurricane frequency: Pleistocene of Curaçao and Bonaire, southern Caribbean. Lethaia 36:273–286

Miller AI (1988) Spatial resolution in subfossil molluscan remains: implications for paleobiological analyses. Paleobiol 14: 91–103

Montaggioni LF (1982) Pleistocene marine depositional environments from Mauritius Island, Indian Ocean. Geobios 15:161–179

Montaggioni LF, Braithwaite CJR (2009) Quaternary coral reef systems: History, development processes and controlling factors. Elsevier, Amsterdam

Moreno-Forero SK, Navas GR, Solano OD (1998) Cryptobiota associated to dead *Acropora palmata* (Scleractinia: Acroporidae) coral, Isla Grande, Columbian Caribbean. Rev Biol Trop 40:29–236

Muscatine LM, Goiran C, Land L, Jaubert J, Cuif JP, Allemand D (2005) Stable isotopes (delta C-13 and delta N-15) of organic matrix from coral skeleton. Proc Nat Acad Sci USA 102:1525–1530

Musso BM (1992) Rates of skeletal degradation following death in three species of *Acropora*. Proc 7th Inter Coral Reef Symp 1:413–418

Nakamori T, Iryu Y, Yamada T (1995) Development of coral reefs of the Ryukyu Islands (Southwest Japan, East China Sea) during Pleistocene sea-level change. Sediment Geol 99:215–231

Neumann AC, Macintyre I (1985) Reef response to sea level rise: keep-up, catch-up or give-up. Proc 5th Inter Coral Reef Symp 3:105–110

Newell ND, Rigby JK, Whiteman AJ, Hickox JE, Bradley JS (1953) The Permian reef complex of the Guadalupe mountain region, Texas and New Mexico. WH Freeman, San Francisco

Nunn PD, Ollier C, Hope G, Rodda P, Omura A, Peltier WR (2002) Late Quaternary sea-level and tectonic changes in northeast Fiji. Mar Geol 187:299–311

Palmer T, Plewes C (1993) Borings and bioerosion in fossils: Geology Today 9:138–142

Pandolfi JM (1992) A palaeobiological examination of the geological evidence of recurring outbreaks of the crown-of-thorns starfish, *Acanthaster planci*. Coral Reefs 11:87–94

Pandolfi JM (1995) Geomorphology of the uplifted Pleistocene lagoon at Henderson Island, Pitcairn Island Group, South Central Pacific. Biol J Linn Soc 56:63–77

Pandolfi JM (1996) Limited membership in Pleistocene reef coral assemblages from the Huon Peninsula, Papua New Guinea: constancy during global change. Paleobiol 22:152–176

Pandolfi JM (2002) Coral reef ecology at multiple spatial and temporal scales. Coral Reefs 21:13–23

Pandolfi JM, Budd AF (2008) Morphology and ecological zonation of Caribbean reef corals: the *Montastraea 'annularis'* species complex. Mar Ecol Prog Ser 369:89–102

Pandolfi JM, Greenstein BJ (1997) Taphonomic alteration of reef corals: effects of reef environment and coral growth form. I: The Great Barrier Reef. Palaios 12:27–42

Pandolfi JM, Jackson JBC (2001) Community structure of Pleistocene coral reefs of Curaçao, Netherlands Antilles. Ecol Monogr 71:49–67

Pandolfi JM, Jackson JBC (2006) Broad-scale patterns in Pleistocene coral reef communities from the Caribbean: Implications for ecology and management. In Aronson RB (Ed) Geological approaches to coral reef ecology. Springer, New York pp 201–236

Pandolfi JM, Jackson JBC (2007) The role of algae. Am Sci 95:1

Pandolfi JM, Lovelock CE, Budd AF (2002) Character release following extinction in a Caribbean reef coral species complex. Evolution 56:479–501

Pandolfi JM, Minchin PR (1995) A comparison of taxonomic composition and diversity between reef coral life and death assemblages in Madang Lagoon, Papua New Guinea. Palaeogeogr Palaeoclimatol Palaeoecol 119:321–341

Pandolfi JM, Tudhope AM, Burr G, Chappell J, Edinger E, Frey M, Steneck R, Sharma C, Yeates A, Jennions M, Lescinsky H, Newton A (2006) Mass mortality following disturbance in Holocene coral reefs from Papua New Guinea. Geology 34:949–952

Pante E, Dustan P (2012) Getting to the point: accuracy of point count in monitoring ecosystem change. J Marine Biol 2012:802875

Pari N, Peyrot-Clausade M, LeCampion-Alsumard T, Hutchings PA, Chazotes V, Golubic S, LeCampion T, Fontaine MF (1998) Bioerosion of experimental substrates on high islands and atoll lagoons (French Polynesia) after two years exposure. Mar Ecol Prog Ser 166:119–130

Parsons-Hubbard K (2003) Molluscan taphofacies in recent carbonate reef/lagoon systems and their application to sub-fossil samples from reef cores. Palaios 20: 175–191

Parsons-Hubbard K, Hubbard D, Tems C, Burkett A (2014) The relationship between modern mollusk assemblages and their expression in subsurface sediment in a carbonate lagoon, St. Croix, US Virgin Islands. In Hembree DI, Platt BF, Smith JJ (eds) Experimental approaches to understanding fossil organisms. Springer, Amsterdam, Topics in Geobiology 41:143–168

Paulay G, Spencer T (1988) Geomorphology, palaeoenvironments and faunal turnover, Henderson Island, S.E. Polynesia. Proc 6th Int Coral Reef Sym 3:461–466

Perry CT (1996) Distribution and abundance of macroborers in an upper Miocene reef system, Mallorca, Spain: implications for reef development and framework destruction. Palaios 11:40–56

Perry CT (2000) Macroboring of Pleistocene coral communities, Falmouth Formation, Jamaica. Palaios 15:483–491

Perry CT (2001) Storm-induced coral rubble deposition: Pleistocene records of natural reef disturbance and community response. Coral Reefs 20:171–183

Perry CT, Edinger EN, Kench PS, Murphy GN, Smithers SG, Steneck RS, Mumby PJ (2012) Estimating rates of biologically driven coral reef framework production and erosion: a new census-based carbonate budget methodology and applications to the reefs of Bonaire. Coral Reefs 31:853–868

Perry CT, Murphy GN, Kench PS, Smithers SG, Edinger EN, Steneck RS, Mumby PJ (2013) Caribbean-wide decline in carbonate production threatens coral reef growth. Nature Communications, 4:1402 I DOI: 10.103

Perry CT, Spencer T, Kench PS (2008) Carbonate budgets and reef production states: a geomorphic perspective on the ecological phase-shift concept. Coral Reefs 27:853–866

Pirazzoli PA, Radtke U, Hantoro WS, Jouannic C, Hoang CT, Causse C, Borel Best M (1991) Quaternary raised coral-reef terraces on Sumba Island, Indonesia. Science 252:1834–1836

Pleydell SM, Jones B (1988) Borings of various faunal elements in the Oligocene-Miocene Bluff Formation of Grand Cayman, British West Indies. J Paleont 62:348–367

Porter JW, Fitt WK, Spero HJ, Rogers CS, White MW (1989) Bleaching in reef corals- physiological and stable isotopic responses. Proc Natl Acad Sci USA 86:9342–9346

Powell EN, Staff GM, Callender WS, Ashton-Alcox KA, Brett CE, Parsons-Hubbard KM, Walker SE, Raymond A (2011) Taphonomic degradation of molluscan remains during thirteen years on the continental shelf and slope of the northwestern Gulf of Mexico. Palaeogeogr Palaeoclimatol Palaeocecol 312:209–232

Precht WF, Aronson RB (2004) Climate flickers and range shifts of reef corals. Front Ecol Environ 2:307–314

Preusser F, Radtke U, Fontugne M, Haghipour A, Hilgers A, Kasper HU, Nazari H, Pirazzoli PA (2003) ESR dating of raised coral reefs from Kish Island, Persian Gulf. Quaternary Sci Rev 22:1317–1322

Rasmussen KA, Brett CE (1985) Taphonomy of Holocene cryptic biotas from St. Croix, Virgin Islands: information loss and preservational biases. Geology 13:551–553

Rasser MW, Riegl B (2002) Holocene coral reef rubble and its binding agents. Coral Reefs 21:57–72

Reuter M, Piller WE (2011) Volcaniclastic events in coral reef and seagrass environments: evidence for disturbance and recovery (Middle Miocene, Styrian Basin, Austria). Coral Reefs 30:889–899.

Reymond CE, Bode M, Renema W, Pandolfi JM (2011) Ecological incumbency impedes stochastic community assembly in Holocene foraminifera from the Huon Peninsula, Papua New Guinea. Paleobiol 37:670–685

Richter C, Wunsch M (1999) Cavity dwelling suspension feeders in coral reefs- a new link in reef trophodynamics. Marine Ecol Prog Ser 188:105–116

Riker-Coleman KE, Gallup CD, Wallace LM, Webster JM, Cheng H, Edwards RL (2006) Evidence of Holocene uplift in east New Britain, Papua New Guinea. Geophys Res Lett 33:L18612

Risk MJ, Sammarco PW, Edinger EN (1995) Bioerosion in *Acropora* across the continental shelf of the Great Barrier Reef. Coral Reefs 14:79–86

Roche RR, Perry CT, Johnson KG, Sultana K, Smithers SG, Thompson AA (2011) Mid-Holocene coral community data as baselines for understanding contemporary reef ecological states. Palaeogeogr Palaeoclimatol Palaeoecol 299:159–167

Roff G, Clark TR, Reymond CE, Zhao J, Feng Y, McCook LJ, Done, TJ, Pandolfi JM (2013) Palaeoecological evidence of a historical collapse of corals at Pelorus Island, inshore Great Barrier Reef, following European settlement. Proc R Soc B 280:20122100

Rogers CS, Miller J, Waara RJ (2000) Tracking changes on a reef in the U S Virgin Islands with videography and SONAR: a new approach. Proc 9th Inter Coral Reef Symp 2:1065–1069

Samimi-Namin K, Riegl B (2012) Raised reef on Larak suggests *Acropora* dominance on the Persian Gulf coral reefs since the Pleistocene. Coral Reefs 31:763

Scheffers SR, Nieuwland G, Bak RPM, van Duyl FC (2004) Removal of bacteria and nutrient dynamics within the coral reef framework of Curaçao (Netherlands Antilles). Coral Reefs 23:413–422

Scheibner C, Speijer RP (2008) Late Paleocene–early Eocene Tethyan carbonate platform evolution- a response to long- and short-term paleoclimatic change. Earth Sci Rev 90:71–102

Schopf TJM (1978) Fossilization potential of an intertidal fauna: Friday Harbor, Washington. Paleobiol 3:261–270

Scoffin TP (1992) Taphonomy of coral reefs: a review. Coral Reefs 11:57–77

Scoffin TP (1993) The geological effects of hurricanes on coral reefs and the interpretation of storm deposits. Coral Reefs 12:203–221
Sheppard CRC, Davy SK, Pilling GM (2009) The biology of coral reefs. Oxford University Press, Oxford
Shinn EA (2004) The mixed value of environmental regulations: do acroporid corals deserve endangered species status? Mar Pollut Bull 49:531–533
Shinn EA, Reich CD, Hickey TD, Lidz BH (2003) Staghorn tempestites in the Florida Keys. Coral Reefs 22: 91–97
Simpson C, Kiessling W, Mewis H, Baron-Szabo RC, Muller J (2011) Evolutionary diversification of reef corals: a comparison of the molecular and fossil records. Evolution 65:3274–3284
Smith LD, Devlin M, Haynes D, Gilmour JP (2005) A demographic approach to monitoring the health of coral reefs. Mar Pollut Bull 51:399–407
Smithers SG, Hopley D, Parnell KE (2006) Fringing and nearshore coral reefs of the Great Barrier Reef: episodic Holocene development and future prospects. J Coastal Res 22:175–187
Spalding MD, Green EP, Ravilious C (2001) World Atlas of Coral Reefs. University of California Press, Oakland
Stafford ES, Leighton LR (2011) Vermeij crushing analysis: a new old technique for estimating crushing predation in gastropod assemblages. Palaeogeogr Palaeoclimatol Palaeocecol 305:123–137
Stanley GD, Helmle KP (2010) Middle Triassic coral growth bands and their implication for photosymbiosis. Palios 25:754–763
Stanley GD, Swart PK (1995) Evolution of the coral-zooxanthenllae symbiosis during the Triassic: a geochemical approach. Paleobiol 21:179–199
Stanley SM (1966) Paleoecology and diagenesis of Key Largo Limestone, Florida. Am Assoc Petr Geol B 50:1927–1947
Stearn CW, Scoffin TP (1977) Carbonate budget of a fringing reef, Barbados. Proc Third Intl Coral Reef Symp 2: 471–477
Strasser A, Strohmenger C (1997) Early diagenesis in Pleistocene coral reefs, southern Sinai, Egypt: response to tectonics, sea-level and climate. Sedimentology 44:537–558
Sugihara K, Nakamori T, Iryub Y, Sasakic K, Blanchon B (2003) Holocene sea-level change and tectonic uplift deduced from raised reef terraces, Kikai-jima, Ryukyu Islands, Japan. Sediment Geol 159:5–25
Susic M, Boto K, Isdale P (1991) Fluorescent humic acid bands in coral skeletons originate from terrestrial runoff. Mar Chem 33:91–104
Suzuki A, Gagan MK, Fabricius K, Isdale PJ, Yukino I, Kawahata H (2003) Skeletal isotope microprofiles of growth perturbations in *Porites* corals during the 1997–1998 mass bleaching event. Coral Reefs 22:357–369
Sweatman H, Delean S, Syms C (2011) Assessing loss of coral cover on Australia's Great Barrier Reef over two decades, with implications for longer-term trends. Coral Reefs 30:521–531
Tager D, Webster JM, Potts DC, Renema W, Braga JC, Pandolfi JM (2010) Community dynamics of Pleistocene coral reefs during alternative climatic regimes. Ecology 91:191–200
Taylor FW, Bloom AL (1977) Coral reefs on tectonic blocks, Tonga island arc. Proc 3rd Int Coral Reef Symp 2:275–281
Taylor FW, Frohlich C, Lecolle J, Strecker M (1987) Analysis of partially emerged corals and reef terraces in the central Vanuatu arc: comparison of contemporary coseismic and nonseismic with Quaternary vertical movements. Geophys Res 92:4905–4933
Taylor FW, Mann P, Bevis MG, Edwards RL, Cheng, H, Cutler KB, Gray SC, Burr GS, Beck JW, Phillips DA, Cabioch G, Recy J (2005) Rapid forearc uplift and subsidence caused by impinging bathymetric features: examples from the New Hebrides and Solomon arcs. Tectonics 24: TC6005
Taylor PD (1990) Preservation of soft-bodied and other organisms by bioimmuration- a review. Paleobiol 33:1–17
Toscano MA, Rodriguez E, Lundberg J (1999) Geologic investigation of the late Pleistocene Jaimanitas Formation: science and society in Castro's Cuba. In Curran HA, Mylroie JE (Eds) Proc. 9th Symposium on the Geology of the Bahamas and other Carbonate Regions, San Salvador, Bahamian Field Station 125–142
Tudhope AW, Chilcott CP, McCulloch MT, Cook ER, Chappell J, Ellam RM, Lea DW, Lough JM, Shimmield GB (2001) Variability in the El Nino-Southern Oscillation through a glacial–interglacial cycle. Science 291:1511–1517
Tudhope AW, Scoffin TP (1984) The effects of *Callianassa* bioturbation on the preservation of carbonate grains in Davies Reef Lagoon, Great Barrier Reef, Australia. J Sedimentary Res 54:1091–1096
Van Valen L (1964) Relative abundance of species in some fossil mammal faunas. Am Nat 98:109–116
Van Woesik R, Franklin EC, O'Leary J, McClanahan TR, Klaus JS, Budd AF (2012) Hosts of the Plio-Pleistocene past reflect modern-day coral vulnerability. Proc Royal Soc B 279:2448–56
Vermeij GJ (1989) Interoceanic differences in adaptation: effects of history and productivity. Mar Ecol Prog Ser 57:293–305
Vermeij GJ (2012) Crucibles of creativity: the geographic origins of tropical molluscan innovations. Evol Ecol 26:357–373
Vermeij GJ, Herbert GS (2004) Measuring relative abundance in fossil and living assemblages. Paleobiol 30:1–4
Veron JEN (1992) Environmental control of Holocene changes to the world's most northern hermatypic coral outcrop. Pacific Sci 46:405–425
Vogel K, Gektidis M, Golubic S, Kiene WE, Radtke G (2000) Experimental studies on microbial bioerosion at Lee Stocking Island, Bahamas and One Tree Island, Great Barrier Reef, Australia: implications for paleoecological reconstructions. Lethaia 33:190–204
Vora, KH, Wagle BG, Veerayya M, Almeida F, Karisiddaiah SM (1996) 1300 km long late Pleistocene-Holocene shelf edge barrier reef system along the western continental shelf of India: occurrence and significance. Mar Geol 134:145–162
Vroom PS, Page KN, Kenyon JC, Brainard RE (2006) Algae-Dominated Reefs. Am Sci 94:430–437
Walbran PD, Henderson RA, Jull AJT, Head MJ (1989) Evidence from sediments of long-term *Acanthaster planci* predation on corals of the Great Barrier Reef. Science 245:847–850
Walker SE (1992) Criteria for recognizing marine hermit crabs in the fossil record using gastropod shells. J Paleont 66:535–558
Webster JM, Wallace L, Silver E, Applegate B, Potts D, Braga RC, Riker-Coleman K, Gallup C (2004) Drowned carbonate platforms in the Huon Gulf, Papua New Guinea. Geochem Geophys Geosyst 5:Q11008
West RR, McKinney FK, Fagerstrom JF, Vacelet J (2011) Biological interactions among extant and fossil clonal organisms. Facies 57:351–374
Willis KJ, Bailey RM, Bhagwat SA, Birks HJ (2010) Biodiversity baselines, thresholds and resilience: testing predictions and assumptions using palaeoecological data. Trends Ecol Evol 25:583–591
Wood R (1999) Reef Evolution. Oxford University Press, Oxford
Woodroffe C, McLean R (1990) Microatolls and recent sea level change on coral atolls. Nature 344:531–534
Woodroffe, CD, Brooke BP, Linklater M, Kennedy DM, Jones BG, Buchanan C, Mleczko R, Hua Q, Zhao J (2010) Response of coral reefs to climate change: expansion and demise of the southernmost Pacific coral reef., Geophys Res Lett 37:L15602
Wright P, Cherns L, Hodges P (2003) Missing molluscs: field testing taphonomic loss in the Mesozoic through early large-scale aragonite dissolution. Geology 31:211–214
Yamano H, Sugihara K, Nomura K (2011) Rapid poleward range expansion of tropical reef corals in response to rising sea surface temperatures. Geophys Res Lett 38:L04601

Yamazaki A, Watanabe T, Urumu T (2011) Nitrogen isotopes of organic nitrogen in reef coral skeletons as a proxy of tropical nutrient dynamics. Geophys Res Lett 38:L19605

Yu KF, Jian-Xin Z, Collerson KD, Shia Q, Chena TG, Wang PX, Liu TS (2004), Storm cycles in the last millennium recorded in Yongshu Reef, southern South China Sea. Palaeogeogr Palaeoclimatol Palaeocecol 210:89–100

Yu KF, Jian-Xin Z, Lawrence MG, Feng Y (2010) Timing and duration of growth hiatuses in mid Holocene massive *Porites* corals from the northern South China Sea. J Quaternary Sci 25:1284–1292

Zamagni J, Mutti M, Kosir A (2012) The evolution of mid Paleocene-early Eocene coral communities: how to survive during rapid global warming. Palaeogeogr Palaeoclimatol Palaeocecol 317:48–65

Zhuravlev AY (2001) Paleoecology of Cambrian reef ecosystems. In Stanley GD (ed) The history and sedimentology of ancient reef systems. Kluwer Academic/Plenum Publishers, New York pp 121–157

Ziebis W, Forster S, Huettel M, Jorgensen BB (1996) Complex burrows of the mud shrimp *Callianassa truncata* and their geochemical impact in the sea bed. Nature 382:619–622

Physical and Biological Drivers of Coral-Reef Dynamics

11

Richard B. Aronson and William F. Precht

Abstract

Coral-reef ecosystems are declining worldwide, compromising their capacity to provide ecosystem services that include feeding hundreds of millions of people and protecting shorelines from erosion. The anthropogenic causes of reef degradation are complex and operate over a broad range of scales and hierarchical levels, but accelerating climate change and its collateral impacts are currently the strongest drivers. Deleterious trends in local-scale, ecological processes that occur within reef communities, such as declining herbivory and increasing eutrophication, generally play a subsidiary role at present, because their effects are overwhelmed by the impacts of climate change on many reefs. That does not mean local-scale ecology is irrelevant. Solving environmental problems at one scale or level will by default leave problems at the other scale as the new primary problems. If humanity is able to control climate change at the global level, then community-level processes will in general become limiting. Both local and global impacts must be mitigated and reversed if we are to save coral reefs.

Keywords

Acropora • Caribbean • Climate change • Coral bleaching • Coral disease • Marine protected areas • MPAs • White-band disease

11.1 Introduction

Marine ecosystems throughout the world ocean have been damaged by human activities, and coral reefs have suffered especially severe impacts (Halpern et al. 2008). The challenge for coral-reef scientists is to determine the strongest causal pathways to degradation. Measuring the relative contributions of proximate and ultimate candidate-causes is not merely an academic exercise; the prescriptions for mitigating and reversing reef degradation differ depending on the scales, hierarchical levels, and identities of those causes. With limited resources available to conserve coral reefs, it is imperative that time, labor, and funds be devoted to corrective measures that will yield the maximum benefits.

Because reefs are geological as well as ecological entities, the physical sciences have been integral to their study from the start. Our understanding of how coral reefs operate strongly emphasizes physical drivers (e.g., Roberts et al. 1992; Hubbard 1997; Montaggioni and Braithwaite 2009). A few basic examples highlight the physical control of biological processes: (1) reef development is limited to latitudes warmer than the 18 °C winter-minimum isotherm (Dana 1843; Johannes et al. 1983; Kleypas et al. 2001); (2) upwelling driven by oceanic gyres restricts reef development off the west coasts of continents (Birkeland 1997; Hubbard 1997), and inimical waters suppress reef

R.B. Aronson (✉)
Department of Biological Sciences, Florida Institute of Technology, 150 West University Boulevard, Melbourne, FL 32951, USA
e-mail: raronson@fit.edu

W.F. Precht
Division of Marine and Coastal Programs, Dial Cordy and Associates, Inc., Miami, FL 33014, USA
e-mail: william.precht@gmail.com

D.K. Hubbard et al. (eds.), *Coral Reefs at the Crossroads*, Coral Reefs of the World 6, DOI 10.1007/978-94-017-7567-0_11

development on smaller scales (Neumann and Macintyre 1985; Hallock and Schlager 1986; Ginsburg and Shinn 1994); (3) antecedent topography and fluctuating sea level determine the growth and form of reefs and their scope for vertical accretion (Darwin 1842; Neumann and Macintyre 1985); (4) light and wave exposure combine with topography to create the biological zonation of reefs (Adey and Burke 1977; Geister 1977; Woodley et al. 1981; Hallock and Schlager 1986; Hubbard 1988; Acevedo et al. 1989; Graus and Macintyre 1989; Murdoch 2007); and (5) climatic fluctuations set the tempo and mode of reef development (Precht and Aronson 2004; Precht and Miller 2007; Toth et al. 2012, 2015). Today geology and paleobiology are helping us to distinguish natural from anthropogenic perturbations of coral reefs and to understand the scales at which those perturbations occur (Aronson 2007). An emerging theme—and the subject of this review—is the overriding influence of physical forcing in the recent, worldwide degradation of reefs.

What is signal to an ecologist is largely noise to a paleontologist. Fossil deposits from shallow, soft-bottom facies are often temporally and spatially averaged, obscuring the record of short-term variability on which the science of ecology still nourishes itself, for better or worse (but see Kidwell 2001, 2007). Time averaging and transport are generally less problematic in interpreting coral-reef deposits than biotas buried in soft sediments, because for coral reefs the benthic assemblages themselves construct their sedimentary fabrics. Coral colonies have longer lifespans than most soft-sediment invertebrates, and their skeletons are to some extent resistant to taphonomic degradation. Furthermore, because coral skeletons are made of calcium carbonate and, in most cases, the skeletal framework and entombing sediments are cemented, corals are often buried and preserved in place and in sequence, or at most subject only to minor transport. Even where reefs are uncemented, the coral assemblages are often autochthonous and sequential (Aronson and Precht 1997; Aronson et al. 2002, 2004, 2005; Wapnick et al. 2004; Greer et al. 2009). Fossil and subfossil reef deposits, therefore, provide an excellent record from which to understand the ecology of coral reefs in times past and thereby discern the time frame and effects of natural and anthropogenic perturbations on modern reefs (Pandolfi 1996; Greenstein et al. 1998; Pandolfi and Jackson 2001, 2006; Precht and Aronson 2006; Greenstein and Pandolfi 2008; Lescinsky et al. 2012; Toth et al. 2012; and many others).

In this chapter we use evidence from fossil and modern coral reefs to review and critically evaluate three related propositions that have attained enormous popularity among coral-reef ecologists: (1) localized human activity, specifically overfishing, has been the primary cause of the decline of coral populations; (2) the cascading, top-down effects of overfishing are currently limiting the recovery of coral assemblages; and, therefore, (3) local management actions are capable of promoting the resilience of reefs to climate change. We contrast these notions with the idea that physical drivers are the primary determinants of reef dynamics at scales not much larger than the scale of the reef or reef system and should be important considerations in management and conservation. We focus on the reefs of Florida, the Bahamas, and the Caribbean (henceforth collectively termed 'the Caribbean'), which are significantly altered (Gardner et al. 2003; Schutte et al. 2010) and for which the historical, ecological, and paleobiological data are complete enough to draw reasonably firm conclusions.

11.2 Causal Connections in the Degradation of Caribbean Reefs

Aronson and Precht (2001a, 2001b, 2006) suggested that larger-scale factors, specifically climate change and coral disease (which is related to climate change), were the primary causes of reef degradation throughout the Caribbean region over the preceding three decades. The elkhorn coral *Acropora palmata* had dominated the reef-crest and shallow fore-reef habitats at 0–5 m depth on windward-facing Caribbean reefs, whereas the staghorn coral *A. cervicornis* had dominated intermediate, fore-reef depths of 5–25 m and some back-reef and lagoonal habitats. A regional outbreak of white-band disease (WBD: Fig. 11.1) was the primary cause of the Caribbean-wide mass mortality of these congeneric corals from the late 1970s to the early 1990s. WBD is an infectious, bacterial syndrome that appears only to affect the acroporids (Gil-Agudelo et al. 2006; Weil et al. 2006; Vollmer and Kline 2008; Kline and Vollmer 2011; Gignoux-Wolfsohn et al. 2012; Sweet et al. 2014). Because in many locations *A. cervicornis* and *A. palmata* were the dominant occupants of reef substratum and the dominant constructors of framework, Aronson and Precht (2001a, 2001b, 2006) concluded that WBD had been the most important cause of coral mortality in the Caribbean in recent decades. Hurricanes, coral bleaching from anomalously high sea temperatures, and additional factors such as corallivory had played subsidiary roles in killing the Caribbean acroporids (ABRT 2005; Gardner et al. 2005). Bleaching and other diseases later killed massive corals, including the formerly abundant, framework-building *Orbicella annularis* species complex (McWilliams et al. 2005; Aronson and Precht 2006; Eakin et al. 2010; Toth et al. 2014; see also Rogers 2008; Miller et al. 2009; Rogers and Miller 2013).

Meta-analyses subsequent to Aronson and Precht (2001a, 2001b) have supported their conclusions (Côté et al. 2005; Alvarez-Filip et al. 2009; Schutte et al. 2010). Furthermore, paleoecological studies of reefs in several locations around

Fig. 11.1 White-band disease on otherwise healthy branches of *Acropora palmata*. Note the characteristic progression of the disease from the bases to the tips of the branches. The photograph was taken at Carysfort Reef in the Florida Keys during the summer of 2004 (Photo credit: William F. Precht)

the Caribbean have demonstrated that the recent mass mortality of acroporids was a novel event in at least the last three millennia (Aronson and Precht 1997; Greenstein et al. 1998; Aronson et al. 2002, 2005; Wapnick et al. 2004; Lescinsky 2012). Epidemiological work has suggested, albeit obliquely, that rising sea temperatures were responsible for the devastating outbreak of WBD in the Caribbean (Kline and Vollmer 2011), providing a link to physical processes. More recently, Randall and van Woesik (2015) linked outbreaks of WBD to increased thermal stress associated with climate change. Outbreaks of some other coral diseases have also been tied to rising temperatures (Rosenberg and Ben-Haim 2002; Selig et al. 2006; Bruno et al. 2007; but see Lafferty et al. 2004).

Jackson et al. (2001), in contrast, asserted in a highly publicized review that the disruption of trophic cascades by overfishing was the most important cause of ecological degradation in shallow-marine environments worldwide. For coral reefs, the scenario was that overfishing reduced herbivory, releasing macroalgae, or seaweeds, to overgrow and otherwise outcompete corals for space (see also Pandolfi et al. 2003). The review by Jackson et al. (2001) was immediately welcomed by conservation groups and the popular media, who touted it as a visionary breakthrough in our understanding of human threats to marine life. With equal rapidity Jackson et al. (2001) drew fire from scientists who pointed out that overfishing was neither the only human assault on marine ecosystems nor necessarily the most significant one. Jackson and colleagues responded that they had never intended to imply a negligible role for other drivers of ecosystem degradation (Peterson et al. 2001). In reality, Jackson et al. (2001) had acknowledged the existence of other factors but had downplayed them.

This group of authors later moderated their stance on overfishing to a more pluralistic view of causality by including sedimentation and nutrient loading from terrestrial sources as another major threat to coral reefs (Bellwood et al. 2004; Kuntz et al. 2005; Jackson 2008). Some of them have recently regressed to their initial stance that overfishing of herbivores, specifically parrotfish, is far and away the primary cause (Jackson et al. 2014).

Terrigenous input has certainly been an important cause of degradation in some situations (Rogers 1990; Cortés 1994; Aronson et al. 2004, 2014; De'ath and Fabricius 2010). On the other hand, one top-down scenario of the impacts of overfishing on coral reefs included speculation of a strong, cascading trophic connection between shark-fishing and a high prevalence of infectious diseases in coral populations (Sandin et al. 2008). That causal chain has not been demonstrated. More convincing are data showing that populations of the corallivorous seastar *Acanthaster planci* are reduced on Pacific reefs that support more intact stocks of predatory fish (Dulvy et al. 2004; McCook et al. 2010).

Claims about the primacy of overfishing are a step backward from Hughes (1994), who argued that overfishing, the regional mass mortality in 1983–1984 of the herbivorous echinoid *Diadema antillarum* from an infectious disease, and direct coral mortality from a hurricane had combined to drive a phase shift from coral to macroalgal dominance on Jamaican reefs. The loss of herbivores has been particularly egregious in Jamaica (Aronson 1990; Hughes et al. 1999; Aronson and Precht 2000), so overfishing was considered an important ingredient in the transition to dominance by macroalgae. Whether or not the construct for Jamaica can be generalized to the rest of the Caribbean is an important question (Côté et al. 2013). In fact, mass coral mortalities permitted macroalgae to rise to dominance opportunistically even in some locations with reasonably intact fish assemblages. It also turns out, surprisingly in hindsight, that only a minority of reefs in the Caribbean actually became dominated by macroalgae (Aronson and Precht

2006; Precht and Aronson 2006; Bruno et al. 2009; Dudgeon et al. 2010; Schutte et al. 2010; Bruno et al. 2014).

There is no doubt that much of the world is overfished and that in some situations overfishing can have drastic, cascading impacts on marine ecosystems. The model of Jackson et al. (2001), that the loss of top predators to overfishing fundamentally alters marine food webs and is the primary impact of human activity, works well for ecosystems with strong top-down trophic connections, such as kelp forests (Estes and Duggins 1995; Shears and Babcock 2003; Estes et al. 2011), but its applicability is far from universal. The overfishing hypothesis has been questioned or refuted for seagrass beds, oyster reefs, pelagic ecosystems, and some kelp forests, as well as coral reefs. In these cases, top-down trophic connections play a minor role in community structure, are too weak to respond substantially to the restoration of fisheries, or are complicated by bottom-up effects and other causes (Boesch et al. 2001; MacKenzie 2007; Waycott et al. 2009; Foster and Schiel 2010; Condon et al. 2012). Like the issue of terrigenous input, the question is not whether overfishing can be important on some coral reefs under some circumstances—because clearly it can—but what is its relative contribution to the overall decline of coral populations and coral reefs, and on what spatio-temporal scales?

Because the two Caribbean species of *Acropora* are now rare to the point of being threatened or endangered (Precht et al. 2004; ABRT 2005; Hogarth 2006; Carpenter et al. 2008; Aronson et al. 2009a, 2009b), accurate knowledge of the timing and causes of their decline is critical to their effective management and conservation (National Marine Fisheries Service 2015). Hughes et al. (2010) presented a history of the causes of decline of the acroporid corals in the Caribbean, which included the following language:

> [T]wo meta-analyses of the loss of structural complexity of Caribbean reefs between 1969 and 2008 [Alvarez-Filip et al. 2009] and of coral cover from 1971 to 2006 [Schutte et al. 2010] have proposed that an *unreported epidemic of white band disease* [emphasis ours] killed off most branching staghorn and elkhorn corals across the region in the1970s. In reality, the loss of coral cover has been highly asynchronous, and disease is only one of many causes of the decline. For instance, cold water killed >90 % of staghorn corals in the Dry Tortugas, Florida in the winter of 1976–77 [Davis 1982]. The collapse of branching acroporids in Jamaica was overwhelmingly because of Hurricane Allen in 1980 [Woodley et al. 1981]. There is only one report of a significant outbreak of white band disease in the Caribbean before 1980, a localized die-off affecting 5 hectares of shallow reef in St. Croix, US Virgin Islands in 1976–1979 [Gladfelter 1982]. In contrast, hurricanes and coral disease were dismissed as causes of the steep decline in coral cover in the Dutch Antilles from 1973 to 1992 [Bak and Nieuwland 1995].

This passage raises several issues. First, Hughes et al. (2010) may be correct that a cold-water event was responsible for the mortality of vast fields of *A. cervicornis* in the Dry Tortugas in 1977. Porter et al. (1982) used photographs of permanent quadrats taken 6 months before and 6 months after the event as evidence. The photographs, unfortunately, do not establish causality and WBD cannot be ruled out as the cause of mortality of *A. cervicornis*. In fact, the before-and-after photographs look suspiciously as though they are displaying mortality from WBD (see especially Fig. 11.2 in the paper) and not bleaching from cold-exposure. Most of the losses of *A. palmata* and *A. cervicornis* throughout the Florida reef tract were from WBD, especially after 1978 (Precht and Miller 2007; references therein).

Second, Hughes et al. (2010) pointed out that mortality from Hurricane Allen in 1980 was the principal cause of the collapse of *Acropora* populations at Discovery Bay and elsewhere along the north coast of Jamaica. Knowlton et al. (1981), however, also noted, "Unusual amounts of tissue exfoliation, resembling that termed 'white band disease' were observed in some colonies of *A. cervicornis* before the hurricane. This exfoliation continued after the storm...," and within 5 months there was a 100-fold decrease in the abundance of living colonies of *A. cervicornis* compared to the population immediately after the storm. Hurricane Allen was clearly a catastrophic disturbance on Jamaican reefs, but it is equally apparent that WBD was critical to the decline of *A. cervicornis* on these reefs. In fact, lagoonal populations of *A. cervicornis* at Discovery Bay were killed outright by WBD, not by Hurricane Allen (Wapnick et al. 2004).

Third, Gladfelter (1982) recognized the devastating effects of WBD on acroporids in St. Croix, spanning the years 1976–1979, prior to Hurricane Allen. Hughes et al. (2010) minimized Gladfelter's work by asserting that the outbreak of WBD was localized to a small area in St. Croix. In contrast, Gladfelter (1982) stated the following:

> Throughout much of its range, *A. palmata* is subject to a necrosis which can cause extensive local mortality of the coral. The author has observed this necrosis (= "white band disease") in the northeastern Caribbean Sea (Virgin Islands, St. Marten, Antigua), Curaçao, [and] Nicaragua (Miskito Cays), and it has been observed in Panama (P. Glynn, pers. comm.) and south Florida (A. Antonius, pers. comm.).

The outbreak of WBD in St. Croix was clearly part of an epidemic that was well underway throughout the Caribbean in the late 1970s, and it was recognized by Gladfelter as a regional phenomenon at the time.

Finally, citing Bak and Nieuwland (1995), Hughes et al. (2010) stated that coral disease was not responsible for the observed mortality of corals in the Netherlands Antilles. Although Bak and Nieuwland (1995) noted that factors such as diseases were unlikely to have been important in structuring the reefs in question, their study was

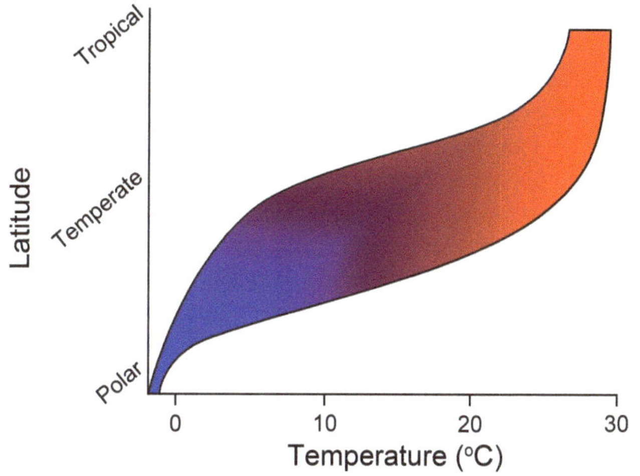

Fig. 11.2 Approximate ranges of annual sea temperatures within shallow-water habitats from polar to tropical latitudes. Narrow ranges have driven the evolution of stenothermy in ectotherms at the poles and in the tropics, whereas temperate ectotherms tend to be eurythermal. The polar range is for Antarctica (Data are from Clarke and Crame 2010 and other sources. Modified from Aronson et al. 2011)

confined to water depths of 10–40 m and their quadrats contained no *Acropora* species. They were, however, careful to note, "White-band disease is practically limited to the *Acropora* species and these are only common at depths shallower than 10 m along these coasts (Bak and Criens 1981; van Duyl 1985)." The two latter references were the same ones Aronson and Precht (2001a, 2001b) cited to describe the strong impact of WBD in the Netherlands Antilles.

Using data from a paleoecological study, Cramer et al. (2012) attempted to dispute the conclusions of Aronson and Precht (2001a, 2001b, 2006): (1) that the most significant losses of acroporid corals in the Caribbean occurred beginning in the late 1970s; and, related, (2) that WBD was the primary cause of the decline of acroporids. Aronson and Precht drew these conclusions based on their compilation of direct observations by a large number of scientists and other informed observers from 31 areas distributed among 16 countries or territories throughout the tropical and subtropical western Atlantic. Cramer et al. (2012) suggested instead that *A. cervicornis* had begun to decline in the Caribbean because of anthropogenic pressure decades before the outbreak of WBD and the frequent occurrence of coral-bleaching events. They based their alternative interpretation on the stratigraphic distribution of subfossil *A. cervicornis* in 18 circular trenches, each 60 cm in diameter and 60–80 cm deep, which they excavated at six stations in one area in one country: Bocas del Toro, Panamá.

Cramer dug three trenches at each of the six stations at Bocas del Toro. Half the sites (nine trenches) were dug in a lagoonal environment and the other half (another nine trenches) were situated 'offshore,' in a near-coastal environment. At the lagoonal stations, which were located in Bahía Almirante and the adjacent Laguna de Chiriquí, three of the nine trenches exhibited a decline of *A. cervicornis* before the regional WBD outbreak (based on radiocarbon dates of *Porites furcata*—not *A. cervicornis*—derived from accelerator mass spectrometry, or AMS), a pattern consistent with the preferred scenario of Cramer et al. (2012). Five lagoonal trenches had little or no *A. cervicornis* at any level, ranging in proportional weight from 0 to 8 % of the total coral material. Radiocarbon dates from the ninth lagoonal trench showed reversals that indicated significant stratigraphic mixing. In the offshore environment, none of the trenches showed evidence of an early decline of *A. cervicornis*. One station, consisting of three trenches, had negligible quantities of *A. cervicornis*. At each of the other two stations, two trenches showed declines of *A. cervicornis* in the 1970s or later (again based on AMS dates from *P. furcata* rather than the focal species), whereas there was no clear pattern in the third trench at any of the stations due to mixing. Not only are the temporal patterns of coral dominance inconsistent, but AMS is inaccurate at the young ages of the corals that were dated, calling into question the age models on which the conclusions were based (Aronson et al. 2014). The claim that the decline of *A. cervicornis* at Bocas del Toro began prior to the outbreak of WBD and subsequent bleaching events is not supported by the data in Cramer et al. (2012), nor by more accurate chronologies from the same area based on ^{210}Pb dating (Aronson et al. 2014).

There can be little doubt that the largest decline in coral cover on Caribbean reefs that occurred in recent decades resulted from the regional mass mortality of *Acropora* spp. from WBD (Gladfelter 1982; Bythell and Sheppard 1993; Aronson and Precht 2001a, 2001b; Schutte et al. 2010), subsequent mass mortalities of massive corals notwithstanding. The disease-induced mass mortality of acroporids that occurred from the late 1970s through the early 1990s was not demonstrably connected to overfishing or changes in land use, and it was more than likely related to warming sea temperatures (Kline and Vollmer 2011; Randall and van Woesik 2015). Why is this important? Cramer et al. (2012) stated, "[Our] results, coupled with increasing evidence that protection from local disturbances may increase reef resilience to climate change (Hughes et al. 2007; Knowlton and Jackson 2008), highlight the importance of managing local impacts such as fishing and land clearing to stem the tide of reef decline." In fact, the primary causes of the decline of *Acropora*, and other Caribbean corals for that matter, operated—and still operate—on larger spatial scales, highlighting the importance of confronting regional and global impacts if we are to save coral reefs.

Below, we evaluate whether local management can promote resilience under current conditions. We emphatically agree that local problems should be addressed, for reasons spelled out at the end of this chapter. Contrary to the accusation of Knowlton and Jackson (2008) that we have been monolithic in our view of the importance of WBD, we have always explicitly subscribed to the pluralistic view of Quinn and Dunham (1983) that ecology seeks to evaluate the relative importance of the many causes underlying an observed pattern. Overestimating and overvaluing our capacity to promote resilience through local action, however, diverts attention and resources from the issue of climate change.

11.3 Indicators of Degradation

McClanahan, Graham et al. (2011) showed that, as reefs in the Indian Ocean degraded, corals were the most resistant components and the last to decline. Planes et al. (2005) reported that when shock waves from nuclear testing extirpated the reef-fish assemblages of Mururoa Atoll in French Polynesia, the living coral assemblages remained intact and the habitat they provided facilitated recolonization of the fish. Likewise, the early losses of reef components in the Caribbean other than corals, such as fish stocks (Jackson and Johnson 2000; Jackson 2008), are not incompatible with the recent loss of acroporids and other coral species to diseases and other causes (cf. Woodley 1992). On reefs where living coral cover has recently declined, however, the loss of that coral and concomitant loss of physical structure have resulted in significant declines in reef fish, independent of any impacts of fishing pressure (Jones et al. 2004; Alvarez-Filip et al. 2009; Paddack et al. 2009).

Our perception of the extent of reef degradation clearly depends on how degradation is defined and which components are considered important from ecological, societal, or other viewpoints. Significant functional degradation is perceived to have occurred earlier if fish stocks are considered most important than if corals are considered to be the signal components of the reefs named for them (Jackson and Johnson 2000; Pandolfi et al. 2003; McClanahan, Graham et al. 2011). The first view, in which the 'health' of a reef hinges on the state of its fish assemblage, implies long periods of latent degradation at the ecosystem level. Undetected losses of resilience are expressed latterly by threshold phenomena in the coral assemblages, including mass mortalities and poor recovery from bleaching events and disease outbreaks. The second view, which focuses on the corals themselves as the bellwethers of reef condition, implies the alternative hypothesis that fish—especially herbivorous fish—are less important to maintaining coral dominance than was previously thought. Herbivory could be critical to recovery in situations in which macroalgae have the potential to monopolize the substratum and suppress populations of juvenile corals (Sammarco 1982; Hughes and Tanner 2000; Mumby 2006; Box and Mumby 2007; Mumby et al. 2007a; Idjadi et al. 2010; Adam et al. 2015). Even in those cases, however, fish are not necessarily the most important herbivores. Echinoids, especially *Diadema antillarum*, are often far more potent herbivores on Caribbean reefs (Sammarco 1982; Edmunds and Carpenter 2001; Idjadi et al. 2006; Idjadi et al. 2010). The latter observation is independent of whether or not overfishing artificially enhanced the abundance of *Diadema* prior to their regional mass mortality (Hay 1984; Carpenter 1986; Precht and Aronson 2006; Sandin and McNamara 2012).

If macroalgae pose a threat to the recovery of coral populations on most reefs, there is little difference in the two views beyond the semantic issue of the point at which a reef is said to be degrading or degraded. The two views differ markedly, however, if macroalgae generally do not threaten coral recovery. Although, as stated above, high abundances of macroalgae can suppress coral recruitment, Bruno et al. (2009, 2014) questioned the proposition that macroalgae dominate most Caribbean reefs in their current state of low coral cover (see also Côté et al. 2005, 2013). The implication is that the decline and recovery of coral populations are largely decoupled from fishing pressure, as has been demonstrated explicitly for several reef systems in the Caribbean (Aronson et al. 2012; Edmunds 2013).

11.4 The Role of Marine Protected Areas

The overfishing hypothesis in its extreme form leads directly to the idea that ecological problems in the sea would largely be solved if only we would control fishing pressure (e.g., Jackson et al. 2014). This presumption provides a clear rationale for continuing to set aside marine protected areas (MPAs) and continuing the protections afforded by existing MPAs, which are designed to control the exploitation of fish stocks. By virtue of their current design, however, MPAs are less effective or wholly ineffective at controlling terrigenous inputs of nutrients and sediments. And by virtue of their scale they do not address the root-causes of climate change: the size of the human population and greenhouse-gas emissions. Climate change is expressed on coral reefs through the direct impacts of increasing sea temperatures, decreasing carbonate saturation states, and rising sea levels, as well as ancillary effects that may include outbreaks of coral disease and increasing intensities of hurricanes (Kleypas et al. 2001; Gardner et al. 2005; Hoegh-Guldberg et al. 2007; Harvell et al. 2009; Anthony et al. 2011; and many others).

Although Mora et al. (2006) concluded that MPAs in their current form do not preserve the trophic cascades of predation and herbivory that putatively maintain coral populations, there are benefits to coral cover of protection from fishing and terrigenous input, as well as benefits of protection from fishing pressure alone (Houk et al. 2010; McCook et al. 2010; Selig and Bruno 2010). Evidence is rapidly mounting, however, that overfishing is not the primary threat to benthic assemblages on coral reefs. Protecting fish stocks does not necessarily reduce the cover of macroalgae, increase coral populations, or preserve or increase the topographic heterogeneity that is critical to maintaining and increasing those fish stocks (McClanahan et al. 2001, 2005; Aronson and Precht 2006; Bood 2006; Idjadi et al. 2006; Vroom et al. 2006; Coelho and Manfrino 2007; Kramer and Heck 2007; Bruno et al. 2009; Myers and Ambrose 2009; Stockwell et al. 2009; Dudgeon et al. 2010; Alvarez-Filip et al. 2011; Lowe et al. 2011; McClanahan, Huntington et al. 2011; Żychaluk et al. 2012; Bégin et al. 2016). The threats of continuing climate change and its collateral impacts loom large, raising questions about the potential of local management alone, or the phenotypic or evolutionary responses of corals and their zooxanthellae (Baker et al. 2008; Sammarco and Strychar 2009; Pandolfi et al. 2011; van Woesik and Jordán-Garza 2011), to reverse or even delay significantly the hemmorhagic damage that is already well underway (Donner et al. 2005; Hoegh-Guldberg et al. 2007; Donner 2009; Hoegh-Guldberg et al. 2011; Toth et al. 2014).

The claim that restoring herbivores will save coral populations by reducing the cover of macroalgae (Aronson 1990; Jackson et al. 2001; Pandolfi et al. 2003; Mumby 2006; Jackson et al. 2014) has been 'augmented' with the idea that MPAs will maintain the resilience of reefs, 'buying time' while we address climate change (Hughes et al. 2003; Bellwood et al. 2004; Hughes et al. 2007; Mumby et al. 2007a, 2007b; Hughes et al. 2010; Edwards et al. 2011). Even the latter concept is of questionable validity (McClanahan et al. 2005; Bood 2006, Graham et al. 2008; McClanahan 2008; Graham et al. 2011; Huntington et al. 2011). Existing MPAs do not enhance the resistance or resilience of reef assemblages to thermal stress (Hoegh-Guldberg and Bruno 2010; Selig et al. 2012). Previous exposure of corals to high-temperature conditions is a far better predictor of the persistence of coral populations during positive thermal anomalies than their status of protection (Thompson and van Woesik 2009; Selig et al. 2012; Grottoli et al. 2014).

Côté and Darling (2010) pointed out that disturbed reef assemblages, which replace more pristine (or less degraded) assemblages following perturbations, are by default more resilient because 'recovery' to those early successional states requires little time and meets with little or no systemic resistance. It is cold comfort to be reminded of the inescapable, thermodynamic reality that the end of marine life will be the most stable state of all. As we labor to prevent that dreadful eventuality from accelerating into the present century, it is well to remember that the most resistant or resilient configurations are not necessarily the most desirable (Rogers 2013).

11.5 Parsimonious Explanations

Millennial-scale physical drivers, including natural trends in climate, often explain the historical limits to the growth and composition of Holocene coral assemblages more simply and more completely than hypotheses of human exploitation and other forms of interference. Fishing and terrigenous input are regional issues, but they are perpetrated and controlled locally. Climate change occurs at the largest spatial scales, but the resultant changes in parameters such as sea temperature and pH act at very small scales. They influence the coral holobiont, its physiological rates, and the microenvironment in which it lays down aragonite crystals, as well as rates of carbonate precipitation and submarine cementation from other biotic and abiotic processes (Kleypas et al. 1999; Macintyre and Aronson 2006; Manzello et al. 2008). Those microscale processes scale up to the level of the reef system and beyond, interacting with such second-order rates as the flux of nutrients into the system and their influence on carbonate deposition and bioerosion. Herbivory, predation, and other rates that ecologists view as critical to the healthy functioning, persistence, and resilience of reef systems (Jackson et al. 2001; Mumby et al. 2007b; Sandin et al. 2008) overprint the impacts of physical processes, driving the trajectories of benthic reef assemblages over a range of relevant time scales (Urban et al. 2012). The challenge is to determine how important those ecological interactions really are on ecological scales of decades to centuries, whether larger, millennial time scales have been more important than ecological scales, and which processes have been important on those millennial time scales.

Just because people were around when acroporid corals ceased building reef framework off present-day Fort Lauderdale 6000 years ago (Lighty et al. 1978; Toscano and Macintyre 2003; Banks et al. 2007) does not mean humans were responsible for their decline. Climatic cooling in the late Holocene was likely the primary cause of reef shutdown off the eastern coast of the Florida Peninsula, and a warming climate is now permitting the northward re-expansion of cold-sensitive coral taxa in the western Atlantic (Precht and Aronson 2004; Precht and Miller 2007; see Greenstein and Pandolfi 2008 for an example from western Australia).

In a similar vein, human activities did not drive branching *Porites* corals to replace *A. cervicornis* ~500 years ago in the shallow zones of the uncemented, lagoonal, rhomboid shoals in Belize. As those reefs grew to sea level, the living coral assemblages relocated themselves to a new physical environment—shallower water—and their species composition changed to produce the observed shallowing-upward sequence (Aronson et al. 1998). The persistence of the coral assemblages and the geomorphology of the reefs are controlled by tectonic events, which occur every few millennia, cause catastrophic slope-failure, and wipe out a substantial proportion of the benthic communities (Aronson et al. 2012).

Some investigators have insisted on anthropogenic causes for the degradation of coral reefs, in spite of evidence that is at best equivocal. Much has been written about a connection between the advent of European agriculture in Barbados and the demise of populations of *Acropora palmata* there, but the supposition that the story is one of nutrient loading from agricultural runoff rests almost entirely on a mild suggestion in a paper by John Lewis (1984). Lewis attributed the late Holocene disappearance of *Acropora palmata* from inshore, fringing reefs along the west coast of Barbados to storm damage, successional processes, and *possibly* terrigenous runoff that resulted from land-clearing and sugar-cane cultivation beginning in the 1600s. Twenty-seven years later, Sala and Jackson (2011, p. 197) had this to say about Lewis's (1984) results from Barbados:

> Circumstantial evidence suggests that the problems in Barbados were due to deforestation of the island for sugarcane and the consequent runoff of sediments and human waste, as well as extreme overfishing to feed the burgeoning population.

Four years before that, however, Macintyre et al. (2007) had commented on Lewis's (1984) interpretation of the late Holocene record of *A. palmata* in Barbados:

> Formerly attributed to human activity, the demise of a bank–barrier reef off southeastern Barbados…is now thought to be largely the result of late Holocene, millennial-scale storm damage.

Macintyre et al. (2007) suggested that the vibrant growth of stands of *A. palmata* in a reef-crest habitat off the south coast had ceased long before the European colonization of Barbados. The primary cause of mortality was physical damage 3000–4500 cal BP, with agricultural runoff possibly accounting for the mortality of remnant colonies 300–400 cal BP (see also Toscano 2016). Roff et al. (2013) echoed this latter model in suggesting that terrigenous runoff combined with climatic perturbation drove a phase shift in a nearshore coral assemblage on the Great Barrier Reef. The loss of *A. palmata* from the south coast of Barbados, millennia before any European influence, could have been part of a regional drawdown of that species (Hubbard et al. 2005; Macintyre et al. 2007).

Lewis's (1984) speculation about agriculture applies to inshore, fringing reefs. It is, therefore, not necessarily incompatible with the conclusions of Macintyre et al. (2007) about a bank–barrier reef further offshore, but populations of *A. palmata* in Barbados were not all killed by runoff. Furthermore, the idea that terrigenous input was to blame cannot fully account for the replacement of *A. palmata* by a vibrant assemblage of massive corals on the inshore reefs (Lewis 1960). A coral assemblage dominated by *Orbicella* spp. and other massive species is precisely what one would expect in a hurricane-dominated environment (Stoddart 1963; Porter et al. 1981; Woodley 1989), such as that envisioned by Macintyre et al. (2007). Overfishing, incidentally, had never been part of Lewis's (1984) original scenario, nor did it figure in the interpretations of Macintyre et al. (2007) or Toscano (2016).

An argument for the runoff hypothesis would also have to explain why Barbados, which is a low, carbonate island, is the only known location in the Caribbean for which such a scenario has been suggested. Jamaica, for example, is a high island with a similar history of colonization and sugar-cane cultivation. The impacts of terrigenous runoff should have been accentuated compared to Barbados, yet there is no evidence for a mass mortality of *A. palmata* in Jamaica (or any other Caribbean island) 300–400 years ago. The best way to test the hypothesis would be through biogeochemical analysis (cf. Aronson et al. 2014).

11.6 Temporal Priority

Which regional or global driver is or was most important in the degradation of modern reefs is largely a consequence of temporal priority. Thermally induced bleaching has been secondarily important in the Caribbean only because bleaching episodes came after much of the *Acropora* had already been killed by white-band disease, in a regional outbreak that itself might have been thermally driven. The situation is reversed in the Indo-Pacific: bleaching has had a greater influence than coral disease because bleaching-induced mortality occurred on a large geographic scale prior to disease outbreaks (Buddemeier et al. 2004). It remains to be seen whether ocean acidification will have the opportunity to damage coral populations and coral reefs (Hoegh-Guldberg et al. 2007; Veron 2008; Hönisch et al. 2012) after the impacts of rising sea temperatures have taken their toll. Some state or rate is the primary limit to the growth of coral populations, the integrity of the assemblages those populations comprise, and the accretion of the reef frameworks on which they perch as a living veneer.

11.7 Conclusion

It should come as no surprise that coral reefs are highly sensitive to climate change. Narrow annual temperature ranges characterize shallow-benthic habitats at tropical and polar latitudes, compared to analogous habitats in the highly seasonal temperate zone (Fig. 11.2). Benthic ectotherms near their latitudinal extremes are adapted to the narrow seasonal temperature ranges to which they are normally exposed. Stenothermy limits their scope for phenotypic or evolutionary adjustment to warming temperatures. The tropical and polar benthos, therefore, are responding earlier and more strongly to warming sea temperatures than are temperate-marine biotas.

All biological interactions play out on the template of the physical environment. For every ecosystem, there is a range of larger scales at which physical drivers trump biotic interactions in determining its trajectory; it is just a matter of scaling up sufficiently to that range. Reefs through Phanerozoic time are no exception (Wood 1999, 2007; Veron 2008; Kiessling and Simpson 2011; Hönisch et al. 2012; Norris et al. 2013). The projected dynamical responses of reefs to large-scale physical drivers, both natural and anthropogenic, are the appropriate foundations of hypotheses against which to test the ecological effects of localized human activities.

Our point about modern coral reefs under human influence is that one does not have to scale up very much at all to discover the scales at which physical controls predominate, because the lower end of the range of scales at which physical drivers strongly influence the biotic milieu, or indeed overpower biological processes, is not very large. For coral populations and coral reefs, physical controls are primary at scales equal to, or only slightly larger than, the scales at which biotic interactions are measured and observed. A good example is the influence of anthropogenically warming temperatures on outbreaks of coral disease. Geological and ecological processes may be disjunct in some marine ecosystems, but for coral reefs they operate on scales that are very similar.

That is not to say that geology equals ecology. Some paleoecologists view the geologically rapid changes in sea level and sea temperature during the Pleistocene glaciations as disturbances in ecological time, precipitating the collapse of reef communities and requiring their subsequent reconstitution in other habitats or locations (Jackson 1992; Pandolfi 1996; Webster et al. 2004; Jackson and Erwin 2006; Pandolfi and Jackson 2006). But even when glacial/interglacial fluctuations were dramatic in geologic time, they were still slow compared to the turnover rates of the corals (see Kleypas 1997). In most cases, coral populations should have been able to alter their spatial and habitat distributions incrementally over long periods, but the endpoints of these incremental shifts displayed in the fossil record have been presumed to represent radical disassembly and reassembly (see discussion in Tager et al. 2010; see also Chap. 7).

Overfishing is a terrible problem with far-reaching consequences. There is more than enough sorrow to go around over the strangulation of marine ecosystems, and we wholeheartedly endorse efforts to protect life in the sea from the grotesque and irrational level of exploitation to which it is being subjected. Even worse for coral reefs, unfortunately, are rising temperatures, ocean acidification, and other potential impacts of climate change, such as predicted future increases in the intensity of hurricanes (see also Glynn 2011). Because these physical drivers operate at a global scale, they exert a powerful influence that is harder to control by a long shot than fishing pressure and terrigenous input, which are the feasible targets of existing MPAs and other local management strategies. Local actions to control fishing and runoff, along with a more strategic, integrative approach to the design and location of marine reserves that accounts for spatial variation in susceptibility to climate change (Riegl and Piller 2003; McClanahan et al. 2008, 2009; Mumby et al. 2011), could prolong the death-throes of coral populations; however, they will keep dying until government and society recognize climate change for the grave threat it is and address it on a geopolitical level. If and when the impacts of climate change can be mitigated or reversed, their impacts will no longer overwhelm local threats. Fishing, terrigenous input, and other localized problems will then more commonly become the limiting factors and will respond more strongly to the appropriate protective measures. Liebig's Law of the Minimum, a nineteenth-century model from agronomy (Hooker 1917), has great value for understanding the challenges facing coral reefs and their human stewards.

The Law of the Minimum describes the serial limitation of different nutrients on crop yields. Liebig used the metaphor of a barrel with staves of different lengths, each stave representing a nutrient. The water-level in the barrel represented crop yield, and the shortest stave represented the limiting nutrient. Adapting the metaphor to coral reefs, the staves of the barrel in Fig. 11.3 represent the many factors that potentially limit the growth of coral populations and accretion of reef frameworks, both of which are represented by the water level. The shortest stave identifies the rate-limiting process. The water can be no higher than that stave, meaning that coral growth or framework accretion can be no greater than allowed by the rate-limiting process.

We have chosen rising global temperatures as the primary limitation in Fig. 11.3, based on the foregoing discussion. If the temperature rise can be slowed or reversed, then warming will cease to be the rate-limiting process. The global-warming stave of the barrel will then be lengthened and will no longer be the shortest one. The shortest stave in the figure will then be disease, which as we have said could be linked to warming seas. Synergistic or antagonistic interactions among the drivers of coral-reef degradation (Mora et al. 2007) mean that lengthening some staves will lengthen or shorten certain other staves as well. If and when regional- and global-scale limitations are adequately addressed, then local-scale factors, such as nutrient loading

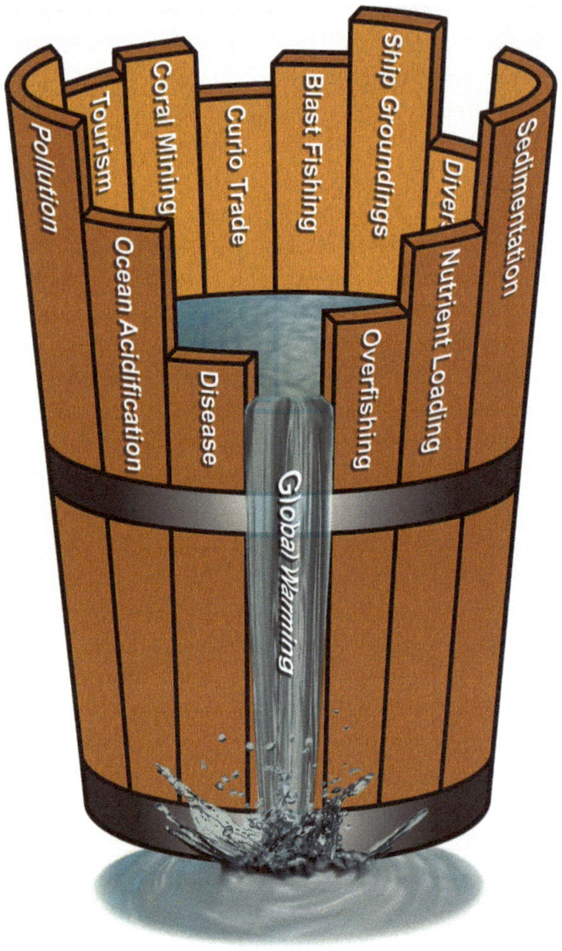

Fig. 11.3 Liebig's Law of the Minimum applied to coral reefs. The lengths of the barrel-staves were set somewhat arbitrarily for illustrative purposes

and overfishing, will serially become the limiting staves. The upshot is that simultaneous actions at local, regional, and global levels are our only hope for saving coral reefs (see also Hoegh-Guldberg and Bruno 2010; Kennedy et al. 2013; Rogers 2013). Clearly, global problems will be more difficult to solve and take longer than local ones, but that is very different from saying we should focus on local management now because it will buy time for us to address the impacts of climate change in the future.

Planning to have reefs around for our children and our children's children to enjoy—meaning that we want to keep them reasonably intact for a little over a century—is just not good enough. We need to conserve reefs on a millennial time scale by fighting climate change on a global spatial scale. It may sound perverse but, considering the jeopardy in which we humans have placed coral reefs, the biosphere, and our very existence, addressing climate change and being left with an exceedingly difficult set of local-scale, ecological problems to attack would be a blessing.

Acknowledgments We thank M. E. S. Bracken, J. F. Bruno, N. L. Hilbun, D. K. Hubbard, L. Kaufman, L. T. Toth, S. Thatje, and R. van Woesik for advice and discussion. R. M. Moody and L. T. Toth drafted the figures. Our research on coral reefs has been supported over the years by the Smithsonian Institution, the National Geographic Society, Northeastern University's Three Seas Program, and the U.S. National Science Foundation (most recently grant OCE-1535007 to R.B.A.). This is contribution number 75 from the Institute for Research on Global Climate Change at the Florida Institute of Technology.

References

ABRT, *Acropora* Biological Review Team (2005) Atlantic *Acropora* Status Review Document. Miami, Florida: Report to National Marine Fisheries Service, Southeast Regional Office, 152 p

Acevedo R, Morelock J, Olivieri RA (1989) Modification of coral reef zonation by terrigenous sediment stress. Palaios 4:92–100

Adam TC, Burkepile DE, Ruttenberg BI, Paddack MJ (2015) Herbivory and the resilience of Caribbean coral reefs: knowledge gaps and implications for management. Mar Ecol Prog Ser 520:1–20

Adey WH, Burke RB (1977) Holocene bioherms of Lesser Antilles: geologic control of development. In: Frost SH, Weiss MP, Saunders JB (eds) Reefs and Related Carbonates—Ecology and Sedimentology. American Association of Petroleum Geologists Studies in Geology 4:67–82

Alvarez-Filip L, Dulvy NK, Gill JA, Côté IM, Watkinson AR (2009) Flattening of Caribbeancoral reefs: region-wide declines in architectural complexity. Proc Soc B 276:3019–3025

Alvarez-Filip L, Gill JA, Dulvy NK, Perry AL, Watkinson AR, Côté IM (2011) Drivers of region-wide declines in architectural complexity on Caribbean reefs. Coral Reefs 30:1051–1060

Anthony KRN, Maynard JA, Diaz-Pulido G , Mumby PJ , Marshall PA, Cao L, Hoegh-Guldberg O (2011) Ocean acidification and warming will lower coral reef resilience. Global Change Biol 17:1798–1808

Aronson RB (1990) Onshore–offshore patterns of human fishing activity. Palaios 5:88–93

Aronson RB (2007) Geological approaches to coral reef ecology. New York: Springer-Verlag, 425 pp

Aronson RB, Bruckner A, Moore J, Precht W, Weil E (2009a) *Acropora cervicornis* (staghorn coral). *IUCN Red List: The IUCN Red List of Threatened Species*, version 2009.1. http://www.iucnredlist.org/details/133381/0.

Aronson RB, Bruckner A, Moore J, Precht W, Weil E (2009b) *Acropora palmata* (elkhorn coral). IUCN Red List: The IUCN Red List of Threatened Species, version 2009.1. http://www.iucnredlist.org/details/133006/0.

Aronson RB, Hilbun NL, Bianchi TS, Filley TR, McKee BA (2014) Land use, water quality, and the history of coral assemblages at Bocas del Toro, Panamá. Mar Ecol Prog Ser 504:159–170

Aronson RB, Macintyre IG, Lewis, SA, Hilbun NL (2005) Emergent zonation and geographic convergence of coral reefs. Ecology 86:2586–2600

Aronson, RB, Macintyre IG, Precht WF, Wapnick CM, Murdoch TJT (2002) The expanding scale of species turnover events on coral reefs in Belize. Ecol Monogr 72:233–249

Aronson RB, Macintyre IG, Wapnick CM, O'Neill MW (2004) Phase shifts, alternative states, and the unprecedented convergence of two reef systems. Ecology 85:1876–1891

Aronson RB, Precht WF (1997) Stasis, biological disturbance, and community structure of a Holocene coral reef. Paleobiology 23:326–346

Aronson RB, Precht WF (2000) Herbivory and algal dynamics on the coral reef at Discovery Bay, Jamaica. Limnol Oceanogr 45:251–255

Aronson RB, Precht WF (2001a) Evolutionary paleoecology of Caribbean coral reefs. In: Allmon WD, Bottjer DJ (eds) Evolutionary paleoecology: the ecological context of macroevolutionary change. New York: Columbia University Press, pp 171–233

Aronson RB, Precht WF (2001b) White-band disease and the changing face of Caribbean coral reefs. Hydrobiologia 460:25–38

Aronson RB, Precht WF (2006) Conservation, precaution, and Caribbean reefs. Coral Reefs 25:441–450

Aronson RB, Precht WF, Macintyre IG (1998) Extrinsic control of species replacement on a Holocene reef in Belize: the role of coral disease. Coral Reefs 17:223–230

Aronson RB, Precht WF, Macintyre IG, Toth LT (2012) Catastrophe and the lifespan of coral reefs. Ecology 93:303–313

Aronson RB, Thatje S, McClintock JB, Hughes KA (2011) Anthropogenic impacts on marine ecosystems in Antarctica. Ann NY Acad Sci 1223:82–107

Bak RPM, Criens SR (1981) Survival after fragmentation of colonies of *Madracis mirabilis*, *Acropora palmata* and *A. cervicornis* (Scleractinia) and the subsequent impact of coral disease. Proc 4th Intl Coral Reef Symp 2:221–228

Bak RPM, Nieuwland G (1995) Long-term change in coral communities along depth gradients over leeward reefs in the Netherlands Antilles. Bull Mar Sci 56:609–619

Baker AC, Glynn PW, Riegl B (2008) Climate change and coral reef bleaching: an ecological assessment of long-term impacts, recovery trends and future outlook. Estuar Coast Shelf Sci 80:435–471

Banks KW, Riegl BM, Shinn EA, Pillar WE Dodge RE (2007) Geomorphology of the Southeast Florida continental reef tract (Miami-Dade, Broward, and Palm Beach Counties, USA). Coral Reefs 26:617–633

Bégin C, Schelten CK, Nugues MM, Hawkins J, Roberts C, Côté IM (2016) Effects of protection and sediment stress on coral reefs in St. Lucia. PLoS ONE 11:e0146855, doi: 10.1371/journal.pone.0146855

Bellwood DR, Hughes TP, Folke C, Nyström M (2004) Confronting the coral reef crisis. Nature 429:827–833

Birkeland C (1997) Geographic differences in ecological processes on coral reefs. In: Birkeland C (ed) Life and death of coral reefs. New York: Chapman and Hall, pp 273–297

Boesch D, Burreson E, Dennison W, Houde E, Kemp M, Kennedy V, Newell RT, Paynter K, Orth R, Ulanowicz R (2001). Factors in the decline of coastal ecosystems. Science 293:1589–1590

Bood, N.D. (2006) Recovery and resilience of coral assemblages on managed and unmanaged reefs in Belize: a long-term study. MS Thesis. Mobile: University of South Alabama

Box SJ, Mumby PJ (2007) Effect of macroalgal competition on growth and survival of juvenile Caribbean corals. Mar Ecol Prog Ser 342:139–149

Bruno JF, Precht WF, Vroom PS, Aronson RB (2014) Coral reef baselines: how much macroalgae is natural? Mar Poll Bull 80:24–29

Bruno JF, Selig ER, Casey KS, Page CA, Willis B, Harvell CD, Sweatman H, Melendy A (2007) Thermal stress and coral cover as drivers of coral disease outbreaks. PLoS Biol 5:e124, doi:10.1371/journal.pbio.0050124

Bruno JF, Sweatman H, Precht WF, Selig ER, Schutte VGW (2009) Assessing evidence of phase shifts from coral to macroalgal dominance on coral reefs. Ecology 90:1478–1484

Buddemeier RW, Kleypas JA, Aronson RB (2004) Coral reefs and global climate change: potential contributions of climate change to stresses on coral reef ecosystems. Arlington, Virginia: Pew Center on Global Climate Change, 44 pp

Bythell J, Sheppard C (1993) Mass mortality of Caribbean shallow corals. Mar Poll Bull 26:296–297

Carpenter RC (1986) Partitioning herbivory and its effects on coral reef algal communities. Ecol Monogr 56:345–363

Carpenter KE Abra M, Aeby G, Aronso RB, Banks S, Bruckner A, Chiriboga A, Cortés J, Delbeek JC, DeVantier L, Edgar GJ, Edwards AJ, Fenner D, Guzmán HM, Hoeksema BW, Hodgson G. Johan O, Licuanan WY, Livingstone SR, Lovell ER, Moore JA, Obura DO, Ochavillo D, Polidoro BA, Precht WF, Quibilan MC, Reboton C, Richards ZT, Rogers AD, Sanciangco J, Sheppard A, Sheppard C, Smith J, Stuart S, Turak E, Veron JEN, Wallace C, Weil E, and Wood E (2008) One-third of reef-building corals face elevated extinction risk from climate change and local impacts. Science 321:560–563.

Clarke A, Crame JA (2010). Evolutionary dynamics at high latitudes: speciation and extinction in polar marine faunas. Phil Trans R Soc B 365:3655–3666

Coelho VR, Manfrino C (2007) Coral community decline at a remote Caribbean island: marine no-take reserves are not enough. Aquat Conserv Mar Freshw Ecosys 17:666–685

Condon RH, Graham WM, Duarte CM, Pitt KA, Lucas CH, Haddock SHD, Sutherland KR, Robinson KL, Dawson MN, Decke MB, Mills CE, Purcell JE, Malej A, Mianzan H, Uye S-I, Gelsich S, Madin. LP2 (2012) Questioning the rise of gelatinous zooplankton in the world's oceans. BioScience 62:160–169

Cortés, J (1994) A reef under siltation stress: a decade of degradation. In: Ginsburg RN (ed.) Proceedings of the Colloquium on Global Aspects of Coral Reefs: Health, Hazards and History. Miami: Rosenstiel School of Marine and Atmospheric Science, University of Miami, pp 240–246

Côté IM, Darling ES (2010) Rethinking ecosystem resilience in the face of climate change. PLoS Biology 8:e1000438, doi:10.1371/journal.pbio.1000438

Côté IM, Gill JS, Gardner TA, Watkinson, AR (2005) Measuring coral reef decline through meta-analyses. Phil Trans R Soc Lond B 360:385–395

Côté IM, Precht WF, Aronson RB, Gardner TA (2013) Is Jamaica a good model for understanding Caribbean coral reef dynamics? Mar Poll Bull 76:28–31

Cramer KL, Jackson JBC, Angioletti CV, Leonard-Pingel J, Guilderson TP (2012) Anthropogenic mortality on coral reefs in Caribbean Panama predates coral disease and bleaching. Ecol Lett 15:561–567

Dana JD (1843) On the temperature limiting the distribution of corals. Am J Sci 45:130–131

Darwin CR (1842) The Structure and Distribution of Coral Reefs. Being the First Part of the Geology of the Voyage of the Beagle, under the Command of Capt. Fitzroy, R.N during the Years 1832 to 1836. London: Smith, Elder and Company, 214 pp

Davis GE (1982) A century of natural change in coral distribution at the Dry Tortugas: a comparison of reef maps from 1881 and 1976. Bull Mar Sci 32:608–623

De'ath G, Fabricius K (2010) Water quality as a regional driver of coral biodiversity and macroalgae on the Great Barrier Reef. Ecol Appl 20:840–850

Donner SD (2009) Coping with commitment: projected thermal stress on coral reefs under different future scenarios. PLoS One 4:e5712, doi:10.1371/journal.pone.0005712

Donner SD, Skirving WJ, Little CM, Oppenheimer M, Hoegh-Guldberg O (2005) Global assessment of coral bleaching and required rates of adaptation under climate change. Global Change Biol 11:1–15

Dudgeon SR, Aronson RB, Bruno JR, Precht WF (2010) Phase shifts and stable states on coral reefs. Mar Ecol Prog Ser 413:201–216

Dulvy NK Freckleton RP, Polunin NVC (2004) Coral reef cascades and the indirect effects of predator removal by exploitation. Ecol Lett 7:410–416

Eakin CM, Morgan JA, Heron SF, Smith TB, Lui G, et al. (2010) Caribbean corals in crisis: record thermal stress, bleaching, and mortality in 2005. PloS ONE 5:e13969, doi: 10.1371/journal.pone.0013969

Edmunds PJ (2013) Decadal-scale changes in the community structure of coral reefs of St. John, US Virgin Islands. Mar Ecol Prog Ser 489:107–123

Edmunds PJ, Carpenter RC (2001) Recovery of *Diadema antillarum* reduces macroalgal cover and increases abundance of juvenile corals on a Caribbean reef. Proc Natl Acad Sci USA 98:5067–5071

Edwards HJ, Elliott IA, Eakin CM, Irikawa A, Madin JS, McField M, Morgan JA, van Woesik R Mumby PJ (2011) How much time can herbivore protection buy for coral reefs under realistic regimes of hurricanes and coral bleaching? Global Change Biol 17:2033–2048

Estes JA, Duggins DO (1995) Sea otters and kelp forests in Alaska: generality and variation in a community ecological paradigm. Ecol Monogr 65:75–100

Estes JA, Terborgh T, Brashares JA, Power ME, Berger J, Bond WJ, Carpenter SR, Essington TE, Hol RD, Jackson JBC, Marquis RJ, Oskanen L, Oskane T, Paine RT, Pikitch EK, Ripple WJ, Sandin SA, Scheffer M, Schoene TW, Shurin JB, Sinclair ARE, Soulé ME, Virtanen R, DA Wardle (2011) Trophic downgrading of Planet Earth. Science 333:301–306

Foster MS, Schiel DR (2010) Loss of predators and the collapse of southern California kelp forests (?): alternatives, explanations and generalizations. J Exp Mar Biol Ecol 393:59–70

Gardner TA, Côté IM, Gill JA, Grant A, Watkinson AR (2003) Long-term region-wide declines in Caribbean corals. Science 301:958–960

Gardner TA, Côté IM, Gill JA, Grant A, Watkinson AR (2005). Hurricanes and Caribbean coral reefs: impacts, recovery patterns, and role in long-term decline. Ecology 86:174–184

Geister J (1977) The influence of wave exposure on the ecological zonation of Caribbean coral reefs. Proc 3rd Intl Coral Reef Symp 1:23–29

Gignoux-Wolfsohn SA, Marks CJ, Vollmer SV (2012) White band disease transmission in the threatened coral, *Acropora cervicornis*. Nature Sci Rep 2:804, doi:10.1038/srep00804

Gil-Agudelo DL, Smith GW, Weil E (2006) The white band disease type II pathogen in Puerto Rico. Rev Biol Trop 54(Suppl 3):59–67

Ginsburg RN, Shinn EA (1994) Preferential distribution of reefs in the Florida reef tract: the past is the key to the present. In Proceedings of the Colloquium on Global Aspects of Coral Reefs: Health, Hazards and History, 1993, Ginsburg RN (ed.), Miami: Rosenstiel School of Marine and Atmospheric Science, University of Miami, pp 21–26

Gladfelter WB (1982) White band disease in *Acropora palmata*: implications for the structure and growth of shallow reefs. Bull Mar Sci 32:639–643

Glynn PW (2011) In tandem reef coral and cryptic metazoan declines and extinctions. Bull Mar Sci 87:767–794

Graham NAJ, McClanahan TR, MacNeil MA, Wilson SK, Polunin NVC, Jennings S, Chabanet P, Clark S, Spalding MD, Letourneur Y, Bigot L, Gazin R, Öhma MC, Garpe KC, Edwards AJ, Sheppard CRC (2008) Climate warming, marine protected areas and the ocean-scale integrity of coral reef ecosystems. PLoS One 3: e3039, doi:10.1371/journal.pone.0003039

Graham NAJ, Nash KL, Kool JT (2011) Coral reef recovery dynamics in a changing world. Coral Reefs 30:283–294

Graus RR, Macintyr IG (1989) The zonation of Caribbean coral reefs as controlled by wave and light energy input, bathymetric setting and reef morphology: computer simulation experiments. Coral Reefs 8:9–18

Greenstein BJ, Curran AH, Pandolfi JM (1998) Shifting ecological baselines and the demise of *Acropora cervicornis* in the western North Atlantic and Caribbean Province: a Pleistocene perspective. Coral Reefs 17:249–261

Greenstein BJ, Pandolfi JM (2008) Escaping the heat: range shifts of reef coral taxa in coastal western Australia. Global Change Biol 14:513–528

Greer L Jackson JE, Curran HA, Guilderson T, Teneva L (2009) How vulnerable is *Acropora cervicornis* to environmental change? Lessons from the early to middle Holocene. Geology 37:263–266

Grottoli AG, Warner ME, Levas SJ, Aschaffenburg MD, Schoepf V, McGinley Baumann MJ, Matsui Y (2014) The cumulative impact of annual coral bleaching can turn some coral species winners into losers. Global Change Biol, doi:10.1111/gcb.12658

Hallock P, Schlager W (1986) Nutrient excess and the demise of coral reefs and carbonate platforms. Palaios 1:389–398

Halpern BS, Walbridge S, Selkoe KA, Kappel CV, Micheli F, D'Agrosa C, Bruno JF, Casey KS, Ebert C, Fox HE, Fujita R, Heinemann D, Lenihan HS, Madin EMP, Perry MT, Selig ER, Spalding M, Steneck R, Watson R (2008) A global map of human impact on marine ecosystems. Science 319:948–952

Harvell D, Altizer S, Cattadori IM, Harrington L, Weil E (2009) Climate change and wildlife diseases: when does the host matter the most? Ecology 90:912–920

Hay ME (1984) Patterns of fish and urchin grazing on Caribbean coral reefs: are previous results typical? Ecology 65:446–454

Hoegh-Guldberg O, Bruno JF (2010) The impact of climate change on the world's marine ecosystems. Science 328:1523–1528

Hoegh-Guldberg O, Mumby PJ, Hooten AJ, Steneck RS, Greenfield P, Gomez E, Harvell CD, Sale PF, Edwards AJ, K. Caldeira, N. Knowlton, C.M. Eakin, R. Iglesias-Prieto, N. Muthiga, R.H. Bradbury, Dubi E, Hatziolos ME (2007) Coral reefs under rapid climate change and ocean acidification. Science 318:1737–1742

Hoegh-Guldberg O, Ortiz JC, Dove S (2011) The future of coral reefs. Science 334:1494–1495

Hogarth WT (2006) Endangered and threatened species: final listing determinations for elkhorn coral and staghorn coral. Fed Reg 71:26852–26872

Hönisch B, Ridgwell A, Schmidt DN, Thomas E, Gibbs SJ, Sluijs A, Zeebe R, Kump L, Martindale RC, Greene SE, Kiessling W, Ries J, Zachos J, Royer DL, Barker S, Marchitto TM Jr, Myoer R, Pelejero C, Ziveri P, Foster GL, Williams B (2012) The geological record of ocean acidification. Science 335:1058–1063

Hooker HD (1917) Liebig's Law of the Minimum in relation to general biological problems. Science 46:197–204

Houk P, Musburger C, Wiles P (2010) Water quality and herbivory interactively drive coral-reef recovery patterns in American Samoa. PLoS One 5:e13913, doi:10.1371/journal.pone.0013913

Hubbard DK (1988) Controls of modern and fossil reef development: common ground for biological and geological research. Proc 6th Intl Coral Reef Symp 1:243–252

Hubbard DK (1997) Reefs as dynamic systems. In: Birkeland C (ed) Life and death of coral reefs. New York: Chapman and Hall, pp 43–67

Hubbard DK, Zankl H, van Heerden I, Gill IP (2005) Holocene reef development along the northeastern St. Croix shelf, Buck Island, US Virgin Islands. J Sed Res 75:97–113

Hughes TP (1994) Catastrophes, phase shifts and large-scale degradation of a Caribbean coral reef. Science 265:1547–1551

Hughes TP, Baird AH, Bellwood DR, Card M, Connoll SR, Folke C, Grosberg R, Hoegh-Guldberg O, Jackson JBC, Kleypas J, Lough JM, Marshall P, Nyström M, Palumbi SR, Pandolfi JM, Rosen B, Roughgarden J (2003) Climate change, human impacts, and the resilience of coral reefs. Science 301:929–933

Hughes TP, Bellwood DR, Folke CS, McCook LS, Pandolfi JM (2007) No-take areas, herbivory and coral reef resilience. Trends Ecol Evol 22:1–3

Hughes TP., Graham NAJ, Jackson JBC, Mumby PJ, Steneck RS (2010) Rising to the challenge of sustaining coral reef resilience. Trends Ecol Evol 25:633–642

Hughes TP, Szmant AM, Steneck R, Carpenter R, Miller S (1999) Algal blooms on coral reefs: what are the causes? Limnol Oceanogr 44:1583–1586

Hughes TP, Tanner JE (2000) Recruitment failure, life histories, and long-term decline of Caribbean corals. Ecology 81:2250–2263

Huntington BE, Karnauskas M, Lirman D (2011) Corals fail to recover at a Caribbean marine reserve despite ten years of reserve designation. Coral Reefs 30:1077–1085

Idjadi JA, Haring RN, Precht WF (2010) Recovery of the sea urchin *Diadema antillarum* promotes scleractinian coral growth and survivorship on shallow Jamaican reefs. Mar Ecol Prog Ser 403:91–100

Idjadi JA, Lee SC, Bruno JF, Precht WF, Allen-Requa L, Edmunds PJ (2006) Rapid phase-shift reversal on a Jamaican coral reef. Coral Reefs 25:209–211

Jackson JBC (1992) Pleistocene perspectives on coral reef community structure. Am Zool 32:719–731

Jackson JBC (2008) Ecological extinction and evolution in the brave new ocean. Proc Natl Acad Sci 105(Suppl 1):11458–11465

Jackson JBC, Donovan MK, Cramer KL, Lam VV (eds) (2014) Status and trends of Caribbean coral reefs: 1970–2012. Gland: Global Coral Reef Monitoring Network, IUCN, 299 pp

Jackson JBC, Erwin DH (2006) What can we learn about ecology and evolution from the fossil record. Trends Ecol Evol 21:322–328

Jackson JBC, Johnson KG (2000) Life in the last few million years. Paleobiology 24 (Suppl):221–235

Jackson JBC, Kirby MX, Berger WH, Bjorndal KA, Botsford LW, Bourque BJ, Bradbury RH, Cooke R, Erlandson J, Estes JA, Hughe TP, Kidwell S, Lange CB, Lenihan HS, Pandolfi JM, Peterson CH, Steneck RS, Tegner MJ. Warner RR (2001) Historical overfishing and the recent collapse of coastal ecosystems. Science 293:629–638

Johannes RE, Wiebe WJ, Crossland CJ, Rimmer DW, Smith SV (1983) Latitudinal limits of coral reef growth. Mar Ecol Prog Ser 11:105–111

Jones GP, McCormick MI, Srinivasan M, Eagle JV (2004) Coral decline threatens fish biodiversity in marine reserves. Proc Natl Acad Sci 101:8251–8253

Kennedy EV, Perry CT, Halloran PR, Iglesias-Prieto R, Schönberg CHL, Wisshak M, Form AU, Carricart-Ganive JP, Fine M, Eakin CM, Mumby PJH (2013) Avoiding coral reef functional collapse requires local and global action. Curr Biol 23:912–918

Kidwell SM (2001) Preservation of species abundance in marine death assemblages. Science 294:1091–1094

Kidwell SM (2007) Discordance between living and death assemblages as evidence for anthropogenic ecological change. Proc Natl Acad Sci USA 104:17701–17706

Kiessling W, Simpson C (2011) On the potential for ocean acidification to be a general cause of ancient reef crises. Global Change Biol 17:56–67

Kleypas J (1997) Modeled estimates of global reef habitat and carbonate production since the last glacial maximum. Paleoceanography 12:533–545

Kleypas JA, Buddemeier RW, Archer D, Gattuso J-P, Langdon, Opdyke, B (1999) Geochemical consequences of increased atmospheric carbon dioxide on coral reefs. Science 284:118–120

Kleypas JA, Buddemeier RW, Gattuso J-P (2001) The future of coral reefs in an age of global change. Intl J Earth Sci (Geol Rdsch) 90:426–437

Kline DI, Vollmer SV (2011) White Band Disease (type I) of endangered Caribbean acroporid corals is caused by pathogenic bacteria. Nature Sci Rep 1:7, doi:10.1038/srep00007

Knowlton N, Jackson JBC (2008) Shifting baselines, local impacts, and global change on coral reefs. PLoS Biology 6:e54, doi:10.1371/journal.pbio.0060054

Knowlton N, Lang JC, Rooney MC, Clifford P (1981) Evidence for delayed mortality in hurricane-damaged Jamaican staghorn corals. Nature 294:251–252

Kramer KL, Heck KL Jr. (2007) Top-down trophic shifts in Florida Keys patch reef marine protected areas. Mar Ecol Prog Ser 349:111–123

Kuntz NM., Kline DI, Sandin SA, Rowher F (2005) Pathologies and mortality rates caused by organic carbon and nutrient stressors in three Caribbean coral species. Mar Ecol Prog Ser 294:173–180

Lafferty KD, Porter, J, Ford SE (2004) Are diseases increasing in the ocean? Ann Rev Ecol Evol Syst 35:31–54

Lescinsky H (2012) Partial mortality and re-sheeting in fossil corals: a disease proxy. Proc 12th Intl Coral Reef Symp 3:1–5

Lescinsky HB, Titus B, Hubbard D (2012) Live coral cover from the fossil record: an example from Holocene reefs of the Dominican Republic. Coral Reefs 31:335–346

Lewis JB (1960) The coral reefs and coral communities of Barbados, WI. Can J Zool 38:1133–1145

Lewis JB (1984) The *Acropora* inheritance: a reinterpretation of the development of ringing reefs in Barbados, West Indies. Coral Reefs 3:117–122

Lighty RG, Macintyre IG, Stuckenrath R (1978) Submerged early Holocene barrier reef south-east Florida shelf. Nature 276:59–60

Lowe PK, Bruno JF, Selig ER, Spencer M (2011) Empirical models of transitions between coral reef states: effects of region, protection, and environmental change. PLoS ONE 6:e26339, doi:10.1371/journal.pone.0026339

Macintyre IG, Aronson RB (2006) Lithified and unlithified Mg-calcite precipitates in tropical reef environments. J Sed Res 76:81–90

Macintyre IG, Glynn PW, Toscano MA (2007) The demise of a major *Acropora palmata* bank–barrier reef off the southeast coast of Barbados, West Indies. Coral Reefs 26:765–773

MacKenzie CL Jr (2007.) Causes underlying the historical decline in eastern oyster (*Crassostrea virginica* Gmelin, 1791) landings. J Shellfish Res 26:927–938

Manzello DP, Kleypas JA, Budd DA, Eakin CM, Glynn PW, Langdon C (2008) Poorly cemented coral reefs of the eastern tropical Pacific: possible insights into reef development in a high-CO_2 world. Proc Natl Acad Sci 105:10450–10455

McClanahan TR (2008) Response of the coral reef benthos and herbivory to fishery closure and the 1998 ENSO disturbance. Oecologia 155:169–177

McClanahan TR, Cinner JE, Graham NAJ, Daw TM, Maina J, Stead SM, Wamukota A, Brown K, Venus V, Polunin NVC (2009) Identifying reefs of hope and hopeful actions: contextualizing environmental, ecological, and social parameters to respond effectively to climate change. Cons Biol 23:662–671

McClanahan TR, Cinner JE, MainaJ, Graham NAJ, Daw TM, Stead, SM, Wamukota A, Brown K, Ateweberhan M, Venus V, Polunin NVC (2008) Conservation action in a changing climate. Cons Lett 1:53–59

McClanahan TR Graham NAJ, McNeil MA, Muthiga NA, Cinner JA, Bruggeman JH, Wilson, SK (2011) Critical thresholds and tangible targets for ecosystem-based management of coral reef fisheries Proc Natl Acad Sci USA 108:17230–17233

McClanahan TR, Huntington BE, Cokos B, (2011) Coral responses to macroalgal reduction and fisheries closure on Caribbean patch reefs. Mar Ecol Prog Ser 437:89–102

McClanahan TR, Maina J, Starge JCr, Herron-Perez P, Dusek E (2005) Detriments to post-bleaching recovery of corals. Coral Reefs 24:230–246

McClanahan TR., Muthiga NA, Mangi S (2001) Coral and algal changes after the 1998 coral bleaching: interaction with reef management and herbivores on Kenyan reefs. Coral Reefs 19:380–391

McCook LJ, Ayling T, Cappo M, Choat JH, Evans RD, De Freitas DM, Heupel M, Hughes TP, Jones GP, Mapstone B, Marsh H, Mills M, Molloy FJ, Pitcher CR, Pressey RL, Russ GR, Sutton S,

Sweatman H, Tobin R, Wachenfeld DR,, Williamson DH (2010) Adaptive management of the Great Barrier Reef: a globally significant demonstration of the benefits of networks of marine reserves. Proc Natl Acad Sci USA 107:18278–18285

McWilliams JP, Côté IM, Gill JA, Sutherland WJ, Watkinson. AR (2005) Accelerating impacts of temperature-induced coral bleaching in the Caribbean. Ecology 86:2055–2060

Miller J, Muller E, Rogers C, Waara R, Atkinson A, Whelan KRT, Patterson M, Witcher B (2009) Coral disease following massive bleaching in 2005 causes 60 % decline in coral cover on reefs in the US Virgin Islands. Coral Reefs 28:925–937

Montaggioni LF, Braithwaite CJR (2009) Quaternary Coral Reef Systems: History, Development Processes and Controlling Factors. Amsterdam: Elsevier, 532 pp

Mora C, Andréfouët S, Costello MJ, Kranenburg C, Rollo A, Veron J, Gaston JK, Myers RA (2006) Coral reefs and the global network of marine protected areas. Science 312:1750–1751

Mora C, Metzger R, Rollo A, Myers RA (2007) Experimental simulations about the effects of overexploitation and habitat fragmentation on populations facing environmental warming. Proc R Soc B 274:1023–1028

Mumby PJ. (2006) The impact of exploiting grazers (Scaridae) on the dynamics of Caribbean coral reefs. Ecol Appl 16:747–769

Mumby PJ, Elliott IA, Eakin CM, Skirving W, Paris CB, Edwards HJ, Enríquez S, Iglesias- Prieto R, Cherubin LM, Stevens JR (2011) Reserve design for uncertain responses of coral reefs to climate change. Ecol Lett 14:132–140

Mumby PJ, Harborne, Williams AJ, Cappel CV, Brumbaugh DR, Micheli F, Holmes KE, Dahlgren CP, Paris CB, Blackwell PG (2007a) Trophic cascade facilitates coral recruitment in a marine reserve. Proc Natl Acad Sci 104:8362–8367

Mumby PJ, Hastings A, Edwards HJ (2007b) Thresholds and the resilience of Caribbean coral reefs. Nature 450:98–101

Murdoch TJT (2007) A functional group approach for predicting the composition of hard coral assemblages in Florida and Bermuda. PhD Dissertation. Mobile: University of South Alabama, 299 pp

Myers MR, Ambrose RF (2009) Differences in benthic cover inside and outside marine protected areas on the Great Barrier Reef: influence of protection or disturbance history? Aquat Conserv Mar Freshw Ecosys 19:736–747

National Marine Fisheries Service (2015) Recovery Plan for Elkhorn *(Acropora palmata)* and Staghorn *(A. cervicornis)* Corals. Prepared by the *Acropora* Recovery Team for the National Marine Fisheries Service, Silver Spring, Maryland.

Neumann AC, Macintyre IG (1985) Reef response to sea level rise: keep-up, catch-up or give-up. Proc 5th Intl Coral Reef Congr 3:105–110

Norris RD, Turner SK, Hull PM, Ridgwell A (2013) Marine ecosystem responses to Cenozoic global change. Science 341:492–498

Paddack M, Reynolds J, Aguilar C, Appledoor R, Beets J, Burkett EW, Chittar PM, Clarke K, Esteves R, Fonseca AC, Forrester GE, Friedlander AM, García-Sais J, González-Sansón G, Jordan LKB, McClellan DB, Miller MW, Molloy PP, Mumby PJ, Nagelkerken I, Nemeth M, Navas-Camacho R, Pitt J, Polunin NVC, Reyes-Nivia MC, Robertson DR, Rodríguez-Ramírez Salas AE, Smith SR, Spieler R, Steele MA, Williams ID, Wormald CL, Watkinson AR, Côté IM (2009) Recent region-wide declines in Caribbean reef fish abundance. Curr Biol 19:590–595

Pandolfi JM (1996) Limited membership in Pleistocene reef coral assemblages from the Huon Peninsula, Papua New Guinea: constancy during global change. Paleobiology 22:152–176

Pandolfi JM, Bradbury RH, Sala E, Hughes TP, K. Bjorndal A, Cooke RG, McArdle D, McClenachan L, Newman MJH, Paredes G, Warner RR, Jackson JBC (2003) Global trajectories of the long-term decline of coral reef ecosystems. Science 301:955–958

Pandolfi JM, Connolly SR, Marshall DJ, Cohen AL (2011) Projecting coral reef futures under global warming and ocean acidification. Science 333:418–422

Pandolfi JM, Jackson JBC (2001) Community structure of Pleistocene coral reefs of Curaçao, Netherlands Antilles. Ecol Monogr 71:49–67

Pandolfi JM, Jackson JBC (2006) Ecological persistence interrupted in Caribbean coral reefs. Ecol Lett 9:818–826

Peterson C, Jackson J, Kirby M, Lenihan H, Bourque B, Bradbury R, Cooke R, Kidwell S (2001) Factors in the decline of coastal ecosystems. Science 293:1590–1591

Planes S, Galzin R, Bablet J-P, Sale PF (2005) Stability of coral reef fish assemblages impacted by nuclear tests. Ecology 86:2578–2585

Porter JW, Battey JF, Smith GI (1982) Perturbation and change in coral reef communities. Proc. Natl Acad Sci 79:1678–1681

Porter JW, Woodley JD, Smith GJ, Neigel JE, Battey JF, Dallmeyer DG (1981) Population trends among Jamaican reef corals. Nature 294:249–250

Precht WF, Aronson RB (2004) Climate flickers and range shifts of reef corals. Front Ecol Environ 2:307–31

Precht WF, Aronson RB (2006) Death and resurrection of Caribbean coral reefs: a palaeoecological approach. In: Côté IM, Reynolds JD (eds) Coral reef conservation. Cambridge: Cambridge University Press. pp 40–77

Precht WF, Miller SL (2007) Ecological shifts along the Florida Reef Tract: the past as the key to the future. In: Aronson RB (ed) Geological approaches to coral reef ecology. New York: Springer-Verlag, pp 237–312

Precht WF, Robbart ML, Aronson RB (2004) The potential listing of *Acropora* species under the US Endangered Species Act. Mar Poll Bull 49:534–536

Quinn JF, Dunham AE (1983) On hypothesis testing in ecology and evolution. Am Nat 122:602–617

Randall CJ, van Woesik R (2015) Contemporary white-band disease in Caribbean corals driven by climate change. Nature Clim Change 5:375–379

Riegl B, Piller WE (2003) Possible refugia for reefs in times of environmental stress. Int J Earth Sci (Geol Rdsch) 92:520–531

Roberts HH, Wilson PA, Lugo-Fernández A (1992) Biologic and geologic responses to physical processes: examples from modern reef systems of the Caribbean–Atlantic region. Cont Shelf Res 12:809–834

Roff G, Clark TR, Reymond CE, Zhao J-X, Feng Y, McCook LJ, Done TD, Pandolfi JM (2013) Palaeoecological evidence of a historical collapse of corals at Pelorus Island, inshore Great Barrier Reef, following European settlement. Proc R Soc B 280, doi: 10.1098/rspb.2012.2100

Rogers CS (1990) Responses of coral reefs and reef organisms to sedimentation. Mar Ecol Prog Ser 62:185–202

Rogers CS (2008) Coral bleaching and disease should not be underestimated as causes of Caribbean coral reef decline. Proc R Soc B 276:197–198

Rogers CS (2013) Coral reef resilience through biodiversity. ISRN Oceanography 2013:739034, doi: 10.5402/2013/739034

Rogers CS, Miller J (2013) Coral diseases cause reef decline. Science 340:1522

Rosenberg E, Ben-Haim Y (2002) Microbial diseases of corals and global warming. Env Microbiol 4:318–326

Sala E, Jackson JBC (2011) Lessons from coral reefs. In: Jackson JBC, Alexander KE, Sala E (eds) Shifting baselines: the past and future of ocean fisheries, Washington, DC: Island Press, pp 193–203

Sammarco PW (1982) Echinoid grazing as a structuring force in coral communities: whole-reef manipulations. J Exp Mar Biol Ecol 61:31–55

Sammarco PW, Strychar KB (2009) Effects of climate change/global warming on coral reefs: adaptation/exaptation in corals, evolution in zooxanthellae, and biogeographic shifts. Environ Bioindic 4:9–45

Sandin SA, McNamara DE (2012) Spatial dynamics of benthic competition on coral reefs. Oecologia 168:1079–1090

Sandin SA, Smith JE, DeMartini EE, Dinsdale EA, Donner SD, Friedlander AM, Konotchik T, Malay M, Maragos JE, Obura D, Pantos O, Paulay G, Richie M, Rohwer F, Schroeder RE, Walsh S, Jackson JBC, Knowlton N, Sala E (2008) Baselines and degradation of coral reefs in the northern Line Islands. PloS One 3(2): e1548, doi:10.1371/journal.pone.0001548

Schutte VGW, Selig ER, Bruno JF (2010) Regional spatio-temporal trends in Caribbean coral reef benthic communities. Mar Ecol Prog Ser 402:115–122

Selig ER, Bruno JF (2010) A global analysis of the effectiveness of marine protected areas in preventing coral loss. PLoS One 5:e9278, doi:10.1371/journal.pone.0009278

Selig ER, Casey KS, Bruno JF (2012) Temperature-driven coral decline: the role of marine protected areas. Global Change Biol 18:1561–1570

Selig ER, Harvell CD, Bruno JF, Willis BL, Page CA, Casey KS, Sweatman H (2006) Analyzing the relationship between ocean temperature anomalies and coral disease outbreaks at broad spatial scales. In: Phinney JT, Hoegh-Guldberg O, Kleypas J, Skirving W, Strong A (eds) Coral reefs and climate change: science and management, Washington, DC: American Geophysical Union, pp 118–128

Shears, NT, Babcock RC (2003) Continuing trophic cascade effects after 25 years of no-take marine reserve protection. Mar Ecol Prog Ser 246:1–16

Stockwell B, Jadloc CRL, Abesamis RA, Alcala AC, Russ GR (2009) Trophic and benthic responses to no-take marine reserve protection in the Philippines. Mar Ecol Prog Ser. 389:1–15

Stoddart DR (1963) Effects of Hurricane Hattie on the British Honduras reefs and cays, October 30–31, 1961. Atoll Res Bull 95:1–142

Sweet MJ, Croquer A, Bythell JC (2014) Experimental antibiotic treatment identifies potential pathogens of white band disease in the endangered Caribbean coral *Acropora cervicornis*. Proc R Soc B 281:20140094, doi: 10.1098/rspb.2014.0094

Tager D, Webster JM, Potts DC, Renema W, Braga JC, Pandolfi JM (2010) Community dynamics of Pleistocene coral reefs during alternative climatic regimes. Ecology 91:191–200

Thompson DM, van Woesik R (2009) Corals escape bleaching in regions that recently and historically experienced frequent thermal stress. Proc R Soc B 276:2893–2901

Toscano MA (2016) Revised paleoenvironmental analysis of the Holocene portion of the Barbados sea-level record: Cobbler's Reef revisited. Coral Reefs. doi:10.1007/s00338-016-1397-0

Toscano MA, Macintyre IG (2003) Corrected western Atlantic sea-level curve for the last 11,000 years based on calibrated ^{14}C dates from *Acropora palmata* framework and intertidal mangrove peat. Coral Reefs 22:257–270

Toth LT, Aronson RB, Cobb KM, Cheng H, Edwards RL, Grothe PR, Sayani HR (2015) Climatic and biotic thresholds of coral-reef shutdown. Nature Clim Change 5:369–374

Toth LT, Aronson RB, Vollmer SB, Hobbs JW, Urrego DH, Cheng H, Enochs IC, Combosch DJ, van Woesik R, Macintyre IG (2012) ENSO drove 2500-year collapse of eastern Pacific coral reefs. Science 337:81–84

Toth LT, van Woesik R, Murdoch TJT, Smith SR, Ogden JC, Precht WF, Aronson RB (2014) Do no-take reserves benefit Florida's corals? 14 years of change and stasis in the Florida Keys National Marine Sanctuary. Coral Reefs 33:565–577

Urban MC, Tewksbury JJ, Sheldon KS (2012) On a collision course: competition and dispersal differences create no-analogue communities and cause extinctions during climate change. Proc R Soc B 279:2072–2080

van Duyl FC (1985) Atlas of the living reefs of Curaçao and Bonaire (Netherlands Antilles). Utrecht: Natuurwetenschappelijke Studiekring voor Suriname en de Nederlandse Antillen [Foundation for Scientific Research in Surinam and the Netherlands Antilles], no. 117, 37 pp

van Woesik R, Jordán-Garza AG (2011) Coral populations in a rapidly changing environment. J Exp Mar Biol Ecol 408:11–20

Veron JEN (2008) Mass extinctions and ocean acidificaiton: geological constraints on biological dilemmas. Coral Reefs 27:459–472

Vollmer SV, Kline DI (2008) Natural disease resistance in threatened staghorn corals. PLoS One 3:e3718, doi:10.1371/journal.pone.0003718

Vroom PS, Page KN, Kenyon JC, Brainard BE (2006) Algae-dominated reefs. Am Sci 94:430–437

Wapnick CM, Prech WF, Aronson RB (2004) Millennial-scale dynamics of staghorn coral at Discovery Bay, Jamaica. Ecol Lett 7:354–361

Waycott M, Duarte CM, Carruthers TJB, Orth RJ, Dennison WC, Olyarnik S, Calladine A, Fourqurean JW, Heck KL Jr., Hughes AR, Kendrick GA, Kenworthy WJ, Short FT, Williams SL (2009) Accelerating loss of seagrasses across the globe threatens coastal ecosystems. Proc Natl Acad Sci 106:12377–12381

Webster JM, Wallace L, Silver E, Potts D, Braga JC, Renema W, Riker-Coleman K, Gallup C (2004) Coralgal composition of drowned carbonate platforms in the Huon Gulf, Papua New Guinea; implications for lowstand reef development and drowning. Mar Geol 204:59–89

Weil E, Smith G, Gil-Agudelo DL (2006) Status and progress in coral reef disease research. Dis Aquat Org 69:1–7.

Wood, R (1999) Reef evolution. Oxford: Oxford University Press, 414 pp

Wood R (2007) The changing fate of coral reefs: lessons from the deep past. In: Aronson RB (ed) Geological approaches to coral reef ecology, New York: Springer-Verlag, pp 3–27

Woodley JD (1989) The effects of Hurricane Gilbert on coral reefs at Discovery Bay. In: Bacon PR (ed.) Assessment of the Economic Impacts of Hurricane Gilbert on Coastal and Marine Resources in Jamaica, Nairobi: UNEP Regional Seas Reports and Studies 110, United Nations Environmental Programme, pp 71–73

Woodley JD (1992) The incidence of hurricanes on the north coast of Jamaica since 1870: are the classic reef descriptions atypical? Hydrobiologia 247:133–138

Woodley JD, Chornesky EA, Clifford PA, Jackson JBC, Kaufman LS, Knowlton N, Lang JC, Pearson MP, Porter JW, Rooney MC, Rylaarsdam KW, Tunnicliffe VJ, Wahle CM, Wulff JL, Curtis AGS, Dallmeyer MD, Jupp BP, Koehl MAR, Neigel J, Sides EM (1981) Hurricane Allen's impact on Jamaican coral reefs. Science 214:749–755

Żychaluk K, Bruno JF, Clancy D, McClanahan TR, Spencer M (2012) Data-driven models for regional coral-reef dynamics. Ecol Lett 15:151–158

Measuring, Interpreting, and Responding to Changes in Coral Reefs: A Challenge for Biologists, Geologists, and Managers

12

Caroline S. Rogers and Jeff Miller

Abstract

What, exactly, is a coral reef? And how have the world's reefs changed in the last several decades? What are the stressors undermining reef structure and function? Given the predicted effects of climate change, do reefs have a future? Is it possible to "manage" coral reefs for resilience? What can coral reef scientists contribute to improve protection and management of coral reefs? What insights can biologists and geologists provide regarding the persistence of coral reefs on a human timescale? What is reef change to a biologist... to a geologist?

Clearly, there are many challenging questions. In this chapter, we present some of our thoughts on monitoring and management of coral reefs in US national parks in the Caribbean and western Atlantic based on our experience as members of monitoring teams. We reflect on the need to characterize and evaluate reefs, on how to conduct high-quality monitoring programs, and on what we can learn from biological and geological experiments and investigations. We explore the possibility that specific steps can be taken to "manage" coral reefs for greater resilience.

Keywords

Monitoring • Random sampling • Marine protected areas • Biodiversity • Connectivity

12.1 Current State of Coral Reefs

For the purposes of this paper, we define "true" coral reefs as rigid, topographically complex structures developed from carbonate accretion by corals and other cementing and calcifying organisms and the product of biological and geological processes. This definition reflects comprehensive discussions in Buddemeier and Hopley (1988), Hubbard (1997), Hubbard et al. (1998) and Kleypas et al. (2001).

C.S. Rogers (✉)
Wetland and Aquatic Research Center, Caribbean Field Station, US Geological Survey, 1300 Cruz Bay Creek, St. John, VI 00830, USA
e-mail: caroline_rogers@usgs.gov

J. Miller
South Florida/Caribbean Inventory and Monitoring Network, US National Park Service, 1300 Cruz Bay Creek, St. John, VI 00830, USA

There is not complete agreement on which "hardbottom" habitats constitute coral reefs. Some would include communities with corals growing on boulders or other non-carbonate pavement, as well as low-relief habitats dominated by gorgonians as reefs, while others would not. However, the lack of consensus among scientists over what constitutes a coral reef, and the total areal extent of reefs in the world, should not interfere with the primary message that coral reefs are important and imperiled, and every effort should be made to reduce the stressors affecting them.

Coral reefs are changing rapidly, and scientists can play a role in communicating with the general public about these changes and what they mean. Coral reefs extend over less than 0.01 % of the marine environment, approximately 250,000 km^2 (Burke et al. 2011). However, they provide many goods and services such as coastal protection, and

D.K. Hubbard et al. (eds.), *Coral Reefs at the Crossroads*,
Coral Reefs of the World 6, DOI 10.1007/978-94-017-7567-0_12

support for tourism and fisheries (Burke et al. 2011) as well as non-economic benefits. Their significance to humans needs to be conveyed more broadly (Burns et al. 2003).

Even though surveys and monitoring have not been conducted everywhere (and using identical methods based on random sampling), there is compelling evidence that reefs worldwide are increasingly stressed (Wilkinson 2008; Burke et al. 2011; Jackson et al. 2014). For obvious reasons, many long-term monitoring programs have focused on reefs that are close to marine field stations or within national parks and other marine protected areas. Has this biased our characterization of the world's coral reefs? If biological monitoring had taken place randomly on reefs distributed throughout the world's oceans, would our assessment of the status of the world's reefs be different? If geologists had cored through randomly selected reefs, would their interpretation of Holocene reef history be different? Conversely, has coral reef degradation been so substantial and pervasive that the lack of random selection of sites for monitoring is not a problem? While we cannot be certain, and we support truly random sampling of coral reefs, we doubt that the prevailing view of the overall degraded condition of the world's reefs would be different if surveys had been done truly randomly--this reflects the seriousness of the current degradation of coral reefs worldwide.

With the genesis of modern scuba diving in the 1940s the ability to document long-term trends on reefs is relatively new. Yet changes in reefs over the past 50 years are quite dramatic, especially given that corals are such slow-growing organisms. While large-scale sample designs increase the ability to detect trends over vast areas, it is best when they are used in combination with smaller, reef-scale monitoring. Our recent access to mesophotic reefs (i.e., those in 30–150 m of water), some of which are less stressed and have coral-cover values similar to nearshore reefs 40 years ago, does not suggest that current threats to reefs overall are any less, or that dramatic declines in nearshore reefs are any less significant. If anything, the increasing knowledge of these "twilight" reefs should heighten the awareness for the need to better understand the connectivity between mesophotic and shallower systems.

The Intergovernmental Panel on Climate Change (2007) concluded that 18 % of the world's coral reefs will likely be "lost" by 2030, but what is meant when we say a reef has been lost? What does this word mean to scientists or to the general public? We can all agree that a coral reef buried by an airport runway is lost. Some people describe reefs as lost if there has been a significant decline in coral cover and an increase in macroalgal cover. However, it would be preferable to be more exact as we consider the spectrum or progression from "threatened" to "lost" in less obvious situations. In general, high coral cover and low algal cover are thought of as desirable.

Perhaps it is more useful to think in terms of the current threats to coral reefs. Burke et al. (2011) estimated that over 90 % of the reefs in the world are already or will be threatened by increasing temperatures, ocean acidification, and local activities and stressors such as fishing, marine-based pollution, physical damage, coastal development, and watershed-based pollution, with about 60 % experiencing high to critical levels of threat. With global stressors such as climate change now added to the mix, it is critical to attempt to mitigate the damage and reverse some of the declines before it is too late. The ultimate goal is to provide rigorous scientific data from monitoring and experimental research that can lead to better management actions that might promote reef recovery, especially given the anticipated effects of global climate change.

12.2 Characterizing Reefs

How should we characterize and describe coral reefs? What are the attributes of degraded, healthy, resistant, resilient or recovering coral reefs? If a group of geologists and biologists were air-lifted to a remote coral reef that they had never seen before, and of which they had no prior knowledge, their assessments of the reef's condition would likely differ. If the reef had 100 % coral cover of several different coral species, and supported numerous large fishes, the scientists would likely agree the reef was in good condition. Similarly, if the reef had no corals, few fishes, and was carpeted by slimy algae, they would agree it was in bad shape. However, if the reef had 25 % coral cover, 25 % cover by macroalgae, and a few large fishes, the biologists might disagree on its current state. Such a snapshot of a coral reef is usually not sufficient to characterize its status. Knowledge of its "ecological history" (sensu Hughes 1989), such as changes in coral and algal cover, changes in fish assemblages, and past disturbances is essential to provide a better sense of its present, and possible future, condition. In addition, the geologists might want to examine several cores from the reef to determine if reef accretion was taking place or whether the reef community differed from that in the more distant past. Whatever the perspective, a baseline is necessary to provide context.

Even when we have the luxury of historical data, we can still argue about what an observed change means or how important it might be in the long run. If there were any documented decreases in coral cover and increases in macroalgal cover, some might conclude that a "phase shift" had occurred (e.g., Dudgeon et al. 2010). Although much discussed, this concept may not be particularly helpful, at least as previously applied. For a shift to occur, coral dominance must give way to algal dominance. However, dominance is too often not defined and not always apparent. Not all losses of coral and increases in algae are shifts that will irrevocably tip the balance from net accretion to losses in calcification, or from calcifying organisms, such as corals and coralline algae, to non-calcifying organisms such as many macroalgal species. It is important to consider if significant losses in coral cover can be reversed. Can a reef that has gone from 80 % coral cover to 80 % algal cover ever return to its initial coral cover? The question is not whether or not a phase shift has occurred—we want to find out if *recovery* is possible and define what that recovery might look like. Although "recovery" is sometimes used in a limited way to refer to a return to an initial level of coral cover, full recovery would also involve the restitution of the preexisting community composition (e.g., relative abundance of coral species) and framework complexity (Johannes 1975; Done 1992).

If our objective is to decide what attributes would indicate a recovering or resilient reef, we need to think in terms of both reef structure and reef function. When biologists speak of structure they are often referring to the composition of the coral reef (the relative abundance of different coral species, algal species, etc.) but, like geologists, they can also be referring to the actual physical structure of the reef—the topographically complex architecture that is the result of deposition of calcium carbonate and cementation. Reef function, on the other hand, refers to *processes* like coral growth, recruitment, nutrient cycling, and calcification.

The most fundamental unit of a coral reef is a coral polyp depositing a limestone (rock) corallite and surviving to grow. Whatever adversely affects this recruitment and calcification process endangers the coral reef. Ultimately, characterizing the status of a reef would require comprehensive knowledge of processes that are more difficult to monitor than just changes in benthic cover. How can we adequately measure changes in the balance of calcification versus bioerosion? Interestingly, Perry et al. (2012) are using basic monitoring data on cover by corals and calcifying algae as well as abundance of substrate-eroding organisms to estimate carbonate budgets that could provide clues to changes in reef structure.

It is challenging to look at changes in structure (including coral composition) but even harder to look at the mix of bioerosion, productivity, recruitment, calcification and the myriad processes that reorganize and redistribute carbonate within the reef over a large spatial scale. Also, we need to understand how these processes can change over time as environmental conditions change.

12.3 Monitoring

Although we acknowledge that reefs were in trouble before most long-term monitoring began (e.g., Jackson 1997), these programs evolved out of a need to quantify the magnitude and rates of change in coral reefs. In this section we discuss the different components of an effective monitoring program, specifically addressing why, where, and how we should monitor coral reefs, and finally what should be done with the results (So what?). We need to carefully consider each of these elements to design effective monitoring strategies.

12.3.1 Why Monitor?

Monitoring has been defined as "the collection and analysis of repeated observations or measurements to evaluate changes in condition and progress toward meeting a management objective" (Elzinga et al. 1998, page 1). Most people value monitoring (and the products, data, results, identified trends) but hope someone else will do it—it is the "Rodney Dangerfield" of coral reef science! Well-designed and carefully implemented monitoring programs are essential for quantifying *changes* on coral reefs. While the characterization of reefs is important, repeated characterization is not monitoring. Monitoring should systematically and consistently measure changes in abundances of organisms, determine ranges in environmental factors, help to reveal possible cause-and-effect relationships, help measure and differentiate the effects of both natural and human-induced stresses, and determine if a specific management action is working. It is essential to state the question you are hoping to answer before you begin your monitoring program. In some cases, the objective of monitoring may not be explicitly stated, but the implicit goal is usually to provide information that can be used to better manage coral reefs. Ideally, the monitoring will be driven by a particular hypothesis. It is possible to "miss the point", that is, to get the right answer to the wrong question, by establishing a monitoring program that is statistically rigorous but ill-conceived. Conversely, challenges arise when comparing or combining reef monitoring data obtained with varying techniques collected in widely different "coral reefs" or coral reef zones to produce regional trends (Jackson et al. 2014).

12.3.2 Where To Monitor?

It has been noted that, "when you are on the wrong train, every stop is the wrong stop" (Stein 1983). Having rigorous, peer-reviewed protocols is critical, but if the monitoring takes place in the wrong locations we are not going to reach our goal.

It is very important to make a distinction between *coral* monitoring and *coral reef* monitoring. The failure to differentiate between these two can lead to confusion and wasted effort. If the question is "What is the percent of coral cover in the Caribbean (or around an island or within a particular national park)?" that is very different from "What are the ecological changes over time on 'true' coral reefs?" Corals can be found growing in seagrass beds, on boulders and in other areas that we would not think of as reefs. If the interest is in documenting coral cover wherever corals occur, then a wider geographic approach (over a larger spatial area) is warranted. However, this broad approach will not provide the necessary ecological data or context for evaluating changes on actual coral reefs that cover a much smaller percent of the overall area.

For example, it is not too helpful for the manager of a marine protected area to learn that coral cover throughout the entire hardbottom area has dropped from 5 % to 4 % or even gone up from 5 % to 7 % over the course of a year. What management action might this provoke? How can one visualize this change and where does one go with this information? However, learning that there were losses of greater than 50 % of the living coral cover on true coral reefs is of great interest to park managers.

Well-developed reefs often will cover a small percent of the bottom, but they are of disproportionate ecological importance as habitats for organisms as well as sinks and sources of the larvae of fishes, corals, and other reef organisms (e.g., Sale 1991). These reefs should be the focus of study, especially within national parks and marine reserves. They are "special places" within "special places" that have been set aside precisely because of their ecological significance. At this spatial scale, the selection of which reefs to study is not and should not be random. Here we focus on monitoring true coral reefs (as defined earlier), not on monitoring of corals in any hardbottom community.

A rigorous statistical approach to sample design is optimal. Until recently few coral-reef monitoring programs used a random approach to sampling (Lewis 2004), but we now have the tools (e.g., GPS, georeferenced maps) and knowledge to do this. Recall, it was only recently, in 1998, that Executive Order 13089 by President Clinton identified—"comprehensive mapping, assessment and monitoring"—as priorities within the Coral Reef Initiative. When monitoring coral reefs, sampling units (i.e., transects or quadrats) should be randomly selected from within the boundaries of the reef or reef zone of interest. Some of these sampling units might not meet the pre-defined criteria, e.g., they may fall in sand or seagrass beds. In this case, they can be rejected and the next randomly selected unit can be chosen.

Caution must be used in extrapolating the results of monitoring to the appropriate area of inference. With random, unbiased selection of a sufficient number of sampling units, the results can be applied to larger spatial areas. The scientific literature is full of the results of monitoring based on haphazard (non-random) sampling in which conclusions are presented as if they applied to an entire reef zone or reef when they are only representative of a few quadrats or transects. The following quote from the 2008 Status of the Reefs report compiled by Wilkinson (2008) is instructive:

> These status assessments and predictions are based on considerable monitoring data using a range of methods, varying from very detailed species level monitoring to rapid monitoring by trained volunteers. However, it is recognized that monitoring in many countries only covers a small and unrepresentative proportion of the reefs, such that the monitoring data are inadequate for a quantitative assessment. In these cases we have relied on qualitative assessments based on the expert opinion of national and visiting scientists, complemented by information from professional dive guides.

Although the focus has been on corals primarily as the major architects of coral reefs, one of the biggest unanswered questions (there are many!) is "What effects will these coral declines have on reef biodiversity, on the reef fishes and other organisms?" [We do not know what happened to fishes when there were extreme losses of elkhorn (*Acropora palmata*) and staghorn (*A. cervicornis*) corals beginning over 25 years ago!] Monitoring fish populations at randomly selected points across a variety of habitats for example, will not help us answer this question. Serious consideration should be given to co-location of sites for monitoring coral cover and reef fish diversity/abundance. Another important question is: "Is coral cover or structural complexity (shelter) a more significant driver of reef fish diversity and abundance?" However, it is also essential to look beyond the boundaries of a particular reef of interest to better understand the *connectivity* among the reef and associated habitats that serve as sources of larvae (see below).

12.3.3 How and How Often?

Monitoring must occur not only in the correct locations but also with appropriate protocols that outline exactly how measurements or samples will be taken and how often. Many different methods are now readily available (e.g., see Rogers et al. 1994; English et al. 1997; Patterson et al. 2008) and will not be reviewed here. In the last two decades, photography (still and video) has become increasingly affordable and valuable. Video provides a permanent visual record, but requires substantial time for processing. Conversely, quadrats, chain transects and line-intercept transects require much less processing after data collection to determine percent cover of organisms, but are time-intensive in the field and provide no permanent visual record. Chain transects and LIDAR provide data on 3-dimensionality/rugosity. LIDAR and other types of remote sensing can cover a much larger spatial scale but lose effectiveness with increasing depth. Remote sensing is not always effective at identifying benthic cover and is most useful for documenting relatively large changes in structure (e.g., Scopoletis et al. 2011). Taking sequential photographs can provide very useful information but it may be difficult to quantify change, depending on the scale (e.g.; aerial vs. in situ). In some cases, the degradation of a reef is so extreme that in situ photographs taken over time are sufficient to indicate that substantial change has occurred. Whatever the method, the critical objective is to reveal problems while there is still time to correct them.

One area where better technology has made a significant difference is in the production of increasingly accurate benthic habitat maps. For example, the progression of maps from Dry Tortugas National Park in Florida (Fig. 12.1) shows the shifting (and better definition) of boundaries as well as greater differentiation and delineation of habitats from the earliest map in 1881 to the most recent in 2010, based on satellite imagery. The new, georeferenced maps provide a basis for both identification of habitats or zones of the greatest interest and selection of appropriate monitoring locations.

Monitoring must also be done at an appropriate frequency. Annual monitoring is often not sufficient. More frequent monitoring has been shown to be essential in revealing the causes of some coral decline. In the absence of major disturbances, coral cover typically changes more slowly than algal cover. Surprisingly, however, many people present single, annual values for macroalgae ignoring the fact that macroalgal cover can vary substantially over short periods of time (days). When major disturbances occur over several months, infrequent monitoring can lead to misinterpretations of the timing and causes of change. For example, someone monitoring permanent transects in August 2005 in the US Virgin Islands and then again in August 2006 could mistakenly conclude that the severe bleaching event which began in September 2005 caused extensive mortality, when, in fact, the coral disease outbreak that began later in 2005 was the actual cause of most of the coral decline (Miller et al. 2009).

12.3.4 So What?

A primary objective of monitoring is to differentiate natural from anthropogenic change to allow identification of actions that can be taken to reduce reef damage. However, this is often not an easy task (Hughes et al. 2011; Sweatman and Syms 2011; Sweatman et al. 2011). Even with several years of monitoring, it can be hard to determine the normal variation in the abundance (cover) of different organisms. Monitoring results will not always be directly applicable (or useful) for local management. To illustrate, the manager of a Marine Protected Area (MPA) should know that over half of the coral within the MPA has succumbed to diseases but will not be able to take specific action to quickly reverse coral cover decline. As seen in the definition of monitoring above, progress towards a management goal is often considered an integral objective. The scales for management actions need to be aligned with the scale of the monitoring and *vice versa*. Regional monitoring may provide regional baseline data, yet regional management actions are rare. Local protective measures within a bay or watershed are more likely to gain public support and produce discrete, measurable results. Public opinion cannot be overlooked with respect to management actions. Whether it is the listing of threatened or endangered species, such as the National Oceanic and Atmospheric Adminstration's recent listing of 20 additional coral species, limits on anchoring, designation of boating/swimming access, or fishing closures, the success of these actions depends on public compliance.

Stressors differ in their essential characteristics, and not all can be categorized as purely natural or anthropogenic. Anchor damage or destruction from a vessel running aground on a reef is very different from chemical or sediment contamination, for example. Stressors that remove living tissue but leave coral skeletons intact differ substantially from those that destroy the physical structure and even the underlying framework of a reef. The effects of some stressors are easier to measure. Bleaching and diseases are much harder to quantify and address than damage from boat anchors and groundings.

Because coral diseases are not clearly either natural or anthropogenic and vary greatly in their temporal and spatial distribution, their global significance can be hard to evaluate. In their report "Reefs at Risk", Burke et al. (2011) compile data on local human-related stressors (the primary focus of the document) and on past thermal stress (bleaching). They address coral disease as a "compounding threat" and include a map showing the global distribution of disease incidence from 1970 through 2011. However, they note that this map is based on an incomplete database. Also, the map does not show disease prevalence or trends in disease occurrence over time.

Coral diseases have had the greatest adverse effects on Caribbean coral reefs in the last 50 years, and they are of increasing concern in the Indo-Pacific (Willis et al. 2004; Raymundo et al. 2008; Weil and Rogers 2011). They are found in all ecosystems, but increased prevalence in some cases can be attributed to release of sewage or other human activities (Bruno et al. 2003). Diseases may also be linked to bleaching which in turn is linked to high seawater temperatures. Given the anticipated increase in seawater temperature associated with climate change, we need to learn more about the relationship among thermal stress, bleaching and disease. Increases in temperatures that lead to bleaching may have both a natural and human component (e.g., Donner et al. 2007).

Often reefs are subjected to several natural and human-related stressors simultaneously, making attempts to manage

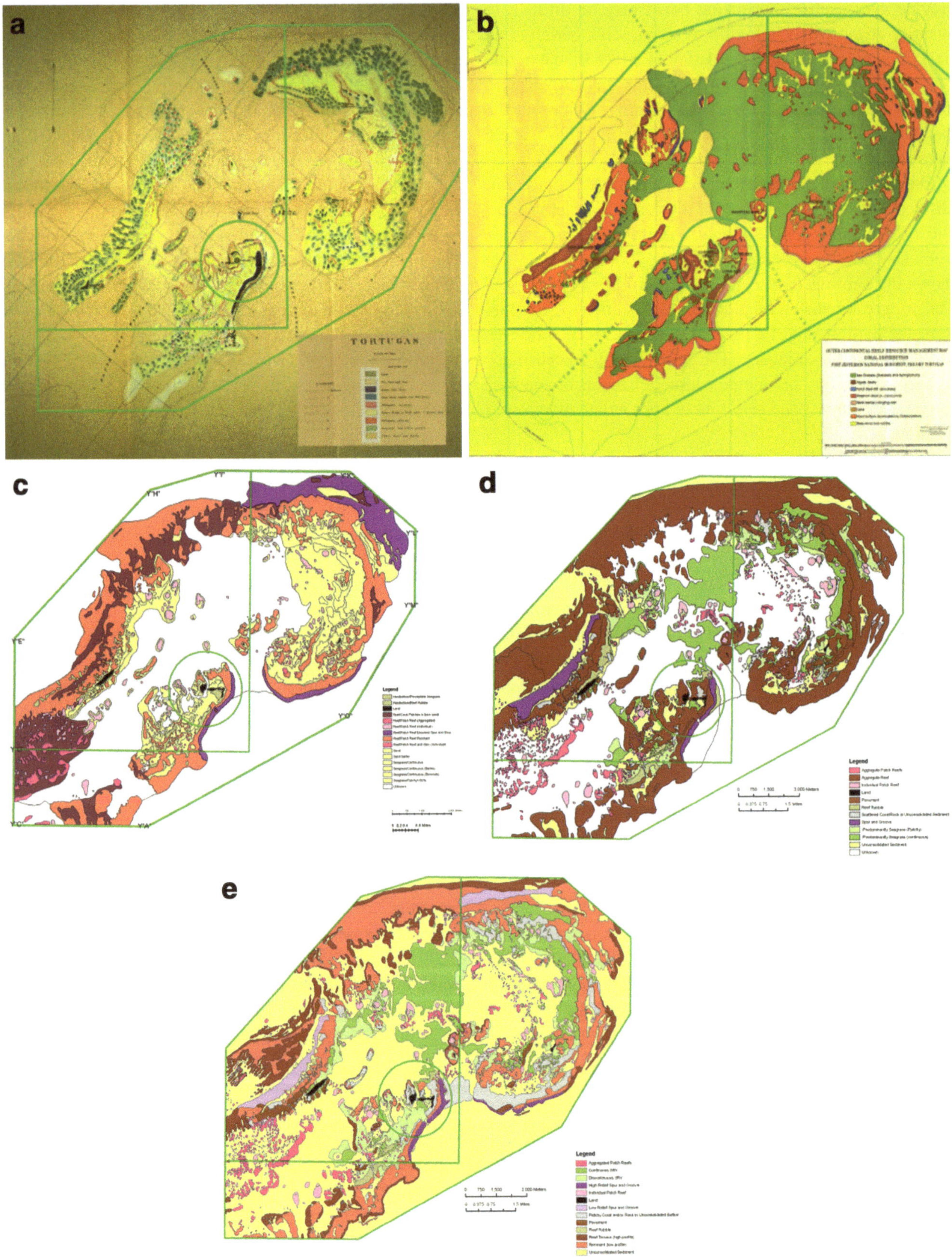

Fig. 12.1 A progression of maps of benthic habitats in Dry Tortugas National Park showing greater accuracy and finer detail with improved mapping technology (see Waara et al. 2011) [Sources for maps **a** Agassiz (1882); **b** Davis (1979); **c** FMRI and NOAA (1998); **d** Avineon (2008); Waara et al. (2011)]

them even more problematic. For example, a heavily overfished reef can be damaged by a hurricane. Explosions of Crown-of-Thorns starfish may reflect natural cycles but also increases in nutrients from agricultural runoff (Brodie et al. 2005; De'ath et al. 2012). The loss of *Diadema antillarum* in the Caribbean (Lessios 1988) is likewise complex regarding both cause and subsequent impact to reef processes. A combination of well-designed long-term monitoring programs and hypothesis-driven experimental research is needed to make progress in sorting out the effects of single and multiple stressors on coral reef organisms and reef processes.

12.3.5 Monitoring: A Few Concluding Thoughts

A primary objective of monitoring programs is to turn data or results into information for better management of reefs. This depends on effective communication among scientists and managers. Many groups have made a strong commitment to high-quality monitoring programs, including but not limited to: The Australian Institute of Marine Science, the Global Coral Reef Monitoring Network, the US National Park Service Inventory & Monitoring Program, the US Geological Survey, the National Oceanic and Atmospheric Administration, the Caribbean Coastal Marine Productivity Program, the Atlantic and Gulf Rapid Reef Assessment program, Reef Check, and the Reef Environmental Education Foundation (Miloslavich et al. 2010; Burke et al. 2011).

These monitoring programs are all good examples of how to turn data into information for improved management of reefs. There is also the need to "monitor the monitoring", to step back and see if adjustments should be made because of changes on the reefs, or monitoring technology. Although methods should be standardized whenever possible (especially when the objective is to compare reefs), and the same methods should be used over time, significant changes in reef structure or the appearance of new stressors may necessitate a shift in methods. For example, as coral cover continues to decline, it is becoming harder and harder to measure, particularly using visual estimates. One will be more accurate observing a decline from 80 % to 40 % coral cover than a decline from 4 % to 2 %, although both reflect a 50 % relative loss.

The support for using a common method and metric (e.g., quadrats to estimate coral cover) does not diminish the need to explore other reef indicators and processes. For example monitoring coral recruitment is necessary for assessing potential coral recovery (Hughes et al. 2000, 2011). Basic monitoring using quadrats and transects is not well suited to documenting trends in recruitment because by the time one can reliably identify recruits on the reef they can be several years old (Vermeij 2006). Conversely, settling plates provide convenient substrates for quantifying recruitment but may not be representative of what is happening on more complex natural reef surfaces. An overall picture of changes on coral reefs requires a combination of different approaches including monitoring and experimental field research.

12.4 Management

The definition of monitoring presented above includes the concept of measuring "progress" towards reaching a management goal. What are the explicit goals of management? Management for sustainable fisheries would presumably differ from management for maintenance of biodiversity. In many cases, the goal will be to conserve or restore a coral reef, i.e., to promote recovery.

There are a number of actions that managers can take locally to increase the likelihood that reefs will be able to persist in the face of local and global stressors. In the early 1990s, many scientists concluded that the most serious threats to reefs were associated with human activities: shoreline development, overfishing, and degraded water quality from sediments and sewage (Ginsburg 1994). Then, with severe bleaching episodes beginning in 1998, the focus shifted more to global stressors and climate change (Wilkinson 1999). In some ways we are back to where we started with an emphasis on managing human activities at a local level (e.g., reducing anchor damage, controlling release of sewage and sediments), while still hoping that international efforts to control greenhouse emissions will become more effective (Hughes et al. 2003; Aronson and Precht 2006; Hoegh-Guldberg et al. 2007; Maynard et al. 2008; Burke et al. 2011; Kelly et al. 2011). Managing local stressors is far more feasible than trying to control global stressors, but even this has not proven to be easy. In spite of all of the uncertainties, it only makes sense to move forward with controlling those stressors that we can control rather than waiting for an international response to climate change.

Monitoring can provide data for models that can be useful for reef management. For example, models have the potential to suggest reef-specific strategies to improve conditions (Wooldridge et al. 2005; Baskett et al. 2010; Mumby et al. 2010). Models can also help to identify the optimal locations for establishing MPAs, e.g., areas with a history of fewer bleaching events (Mumby et al. 2010). Many of the existing models focus on control of macroalgae, reflecting the importance of algae in restricting coral recruitment. Some models suggest that there are thresholds in the abundance of some reef organisms such as herbivorous fishes that must be exceeded if coral cover is to remain level or increase. Many models focus on abundance of herbivores because of their role in controlling macroalgae and some indicate that even small increases in herbivore numbers can have significant effects (Wooldridge et al. 2005). However, further research is needed to determine if thresholds can be identified for specific reef processes in ways that can inform management.

12.4.1 Marine Protected Areas

The establishment of MPAs is one of the most promising management actions that can be taken. When we say that an area is "protected" we are implying that it is less subject to human activities that adversely affect the organisms within it. Sometimes the goal of a monitoring program is to evaluate the effectiveness of an MPA. However, designing such a monitoring program can be challenging because perfect controls or even reference areas may not exist. The MPA and non-protected area may differ in several characteristics, such as distance from shore, depth, and reef zonation. Another complicating factor is that MPA boundaries are often politically rather than ecologically derived. They are often unmarked, further complicating results as users are often unaware of management restrictions. Public education regarding MPA regulations may be lacking or ineffective, and evaluation of user compliance with regulations, integral for an accurate "inside/outside" comparison, is often not done (Claudet and Guidetti 2010).

Marine protected areas, particularly no-take marine reserves, have been shown to increase the abundance of targeted species and sustain or, in a few cases, increase cover of corals and other non-harvested species (Lester et al. 2009; Selig and Bruno 2010). It is harder to find evidence of increases in coral cover than increases in abundance of previously harvested fishes. Protecting fish does not necessarily reduce macroalgal cover, increase coral abundance, or preserve or increase topographic heterogeneity that is critical to maintaining or increasing fish abundance (e.g., Aronson and Precht 2006; Alvarez-Filip et al. 2011; Huntington et al. 2011). Analyses of the effectiveness of MPAs may be misleading if they include areas where fishing and other extractive uses are allowed or areas with ineffective enforcement of strict regulations.

Selig and Bruno (2010) found no change in coral cover over time across all reefs within MPAs over about four decades (1969–2006) but a decline in cover on unprotected reefs. Their analysis did not conclude that MPAs (encompassing areas with different levels of enforcement and protection) would lead to increases in coral abundance. It is also important to note that their study did not incorporate significant losses in coral cover following the Caribbean bleaching event in 2005 and the subsequent disease outbreak (Miller et al. 2009).

Networks of MPAs, if well-designed, can result in more benefits than single protected areas—with the whole being greater than the sum of its parts. They can include sources of larvae to replenish reef organisms and areas with stronger currents and upwelling, bringing cooler temperatures. However, variation in stressors as well as the heterogeneity of reefs even over short distances can make it difficult to predict where the greatest protection can be realized. Where it is feasible to design networks of marine reserves, every effort should be made to protect areas that are likely to survive future climate-driven changes, although this is certainly not very straightforward (McLeod et al. 2009).

12.5 Biodiversity as a Management Goal

Networks of marine reserves that protect the connections among coral reefs, seagrass beds, and mangroves have the potential to maintain biodiversity. There is an urgent need for more information on the biodiversity of coral reefs and on how different reef species will respond to the combination of local and global stressors.

It is estimated that over 90 % of the species inhabiting coral reefs have yet to be discovered (Reaka-Kudla 1997; Ausubel et al. 2010). We need to have a better understanding not only of how organisms might change in their distribution and abundance, but how they may or may not adapt to changes in climate. Also, the complex interactions among different organisms may themselves change over time (e.g., Harley 2011). Although Harley (2011) does not specifically mention corals or reefs, his conclusions on interactions among species and how these could react to climate change seem relevant. What happens to other reef organisms when coral cover or diversity is reduced?

Coral reef species differ in their responses to local stressors and to environmental factors that are anticipated to change with changing climate. For example, some coral species may not be as affected by increasing temperatures in terms of their physiology, larval development and survival. Weil and Vargas (2010) note the importance of learning more about the reproductive biology of major reef-building corals and how increasing water temperatures, whether or not they lead to bleaching, could affect fecundity and larval survival. Corals and other reef species also vary in their response to acidification (Pandolfi et al. 2011). More information on the variation in responses of reef organisms to local and global stressors, only briefly referred to here, must come from experimental research.

It is possible that the complexity and biodiversity of coral reefs, one of their core characteristics, can help to ensure that they have a future. In other words, biodiversity, and effectively managing for biodiversity, may confer resilience. Because reefs are now so degraded, there has been considerable discussion about managing for "resilience". Humans are clearly reducing the resilience of reefs (Nystrom et al. 2000).What is resilience and can we manage for it?

12.6 Coral Reef Resilience

Resilience has been defined as "a measure of the persistence of systems and of their ability to absorb change and disturbance and still maintain the same relationships between populations or state variables" (Holling 1973, page 14). In short, resilience is the ability to rebound. Although the word resilience is sometimes used to encompass "resistance," greater clarity is achieved by making a distinction between the two (Tang 2001; West and Salm 2003). When applied to coral reefs, resistance can refer to the ability of the entire reef or of individual corals and other organisms to remain unaffected by a stress, and resilience can be thought of as the capacity for the reef to recover after disturbance or stress.

In general, few examples of significant recovery can be documented, although some reefs have seen increases in coral cover primarily after bleaching events (Diaz-Pulido et al. 2009; Graham et al. 2011). Reefs with faster-growing corals, cooler water temperatures, less macroalgae, and sources of larvae are more likely to be resilient (e.g., see McLeod et al. 2009). Some scientists have argued that it is better to manage for a reduced number of more resistant reef organisms than to try to restore original species assemblages (Cote and Darling 2010). However, given how little is known about the tolerances of different species and how difficult it is to predict future environmental conditions, managing for a resistant reef with less biodiversity rather than a resilient, diverse reef seems counterproductive. A major question is the degree to which human-caused changes on modern reefs are unprecedented, jeopardizing reef resilience (Hughes et al. 2003).

The ability of coral reefs to recover after disturbances could depend on their connectivity not only to other reefs but also to seagrass beds and mangroves, systems which are often neglected or understudied (Rogers 2013). Sources of larvae are needed to replenish depleted populations of corals, fish and other organisms. Mumby and Hastings (2008) used models to show that reefs near mangroves were better able to recover from intense storms than those that were not linked to mangroves. In addition, these systems (and the connections or linkages between and among them) may respond differently to stressors associated with global climate change. Sea level rise could be more of a threat to mangroves than to coral reefs, which could actually benefit (Cubit 1985; Hallock 1997). It is important to keep in mind that changing climate has many components (sea level rise, alterations of current patterns, ocean acidification, increasing temperatures) and marine ecosystems can respond differently to each of these.

12.7 Predicting Coral Reef Change: Back to the Future

Given the complexity of coral reefs, it is not surprising that an accurate evaluation of their potential for recovery or for persistence will require efforts by scientists in many different disciplines—including biology, geology, and physical oceanography. Biology and geology are more closely interrelated in studies of coral reef ecosystems than in any other ecosystems. What can biologists learn from geologists and *vice versa*? (see Box 12.1 for an example based on Caribbean *A. palmata*). Biologists can look to the fossil record and patterns therein to form hypotheses for experimental research. Geologists can learn from biologists about the environmental constraints to coral growth. Working more closely with each other could help advance our knowledge of reefs at this critical crossroads. One hundred years from now what will we find? Some scientists have predicted reefs could collapse into rubble as early as 2050 if CO_2 concentrations reach 560 ppm (Silverman et al. 2009; Erez et al. 2011). Are reefs collapsing anywhere now? What can geologists tell us about this and where it is likely to happen? What is the time frame?

Box 12.1 The Evolving Story of *Acropora palmata*

An examination of the history of research on *Acropora palmata* provides interesting insights into the challenges involved in documenting reef change and predicting the future for coral reefs. *Acropora palmata* is without doubt one of the most significant corals on Caribbean reefs. With its large size and intricate, branching morphology, it creates a complex architecture for the shallow zones seaward of the reef crests of fringing and barrier reefs. Before the 2005 Caribbean bleaching event followed by a disease outbreak that led to an average loss of 60 % live coral cover (of primarily massive species) in the Northeastern Caribbean (Miller et al. 2009), white band disease (WBD) was probably the single greatest cause of coral mortality on shallow coral reefs in the Caribbean. White band disease and hurricanes are thought to have caused mortality of over 90 % of the *A. palmata* populations at several sites throughout the region beginning in the late 1970s and continuing through at least the 1980s. Evidence that WBD was the primary cause of this extensive mortality comes from a variety

(continued)

Box 12.1 (continued)

of sources, ranging from anecdotal observations to a few cases of well-documented research on the disease as it advanced through zones dominated by this species (Rogers 1985; Aronson and Precht 2001a, 2001b). Careful scrutiny of the literature reveals very few quantitative and/or definitive studies of declines in *A. palmata*, and these are from only a few locations (more data are available for *A. cervicornis*.) In some cases, scientists returned to sites they had visited 10 or more years previously and found piles of rubble and/or stands of dead *A. palmata* colonies but did not have direct proof that WBD was the cause of the losses (e.g., Panama: Ogden and Ogden 1994; Bahamas: Greenstein et al. 1998). Although it is likely that disease caused the observed mortality, few people have actually documented the disease progressing through a reef. The most definitive records come from St. Croix (Gladfelter 1982), Belize (Aronson and Precht 1997), and the British Virgin Islands (Davis et al. 1986; Bythell and Sheppard 1993). From 1976 to 1988, Gladfelter (1991) noted a drop from 85 % to 5 % at two sites at Buck Island, St. Croix, from WBD and Hurricanes David and Frederic (1979). There was a further loss from 5 % to less than 1 % attributable to Hurricane Hugo in 1989 (Gladfelter 1991). Data from monitoring along with anecdotal observations (see Bruckner 2003) documented the low population levels in the Caribbean that eventually led to the listing of this species as threatened under the US Endangered Species Act in 2006 (Hogarth 2006) and as critically endangered on the International Union for Conservation of Nature Red List in 2008.

Surprisingly, data that would indicate the effects of the loss of *A. palmata* on fishes or other organisms are not available. Perhaps this is because some scientists focus more on corals and others focus more on fishes! Co-location of monitoring at permanent reference sites has the highest potential to provide useful ecological data. Fish abundance and diversity reflect the abundance (cover) of coral but also the presence of complex structure that provides habitat: even dead coral can provide shelter. Collapse of *A. palmata* from hurricanes flattens the topography and makes an area less likely to support large fish populations. In the absence of hurricanes, how long does it take for *A. palmata* to break down from boring sponges or other bioeroders?

Currently we lack sufficient quantitative data for the wider Caribbean to state whether *A. palmata* recovery is occurring substantially or to provide a baseline for future evaluation. Information on distribution and abundance comes from Aronson and Precht (2001a, 2001b), Bruckner (2003), Precht and Aronson (2006) and the *Acropora* Biological Status Review Team (2005), but there are few records since 2000. Given the species' status as threatened/endangered, the low number of current monitoring programs focused on this species is surprising. No studies document significant recovery to levels of abundances or densities characteristic of the 1970s, and only a few studies indicate limited recovery (Zubillaga et al. 2008; Rogers and Muller 2012). It is possible that *A. palmata* will recover more quickly than *Orbicella annularis* and *O. faveolata* which have been disproportionately affected by diseases in the last decade (Edmunds and Elahi 2007; Miller et al. 2009; Rogers et al. 2009; Bruckner 2012) although they continue to be the most abundant species on many reefs in the USVI (Fig. 12.2).

Disease continues to affect *A. palmata* populations, although white pox (white patch) disease is now more prevalent than the more virulent WBD, at least at some sites (Mayor et al. 2006; Rogers and Muller 2012). Problems in differentiating white pox from other "white" diseases make an accurate assessment difficult. Some *A. palmata* colonies exhibit signs of white pox, white band, and other un-described diseases simultaneously. Climate change is expected to increase the frequency of bleaching episodes, and these may be linked to disease outbreaks. In 2005, *A. palmata* bleached for the first time on record in the USVI, and bleached colonies in Hawksnest Bay, St. John, had more disease than unbleached ones (Muller et al. 2008). In 2010, there was moderate to severe bleaching of many coral species but not of *A. palmata*.

Some have questioned the value of monitoring, stating that there is little point in continuing to monitor declines in coral cover or abundance. However, monitoring provides a quantitative basis to determine if recovery is taking place, either as part of a natural cycle or in response to a specific management action. The best evidence of *A. palmata* recovery would come from multi-year studies showing all of the following: an increase in the overall amount of living tissue of this species, growth of existing colonies, and an increase in the number of small corals arising from

(continued)

Box 12.1 (continued)

Fig. 12.2 An *A. palmata* colony growing on top of a dead *Orbicella annularis* colony

Fig. 12.3 Two *A. palmata* zones in St. John, US Virgin Islands: (**a**) Newfound Reef: Little live coral is present in this zone that was probably decimated by white band disease and storms. (**b**) Hawksnest Reef: *A. palmata* grows in high density here

sexual recruitment. Some *A. palmata* zones continue to have little to no *A. palmata* cover while others have high densities of the species (Fig. 12.3).

Although WBD was first noted over 30 years ago, no specific cause of the disease has been identified, partly because of the difficulty in culturing bacteria and because both the coral host and associated symbionts can be involved. Recently Kline and Vollmer (2011) have experimentally shown that the causative agent for WBD is probably pathogenic

(continued)

Box 12.1 (continued)
bacteria. However, it is unlikely we will ever know why these bacteria triggered the disease beginning in the late 1970s. In 1977, Gladfelter observed that 5 ha of Tague Bay Reef, off the northeast coast of St. Croix, was affected by WBD and remarked that there was little evidence of any link with human disturbance (Gladfelter 1982).

Clearly, linking coral disease to human activities would be vitally important as it provides a basis for management action. White pox in Florida appears to be associated with human sewage (Sutherland et al. 2010, 2011), but that association has not been made for what appears to be the same disease in St. John (Polson et al. 2009; May et al. 2010). The listing of *A. palmata* and the closely related *A. cervicornis* has focused more attention on these species. At Buck Island Reef National Monument, St. Croix, managers are proposing "boat free zones" to eliminate anchor damage and reduce the possibility of vessels running aground on shallow stands of these species.

What insights does the geological record provide? Van Woesik et al. (2012) explore the vulnerability of modern corals specifically to thermal stress by examining extinctions in the Plio-Pleistocene. They developed "resilience scores" based on biological traits such as type of symbionts, calcification rate, and colony size as well as on biological processes such as sexual recruitment and colony re-growth for several Caribbean taxa. On a scale of 5 to −6 (most to least resilient), *A. palmata* had a score of 2 and *O. annularis* complex a score of 4. Above we suggested that *A. palmata* might recover more quickly than *O. annularis* based on consideration of other stressors including susceptibility to diseases.

Hubbard (2009) describes how the perspective on *A. palmata* reefs in the Caribbean shifted with an increase in the number of cores that were drilled in reefs. This story underscores the importance of having sufficient data to provide a basis for extrapolating findings to larger spatial scales. He notes "For geologists, three cores from Lang Bank [St. Croix] described at the 1977 ISRS meeting in Miami, Florida set the direction of the coral-reef discussion for the next three decades." These cores indicated a gap in reef accretion at a time when *A. palmata* reefs would have been expected to keep up with sea level rise. Additional cores in St. Croix, Puerto Rico and Florida suggested that reef accretion had continued during these time periods (Hubbard 2009). However, they also pointed out an absence of *A. palmata* reef accretion for unknown reasons between 7000 and 6000 Cal BP (calibrated years before present) and again at c. 3000 Cal PB. Hubbard (2009) asks if this regional decrease in reef accretion provides some clues into the more recent decline of this species. It is interesting to consider if other widespread disease epidemics led to the hiatus in coral growth during these two time periods. The dynamic between sea level rise and *A. palmata* reef accretion described in Hubbard (2009) may be pertinent to predicting the future of some reefs as sea level continues to rise.

Given how difficult it is to get cores from reefs and interpret them, have enough been taken to provide an adequate picture of reef changes and reef history? It would be interesting to look back in time to see what the geological record can (or can't) tell us about connectivity among reefs, seagrass beds and mangroves, and how these ecosystems have changed in spatial relationship to each other. Decades into the future, geological investigations may be able to tell us if our management actions have been successful. One hundred years from now, will we be able to tell from the brief geological signature if reefs were effectively protected within MPAs?

There are constraints and challenges in predicting what to expect with climate change (e.g., Kleypas 2007), and biologists and geologists can both contribute to reducing the uncertainty through monitoring and experimental research. Climate change is sometimes described as the greatest "single" threat to coral reefs. However, climate change is comprised of different components (e.g., sea level rise, ocean acidification, higher seawater temperatures, more frequent and intense storms), and these should be considered separately. Major bleaching episodes are likely to become more frequent (Hoegh-Gulberg 2011). Can the geological record provide different signatures for coral bleaching and disease, and indicate how the two are related? The greater precipitation expected with changing climate could increase runoff of sediment into waters overlying coral reefs, a pattern that can be discerned in geological cores.

One of the most fundamental questions is: What can the fossil record tell us about the past, present and future of corals reefs? Pandolfi (2011) observed that paleoecological studies can help put the current status of coral reefs into perspective and provide insights into the reefs of the future. For example, such studies can indicate how disturbances and environmental changes in the past have influenced processes that affect species diversity through time (Precht and Miller

2007). Variation in coral growth rates, the susceptibility of different coral species to extinction, and overall reef accretion can all be put into context. Even processes such as predation, herbivory and competition can be studied through the fossil record (see Chap. 10). Pandolfi (2011) notes it has been challenging to answer a key question in ecology: How is biodiversity maintained in ecological communities? The geological record may help to explain why biodiversity "hotspots" form. The fossil record can tell us not only about corals but also about other reef organisms that contribute to the complexity of the reef.

There is the potential for the geological record to provide many clues into the future persistence of coral reefs. However, this record will provide more information on some aspects of climate change (warming temperatures, sea-level rise) than others (rates and magnitude of CO_2 rise) [Pandolfi and Greenstein 2007; Solomon et al. 2007].

12.8 Conclusions

Coral reefs are clearly more than just hazards to navigation (Columbus 1492). They enrich and protect human life. Geologists and biologists can provide evidence of the ecosystem services that are associated with these beautiful and threatened ecosystems and make recommendations for more effective management. Understanding and predicting future changes and the very survival of coral reefs will require continued long-term monitoring as well as biological/geological experiments and investigations.

Interpretation of monitoring results needs to be done with full awareness of the context in which they were obtained. Meta-analysis combining numerous regional monitoring studies with disparate objectives and methods can identify large-scale trends, but often errantly extrapolate findings of monitoring studies beyond the areas of inference for which they were originally designed.

Considering the relatively recent advent of underwater exploration, we've learned much in a short period of time, but for a system that is the poster-child for "geologic time" we must acknowledge that this is a 'work in progress'. Changes in global climate, declines in coral cover from habitat loss/disease, and rearrangement of trophic assemblages due to overfishing are examples of manipulations currently taking place on coral reefs, for which the effects are still unknown. In many places we've moved from convincing managers that monitoring was necessary, to engaging managers in discussions on the merits of stratified random versus haphazard sampling and understanding monitoring results.

Acknowledgements We thank Dr. Dennis Hubbard for inviting us to submit this chapter and for stimulating exhausting but productive discussions and two USGS scientists for reviewing an earlier draft. Dr. Bill Precht's review was particularly insightful and constructive. We also wish to thank Julia Dileo and Lee Richter for their helpful comments.

References

Acropora Biological Status Review Team (2005) Atlantic *Acropora* Status Review Document. Report to National Marine Fisheries Service, Southeast Regional Office. March 3, 2005. 152 p + App.

Agassiz A (1882) Explorations of the surface fauna of the Gulf Stream, under the auspices of the United States Coast Survey. II. The Tortugas and Florida reefs. Mem Amer Acad Arts Sci Centennial, II: 107–134

Alvarez-Filip L, Cote IM, Gill JA, Watkinson AR, Dulvy NK (2011) Region-wide temporal and spatial variation in Caribbean reef architecture: is coral cover the whole story? Global Change Biology 17: 2470–2477 doi: 2410.1111/j.1365-2486.2010.02385.x

Aronson RB, Precht WF (1997) Stasis, biological disturbance, and community structure of a Holocene coral reef. Paleobiology 23: 326–346

Aronson RB, Precht WF (2001a) White-band disease and the changing face of Caribbean coral reefs. Hydrobiologia 460: 25–38

Aronson RB, Precht WF (2001b) Evolutional paleoecology of Caribbean coral reefs. In: Allmon WD, Bottjer DJ (eds) Evolutionary paleoecology: The ecological context of macroevolutionary change. Columbia University Press, NY, pp 171–233

Aronson RB, Precht WF (2006) Conservation, precaution, and Caribbean reefs. Coral Reefs 25: 441–450

Ausubel, JH, Crist, DT, Waggoner, PE (eds) (2010) First Census of Marine Life 2010: Highlights of a decade of discovery. Census of Marine Life. Washington, D.C. 64 pp

Avineon, Inc (2008) DRTO_coral\crdublin\photo\FWRI\Tortugas \Final_Inspection\shapefiles\ DRTO_coral.shp. Clearwater, FL

Baskett M, Nisbet R, Kappel C, Mumby P, Gaines S (2010) Conservation management approaches to protecting the capacity for corals to respond to climate change: a theoretical comparison. Global Change Biology 16: 1229–1246

Brodie J, Fabricius K, De'ath G, Okaji K (2005) Are increased nutrient inputs responsible for more outbreaks of crown-of-thorns starfish? An appraisal of the evidence. Mar Poll Bull 51: 266–278

Bruckner AW (ed) (2003) Proceedings of the Caribbean *Acropora* workshop: Potential application of the U.S. Endangered Species Act as a conservation strategy. NOAA Tech Memo NMFS-OPR-24, Silver Spring, MD, 199 pp

Bruckner AW (2012) Factors contributing to the regional decline of *Montastraea annularis* (complex). Proc 12th International Coral Reef Symposium: 5p

Bruno JF, Petes LE, Harvell CD, Hettinger A (2003) Nutrient enrichment can increase the severity of coral diseases. Ecol Lett 6: 1056–1061

Buddemeier RW, Hopley D (1988) Turn-ons and turn-offs: causes and mechanisms of the initiation and termination of coral reef growth, Proc. 6th Intl Coral Reef Symp. 1:253–261

Burke L, Reytar K, Spalding MD, Perry A (2011) Reefs at Risk Revisited. World Resources Institute, Washington, D.C. 114 pp

Burns TW, O'Connor DJ, Stocklmayer SM (2003) Science communication: a contemporary definition. Public Understanding Sci 12: 183–202.

Bythell JC, Sheppard C (1993) Mass mortality of Caribbean shallow water corals. Marine Pollution Bulletin 26: 296–297

Claudet J, Guidetti P (2010) Improving assessments of marine protected areas. Aquatic Conservation: Marine and Freshwater Ecosystems 20: 239–242

Columbus C (1492) Journal of New World Pillage, Plunder and Exploitation 1: 1–21

Cote I, Darling E (2010) Rethinking ecosystem resilience in the face of climate change. PloS biology 8 (7): e1000438. doi:10.1371

Cubit JD (1985) Possible effects of recent changes in sea level on the biota of a Caribbean reef flat and predicted effects of rising sea levels. Proc 5th International Coral Reef Symposium 3: 111–118

Davis GE (1979) Outer continental shelf resource management map, coral distribution Fort Jefferson National Monument, the Dry Tortugas. US Dept of Interior, Bureau of Land Management, Outer Continental Shelf Office, 500 Camp Street, New Orleans, LA 70130

Davis M, Gladfelter E, Lund H, Anderson M (1986) Geographic range and research plan for monitoring white band disease. Biosphere Reserve Research Report No. 6. National Park Service. 28 pp

De'ath G, Fabricius KE, Sweatman H, Puotinen M (2012) The 27-yr decline of coral cover on the Great Barrier Reef and its causes. PNAS doi:10.1073/pnas.1215836109

Diaz-Pulido G, McCook LJ, Dove S, Berkelmans R, Roff G, Kline DI, Weeks S, Evans RD, Williamson DH, Hoegh-Guldberg O (2009) Doom and Boom on a Resilient Reef: Climate Change, Algal Overgrowth and Coral Recovery. PLoS ONE 4(4): e5239. doi:10.1371/journal.pone.0005239

Done T (1992) Phase shifts in coral reef communities and their ecological significance. Hydrobiologia 247: 121–132

Donner S, Knutson T, Oppenheimer M (2007) Model-based assessment of the role of human-induced climate change in the 2005 Caribbean coral bleaching event. PNAS 104: 5483–5488

Dudgeon S, Aronson R, Bruno J, Precht W (2010) Phase shifts and stable states on coral reefs. Marine Ecology Progress Series 413: 201–216

Edmunds PJ, Elahi R (2007) The demographics of a 15-year decline in cover of the Caribbean reef coral *Montastraea annularis*. Ecological Monographs 77: 3–18

Elzinga C, Salzar D, Willoughby J (1998) Measuring and monitoring plant populations. Bureau of Land Management. Technical Ref 1730–1. 477 pp

English S, Wilkinson C, Baker V (1997) Survey Manual for Tropical Marine Resources. Townsville. 390 pp

Erez J, Reynaud S, Silverman J, Schneider K, Allemand D (2011) Coral calcification under ocean acidification and global change. In: Dubinsky Z, Stambler N (eds.) Coral reefs: an ecosystem in transition. Springer, Dordrecht, pp 151–176

Florida Marine Research Institute (FMRI) and National Oceanic and Atmospheric Administration (NOAA) (1998) Benthic habitats of the Florida Keys. FMRI Technical Report TR-4. 59 pp.

Ginsburg RN (ed.) (1994) Proceedings of the colloquium on global aspects of coral reefs: health, hazards and history, 1993. Rosenstiel School of Marine and Atmospheric Science, University of Miami

Gladfelter WB (1982) White-band disease in *Acropora palmata*: implications for the structure and growth of shallow reefs. Bull Mar Sci 32: 639–643

Gladfelter WB (1991) Population structure of *Acropora palmata* on the windward forereef, Buck Island National Monument; seasonal and catastrophic changes 1988–1989. Chapter 5. Ecological studies of Buck Island Reef National Monument, St. Croix, US Virgin Islands: a quantitative assessment of selected components of the coral reef ecosystem and establishment of long term monitoring sites. Part 1. NPS Coral Reef Assessment Program. 22 pp

Graham NAJ, Nash KL, Kool JT (2011) Coral reef recovery dynamics in a changing world. Coral Reefs 30: 283–294

Greenstein BJ, Curran HA, Pandolfi JM (1998) Shifting ecological baselines and the demise of *Acropora cervicornis* in the western North Atlantic and Caribbean province: a Pleistocene perspective. Coral Reefs 17: 249–261

Hallock P (1997) Reefs and reef limestones in earth history. In: Birkeland C (ed) Life and Death of Coral Reefs. pp 13–42.

Harley CDG (2011) Climate change, keystone predation, and biodiversity loss. Science 334: 1124–1127

Hoegh-Gulberg O (2011) The impact of climate change on coral reef ecosystems. In: Dubinsky Z, Stambler N (eds) Coral reefs: an ecosystem in transition. Springer, Dordrecht (pp 391–403)

Hoegh-Guldberg O, Mumby P, Hooten A, Steneck R, Greenfield P, Gomez E, Harvell C, Sale P, Edwards A, Caldeira K, Knowlton N, Eakin C, Iglesias-Prieto R, Muthiga N, Bradbury R, Dubi A, Hatziolos M (2007) Coral Reefs Under Rapid Climate Change and Ocean Acidification. Science 318: 1737–1742

Hogarth WT (2006) Endangered and threatened species: final listing determinations for elkhorn coral and staghorn coral. Fed Regist 71: 26852–26861

Holling CS (1973) Resilience and stability of ecological systems. Ann Rev Ecol Syst 4: 1–23

Hubbard D (1997) Reefs as dynamic systems. In: Birkeland C (ed.) Life and Death of Coral Reefs. Chapman & Hall, pp 43–67

Hubbard DK (2009) A new Caribbean reef model: a view from ye shoulders of giants. Proc 11th Int Coral Reef Symp 1: 2–6

Hubbard DK, Burke RB, Gill IP (1998) Where's the reef: the role of framework in the Holocene. Carbonates and Evaporites 13: 3–9

Hughes TP (1989) Community structure and diversity of coral reefs: the role of history. Ecology 70: 275–279

Hughes TP, Baird AH, Bellwood DR, Card M, Connolly SR, Folke C, Grosberg R, Hoegh-Guldberg O, Jackson JBS, Kleypas JA, Lough JM, Marshall P, Nystrom M, Palumbi SR, Pandolfi JM, Rosen B, Roughgarden J (2003) Climate change, human impacts, and the resilience of coral reefs. Science: 929–933

Hughes TP, Baird AH, Dinsdale EA, N.A. M, Pratchett MS, Tanner JE, Willis BL (2000) Supply-side ecology works both ways: the link between benthic adults, fecundity, and larval recruits. Ecology 81: 2241–2249

Hughes TP, Bellwood DR, Baird AH, Brodie J, Bruno JF, Pandolfi JM (2011) Shifting base-lines, declining coral cover, and the erosion of reef resilience: comment on Sweatman et al. (2011). Coral Reefs 30: 653–660

Huntington BE, Karnauskas M, Lirman D (2011) Corals fail to recover at a Caribbean marine reserve despite ten years of reserve designation. Coral Reefs doi 10.1007/s00338-011-0809-4

IPCC (2007) Climate Change 2007: Impacts, adaptation and vulnerability. Contribution of Working Group II to the Fourth Assessment Report of the Intergovernmental Panel on Climate Change. Parry ML, Canziani OF, Palutikof JP, van der Linden PJ, Hanson CE (eds) Cambridge University Press, Cambridge, UK. 976 pp

Jackson JBC (1997) Reefs since Columbus. Coral Reefs 16, Suppl: S23-S32

Jackson JBC, Donovan MK, Cramer KL, Lam VV (eds) (2014) Status and trends of Caribbean coral reefs: 1970–2012. Global Coral Reef Monitoring Network, IUCN, Gland, Switzerland. 306 pp

Johannes RE (ed) (1975) Pollution and degradation of coral reef communities. Elsevier Scientific Publ, New York pp 1–12

Kelly RP, Foley MM, Fisher WS, Feely RA, Halpern BS, Waldbusser GG, Caldwell MR (2011) Mitigating local causes of ocean acidification with existing laws. Science 332: 1036–1037

Kleypas JA (2007) Constraints on predicting coral reef response to climate change. In: Aronson RB (ed) Geological approaches to coral reef ecology. Ecological studies 192. Springer, New York, pp 386–424

Kleypas JA, Buddemeier RW, Gattuso J-P (2001) The future of coral reefs. Inst J Earth Sciences (Geol Rundsch) 90: 426–437

Kline D, Vollmer SV (2011) White band disease (type 1) of endangered Caribbean acroporid corals is caused by pathogenic bacteria. Scientific Reports 1: doi:10.1038/srep00007

Lessios HA (1988) Mass mortality of *Diadema antillarum* in the Caribbean: what have we learned? Ann Rev Ecol Syst 19: 371–393

Lester SE, Halpern BS, Grorud-Colvert K, Lubchenco J, Ruttenberg BI, Gaines SD, Airam S, Warner RR (2009) Biological effects within no-take marine reserves: a global synthesis. Mar Ecol Prog Ser 384: 33–46

Lewis JB (2004) Has random sampling been neglected in coral reef faunal surveys? Coral Reefs 23: 192–194

May LA, Avadanei AR, Rogers CS, Miller J, Woodley CM (2010) Microbial community analysis of *Acropora palmata* mucus swabs, water and sediment samples from Hawksnest Bay, St. John, U.S. Virgin Islands. NOAA Technical Memorandum NOS NCCOS 123 and NOAA Technical Memorandum Coral Reef Conservation Program 14

Maynard JA, Baird AH, Pratchett MS (2008) Revisiting the Cassandra syndrome: the changing climate of coral reef research. Coral Reefs 27: 745–749

Mayor PA, Rogers CS, Hillis-Starr ZM (2006) Distribution and abundance of elkhorn coral, *Acropora palmata*, and prevalence of white-band disease at Buck Island Reef National Monument, St. Croix, US Virgin Islands. Coral Reefs 25:239–242

McLeod E, Salm R, Green A, Almany J (2009) Designing marine protected area networks to address the impacts of climate change. Front Ecol Environ doi: 10.1890/070211

Miller J, Muller E, Rogers C, Waara R, Atkinson A, Whelan KRT, Patterson M, Witcher B (2009) Coral disease following massive bleaching in 2005 causes 60 % decline in coral cover on reefs in the US Virgin Islands. Coral Reefs 28: 925–937

Miloslavich P, Diaz JM, Klein E, Alvarado JJ, Dıaz C, Gobin J, Escobar-Briones E, Cruz-Motta JJ, Weil E, Cortes J, Bastidas AC, Robertson R, Zapata F, Martın A, Castillo J, Kazandjian A, Ortiz M (2010) Marine Biodiversity in the Caribbean: Regional Estimates and Distribution Patterns . PLoS One 5: e11916

Muller EM, Rogers CS, Spitzack AS, van Woesik R (2008) Bleaching increases likelihood of disease on *Acropora palmata* (Lamarck) in Hawksnest Bay, St John, US Virgin Islands. Coral Reefs 27:191–195

Mumby PJ, Elliott IA, Eakin CM, Skirving W, Paris CB, Edwards HJ, Enriquez S, Iglesias-Prieto R, Cherubin LM, Stevens JR (2010) Reserve design for uncertain responses of coral reefs to climate change. Ecology Letters doi: 10.1111/j.1461-0248.2010.01562

Mumby PJ, Hastings A (2008) The impact of ecosystem connectivity on coral reef resilience. J Applied Ecology 45: 854–862

Nystrom M, Folke C, Moberg F (2000) Coral reef disturbance and resilience in a human-dominated environment. TREE 15: 413–417

Ogden JC, Ogden NB (1994) The coral reefs of the San Blas islands; revisited after 20 years. In: Ginsburg, RN (compiler) Proceedings of the Colloquium on Global Aspects of Coral Reefs: Health, Hazards, and History, 1993. RSMAS, University of Miami. pp 267–272

Pandolfi JM (2011) The Paleoecology of coral reefs. In: Dubinsky Z, Stambler N (eds.) Coral Reefs: an Ecosystem in Transition. Springer, Dordrecht, pp 13–24

Pandolfi JM, Connolly SR, Marshall DJ, Cohen AL (2011) Projecting coral reef futures under global warming and ocean acidification. Science 333: 418–422

Pandolfi JM, Greenstein B (2007) Using the past to understand the future: palaeoecology of coral reefs. In: Johnson JE, Marshall PA (eds) Climate change and the Great Barrier Reef. Great Barrier Reef Marine Park Authority and the Australian Greenhouse Office, Townsville, Australia, pp 717–744

Patterson ME, Atkinson AJ, Witcher BD, Whelan KRT, Miller WJ, Waara RJ, Patterson JM, Ruttenberg BI, Davis AD, Urgelles R, Shamblin RB (2008) South Florida/Caribbean Network vital signs monitoring plan. Natural Resource Report NPS/SFCN/NRR-2008/063. National Park Service, Fort Collins, Colorado

Perry C, Edinger EN, Kench PS, Murphy G, Steneck RS, Smithers SG, Mumby PJ (2012) Estimating rates of biologically driven coral reef framework production and erosion: a new census-based carbonate budget methodology and applications to the reefs of Bonaire. Coral Reefs 31: 853–868

Polson SW, Higgins JL, Woodley CM (2009) PCR-based assay for detection of fourcoral pathogens. Proc 11th Int Coral Reef Symp 1:251–255

Precht WF, Aronson R (2006) Death and resurrection of Caribbean coral reefs: a paleoecological perspective. Conservation Biology Series-Cambridge 13: 40–77

Precht WF, Miller SL (2007) Ecological shifts along the Florida Reef Tract: the past as a key to the future. In: Aronson RB (ed) Chapter 9: 237–312

Raymundo LJ, Couch CS, Harvell CD (eds) (2008) Coral disease handbook. Guidelines for assessment, monitoring and management. GEF-CRTR program. Australia, 121 pp

Reaka-Kudla M (1997) Chapter 7. The global biodiversity of coral reefs: a comparison with rain forests. In: Reaka-Kudla M, Wilson D, Wilson E (eds) Biodiversity II: Understanding and protecting our biological resources. Joseph Henry Press, Washington, DC pp 83–108

Rogers CS (1985) Degradation of Caribbean and Western Atlantic coral reefs and decline of associated fisheries. Proc 5th Intl Coral Reef Symp 6:491–496

Rogers CS (2013) Coral reef resilience through biodiversity. ISRN Oceanography 2013: article ID 739034.

Rogers CS, Garrison G, Grober R, Hillis Z-M, Franke MA (1994) Coral reef monitoring manual for the Caribbean and Western Atlantic, 107 pp

Rogers CS, Muller E, Spitzack T, Miller J (2009) Extensive coral mortality in the US Virgin Islands in 2005/2006: A review of the evidence for synergy among thermal stress, coral bleaching and disease. Caribb J of Sci 45: 204–214

Rogers CS, Muller EM (2012) Bleaching, disease and recovery in the threatened scleractinian coral *Acropora palmata* in St. John, US Virgin Islands: 2003–2010. Coral Reefs 31: 807–819

Sale PF (ed) (1991) The ecology of fishes on coral reefs. Academic Press, Inc, New York (754)

Scopoletis J, Andrefouet S, Phinn S, Done T, Chabanet P (2011) Coral colonisation of a shallow reef flat in response to rising sea level: quantification from 35 years of remote sensing data at Heron Island, Australia. Coral Reefs 30: 951–965

Selig ER, Bruno JF (2010) A global analysis of the effectiveness of marine protected areas in preventing coral loss. PLoS ONE 5(2): e9278. doi:10.1371/journal.pone.0009278

Silverman J, Lazar B, Cao L, Caldeira K, Erez J (2009) Coral reefs may start dissolving when atmospheric CO_2 doubles. Geophysical Research Letters 36. doi:10.1029/2008GL036282

Solomon S, Qin D, Manning M, Chen Z, Marquis J, Averyt KB, Tignor M, Miller HL (eds) (2007) IPCC, 2007. Climate change 2007: the physical science basis, contribution of Working Group 1 to the Fourth Assessment Report of the Intergovernmental Panel on Climate Change. Cambridge University Press

Stein, Benjamin (1983) Ludes: A Ballad of the Drug & the Dream. New York: St. Martin's/Marek. pp 248.

Sutherland KP, Porter JW, Turner JW, Thomas BJ, Looney EE, Luna TP, Meyers MK, Futch JC, Lipp EK (2010) Human sewage identified as likely source of white pox disease of the threatened Caribbean elkhorn coral, *Acropora palmata*. Environ Microbiol 12: 1122–1131

Sutherland KP, Shaban S, Joyner JL, Porter JW, Lipp EK (2011) Human pathogen shown to cause disease in the threatened elkhorn coral *Acropora palmata*. PLoS ONE 6(8): e23468. doi:10.1371/journal.pone.0023468

Sweatman H, Delean S, Syms C (2011) Assessing loss of coral cover on Australia's Great Barrier Reef over two decades, with implications for longer-term trends. Coral Reefs 30: 521–531

Sweatman H, Syms C (2011) Assessing loss of coral cover on the Great Barrier Reef: A response to Hughes et al. (2011). Coral Reefs 30: 661–664

Tang C (2001) Chapter 5. Stability in ecological and paleoecological systems: variability at both short and long timescales. In: Allmon W, Bottjer D (eds) Evolutionary paleoecology: The ecological context of macroevolutionary change. Columbia University Press, New York, pp 63–81

van Woesik R, Franklin EC, O'Leary J, McClanahan T, Klaus JS, Budd AF (2012) Hosts of the Plio-Pleistocene past reflect modern-day coral vulnerability. Proc Royal Society B doi: 10.1098/rspb.2011.2621

Vermeij MJA (2006) Early life-history dynamics of Caribbean coral spescies on artificial substratum: the importance of competition, growth and variation in life-history strategy. Coral Reefs 25: 59–71

Waara, RJ, Patterson JM, Atkinson AJ, Estep AJ (2011) Development and policy applications of the 2010 benthic habitat map for Dry Tortugas National Park. Natural Resource Technical Report. NPS/SFCN/NRTR—2011/474. National Park Service, Fort Collins, Colorado

Weil E, Rogers CS (2011) Coral reef diseases in the Atlantic-Caribbean. In: Dubinsky Z, Stambler N (eds) Coral reefs: an ecosystem in transition. Springer, Dordrecht, pp 465–491

Weil E, Vargas WL (2010) Comparative aspects of sexual reproduction in the Caribbean coral genus *Diploria* (Scleractinia: Faviidae). Mar Biol 157: 413–426

West JM, Salm RV (2003) Resistance and resilience to coral bleaching: implications for coral reef conservation and management. Cons Biol 17:956–967

Wilkinson C (1999) Global and local threats to coral reef functioning and existence: review and predictions. Mar Freshwater Research 50: 867–878

Wilkinson CR (2008) Status of coral reefs of the world: 2008. Global Coral Reef Monitoring Network, and Reef and Rainforest Research Centre, Townsville, 296 pp

Willis BL, Page CA, Dinsdale EA (2004) Coral disease on the Great Barrier Reef. In: E. Rosenberg E, Loya Y (eds) Coral Health and Disease. Springer-Verlag. Berlin, pp 69–104

Wooldridge SA, Done TJ, Berkelmans R, Jones R, Marshall P (2005) Precursors for resilience in coral communities in a warming climate: a belief network approach. Mar Ecol Prog Ser 295: 157–169

Zubillaga AL, Marquez LM, Croquer A, Bastidas C (2008) Ecological and genetic data indicate recovery of the endangered coral *Acropora palmata* in Los Roques, Southern Caribbean. Coral Reefs 27: 63–72

Index

D.K. Hubbard et al. (eds.), *Coral Reefs at the Crossroads*,
Coral Reefs of the World 6, DOI 10.1007/978-94-017-7567-0

FSC
www.fsc.org
MIX
Papier | Fördert
gute Waldnutzung
FSC® C083411

Kolibri 360 GmbH
Ferdinand-Jühlke-Straße 7
99095 Erfurt, Deutschland
produktsicherheit@kolibri360.de